Grasslands:
Developments Opportunities Perspectives

Edited by
S.G. Reynolds and J. Frame

Food and Agriculture Organization of the United Nations
Rome

CRC Press
Taylor & Francis Group
Boca Raton London New York

CRC Press is an imprint of the
Taylor & Francis Group, an **informa** business
A SCIENCE PUBLISHERS BOOK

CRC Press
2385 NW Executive Center Drive, Suite 320, Boca Raton FL 33431

and by CRC Press
4 Park Square, Milton Park, Abingdon, Oxon, OX14 4RN

First issued in hardback 2019

CRC Press is an imprint of Taylor & Francis Group, LLC

ISBN: 978-1-138-42712-9 (hbk)
ISBN: 978-1-57808-359-6 (pbk)

Visit the Taylor & Francis Web site at
http://www.taylorandfrancis.com

and the CRC Press Web site at
http://www.crcpress.com

world. Reliable techniques now exist, mainly due to innovations in recent decades, but there is still scope and need to limit losses in nutritive value during ensiling, and to reduce costs and resource inputs. In tropical and subtropical countries, silage making is forecast to increase as suitable techniques are developed to handle the range of forages and crop by-products available. Haying systems are still important in many parts of the world, with methods ranging from traditional harvesting by hand, through highly mechanized round, square or densified bales, to pelleting systems. However, in spite of past technical developments, weather during hay making is still the major controlling factor in its success.

In the *Grass-based Systems and Organic Production* section, the three main drivers forecast to affect milk and meat production are a general increase in global demand, particularly from populous China and Southeast Asia; public concern about the environmental impact of agricultural practices; and consumer concern about animal welfare, food quality and food safety. In these contexts, grass-based systems are likely to be viewed favourably. Fuelled by these concerns, interest in and demand for organic food production has increased dramatically in recent years. Thus the timely chapter on organic farming (also called biological or ecological agriculture) deals with its principles, practices and potential. Statutory support schemes to encourage conversion from conventional to organic farming have been initiated in many countries, but its growth is dependent on the willingness of the consumer to pay a premium for the products. The New Zealand model of pastoral resource use serves as an example of efficient conventional milk and meat production through its emphasis on effective use of grazed forage, the production of which is underpinned by the legume, white clover. The model highlights the dynamic nature of the pastoral industries in their flexible response to export market conditions and opportunities.

In *Climatic Changes, Biodiversity and Biotechnology*, global climate change, with higher temperatures and more intense precipitation events, is foreseen to require grassland management modifications to counteract both above- and below-ground ecological consequences. At present, the ability to predict and prepare for change is limited and so it is necessary to develop proactive strategies and technologies for the future, especially to take the multi-factor nature of global change into account. Changes in land use and ground cover, in response to pressures for food and fibres, are to blame for loss of biodiversity in grasslands and associated species extinction rates. Grassland management needs to be adjusted, with maintenance of biodiversity being an objective in addition to forage production and animal output. A review of modern biotechnological approaches to grass and legume varietal improvement is presented. Though concentrating on temperate forage species, potential applications of biotechnology are germane across other climatic ranges. Progress in genome analysis, mapping, marker-assisted selection, introgression and transformation is summarized. It is expected that the end result will be a flow of novel forage varieties, not least with improved nutritional quality parameters.

The section, *Geographical Information Systems (GIS)*, considers adaptation mapping of forage species based on GIS and the application of remote sensing technology to grassland resource management. These technologies are particularly suitable for large countries or regions with diverse environments. The thrust of the GIS chapter is that better matching of plants to site conditions would enhance economic returns and reduce environmental hazards associated with suboptimal performance. Essentially, the matching is done by integrating a matrix of plant species characteristics into a database of climatic and soil criteria. The application of remote sensing technology for grassland resource management and livestock production is exemplified by its use in China, where changes in grassland production, utilization, degradation and nature conservation value have been assessed.

Other practical offshoots include snow cover and depth gauging, grassland fire incidence, drought monitoring and land use changes.

In *Farmer and Pastoralist Participation*, the development and impact of new forages, together with opportunities, are discussed in relation to smallholder livestock systems in Southeast Asia. Another presentation discusses the role of pastoralists in the development of rangeland in West Asia and North Africa. Against a background of increasing demand for livestock products and degradation of traditional feed resources, which include natural grazing and crop or tree by-products, farmers in Southeast Asia are having to adjust to the new idea of creating improved grass and legume swards. Success requires that they become involved early in the species and varieties testing and selection process together with extensionists and researchers. A notable point is the high potential of forage use for non-ruminants, especially pigs, poultry and fish. In countries of West Asia and North Africa, centralised rehabilitation and grazing schemes to increase the productivity of the steppe rangelands have resulted in degradation of the resource base, including its biodiversity. Future success is adjudged to rest upon greater involvement of pastoralists and their associated institutions, that is, using a 'bottom up' rather than a 'top down' approach.

The final section, entitled *Regional Developments*, discusses developments from contrasting environments, though mainly in the Southern Hemisphere. In Southeast Asia, the integration of forage, tree crops and livestock is debated in relation to its potential for a sustainable ruminant industry. It is concluded that further advances will hinge upon multilateral partnership among research institutes, government agencies and plantation entrepreneurs. In eastern Africa, over 80 percent of milk produced in improved dairying enterprises comes from intensively managed forage-based smallholdings. Increased human population and decreased farm size has led to this increased intensification. The discourse on the relationship in Patagonia, Argentina, between grassland sheep farming and rural depopulation notes that in recent decades adverse economic conditions have led to a 50 percent decline in sheep stocks, lack of capital investment, and consequent human depopulation. The solution posited to solve these problems lies in significant infrastructural changes, sustainable utilization of the grasslands, production of diverse high-quality products, and improved living conditions for the rural population. To achieve these targets, public funding and special technical assistance would be necessary. The discussion on Australian temperate grasslands illustrates how management progressed from an exploitive phase to a reactive phase. While the former saw the rapid expansion of livestock farming, this led to overexploitation, characterized by expansion into new areas and by pasture renovation through introduction of improved grassland species. It is concluded that future management will require a pro-active approach based on the development of sustainable and environmentally friendly farming systems, supported by research and extension programmes. The future for savannah and tropical grasslands is discussed from a Latin American perspective. Grassland-based animal production has increased, with milk production in particular benefiting from advances in technology. However, land and grassland degradation have also become evident and appropriate ameliorative measures are emerging. The industry is becoming more market-oriented than before, including diversification into a greater range of final products. While some societal and environmental concerns are arising, the signs point to the development of a vibrant pastoral industry.

S.G. Reynolds and J. Frame
Rome, 2004

REFERENCES

Allen-Diaz, B. 1996. Rangelands in a changing climate; impacts, adaptations, and mitigation. pp. 131–158. *in:* R.T. Watson, M.C. Zinyowera and R.H. Moss (eds). *Climate Change 1995: Impacts, Adaptions, and Mitigation of Climate Change: Scientific-Technical Analyses.* Contribution of Working Group II to the Second Assessment Report for the Intergovernmental Panel on Climate Change. Cambridge, UK, and New York, NY: Cambridge University Press.

Anon. 2003. Livestock, Environment and Development Initiative (LEAD) – an Inter-institutional project with secretariat based in FAO. Meeting the Challenge. See: http://lead-en.virtualcentre.org/en/wha/meeting/default.htm.

CBD [Convention on Biological Diversity]. 2002. Dry and Sub-humid Lands Biodiversity Definitions; Grasslands and savannah ecosystems. Convention on Biological Diversity, United Nations Environment Programme (UNEP). See: http://www.biodiv.org/programmes/areas/dryland/definitions.asp.

Diouf, J. 2002. Mobilizing the political will and resources to banish world food hunger. Foreword by Director-General, FAO, to: The World Food Summit – five years later. Technical Background Document, 2002. 105p.

IPCC [Intergovernmental Panel on Climate Change]. 2001. Climate Change 2001: Impacts, Adaptation and Vulnerability – Contribution of Working Group II to the Third Assessment Report of IPCC. UNEP/WMO. 1 000p.

Peeters, A. 2004. *Wild and Sown Grasses. Profiles of a temperate species selection; Ecology, biodiversity and use.* Rome: FAO, and London: Blackwell.

Steinfeld, H., de Haan, C. & Blackburn, H. 1997. Livestock-environment interactions. FAO. 56p.

UNEP. 2002. Global Environment Outlook 3 (GEO-3). 416p.

UNFPA. 2001. The State of World Population 2001. See: http://www.unfpa.org/swp/2001/english/

White, R., Murray, S. & Rohweder, M. 2000. Pilot analysis of global ecosystems (PAGE): Grassland ecosystems. World Resources Institute (WRI). 100p.

WRI [World Resources Institute]. 2000. World Resources 2000-2001. People and Ecosystems: The Fraying Web of Life. World Resources Institute, Washington DC, USA. 400pp.

ACKNOWLEDGEMENTS

Several persons assisted in identifying potential authors for papers presented in this book; thanks in particular to Bill Lauenroth of Colorado State University, Josef Nösberger of the Swiss Federal Institute of Technology, Richard Snowball of the Australian Trifolium Genetic Resource Centre and Len 't Mannetje, formerly of Wageningen University. Thanks also to Mary Reynolds, R. Daniel and B. Trelawney for assistance and to Jim Suttie for commenting on earlier versions of several papers and for assisting with the finalization and editing of the East African paper. John Latham and Jonathan Robinson kindly provided comments on several draft chapters.

Some of the ideas presented in the paper by Assoc. Prof. Max Shelton were developed when he was visiting scientist with the FAO Grassland and Pasture Crops Group. The paper by Wong *et al*. originated from a presentation at the 7th Meeting of the [FAO-sponsored] Grazing and Feed Resources Working Group for Southeast Asia, held in Manado, North Sulawesi, Indonesia, 2-7 July 2001.

The copyright holders, Blackwell Publishing, Ames, Iowa, and the authors kindly agreed to the use of Figure 13.1 in this publication.

Photographs, unless otherwise acknowledged, are by the authors and editors. Particular thanks to Angelo Loi, Brian Dear, Ismail Tajuddin, Marcus Halling and Marzio Marzot for providing photographs and for agreement to their use in this publication.

The cover was designed by Studio Bartoleschi, Rome. Lucie Herzigova assisted with finalization of a number of the Figures. Cover photographs are by the editors, Duane McCartney and Marzio Marzot. Final editing for consistency of language and style, and preparation for publication, was by Thorgeir Lawrence.

CONTENTS

SECTION IV – CLIMATE CHANGE, BIODIVERSITY AND BIOTECHNOLOGY

SECTION V – GEOGRAPHICAL INFORMATION SYSTEMS (GIS)

SECTION VI – FARMER AND PASTORALIST PARTICIPATION

SECTION VII – REGIONAL DEVELOPMENTS

LIST OF CONTRIBUTORS

Abberton, M.T., Institute of Grassland and Environmental Research, Plas Gogerddan, Aberystwyth SY23 3EB, Wales, United Kingdom.

Aguiar, Martín R., Cátedra de Ecología - IFEVA, Facultad de Agronomía, Universidad de Buenos Aires, San Martin 4453, Buenos Aires C1417DSQ, Argentina.

Baars, Ton, Louis Bolk Institute, Hoofdstraat 24, 3972 LA Driebergen, The Netherlands.

Borrelli, Pablo, OVIS XXI, Santa Fe 2843 10□B, 1425-Buenos Aires, Argentina.

Cameron, Keith, Lincoln University, Canterbury, New Zealand.

Chapman, D., Institute of Land and Food Resources, University of Melbourne, Victoria, Australia.

Chen, C.P., 78, Jalan Reko, Taman Sri Reko, Kajang 43000, Selangor Darul Ehsan, Malaysia.

Cibils, Andrés, Dept. of Animal and Range Sciences, New Mexico State University, Las Cruces, NM 88003, USA.

Clark, David, Dexcel Co. NZ, Hamilton, New Zealand.

Condron, Leo, Lincoln University, Canterbury, New Zealand.

Coop, L., Department of Entomology, Oregon State University, Corvallis OR 97331, USA

Daly, C., Spatial Climate Analysis Service, Oregon State University, Corvallis OR 97331, USA

de F. Carvalho, P., Federal University of Rio Grande do Sul, Agronomy Faculty, C.P. 776, 91501-970 Porto Alegre, RS, Brazil

do Valle, C. B., Embrapa Beef Cattle Centre, C. P. 154, 79002-970 Campo Grande, MS, Brazil

Frame, J., Ard Choille, 13, St Vincent Crescent, Alloway, Ayr KA7 4QW, Scotland, United Kingdom.

Fraser, Tom, Agresearch Ltd. NZ, Hamilton, New Zealand.

Gabunada, F., Leyte State University, 6521-A Baybay, Leyte, The Philippines

Hannaway, D.B., Department of Crop and Soil Science, Oregon State University, Corvallis, OR 97331, USA

Hedley, Mike, Massey University, Palmerston North, New Zealand.

Hodgson, John, Massey University, Palmerston North, New Zealand.

Holmes, Colin, Massey University, Palmerston North, New Zealand.

Horne, P., Lao-CIAT Forages and Livestock Systems Project, PO Box 6766, Vientiane, Lao PDR.

Humphreys, M.O., Institute of Grassland and Environmental Research, Plas Gogerddan, Aberystwyth SY23 3EB, Wales, United Kingdom.

Humphreys, M.W., Institute of Grassland and Environmental Research, Plas Gogerddan, Aberystwyth SY23 3EB, Wales, United Kingdom.

Jank, L., Embrapa Beef Cattle Centre, C.P. 154, 79002-970 Campo Grande, MS, Brazil.

Kemp, D.R., University of Sydney, Centre for Rural Sustainability, Leeds Parade, Orange, New South Wales, Australia 2800.

Kemp, Peter, Massey University, Palmerston North, New Zealand.

Laidlaw, A.S., Department of Applied Plant Science, Queen's University of Belfast, Plant Testing Station, 50 Houston Road, Crossnacreevy, Belfast BT6 9SH, Northern Ireland, United Kingdom.

Lucas, Richard, Lincoln University, Canterbury, New Zealand.

McCartney, Duane, Agriculture and Agri-Food Canada Research Centre, Lacombe, Alberta, Canada.
Michalk, D. L., NSW Agriculture, Orange Agricultural Institute, Forest Rd, Orange, New South Wales, Australia 2800.
Moog, F., Bureau of Animal Industry, Department of Agriculture, Visayas Avenue, Diliman, Quezon City, The Philippines.
Moot, Derrick, Lincoln University, Canterbury, New Zealand.
Morgan, Jack A., Rangeland Resources Research Unit, USDA-Agricultural Research Service Crops Research Laboratory, 1701 Center Ave., Fort Collins CO 80526, USA.
Morris, P., Institute of Grassland and Environmental Research, Plas Gogerddan, Aberystwyth SY23 3EB, Wales, United Kingdom.
Morris, Steve, Massey University, Palmerston North, New Zealand.
Nicholas, Phillipa, Institute of Rural Studies, University of Wales, Aberystwyth, United Kingdom.
Orodho, Apollo Bwonya, P.O. Box 1667, Kitale, Kenya.
Phengsavanh, P., Lao-CIAT Forages and Livestock Systems Project, PO Box 6766, Vientiane, Lao PDR.
Pollock, C.J., Institute of Grassland and Environmental Research, Plas Gogerddan, Aberystwyth SY23 3EB, Wales, United Kingdom.
Rae, Jon, School of the Environment, Cockcroft Building, Brighton University, Moulsecoomb, Brighton BN2 4GJ, United Kingdom.
Roothaert, R., CIAT, ILRI-Ethiopia, P.O. Box 5689, Addis Ababa, Ethiopia.
Shadbolt, Nicola, Massey University, Palmerston North, New Zealand.
Sheath, Gavin, Massey University, Palmerston North, New Zealand.
Shelton, H.M., Faculty of Natural Resources, Agriculture, and Veterinary Science, University of Queensland, Australia.
Stür, W., 22 Seventh Avenue, Windsor, Queensland 4030, Australia.
Sulas, Leonardo, C.N.R.- Istituto per il Sistema Produzione Animale in Ambiente Mediterraneo, Via E. De Nicola, 07100 Sassari, Italy.
Valentine, Ian, Massey University, Palmerston North, New Zealand.
Vera, Raúl R., 2 Norte 443, Viña del Mar, Chile.
Waghorn, Garry, Agresearch Ltd. NZ, Hamilton, New Zealand.
Wei, Y., Meteorological Institute of Inner Mongolia, Hohhot 010051, People's Republic of China.
Wilkins, Roger J., Institute of Grassland and Environmental Research, North Wyke, Okehampton, United Kingdom, and University of Plymouth, Seale-Hayne Faculty, Newton Abbot, United Kingdom.
Wong, C.C., Rice and Industrial Crops Research Centre, MARDI, P.O.Box 12301 GPO, 50774 Kuala Lumpur, Malaysia.
Woodfield, Derek, Agresearch Ltd. NZ, Hamilton, New Zealand.
Wright, Iain, Macaulay Institute, Craigiebuckler, Aberdeen AB15 8QH, Scotland, United Kingdom.
Younie, David, SAC Plant & Crop Science Division, Craibstone Estate, Bucksburn, Aberdeen AB21 9YA, Scotland, United Kingdom.
Zhibiao, Nan, College of Pastoral Agriculture Science and Technology, Lanzhou University, Gansu Grassland Ecological Research Institute, P. O. Box 61, Lanzhou 730021, People's Republic of China.

Section I
FORAGE GERMPLASM

CHAPTER 1

Prospects for Temperate Forage Legumes

J. Frame[1] and A.S. Laidlaw[2]

ABSTRACT

Forage legumes are playing an expanded and invaluable role in the nitrogen economy, in animal productivity and in sustainability of temperate grasslands, and advances in their technology are foreseen. The large pool of genetic variability among the genera and their species is being exploited further by the development of cultivars adapted to different edaphic, climatic and biotic conditions, though not forgetting the value of already adapted landraces. Some lesser species have shown promise as pioneer swards for difficult soils, degraded land and extreme climes, though sufficiency of certified seed supplies can be a problem. The N_2-fixing capacity of forage legumes is discussed in terms of rhizobial efficiency, N transference to associated companion grasses, provision of on-farm protein, and N supply to subsequent arable crops. Individual animal performance is enhanced by high intake and nutritive value characteristics of legume-rich diets. Condensed tannins, present in certain legumes, e.g. birdsfoot trefoil (*Lotus corniculatus*), have benefited ruminant animal nutrition through protein protection and bloat prevention and reduction of internal parasites. Anti-quality factors associated with some legumes, e.g. oestrogens in red clover (*Trifolium pratense*), are also noted, together with preventative and control measures. Macro- and micro-economic aspects of forage legume use in grassland are dealt with briefly. Environmental issues, especially concerning nitrate leaching and emissions of ammonia and nitrous oxide, which are featuring more strongly in grassland farming, are addressed, as is the influence on legumes of global warming and accompanying elevated carbon dioxide in the atmosphere. Finally some alternative uses of legumes are noted.

[1]Ard Choille, 13 St Vincent Crescent, Alloway, Ayr KA7 4QW, Scotland, United Kingdom.

[2]Department of Applied Plant Science, Queen's University of Belfast, Plant Testing Station, 50 Houston Road, Crossnacreevy, Belfast BT6 9SH, Northern Ireland, United Kingdom.

INTRODUCTION

The legume plant family—Fabaceae—is one of the largest plant families in the world, with the subfamily Papilionoideae containing some widely used forage species. The genera *Trifolium* and *Medicago* are particularly prominent in sustainable farming systems in temperate climes. However, relatively few species in the sub-family have been exploited in grassland farming, although alfalfa or lucerne (*Medicago sativa*), white clover (*Trifolium repens*) and red clover (*T. pratense*) play an enormously valuable role in grazing and conservation. The vast pool of genetic variability offers further scope for developing cultivars adapted to a wider range of edaphic, climatic and biotic conditions, including marginal and degraded lands, drylands and winter-cold zones. Genetic manipulation and other biotechnology techniques complement conventional methods in offering opportunities for major advances in plant breeding.

The benefits of forage legumes have been propounded firstly in terms of their contribution to the nitrogen (N) economy of grasslands and, secondly, their superior feeding value for ruminant livestock nutrition compared with grasses. For plant growth and development in general, N acquisition and supply are second only to photosynthesis in importance. Biological N_2-fixation accounts for about 65 percent of the N used in world agriculture (Vance, 1997), of which a high proportion is rhizobially fixed by forage legumes in grasslands. Previously, Hauck (1988) estimated that about 50 million tonnes are fixed annually by forage legumes. The efficiency of the plant-*Rhizobium* symbiosis is therefore a key factor in legume technology and N supply for soil fertility. The main routes of transfer of fixed N to associated grasses are through the grazing animal and by decomposition of surface litter and subterranean organs, while subsequent arable crops benefit from the residual N left in the soil.

For animal nutrition, the advantages of forage legumes over grasses are expressed through an amalgam of inherent nutritive value, nutrient availability and voluntary food intake. Animal products from legume-rich swards are perceived by the consumer as being 'natural', a perception associated with the products of organic farming systems, in which legumes figure prominently. In certain legumes, there are also some anti-quality factors that can cause bloat, or reduce reproductive performance in ruminants, but these adverse effects are preventable or controllable by strategic or tactical husbandry measures.

The agronomy and management requirements of the main forage legumes used for grazing and conservation have been well researched and a bank of knowledge is available to draw upon (Frame, Charlton and Laidlaw, 1998). Practical management guidelines are available to farmers, but interactive development and advisory work, together with decision-support, are still necessary, particularly where forage legume use is being introduced or expanded. Environmental aspects have also assumed greater importance. There is less information about the lesser-used forage legumes, but there are several world centres of excellence carrying out pioneering research and development to expand the roles of lesser-known forms. There is also increased interest in, and appraisal of, the alternative and sometimes quite novel uses to which forage legumes could be put.

Temperate forage legumes play, and will continue to play, an invaluable role in sustainable grassland agriculture, encompassing cold to warm climes. The relevant factors in relation to their future use are reviewed below, and recent advances that support their development are discussed, as are some limitations.

THE PLANT RESOURCE

A key feature of forage legumes is their intrageneric adaptation to different environmental and management conditions. Species of the genera *Trifolium, Medicago, Vicia* and *Melilotus* in particular (Plate 1.1) have a wide latitudinal range from northern and southern cool temperate regions to high altitude areas with dry, cold winters in subtropical and tropical regions (FAO, 1988). In addition, a large pool of genetic variability among and within species is available to plant breeders for the development of new cultivars, whether by conventional techniques or by innovative use of biotechnology.

Plate 1.1 *Melilotus officinalis* (top left), *Medicago sativa* (top right), *Vicia sativa* (bottom left) and *Trifolium pratense* (bottom right).
[Photographs: Dr Marcus Halling, Swedish University of Agriculture Sciences, Ultuna Campus, Uppsala, Sweden and Dr Stephen Reynolds].

This genetic diversity is still not fully tapped, and coupled with a greater understanding of the physiological basis of variation in yield and persistence, is leading to continued substantial improvements in cultivars (Rhodes and Ortega, 1996). The sheer number of species in some genera is sometimes forgotten. For example, in the genus *Trifolium*, there are about 250 species, and yet only a dozen or so are currently important in agriculture.

Concerning biotechnology, the use of transformation techniques to insert specific traits into arable and food crops has provoked mixed reactions from the public, ranging, for example, from a high level of acceptance in North America to hostility and mainly rejection in Western Europe. European scientists are partly at fault because of poor dissemination to the public of information concerning the work and its potential value in the human food chain, while aggressive marketing by multinational food companies without pre-consultation with customers has also been a major factor. Arguably, a somewhat similar scenario exists for forages, even though they have to be converted by animals into animal products. Such public concerns will have to be addressed, else the development of improved cultivars of forage legumes will be slowed. Another limiting factor identified is the difficulty of recouping costs and generating sufficient profitability from seed sales, even of improved cultivars, because of farmer reaction against enhanced seed prices, mainly since forage production cost per se is only one of the input costs for livestock farmers (Quesenberry and Casler, 2001).

Exploiting new germplasm

Many countries have plant improvement programmes based on existing collections of germplasm, but world grasslands are still being scoured for new germplasm. Such work is necessary primarily to develop improved cultivars of species already in use in traditional farming systems, but secondarily to produce cultivars for new systems that are evolving, and also to identify germplasm suitable for edaphic and climatic conditions for which existing cultivars are or may be unsuited. The last case has arisen in countries where population and economic pressures are increasingly forcing arable cropping and direct food production to be given priority on the better quality land. Thus, ruminant production has been forced on to marginal and poor, infertile lands. This situation can clearly be seen in the Eastern European steppes, the South American pampas and North American prairies (Riveros, 1993). Other vast grazing areas noted by Riveros, and which conceivably could benefit from the introduction of legumes or the improved presence of existing forage legumes, include the steppes that stretch from Mongolia into Europe, the Tibet-Qinghai plateau and adjacent Himalayan lands, the Australian grasslands, and the semi-arid lands of the Mediterranean region.

Of the world's grasslands–circa 3 500 million hectares–only about 500 million hectares are estimated to be in the high and medium land-use categories, while the remaining 3 000 million hectares are in low and zero categories (Buringh and Dudal, 1987). About 30 percent of the world's land surface is semi-arid to arid grazed rangelands, typically characterized by low productivity (because of water or nutrient limitation, or both), variable rainfall, complex natural vegetation, and limited scientific attention (Stafford Smith, 1998). Globally, annual rainfall is more variable at lower latitudes, or in regions with low annual rainfall or that are subject to the El Niño-southern oscillation phenomenon (Nicholls and Wong, 1990). Improving selected areas of rangeland by introducing forage legumes and utilizing these areas strategically to complement natural pastures is one option, while another option is the wider introduction of legumes by various partial seeding techniques ranging from surface seeding on open swards to direct drilling (also known as sod seeding) on denser swards. Such improvements have to be sustainable in the long term within the local ecological, production and socio-economic contexts (Riveros, 1993).

Within temperate regions, climate, soils and biotic pressures may limit the choice of forage legume species that can be grown. In the EU project, LEGSIL, in which five temperate forage legumes were grown with and without companion grass at three sites in each of four northern European countries, relative performance of the legumes varied due to strong genotype × environment interactions (Halling *et al.*, 2002). For example, yields of alfalfa and galega were the most sensitive to environment, while white clover was the least. Red clover tended to perform relatively well at all sites, especially in the first year, whereas yields of lotus were consistently poor. Taking management into account might further influence genotype × environment interactions.

Some caution is relevant when considering the widespread introduction of potentially productive new forage species. Firstly, there is the danger of genetic erosion of existing adapted landraces (Rebuffo and Abadie, 2001). Secondly, there is an example where transfer of successful Australian pasture technology based on annual subterranean clover (*T. subterraneum*) and application of phosphorus (P) did not have as significant an effect in other Mediterranean-climate-type regions (Seligman, 1998). This was because of socio-economic conditions, often related to land tenure patterns or to subsidization of feed grains and concentrates. Even in southern Australia, the pasture:crop ratio of land has fallen from 3:1 to less than 1:1 mainly as a consequence of low wool prices in relation to grain crops and the availability of cheap N fertilizer. Thirdly, there are often indigenous legumes well adapted to local conditions, but which may need improved nutrition and a more effective *Rhizobium* for N_2 fixation in order to perform better. In Mediterranean regions, local legume ecotypes generally outperformed improved Australian cultivars (Falcinelli *et al.*, 1993). As a general comment, development of local ecotypes, even if regarded as lesser species, may be more appropriate and rewarding than attempting to adapt existing major species to fit all conceivable agro-ecological environments.

Suitable legume species are also required for situations where land has become degraded. Overgrazing, excessive use of fire, agricultural disturbance, fragmentation and urbanization are causal factors that have led to increasing degradation of grasslands (Reynolds, 2001). In China, for example, about 34 percent of all rangelands are moderately to severely degraded, and some 90 percent are degraded to some degree (Miller, 2001). Again, it can be emphasized that the introduction of forage legumes is only one element in a grassland system that has to be economically, socially and ecologically sustainable. In parts of southern Australia, soil acidification and salinity have damaged or are threatening millions of hectares because of incomplete use of soil nitrogen and water by farming systems based on annual plants (subterranean clover, medics (*Medicago* spp.), cereals) (Cocks and Bennett, 1999).

Alfalfa

There is a wide range of existing cultivars with differing characteristics, but there is opportunity to make further breeding advances, not least because of ease of use of new biotechnologies (Bouton, 2001). Biotechnology tools in the form of molecular DNA markers and somatic hybridization, and plant transformation technologies involving the insertion of specific novel genes of one species into another, are being increasingly used in alfalfa plant breeding. For example, cultivars bred for increased resistance to potato leaf hopper (*Empoacea fabae*) by introgressing trichomes with glands on the leaf from two other *Medicago* species are now widely used in the USA (McCaslin, 1994).

Improved yield and persistence, multiple resistance to many diseases and pests, and cold temperature dormancy to improve cold hardiness are typical ongoing objectives. There have also been advances in tolerance to grazing without sacrificing yield and quality (Smith *et al.*, 2000) using a 'Standard Test Protocol' devised for alfalfa breeding and cultivar evaluation, a protocol that involves intensive grazing in a continuous stocking regime. In Canada, a cultivar has been developed—'AC Grazeland'—with reduced potential for causing bloat because of a low initial rate of digestion (Coulman *et al.*, 2000).

Other potentially promising developments include the use of alfalfa in low-input, extensively managed pastures, and its adaptation to hostile environments such as saline or acid, aluminium (Al)-toxic soils. This adaptation, together with resistance to high disease and pest pressure, and selection among non-autumn-dormant, winter-active germplasm could make alfalfa a viable forage for animal production in the tropics and subtropics (Bouton, 2001).

White clover

White clover has been improved by scientific breeding programmes since the late 1920s, and over 300 cultivars were listed by Caradus and Woodfield (1996), though the seed supplies of many are scarce. The rate of improvement of clover yield and content in the sward increased by 0.6 percent per year over a period of six decades (Woodfield and Caradus, 1994), but using cultivars adapted to New Zealand conditions, further work there indicated that genetic gains exceeded 1 percent per year for yield, with more rapid improvement in the past decade—4.4 percent per year under sheep grazing and 2.5 percent under cattle grazing (Woodfield, 1999). There is an annual world consumption of 8 000-10 000 tonne of seed (Mather, Melhuish and Herlihy, 1996) and the cultivar 'Grassland Huia', which has dominated the market for many years, is gradually being replaced by improved cultivars, particularly from the United Kingdom and from New Zealand. Table 1.1 shows selected features of New Zealand-bred cultivars.

Breeding objectives include improved compatibility with grasses, increased resistance to pests and diseases, and better adaptation to drought. Success has been achieved in combining spring growth and improved cold hardiness (Rhodes and Ortega, 1996). These two attributes had been previously considered mutually exclusive and so this breakthrough augurs well for similar developments in other forage legume species. In tandem with the development of new traits in white clover (and in other species), current evaluation protocols for National and Recommended List purposes should be more comprehensive than the current, fairly rigid, management regimes.

Red clover

Major breeding aims for red clover have included improved persistence beyond the normal longevity of two to three years, greater resistance to pests and diseases, and increased tolerance of grazing. The increased longevity of newly released red clover cultivars is partly due to improved disease resistance (Smith, 1997). Semiprostrate cultivars that reproduce by seed, but also vegetatively by stolons and daughter plants, have shown persistence under grazing (Smith and Bishop, 1993; Hyslop, Kemp and Hodgson, 1999).

Table 1.1 The agronomic positioning of white clover cultivars in New Zealand.

Cultivar and type	Land class or climatic adaptation	Stock class or management	Key seasonal performance (rel. to Huia)	Pest and disease tolerance
Prop Small	North Island dry hill country	set-stock sheep	winter/spring	leaf rust(2)
Tahora Small	North Island moist and dry hill country, and dry lowland	set-stock sheep	autumn	leaf rust
Prestige Medium/small	Hill and rolling country northern North Island, and dry lowland	set-stock and rotational sheep	winter/ spring/ summer	stem nematode(3), leaf rust, clover flea(4)
Demand Medium/small	Hill and rolling country South Island, and moist lowland	set-stock and rotational sheep	spring/ summer/ autumn	
Huia Medium	General	rotational sheep	(mid-spring)	
Sustain Medium /large	Rolling-flat	rotational sheep/ cattle/dairy	autumn/ winter/summer	stem nematode
Pitau Medium /large	Flat, warm winter	rotational cattle/ dairy	winter	
Challenge Large	Flat, warm winter	rotational cattle/ dairy	autumn/ winter	stem nematode, leaf rust
Kopu Large/ladino	Flat, North Island	rotational dairy	winter	stem nematode
Le Bons Large	Flat, North and South Island	rotational dairy	winter	
Aran(1) Very large	Flat, North Island	rotational dairy	summer/autumn	

NOTES: (1) Irish cultivar included for morphological comparison. (2) *Uromyces trifolii.* (3) *Ditylenchus dipsaci.* (4) *Sminthurus viridis.*
SOURCE: Adapted from Caradus, Hay and Woodfield, 1996.

In the USA, a seed crop from the aftermath growth of a conservation crop is commonly harvested for home use or inter-farm sales and so the financial rewards for marketing new cultivars are low. Thus breeding efforts have remained modest and mainly in the hands of publicly-funded institutes, in contrast to the vibrant and profitable breeding of alfalfa that is almost totally in the hands of private seed companies (McKersie, 1997). There has been some resurgence in breeding effort in Europe partly in response to red clover's high yield and nutritive value potential, and partly due to the increase in popularity of organic farming, where red clover can be used for green manure or as a break crop in arable-grassland rotations. High-yielding cultivars with improved persistence have been developed in Switzerland (Decamps *et al.*, 2001).

Subterranean clover

Annual subterranean clover is a classic example of how a species from one world region, the Mediterranean Basin in this case, has proved of enormous value to another region, southern Australia, where it was introduced by accident in the late nineteenth century and subsequently became a key component in farming systems. It begs the question of what other species, as yet unrealized, could play a similar role somewhere in the world. The majority of subterranean clover cultivars are of the subspecies *subterraneum*, but cultivars of the subspecies *yannicum* can tolerate poorer drained, acidic soils while those of the subspecies *brachycalycinum* are more suited to neutral and alkaline soils. An important breeding objective is the introduction of different flowering times so that cultivar maturities can fit in with the different onsets of summer drought, since seed development and burr burial in the soil are required for annual regeneration. A good degree of hardseededness is necessary as premature seed softening and unseasonal rains can lead to germination and seedling dieback before sustained rainfall arrives. Improved seedling vigour, better yield, more resistance to pests and diseases, and a low oestrogen content are other current breeding aims.

Birdsfoot trefoil

Birdsfoot trefoil is sown in mixtures for grazing but is also used, though to a lesser extent, for conservation. It is a pioneer legume for marginal infertile land that is less favourable for the more productive forage legumes. Nearly 800 000 ha are sown annually, e.g. in northeast USA, southeast Canada and southern Latin America (Blumenthal and McGraw, 1999) but its use could be extended to other less-developed grassland regions. It has proved suitable for reclamation and revegetation of disturbed sites – after mining, for example (Belesky, 1999). Its drought resistance (Peterson, Sheaffer and Hall, 1992) and tolerance of saline soils (Schachtman and Kelman, 1991) are attributes capable of being more widely exploited.

The vast genetic resources of the *Lotus* genus have not been fully tapped, with only birdsfoot trefoil (*Lotus corniculatus*) receiving significant breeding efforts in the past (Beuselinck, 1999) and none of the cultivars carries genes from related or distant *Lotus* species (Papadopoulos and Kelman, 1999). Therefore, there is opportunity to overcome the main shortcomings, namely slow establishment, lack of persistence (partly due to diseases and pests), and seed production. Rhizomatous growth of Moroccan genotypes has been transferred to a domesticated cultivar with the object of increased persistence under grazing (Beuselinck and Steiner, 1996).

Lesser forage legumes

Many lesser-used legumes are under evaluation and selection, sometimes for the first time, in countries in different parts of the world. Sainfoin (*Onobrychis viciifolia*) is drought tolerant but responds to irrigation on shallow soils; it is suited to infrequent cutting for hay or silage in dryland areas (Miller and Hoveland, 1995). Alsike clover (*Trifolium hybridum*) is an alternative to red clover for cold, infertile acidic soils and may be regarded as a pioneer-type legume. Stoloniferous strawberry clover (*T. fragiferum*) is adapted to wet saline or alkaline conditions and is highly tolerant of grazing. Rhizomatous Caucasian clover (*T. ambiguum*) is resistant to summer drought and winter cold and is also highly tolerant of grazing—and a notable source of honey (Taylor and Smith, 1998); one important breeding objective is to develop a white clover × Caucasian clover hybrid that combines their key characteristics. Sulla (*Hedysarum coronarium*) is a short-lived perennial or biennial useful for silage, hay, green feed or grazing; 250 000 ha are grown in southern Italy (Martiniello and Ciola, 1994) and it is being evaluated in other drylands, such as the Iberian peninsula. Greater lotus (*Lotus uliginosus*), is adapted to wet infertile soils, and a few cultivars have been released in Australasia and the USA. Selection for improved establishment tolerance of grazing, competitiveness with companion grasses and optimal content of condensed tannins (CT) are among the breeding aims (Blumenthal *et al.*, 1993). Greater lotus has also been used in silvipastoral systems as understorey in pine (*Pinus radiata*) plantations, where the forage is grazed and its N_2-fixation ability is harnessed to boost tree growth. Narrow-leaved birdsfoot trefoil (*Lotus glaber*) has proved suitable for saline and alkaline soils, and for swampy conditions (Vigniola, Fernandez and Maciera, 1999). Fodder galega (*Galega orientalis*) has shown promise for persistent high production in northern, winter-cold latitudes, though it does not reach full yielding ability until the stands are in their third year, and low forage digestibility is still a shortcoming (Raig *et al.*, 2001).

A number of annual clovers, mainly fast-establishing cool-season types, are adapted to differing soil conditions and managements – arrowleaf (*Trifolium vesiculosum*), crimson (*T. incarnatum*), rose (*T. hirtum*) and ball (*T. nigrescens*) clovers for example (Hoveland and Evers, 1995; Pederson, 1995). There is potential for the use of annual *Lotus* spp. in relation to Al, Mn and salt tolerance (Kelman, 1994), an example being *L. subbiflorus* cv. El Rincón, which is used in Uruguay (Asuaga, 1994). Vetches (*Vicia* spp.), forage pea (*Pisum sativum*), biennial sweetclover (*Melilotus* spp.) also have proven roles under various ecological conditions. For example, forage peas sown pure as a short-term crop can produce high yields in 14-18 weeks in both winter-cold zones (Sparrow, Cochran and Sparrow, 1993) and semi-arid zones (Caballero, Goicoechea and Hernaiz, 1995). Recent plant improvement programmes in southern Australia have identified a number of annual legume species adapted to degraded soils where subterranean clover no longer fits the bill; for example, yellow serradella (*Ornithopus compressus*), pink serradella (*O. sativus*) and biserrula (*Biserrula pelecinus*) for sandy, infertile, acid soils, and balansa clover (*Trifolium michelianum*) and Persian clover (*T. resupinatum*) for waterlogged and mildly saline soils (Norman *et al.*, 2000).

Tagasaste (*Chamaecytisus proliferus*), a perennial shrub or small tree that originates from the Canary Islands, has become naturalized and utilized in several warm, temperate regions, e.g. those with or near to Mediterranean-type climates (Wiley *et al.*, 1994) and has the ability to fix large quantities of N (Unkovich, 2000). Perhaps the best lesson to be drawn from tagasaste is the need to evaluate other shrubs that could conceivably have a role in temperate areas for forage, either grazed or cut and carried.

Research and development on the above and other lesser species are taking place in several countries, though fairly limited in extent, to judge from published papers. Nevertheless, the information will assist other countries with similar environmental conditions. Several drawbacks are associated with the exploitation of lesser legumes: insufficient resources from official sources for development; time lag between initial development and commercialization; limited availability of seed; and possible lack of knowledge initially for advisory services to spread the necessary technology.

Limitation of forage (and crop) production due to salinization on irrigated land is an increasing world problem and so salt tolerance is becoming a priority objective for plant breeders in a number of regions. Drought tolerance is also becoming an increasingly important objective: it has been estimated that vegetation growth on 40 percent of the area of the world's grazing lands is limited by moisture supply (Spedding, 1996). Detailed databases of the characteristics of temperate forage legumes have a role to play in identifying potential candidates for specific edaphic, climatic or biotic conditions.

SEED PRODUCTION

The greatest advances in forage legume (and grass) cultivar improvement and hence production of certified seed started in the early twentieth century. Government Seed Acts and stringent seed regulations followed, and with the upsurge in plant breeding, seed production research and development of new technologies, average seed yields steadily improved. Nevertheless seed yields from individual species are still highly variable. For example, in New Zealand, which produces half of the world's annual output of white clover, seed yields in the 1990s ranged from 100 to 1000 kg ha^{-1}, with a national average of 300 kg ha^{-1}, though skilled growers achieve 600 kg ha^{-1} (Mather, Melhuish and Herlihy, 1996). These findings, which are undoubtedly replicated for legume species elsewhere, indicate the scope for improvement in seed production technology, although random factors such as local weather or attack by diseases or pests can obviously affect outcomes. Attempts to widen the range of suitable regions or local areas capable of seed production have had mixed success, mainly because of lack of grower expertise. Apart from the growing of seed crops, adequate harvesting, cleaning, drying and storage facilities are necessary, since commercial markets can be limited for crops that fail to meet the standards for certification.

There is widespread commerce in uncertified seed of red clover and birdsfoot trefoil in the USA and arguably this is holding back advances in the traits of these species since breeding of proprietary cultivars is restricted on account of lack of commercial returns. This is in contrast to alfalfa, where there is a high rate of release of improved proprietary cultivars (Barnes and Sheaffer, 1995) and 35-40 tonnes of seed are produced annually, much of it for export. Certified seed of birdsfoot trefoil cultivars comprises the bulk of production in Canada (Beuselinck, 1997).

Improved forage legume cultivars developed in countries that do not have suitable climates or skills for multiplication, typically sub-contract growers in the traditional and favourable seed producing regions. For example, white clover cultivars developed in the United Kingdom are multiplied in specialist seed growing areas in New Zealand and northwestern USA. Market demands for most of the major forage legumes are likely to be met as before, but the question is how well will demands be met for seed of lesser legume species that fit specific ecological niches. Seed production of these lesser legumes may have to be carried out at official institutes or else officially subsidized (a conclusion that can also be drawn for secondary or lesser grasses that may be suitable in mixture with the legumes for specific situations).

CONTRIBUTION TO GRASSLAND NITROGEN ECONOMY

In world agriculture, it has been estimated that 74 million tonnes of N was supplied to crops and grassland as fertilizer in 1994, and 45 million tonnes of N supplied from N_2-fixation, indicating that about 38 percent of the world crop and grassland demand for N, excluding recycled animal waste, is met from biological N_2-fixation (van der Hoek, 1998). Output of N was estimated to be 49 million tonnes in crops and 83 million tonnes in consumed forage, i.e. 47 percent and 63 percent, respectively, of N output. Assuming these proportions apply to fixed N_2 utilized by crops and forage, grassland receives approximately 28 million tonnes of fixed N per annum, most of which will be derived from forage legumes. This N has an important role to play in the provision of home-grown forage protein (e.g. from legume monocultures), N transfer to grass (e.g. in grass-legume swards) and N supply to a subsequent arable crop (e.g. from ploughed-in red clover). Reliance on N_2-fixation varies widely among regions of the world. In Europe and North America, significant amounts of fertilizer N are applied to grassland, but in the rest of the world, much of N supply to grassland is derived from recycling of ingested N by the grazing animal, redistribution by application of animal manures from housed or tethered animals, or by N_2-fixation. There is also a question of how long the use of non-renewable energy for fertilizer N manufacture can be sustained, the obvious response being either reduced use of N fertilizer—with the corollary of more N_2 fixation being required—or else finding renewable sources of energy for N fertilizer manufacture.

In many systems with annual fertilizer applications of about 150-200 kg N ha^{-1}, production responses of grass-legume to N fertilizer are much lower than those of grass in monoculture, due mainly to direct replacement of the legume by N-stimulated grass growth. Thus N fertilizer is used consciously as an alternative to relying on N fixed by legumes in such systems.

Nitrogen fixation

In order to implement an N management programme for a production system involving forage legumes, it is important to estimate the likely contribution of the legume to the N economy of the system. However, a fundamental problem is the sheer variation in annual N_2 fixation that can occur for various reasons, such as amount of legume, its vigour of growth and the efficiency of the N_2 fixation process. Estimates have therefore ranged up to 600 kg N ha^{-1}, but are most commonly in the 150-250 kg N ha^{-1} range. Most of the estimates were derived from the amount of N_2 fixed in harvested herbage, and are probably underestimates by 19-25 percent (Høgh-Jensen and Kristensen, 1995).

Actual N_2 fixation rates in legume monocultures are relatively easily interpreted on a unit area basis. However, when in association with a companion species, usually a grass, comparison is more difficult, due to both the relative biomass of the legume being less than in monoculture, and the contribution of legume to the N economy of the accompanying species having to be taken into account. Generally, the proportion of total legume N derived from N_2-fixation (Ndfa) is usually 0.7 to 0.9, e.g. Vinther and Jensen (2000) for white and red clovers, Peoples *et al*. (1998) and Dear *et al*. (1999) for subterranean clover, Trystman and Smith (1998) for berseem and arrowleaf clovers, and Farnham and George (1994) for birdsfoot trefoil. Therefore, most of the variation in the contribution of fixed N among species is probably due to variation in dry matter yield. Nevertheless, there is potential to increase Ndfa, since the highest estimates may be at least 50 percent above that of the lowest in the range. The most obvious means of increasing Ndfa is to minimize the amount of soil mineral N. Companion grasses, albeit under strict control by management, have a role in this.

When estimating N_2 fixed in grazed swards, the proportion of N derived from fixation has to be adjusted as grazing reduces the potential of legumes to contribute fixed N to a sward. For example an Ndfa of 0.64 has been measured for white clover in grazed swards compared with 0.79 for an ungrazed sward (Eriksen and Høgh-Jensen, 1998), since recycled N from the grazing animal increases soil mineral N concentration and reduces the effectiveness of the nitrogen fixing enzyme, nitrogenase.

Improving nitrogen fixation

Selecting for improvements in nitrogen fixation efficiency has met with limited success. Alfalfa genotypes selected for improvement in various components of the fixation process under controlled environments and glasshouses (improved nodulation rate, increased root and nodule mass, higher nodule activity and nitrogen content of the dry matter) have not translated into major improvements in N_2 fixation under field conditions (Teuber and Phillips, 1988). Limitations in N_2 fixation most likely to be overcome are those associated with the rhizobium, either by selection from natural strains or by biotechnology. Surveys may identify strains that could be both highly effective and competitive against other strains and be exploited as inoculants. While the use of biotechnology is unravelling the complex interaction within the host plant-*Rhizobium/Bradyrhizobium* microbe symbiosis, at least 100 genes each from the micro-organism and the host plant must function in unison to obtain effective N_2 fixing root nodules, and so producing new strains is a difficult process (Vance, 1997).

Nitrogenase enzyme activity in the root nodule is more energy demanding than inorganic nitrogen assimilation within the legume plant. Thus, improvement in the efficiency of energy utilization in the fixation process offers potential to increase the N_2 fixation impact in forage legumes.

Nitrogen transfer

Soil mineral N mineralized from N in senescing legume roots and nodules is readily taken up by grass as grasses have a higher affinity for soil mineral N than the accompanying legume. If transfer is omitted in estimates of N input by legumes, contribution of legumes to N economy of a sward could be markedly underestimated. Potential routes of transfer are presented in Table 1.2; the main routes in both cut and grazed swards are death and decay of subterranean parts and litter.

Table 1.2 Routes of N 'transfer'.

Main routes:	Decomposition of litter, unharvested herbage and subterranean organs
	Grazing animal
Other routes:	Root-feeding fauna
	Leakage from roots and nodules
	Vesicular arbuscular fungal hyphal connections
	Nitrous oxide (N_2O), ammonia (NH_3) from legume

SOURCE: Summarized from Laidlaw and Teuber, 2001.

Transfer of fixed N from white or red clover to grass can form more than one-third of total fixed N (Frame, Charlton and Laidlaw, 1998). In alfalfa-grass mixtures, less N is generally transferred than in other legume-grass mixtures.

Grazing increases the rate of transfer of fixed N_2, as regular defoliation increases the rate of turnover of nodules and much of the 'harvested' N passing though the animal is potentially and readily mineralizable. Thus, in grazed swards, relatively high combined transfer of legume N to grass has been measured. In New Zealand, Ledgard (1991) estimated that about 20 percent of fixed N was transferred by dairy cows grazing a grass-white clover sward and a further 25 percent was transferred underground.

As N_2 fixation rate per unit weight of legume and the proportion of fixed and legume N transferred vary widely, and as the amount transferred will influence the vigour with which grass will compete with the legume, the long-term balance between grass and legume is not easily predicted.

At extreme soil (low or high) mineral N concentrations, one or other of the grass and clover components is excluded, but intermediate levels do not necessarily guarantee co-existence (Schwinning and Parsons, 1996a,b). Co-existence in the area as a whole depends on varying soil mineral N concentration, influenced by the plants but operating against a background of varying

received is an all-important factor, and this is achieved by using all resources to best effect. In the recent past in the United Kingdom, grass-legume systems have compared favourably in economic terms with grass-N fertilizer systems when the outputs of the two were similar (Leach *et al.*, 2000) and also with grass receiving about 200 kg N ha^{-1} annually, but not with higher rates (300-400 kg N ha^{-1}) due to a lower stock carrying capacity (Doyle and Bevan, 1996). In most economic models, account is not taken of the generally higher risk or higher standard of management required when relying on forage legumes rather than N fertilizer.

ROLE OF FORAGE LEGUMES IN ENVIRONMENTAL ISSUES

Nitrogen loss

Nitrogen loss from ruminant systems contributes to three environmental concerns, namely nitrate (NO_3^-) leaching, with implications for contaminating drinking water; emission of ammonia (NH_3), affecting the chemistry of the atmosphere; and nitrous oxide (N_2O) production, a greenhouse gas resulting from denitrification in the soil. Although legumes do not input inorganic N directly into the ecosystem in which they are growing, their high N content contributes to the higher-than-desired rate of N leaching when grazed due to the N in urine from grazing animals nitrifying to nitrate (McDuff, Jarvis and Roberts, 1990). The stock carrying capacity of the sward therefore seems to be a more important determinant of nitrate leached than sward type (Cuttle, Scurlock and Davies, 1998; Hooda *et al.*, 1998). However, there is now evidence to suggest that the ecosystem of a mixed grass-white clover sward can adjust to the high soil mineral N concentrations that inevitably develop in such an N-rich system by reducing N_2-fixation. With time, the level of leaching becomes relatively low and independent of the capacity of the soil to supply mineral N (Loiseau *et al.*, 2001).

Theoretically, due to the rate of cycling in a grazed grass-/white clover system, the required input of N from the legume to sustain animal production equivalent to a grass sward receiving about 200 kg N ha^{-1} may be much lower than is usually expected (Parsons *et al.*, 1991), relatively high rates of animal production being achieved from grass-white clover sward containing 5-15 percent white clover in above-ground herbage (Orr *et al.*, 1990; Steen and Laidlaw, 1995). It has been calculated by Jarvis, Wilkins and Pain (1996) that, provided enough white clover is present to fix just over 70 kg N ha^{-1} $annum^{-1}$, it could sustain milk production equivalent to 80 percent of that from a system receiving 250 kg N ha^{-1} and with a reduction in N losses of almost 60 percent. However, this assumes that a given grass-clover balance can be maintained from year to year, whereas the grass-legume relationship is dynamic, and so it is unreasonable to expect a steady balance and amount of N_2-fixation to be maintained. Clearly, more information is required on the dynamics of N in grazing systems in general and grass-legume systems in particular.

Strategies to reduce the potentially high rates of N that can be lost from legume or grass-legume crops when ploughed up include avoiding grazing before ploughing to reduce the mineralization rate of aboveground herbage and removal of as much herbage as possible before ploughing to reduce the pool of mineralizable N (Davies, Vinten and Smith, 1996). Ploughing in spring rather than autumn is another suggested strategy, as N leaching is high during winter from autumn-ploughed ground, even if sown with a winter cereal (Høgh-Jensen, 1996; Djurhuus and Olsen, 1997).

Restrictions on N fertilizer usage are either being considered or are already in force in some countries. Examples within the EU include the designation of Nitrate Vulnerable Zones, where N fertilizer use is capped, or limits on N usage in some Environmentally Sensitive Area schemes are prescribed. Imposition of a tax on N fertilizer is another strategy being considered in some countries. While these would seem to be opportunities for increased use of legumes, the lack of unequivocal evidence in favour of legumes reducing environmental damage relative to N fertilizer prevents positive advocacy of legumes as substitutes for N fertilizer. This is exemplified in a theoretical analysis of the impact of an N fertilizer tax on cattle production systems in Denmark (Hasler, 1998). Considering the possibility of 'crop substitution', where N-fertilized grass could be replaced by grass-legume, it is concluded that nitrate leaching would not be reduced. Until unequivocal evidence is presented to the contrary, it seems unlikely legumes will be specified as alternatives in policies underpinning nitrate reduction programmes.

Environmental change

In general, it is expected that as atmospheric CO_2 content increases, white clover content in mixtures with perennial ryegrass will increase. Almost doubling CO_2 content increased yields of grass-white clover miniswards by 12-29 percent and clover content by 7-10 percent during the first five months after sowing (Schenk, Jaeger and Weigel, 1997). Doubling CO_2 concentration increased subterranean clover yields by 31 percent when grown with phalaris (*Phalaris aquatica*) (Lilley *et al.*, 2001). Cocksfoot responded better than red clover to elevated CO_2 concentration in a stress-free environment, while the benefits were reversed when mineral N was limiting (Meier and Fuhrer, 1997).

The increase in temperature that is likely to accompany elevated CO_2 concentration is predicted to reduce the beneficial effect of CO_2 elevation on subterranean clover (Lilley, Bolger and Gifford, 2001). From a modelling study, a combination of CO_2 elevated by 50 percent and mean temperature of 2°C was predicted to increase white clover content from 31 to 46 percent (mean over four sites). Although global warming would increase growth (and hence CO_2 uptake), it would also increase C turnover in the soil (Ross *et al.*, 1996) and the low phenolic content of legume roots will stimulate decomposition at higher temperatures (Martens, 2000), negating transient benefits of higher C fixation.

Although the impact of environmental change on dry matter production and morphological development has been well studied in some forage legumes, the effect on nutritional aspects has been less well covered. Other than at the beginning of the season, crude protein (CP) was increased and fibre content decreased in a perennial ryegrass-white clover mixture due to CO_2 elevation; there was a positive effect on the Ca:P ratio and the content of most elements increased (Schenk, Jaeger and Weigel, 1997). Condensed tannins in *Lotus corniculatus* increased with doubling of the CO_2 concentration, but declined with rising temperature or slight drought stress (Carter, Theodorou and Morris, 1995).

As already stated, some areas are expected to become drier than at present as CO_2 levels increase. Therefore it would be assumed that, as many of the forage legumes are relatively sensitive to drought, they would be at a disadvantage in drier areas as the climate changes. However, alfalfa has been shown to increase its drought tolerance under elevated CO_2 due to increased net photosynthesis resulting in higher biomass, more nodules and a higher N content (De Luis, Irigoyen and Sanchez-Diaz, 1999).

Taking these studies as a whole, the balance of evidence suggests that legumes will generally be no more adversely affected by climate change than other forage species. However, much will depend on the rate of increase in temperature or decrease in precipitation relative to the rate of change in CO_2, due to the independent, sometimes conflicting, influences these exert on legume growth and development.

ALTERNATIVE USES

Forage legumes have a number of biological advantages beyond their specific role in grassland farming. Inclusion of red clover as a living understorey with wheat in a 'no-till' soya bean-wheat-maize rotation increased soil temperature for the establishing maize sown into the wheat stubble and germination of the maize crop was advanced by 3-4 days (Drury *et al.*, 1999). The clover can reduce the possibility of excess water presence and provide less residue to hinder the developing maize than other living mulches. Potatoes following red clover may derive benefit from the development of endophyte bacterial associations in the soil encouraged by the clover crop (Sturtz and Christie, 1998).

White clover and some of the annual clovers, e.g. crimson clover, have been undersown in spaced arable crops such as maize to protect the soil from erosion and minimize soil damage from harvesting operations. White clover, pure-sown or in mixture with grasses, acts as a protective ground cover against rain runoff erosion and as a soil stabilization plant, where its stoloniferous growth is advantageous, and as a weed suppressing, living mulch in orchards and vineyards (Parente and Frame, 1993); subterranean clover is also a suitable species as a cover crop, though cold tolerance is needed for sub-Mediterranean and continental environments with low winter temperatures, e.g. in Italy (Pecetti, Carroni and Piano, 2000).

Other benefits of legumes to soil include increase in soil aggregate stability, aided by their association with mycorrhizal fungi; the hyphae and decomposing legume roots bind soil particles into stable aggregates and the relatively rapid turnover of roots provides substrates for microbes to produce polysaccharides that also contribute to improving soil structure (Miller and Jastrow, 1996; Misra *et al.*, 1999).

Because of its high growth rate, alfalfa has potential as a biomass crop for energy, a raw material for the production of biodegradable plastics or as a substrate for production of fermentation products such as lactic acid or alcohol (Talamucci, 1994). Protein, carotene and xanthophylls in extracted juice from fractionation are other possible products. Alfalfa is also under investigation as a bioremediation system for removal of harmful nitrates, and as a source of industrial enzymes (Bouton, 2001).

Agricultural legumes enhance the beauty of the landscape, especially with their different flower colours. The contrast of colours is also utilized for roadside vegetation covers where natural reseeding of annual clovers or the spread by stolons of perennial clovers are valuable assets. Their seed and forage are also important food sources for various forms of wildlife. Several lesser legumes are recommended for use in wild flora or nature conservation mixtures that thrive on infertile soils and become havens for wildlife. Some legumes are sold as ornamental plants for gardens, e.g. birdsfoot trefoil, well leaf-marked selections or 'lucky' four-leaved white clovers. The leaves of the latter are also incorporated into various gift items including jewellery. Other species could also fulfil the role of novelty garden ornamentals.

In terms of human nutrition, the production of honey by bees is of tremendous value, particularly from white clover and some *Medicago* and *Lotus* species. The leaves of white clover can be used as a salad or a vegetable while a 'wine' can be made from clover flowerheads, such as white clover. Sprouted alfalfa seeds are another delicacy. There is a new wave of interest in exploiting secondary compounds of plants in general by the medical and pharmaceutical industry, and legumes are being screened as part of this. One example is the extraction of isoflavones (genistein, diadzen, biochanin A, formononetin) from red clover as dietary food supplements.

CONCLUDING REMARKS

It is foreseen that the role of forage legumes in sustainable farming systems in temperate regions will increase, certainly in the long term if not the short term. In the countries of the EU, land-use polices have changed from a production orientation to environmental protection, landscape enhancement and nature conservation. Reduction in N fertilizer is a major strand in such policies and this creates opportunity for increased use of forage legumes, particularly but not exclusively in stock extensification policies. In other countries, e.g. in eastern Europe and the Baltic countries, and in less developed grasslands, such as in some of the world's rangelands, the use of forage legumes will depend on the availability and cost of N fertilizer in relation to end-product prices and in some cases the availability and cost of legume seed supplies. Development of indigenous forage legumes should be a major aim for rangelands, thus avoiding complete reliance on cultivars developed elsewhere that may not be totally adapted to the agro-ecological conditions. Where land has degraded, e.g. due to salinization or acidity, a wider spectrum of legumes than before is necessary.

Where grassland is intensively used, e.g. on farms with limited land area and highly N-fertilized grass, forage legume use is less likely, particularly if the N fertilizer is cost effective. On such farms strong measures to prevent environmental pollution will be instituted, as illustrated by the legislation in the Netherlands to prevent N and P pollution of watercourses and watertables. In this respect, zero grazing (cut-and-carry) systems offer more control than grazing vis-à-vis recycling of excretal N and P, and also more control by management of the grass-legume balance in mixed swards, a stability that is difficult to achieve in long-term grazed swards. Zero grazing also offers the chance to grow and utilize separate grass and forage legume monocultures, for which management is simpler than for mixed swards. One example of a successful legume monoculture is the alfalfa-maize conservation system of dairying in North America. The increased costs of zero compared with natural grazing has to be offset by sufficiently high commodity prices. High grain prices, perhaps because of poor harvests, allied to a decline in milk prices, could encourage a greater interest in grassland-based farming. Certainly, as the global marketplace becomes more open and world competition increases, low unit cost of commodity products will be increasingly necessary.

The scenarios above require exploitation of existing knowledge and it is advantageous that there is a considerable body of information on forage legumes from which to draw. There are also some world centres of excellence advancing forage legume technology (Sulas, 2000; Norman *et al.*, 2000). Nevertheless, because of the scope envisaged for their wider use, a great deal of further research and development, particularly systems development as opposed to component-part research, is required. This need is greatest where the legume ethos for grasslands is relatively new. The resource needs to develop forage legume technology further, call for more international cooperation and networking than hitherto to share costs, allocate areas of responsibility and replicate research in different environments.

REFERENCES

Anderson, G.C., Fillery, I.R.P., Dolling, P.J. & Asseng, S. 1998. Nitrogen and water flows under pasture-wheat and lupin-wheat rotations in deep sands in Western Australia. 1. Nitrogen fixation in legumes, net nitrogen mineralisation and utilization of soil-derived nitrogen. *Australian Journal of Agricultural Research,* 49: 329-343.

Asuaga, A. 1994. *Lotus subbiflorus* cv. El Rincón, a new alternative for extensive improvements of natural pastures. pp. 147-150, *in: Proceedings of the 1st International Lotus Symposium.* St. Louis, Missouri, 22-24 March 1994. Columbia Press, University of Missouri.

Barnes, D.K. & Sheaffer, C.C. 1995. Alfalfa. pp. 205-216, *in:* R.F. Barnes, D.A. Miller and C.J. Nelson (eds). *Forages* 5th edition. Vol. 1. *An introduction to grassland agriculture.* Ames, Iowa: Iowa State University Press.

Barry, T.N., McNeill, D.M. & McNabb, W.C. 2001. Plant secondary compounds; their impact on forage nutritive value and upon animal production. pp. 445-552, *in:* J.A. Gomide, W.R.S. Mattos and S. Carneiro da Silva (eds). *Proceedings of the XIX International Grassland Congress.* São Pedro, Brazil, 11-21 February 2001. FEALQ, Piracicaba.

Bax, J.A. & Browne, I. 1995. The use of clover on dairy farms. Milk Development Council, Cirencester *Research Summary.* 20p.

Beever, D.E. & Thorp, C. 1996. Advances in the understanding of factors influencing the nutritive value of legumes. pp. 194-207, *in:* D. Younie (ed). *Legumes in sustainable farming systems. British Grassland Society Occasional Symposium,* No.30.

Belesky, D.P. 1999. Lotus species used in reclamation, renovation and revegetation. pp. 133-143, *in:* P.R. Beuselinck (ed). *Trefoil: the science and technology of lotus.* ASA/CSSA, Madison, Wisconsin. *CSSA Special Publication,* No.28.

Beuselinck, P.R. 1997. *Lotus corniculatus* L. in North America. pp. 351-360, *in:* D.T. Fairey and J.G. Hampton (eds). *Forage seed production.* Vol. 1. *Temperate species.* Wallingford, UK: CAB International.

Beuselinck, P.R. (ed) 1999. *Trefoil: the science and technology of Lotus.* ASA/CSSA, Madison, Wisconsin. *CSSA Special Publication,* No.28.

Beuselinck, P.R. & Steiner, J.J. 1996. Registration of 'ARS-2620' birdsfoot trefoil. *Crop Science,* **36:** 1414.

Blumenthal, M.J. & McGraw, R.L. 1999. Lotus adaptation, use and management. pp. 97-119, *in:* P.R. Beuselinck (ed). *Trefoil: the science and technology of lotus.* ASA/CSSA, Madison, Wisconsin. *CSSA Special Publication,* No.28.

Blumenthal, M.J., Kelman, W.M., Lolicato, S., Hare, M.D. & Bowman, A.M. 1993. Agronomy and improvement of *Lotus*: a review. pp. 74-85, *in:* D.L Michalk., A.D. Craig and W.J. Collins (eds). *Alternative pasture legumes 1993.* Department of Primary Industries, South Australia. *Technical Report,* No. 219.

Bouton, J.H. 2001. Alfalfa. pp. 545-547, *in:* J.A. Gomide, W.R.S. Mattos and S. Carneiro da Silva (eds). *Proceedings of the XIX International Grassland Congress.* São Pedro, Brazil, 11-21 February 2001. FEALQ, Piracicaba.

Buringh, P. & Dudal, R. 1987. Agricultural land use in space and time. *In:* M.G. Wolman and F.G.A. Fournier (eds). *Land transformation in agriculture.* Scope vol. 32. Chichester, UK: John Wiley & Sons.

Caballero, R., Goicoechea, E.L. & Hernaiz, P.J. 1995. Forage yields and quality of common vetch and oat sown at varying seeding ratios and seeding rates of vetch. *Field Crops Research,* **41**: 135-140.

Caradus, J.R. & Williams, W.M. 1989. Breeding for legume persistence in New Zealand. pp. 523-539, *in:* G.C. Marten *et al.* (eds). *Persistence of Forage Legumes.* Proceedings of a Trilateral Workshop. Honolulu, Hawaii, 17-22 July 1988. ASA, Madison, Wisconsin.

Caradus, J.R. & Woodfield, D.R. 1996. World checklist of white clover varieties. *New Zealand Journal of Agricultural Research,* **40**: 115-206.

Caradus, J.R., Hay, R.J.M. & Woodfield, D.R. 1996. The positioning of white clover cultivars in New Zealand. pp. 45-49, *in:* D.R. Woodfield (ed). *White clover: New Zealand's competitive edge.* New Zealand Grassland Association *Grassland Research and Practice Series,* No. 6.

Caradus, J.R., MacKay, A.C., Woodfield, D.R., Van Den Bosch, J. & Wewala, S. 1989. Classification of a world collection of white clover cultivars. *Euphytica,* **42**: 183-196.

Carter, E., Theodorou, M.K. & Morris, P. 1995. Responses of *Lotus corniculatus* to environmental change. 2. Effect of elevated CO_2, temperature and drought on tissue digestion in relation to condensed tannin and carbohydrate accumulation. *Journal of the Science of Food and Agriculture,* **79**: 1431-1440.

Chapman, D.E., Parsons, A.J. & Schwinning, S. 1996. Management of clover in grazed pastures: expectations, limitations and opportunities. pp. 55-64, *in:* D.R. Woodfield (ed). *White clover: New Zealand's competitive edge.* New Zealand Grassland Association *Grassland Research and Practice Series,* No. 6.

Cocks, P.S. & Bennett, S.J. 1999. Introduction: Role of pasture and forage legumes in Mediterranean farming systems. pp. 9-19, *in:* S.J. Bennett and P.S. Cocks (eds). *Genetic resources of Mediterranean pasture and forage legumes.* Dordrecht, The Netherlands: Kluwer Academic Publishers.

Coulman, B., Goplen, B., Majak, W., McAllister, T., Cheng, K.-J., Berg, B., Hall, J., McCartney, D. & Acharya, S. 2000. A review of the development of a bloat-reduced alfalfa cultivar. *Canadian Journal of Plant Science,* **80**: 487-491.

Cuttle, S.P., Scurlock, R.V. & Davies, B.M.S. 1998. A 6-year comparison of nitrate leaching from grass/clover and N-fertilized grass pastures grazed by sheep. *Journal of Agricultural Science* (Cambridge), **131**: 39-50.

Davies, M.G., Vinten, A.J.A. & Smith, K.A. 1996. The mineralisation and fate of nitrogen following the incorporation of grass and grass/clover swards. pp. 133-134, *in:* D. Younie (ed). *Legumes in sustainable farming systems. British Grassland Society Occasional Symposium,* No. 30.

Dear, B.S., Cocks, P.S., Peoples, M.B., Swan, A.D. & Smith, A.B. 1999. Nitrogen fixation by subterranean clover (*Trifolium subterraneum* L.) growing in pure culture and in mixtures with varying densities of lucerne (*Medicago sativa* L.) or phalaris (*Phalaris aquatica* L.). *Australian Journal of Agricultural Research,* **50**: 1047-1058.

Decamps, C., Collignon, A., Van Der Vennet, D., Lambert, R., Toussaint, B. & Peeters, A. 2001. Persistance, qualité et rendement de variétiés de trèfle violet (*Trifolium pratense* L.) dans des associations fourragères en Moyenne et Haute Belgique [Persistency, quality and yield of red clover (*Trifolium pratense* L.) varieties in forage mixtures in Mid and High Belgium]. Laboratoire d'Écologie des Prairies/Centre Provincial d'Information Agricole, Michamps, Bastogne. *Research Bulletin.* 27p.

del Pozo, M., Wright, I.A, Whyte, T.K. & Colgrove, P.M. 1996. Effects of grazing by sheep or goats on sward composition in ryegrass/white clover pasture and on subsequent performance of weaned lambs. *Grass and Forage Science,* **51**: 142-154.

De Luis, I., Irigoyen, J.J. & Sanchez-Diaz, M. 1999. Elevated CO_2 enhances plant growth in droughted N_2-fixing alfalfa without improving water status. *Physiologia Plantarum,* **107**: 84-89.

Djurhuus, J. & Olsen, P. 1997. Nitrate leaching after cut grass/clover leys as affected by time of ploughing. *Soil Use and Management,* **13**: 61-67.

Doyle, C.J. & Bevan, K. 1996. Economic effects of legume-based grassland systems. *British Grassland Society Occasional Symposium,* No. 30.

Drury C.F., Tan, C.-S., Welacky, T.W., Oloya, T.O., Hamill, A.S. & Weaver, S.E. 1999. Red clover and tillage influence on soil temperature, water content and corn emergence. *Agronomy Journal,* **91**: 101-108.

Eriksen, J. & Høgh-Jensen, H. 1998. Variations in the natural abundance of ^{15}N in ryegrass/white clover shoot material as influenced by cattle grazing. *Plant and Soil,* **205**: 67-76.

Falcinelli, M., Veronesi, F., Russi, L. & Pollodori, P. 1993. Persistence and productivity of some forage varieties and landraces of different origin grown in central Italy. pp. 266-268, *in:* M.J. Baker (ed). *Proceedings of the XVII International Grassland Congress, New Zealand/Australia.* Palmerston North, NZ: New Zealand Grassland Association and others.

FAO [Food and Agriculture Organization of the United Nations]. 1988. *Tropical forage legumes.* Prepared by P.J. Skerman, D.G. Cameron and F. Riveros. Rome: FAO.

Farnham, D.E. & George, J.R. 1994. Dinitrogen fixation and nitrogen transfer in birdsfoot trefoil and orchard grass communities. *Agronomy Journal,* **86**: 690-696.

Frame, J., Charlton, J.F.L. & Laidlaw, A.S. 1998. *Temperate forage legumes.* Wallingford, UK: CAB International.

Frankow-Lindberg, B.E., Danielson, D.A. & Moore, C.A. 1996. The uptake of white clover technology in farm practice. *FAO/REUR Technical Series,* No. 42: 37-43. FAO, Rome.

Granstedt, A. 2000. Ecological grassland management in the Nordic countries. pp. 27-33, *in:* R. Viiralt (ed). *Conventional and ecological grassland management: comparative research and development.* Proceedings of an International Symposium, Tartu, Estonia, 4-6 July 2000.

Halling, M., Hopkins, A., Nissinen, O., Paul, C., Tuori, M. & Soelter, U. 2002. Forage legumes – productivity and composition. pp. 5-15, *in:* R.J. Wilkins and C. Paul (eds). *Legume silages for animal production – LEGSIL.* Braunschweig, Germany: FAL Agricultural Research.

Harris, S.L., Clark, D.A., Auldist, M.J., Waugh, C.D. & Laboyrie, P.G. 1997. Optimum white clover content for dairy pastures. *Proceedings of the New Zealand Grassland Association,* **59**: 29-33.

Hasler, B. 1998. Analysis of environmental policy measures aimed at reducing nitrogen leaching at the farm level. *Environmental Pollution,* **102**: 749-754.

Hauck, R.D. 1988. A human ecosphere perspective of agricultural nitrogen cycling. pp. 3-19, *in:* J.R. Wilson (ed). *Advances in nitrogen cycling in agricultural ecosystems.* Wallingford, UK: CAB International.

Hill, M.J., Hockney, M.J., Mulcahy, C.A. & Rapp, G. 1995. The effect of hydrogen cyanide potential (HCNp) and sward morphology on the relative acceptability to sheep of white clover and Caucasian clover herbage. *Grass and Forage Science,* **50**: 1-9.

Høgh-Jensen, H. 1996. Nitrogen recovery after clover/ryegrass leys. pp. 130-132, *in:* D. Younie (ed). *Legumes in sustainable farming systems. British Grassland Society Occasional Symposium,* No. 30.

Høgh-Jensen, H. & Kristensen, E.S. 1995. Estimation of biological N_2-fixation in a clover-grass system by the ^{15}N dilution method and the total-N difference method. *Biological Agriculture and Horticulture,* **11**: 203-219.

Hooda, P.S., Moynagh, M., Svoboda, I.F. & Anderson, H.A. 1998. A comparative study of nitrate leaching from intensively managed monoculture grass and grass-clover pastures. *Journal of Agricultural Science* (Cambridge), **131**: 267-275.

Hoveland, C.S. & Evers, G.W. 1995. Arrowleaf, crimson and other annual clovers. pp. 249-260, *in:* Barnes, R.F., Miller, D.A. and Nelson, C.J. (eds) *Forages* 5th edition, Vol. 1, *An Introduction to Grassland Agriculture*. Iowa State University Press, Iowa.

Hyslop, M.G., Kemp, P.D. & Hodgson, J. 1999. Vegetatively reproductive red clovers (*Trifolium pratense* L.): an overview. *Proceedings of the New Zealand Grassland Association*, **61**: 121-126.

Jarvis, S.C., Wilkins, R.J. & Pain, B.F. 1996. Opportunities for reducing the environmental impact of dairy farm managements: a systems approach. *Grass and Forage Science*, **51**: 21-31.

Jones, L. & Clements, R.O. 1993. Development of a low-input system for growing wheat (*Triticum vulgare*) in a permanent understorey of white clover. *Annals of Applied Biology*, **123**: 109-119.

Kelman, W.M. 1994. Lesser known *Lotus* species. pp. 86-87, *in:* D.L. Michalk, A.D. Craig and W.J. Collins (eds). *Alternative pasture legumes 1993*. Department of Primary Industries, South Australia. *Technical Report*, No.219.

Laidlaw, A.S. & Teuber, N. 2001. Temperate forage grass-legume mixtures: advances and perspectives. pp. 85-92, *in:* J.A. Gomide, W.R.S. Mattos and S. Carneiro da Silva (eds). *Proceedings of the XIX International Grassland Congress*. São Pedro, Brazil, 11-21 February, 2001. FEALQ, Piracicaba.

Leach, K.A., Bax, J.A., Roberts, D.J. & Thomas, C. 2000. The establishment and performance of a dairy system based on perennial ryegrass-white clover swards compared with a system based on nitrogen fertilized grass. *Biological Agriculture and Horticulture*, **17**: 202-217.

Leake, A. 2001. Integration: farm and the environment. *Biologist*, **48**: 159-162.

Ledgard, S.F. 1991. Transfer of fixed nitrogen from white clover to associated grasses in swards grazed by dairy cows estimated using ^{15}N methods. *Plant and Soil*, **131**: 215-223.

Ledgard, S.F., Penno, J.W. & Sprosen, M.S. 1999. Nitrogen inputs and losses from clover/grass pastures grazed by dairy cows, as affected by nitrogen fertilizer. *Journal of Agricultural Science* (Cambridge), **132**: 215-225.

Lilley, J.M., Bolger, T.P. & Gifford, R.M. 2001. Productivity of *Trifolium subterraneum* and *Phalaris aquatica* under warmer high CO_2 conditions. *New Phytologist*, **150**: 371-383.

Lilley, J.M., Bolger, T.P., Peoples, M.B. & Gifford, R.M. 2001. Nutritive value and the nitrogen dynamics of *Trifolium subterraneum* and *Phalaris aquatica* under warmer, high carbon dioxide conditions. *New Phytologist*, **150**: 385-395.

Loiseau, P., Carrere, P., Lafarge, M., Delpy, R. & Dublanchet, J. 2001. Effect of soil N and urine N on nitrate leaching under pure grass, pure clover and mixed grass/clover swards. *European Journal of Agronomy*, **14**: 113-121.

Martens, D.A. 2000. Plant residue biochemistry regulates soil carbon cycling and carbon sequestration. *Soil Biology and Biochemistry*, **32**: 361-369.

Martiniello, P. & Ciola, A. 1994. The effect of agronomic factors on seed and forage production in perennial legumes sainfoin (*Onobrychis viciifolia* Scop.) and French honeysuckle (*Hedysarum coronarium* L.). *Grass and Forage Science*, **49**: 121-129.

Mather, R.D.J., Melhuish, D.T. & Herlihy, M. 1996. Trends in the global marketing of white clover cultivars. pp. 7-14, *in: White Clover: New Zealand's Competitive Edge. Grassland Research and Practice Series*, No. 6. New Zealand Grassland Association, Palmerston North.

McCaslin, M. 1994. Breeding for potato leafhopper resistance in alfalfa. p. 61, *in: Report of 34th North American Alfalfa Improvement Conference*. Guelph, Canada, 10-14 July 1994.

McDuff, J.H., Jarvis, S.C. & Roberts, D.H. 1990. Nitrates: leaching from grazed grassland systems. pp. 405-410, *in:* R. Calvet (ed). *Nitrates, agriculture, water*. Paris: INRA.

McKersie, B.D. 1997. Improving forage production systems using biotechnology. pp. 3-21, *in:* B.D. McKersie and D.C.W. Brown (eds). *Biotechnology and the improvement of forage legumes*. Wallingford, UK: CAB International.

McMahon, L.R., Majak, W., McAllister, T.A., Hall, J.W., Jones, G.A., Popp, J.D. & Cheng, K.J. 1999. Effect of sainfoin on *in vitro* digestion of fresh alfalfa and bloat in steers. *Canadian Journal of Animal Science*, **79**: 203-212.

McMahon, L.R., McAllister, T.A., Berg, P.B., Majak, W., Acharya, S.N., Popp, J.D., Coulman, B.E., Wang, Y. & Cheng, K.-J. 2000. A review of the effects of forage condensed tannins on ruminal fermentation and bloat in grazing cattle. *Canadian Journal of Plant Science*, **80**: 469-480.

Meier, M. & Fuhrer, J. 1997. Effect of elevated CO_2 on orchard grass and red clover grown in mixture at two levels of nitrogen or water supply. *Environmental and Experimental Biology*, **38**: 251-262.

Miller, D. 2001. Looking back to move ahead: integrating indigenous nomadic knowledge into the modern range profession in China. Paper presented at the *Society for Range Management Annual Meeting, Session on Rangeland Professionals and Society: Future Directions*. Kailua-Kona, Hawaii, USA, 20 February 2001.

Miller, D.A. & Hoveland, C.S. 1995. Other temperate legumes. pp. 273-281, *in:* R.F. Barnes, D.A. Miller and C.J. Nelson (eds). *Forages*. 5th edition. Vol. 1. *An introduction to grassland agriculture*. Ames, Iowa: Iowa State University Press.

Miller, R.M. & Jastrow, J.D. 1996. Contribution of legumes to the formation and maintenance of soil structure. pp. 105-112, *in:* D. Younie (ed). *Legumes in sustainable farming systems. British Grassland Society Occasional Symposium,* No.30.

Misra, A.K., Daniel, H., Till, R. & Blair, G.J. 1999. Effect of long-term crop rotations and rewetting of soil on stability. *Soil Use and Management,* **15**: 254-255.

Morris, P. & Robbins, M.P. 1997. Manipulating condensed tannins in forage legumes. pp. 147-174, *in:* B.D. McKersie and D.C.W. Brown (eds). *Biotechnology and the improvement of forage legumes*. Wallingford, UK: CAB International.

Nicholls, N. & Wong, K.K. 1990. Dependence of rainfall variability on mean rainfall, latitude and the southern oscillation. *Journal of Climate,* **3**: 163-170.

Norman, H.C., Ewing, M.A., Loi, A., Nutt, B.J. & Sandral, G.A. 2000. The pasture and forage industry in the Mediterranean bioclimates of Australia. *Cahiers Options Méditerranéennes,* **45**: 469-476.

Orr, R.J., Parsons, A.J., Penning, P.D. & Treacher, T.T. 1990. Sward composition, animal performance and the potential production of grass/white clover swards continuously stocked with sheep. *Grass and Forage Science,* **45**: 325-336.

Papadopoulos, Y.A. & Kelman, W.M. 1999. Traditional breeding of *Lotus* species. pp. 187-198, *in:* P.R. Beuselinck (ed). *Trefoil: the science and technology of lotus*. ASA/CSSA, Madison, Wisconsin. *CSSA Special Publication,* No.28.

Parente, G. & Frame, J. 1993. Alternative uses of white clover. *FAO/REUR Technical Series,* No.29: 30-36.

Parsons, A.J., Orr, R.J., Penning, P.D. & Lockyer, D.R. 1991. Uptake, cycling and fate of nitrogen in grass-clover swards continuously grazed by sheep. *Journal of Agricultural Science* (Cambridge), **116**: 47-61.

Pecetti, L., Carroni, A.M. & Piano, E. 2000. Cold tolerance of subterranean clover in a continental-climate environment. *Cahiers Options Méditerranéennes,* **45**: 43-46.

Pederson, G.A. 1995. White clover and other perennial clovers. pp. 227-236, *in:* R.F. Barnes, D.A. Miller and C.J. Nelson (eds). *Forages*. 5th edition. Vol. 1. *An introduction to grassland agriculture*. Ames, Iowa: Iowa State University Press.

Peoples, M.B., Gault, R.R., Scammell, G.J., Dear, B.S., Virguoa, J., Sandral, G.H., Paul, J., Wolf, E.C. & Angus, J.F. 1998. Effects of pasture management on the contributions of fixed N to the N economy of ley farming systems. *Australian Journal of Agricultural Research,* **49**: 459-474.

Peterson, P.R., Sheaffer, C.C. & Hall, M.H. 1992. Drought effects on perennial forage legume yields and quality. *Agronomy Journal,* **84**: 774-779.

Quesenberry, K.H. & Casler, M.D. 2001. Achievements and perspectives in the breeding of temperate grass and legumes. pp. 517-524, *in:* J.A. Gomide, W.R.S. Mattos and S. Carneiro da Silva (eds). *Proceedings of the XIX International Grassland Congress*. São Pedro, Brazil, 11-21 February 2001. FEALQ, Piracicaba.

Rae, A.N. 2001. The role of grasslands in world food trade. p. 1017, *in:* J.A. Gomide, W.R.S. Mattos and S. Carneiro da Silva (eds). *Proceedings of the XIX International Grassland Congress*. São Pedro, Brazil, 11-21 February 2001. FEALQ, Piracicaba.

Raig, H., Nõmmsalu, H., Meripõld, H. & Metlitskaja, J. 2001. Fodder Galega. Estonian Research Institute of Agriculture, Saku.

Rebuffo, M. & Abadie, T. 2001. Genetic resources for temperate areas – achievements and perspectives. pp. 469-476, *in:* J.A. Gomide, W.R.S. Mattos and S. Carneiro da Silva (eds). *Proceedings of the XIX International Grassland Congress*. São Pedro, Brazil, 11-21 February 2001. FEALQ, Piracicaba.

Reynolds, S.G. 2001. Sustainable development of grassland ecosystems: two case studies from China. Paper presented at *The International Symposium on Sustainable Development of Grassland Ecosystems (SDGE-2001)*, Xilinhot, China, 27-30 August 2001. 14p.

Rhodes, I. & Ortega, F. 1996. Progress in forage breeding. pp. 62-71, *in:* D. Younie (ed). *Legumes in sustainable farming systems. British Grassland Society Occasional Symposium,* No. 30.

Riveros, F. 1993. Grasslands for our world. pp. 15-20, *in:* M.J. Baker (ed). *Proceedings of the XVII International Grassland Congress*. New Zealand and Australia. Palmerston North, NZ: New Zealand Grassland Association and others.

Ross, D.J., Saggar, S., Tate, K.R., Feltham, C.W. & Newton, P.C.D. 1996. Elevated CO_2 effects on carbon and nitrogen cycling in grass/clover turves of a Psammaquent soil. *Plant and Soil,* **182**: 185-198.

Schachtman, D.P. & Kelman, W.M. 1991. Potential of *Lotus* germplasm for the development of salt, aluminium and manganese tolerant pasture plants. *Australian Journal of Agricultural Research,* **42**: 139-149.

Schenk, U., Jaeger, H-J., & Weigel, H-J. 1997. The response of perennial ryegrass/white clover miniswards to elevated atmospheric CO_2 concentrations: Effects on yield and fodder quality. *Grass and Forage Science,* **52**: 232-241.

Schwinning, S. & Parsons, A.J. 1996a. Analysis of the coexistence mechanisms for grasses and legumes in grazed systems. *Journal of Ecology,* **84**: 799-813.

Schwinning, S. & Parsons, A.J. 1996b. A spatially explicit population model of stoloniferous N-fixing legumes in mixed pasture with grass. *Journal of Ecology,* **84**: 815-826.

Seligman, N.G. 1998. Management of Mediterranean grasslands. pp. 359-391, *in:* J. Hodgson and A.W. Illius (eds). *The ecology and management of grazing systems.* Wallingford, UK: CAB International.

Smith, R.R. 1997. Increased longevity of red clover through selection for persistence and disease resistance. Vol. 1, pp. 4.19-4.20, *in: Proceedings of the XVIII International Grassland Congress.* Winnipeg, Saskatoon, Canada. 8-19 June 1997.

Smith, R.S. & Bishop, D.J. 1993. Astred – a stoloniferous red clover. pp. 421-423, *in:* M.J. Baker (ed). *Proceedings of the XVII International Grassland Congress.* New Zealand and Australia. Palmerston North, NZ: New Zealand Grassland Association and others.

Smith, S.R., Jr., Bouton, J.H., Singh, A. & McCaughey, W.P. 2000. Development and evaluation of grazing-tolerant alfalfa cultivars: A review. *Canadian Journal of Plant Science,* **80**: 503-512.

Sparrow, S.D., Cochran, V.L. & Sparrow, E.B. 1993. Herbage yield and nitrogen accumulation by seven legume crops on acid and neutral soils in a subarctic environment. *Canadian Journal of Plant Science,* **73**: 1037-1045.

Spedding, C.R.W. 1996. *Agriculture and the citizen.* London: Chapman & Hall.

Stafford Smith, D.M. 1998. Management of rangelands: Paradigms at their limits. pp. 325-357, *in:* J. Hodgson and A.W. Illius (eds). *The ecology and management of grazing systems.* Wallingford, UK: CAB International.

Steen, R.W.J. & Laidlaw, A.S. 1995. The effect of fertilizer nitrogen on the stock-carrying capacity of ryegrass/white clover swards. *Irish Journal of Agricultural and Food Research,* **34**: 123-132.

Stewart, T.A. & Haycock, R.E. 1984. Beef production from low N and high N S24 perennial ryegrass/Blanca white clover swards – a six-year farmlet scale comparison. *Research and Development in Agriculture,* **1**: 103-113.

Sturtz A.V & Christie B.R. 1998. The potential benefits from cultivar-specific red clover-potato crop rotations. *Annals of Applied Biology,* **133**: 365-373.

Sulas, L. (ed). 2000. Legumes for Mediterranean forage crops, pastures and alternative uses. *Cahiers Options Méditerranéennes,* **45**: 1-483.

Talamucci, P. 1994. Lucerne's role in farming systems, technical itineraries and managements for different uses in diverse physical and socio-economic environments. *FAO/REUR Technical Series,* No.36: 6-17.

Tava, A., Forti, D. & Odoardi, M. 1992. Alfalfa saponins: isolation, chemical characterisation and biological activity against insects. pp. 283-288, *in:* P. Rotili and L. Zannone (eds). *The Future of Lucerne – Biotechnology Breeding and Variety Constitution.* Proceedings of the X International Conference of the EUCARPIA *Medicago* spp. Group, Lodi, Italy, 15-19 June 1992. Instituto Sperimentale per le Colture Foraggere, Lodi.

Taylor, N.L. & Smith, R.R. 1998. Kura clover (*Trifolium ambiguum* M.B.) breeding, culture and utilization. *Advances in Agronomy,* **63**: 153-178.

Teuber, L.R. & Phillips, D.A. 1988. Influences of selection method and N environment on breeding alfalfa for increased forage yield and quality. *Crop Science,* **28**: 599-604.

Tiffin, P. & Hesterman, O.B. 1998. Response of corn grain yield to early and later killed red clover green manure and subirrigation. *Journal of Production Agriculture,* **11**: 112-121.

Trystman, M. & Smith, M.F. 1998. Comparison of methods used for determining nitrogen fixation in annual clovers. *Applied Plant Science,* **12**: 20-23.

Unkovich, M.J. 2000. Nitrogen isotope fractionation in the fodder tree legume tagasaste (*Chamaecytisus proliferus*) and assessment of N_2-fixation inputs on deep sandy soils of Western Australia. *Australian Journal of Plant Physiology,* **27**: 921-929.

van der Hoek. 1998. Nitrogen efficiency in global animal production. *Environmental Pollution,* **102** (Supplement 1): 127-132.

Vance, C.P. 1997. Nitrogen fixation capacity. pp. 375-407, *in:* B.D. McKersie and D.C.W. Brown (eds). *Biotechnology and the improvement of forage legumes.* Wallingford, UK: CAB International.

Vigniola, O.R., Fernandez, O.N. & Maciera, N.O. 1999. Flooding tolerance in five populations of *Lotus glaber* Mill. Syn. *Lotus tenuis* Waldst & Kit ex Willd. *Australian Journal of Agricultural Research,* **50**: 555-559.

Vinther, F.P. & Jensen, E.S. 2000. Estimating legume N_2 fixation in grass-clover mixtures of a grazed organic cropping system using two ^{15}N methods. *Agriculture Ecosystems and Environment,* **78**: 139-147.

Waghorn, G.C., Reed, J.D. & Ndlovu, L.R. 1997. Condensed tannins. pp. III: 153-166, *in: Proceedings of the XVIII International Grassland Congress.* Winnipeg, Saskatoon, Canada, 8-19 June 1997.

Waghorn, G.C., Shelton, I.D. & Thomas V.J. 1989. Particle breakdown and rumen digestion of fresh ryegrass (*Lolium perenne* L.) and lucerne (*Medicago sativa* L.) fed to cows during a restricted feeding period. *British Journal of Nutrition,* **61**: 409-423.

Wheeler, J.L. & Vickery, P.J. 1989. Variation in HCN potential among cultivars of white clover (*Trifolium repens*). *Grass and Forage Science,* **44**: 107-109.

Wiley, T.J., Oldham, C.M., Allen, G. & Wiese, T. 1994. *Tagasaste.* Department of Agriculture, Western Australia. *Bulletin,* 4291. 23p.

Wilman, D., Mtengeti, E.J. & Moseley, G. 1996. Physical structure of 12 forage species in relation to rate of intake by sheep. *Journal of Agricultural Science* (Cambridge), **126**: 277-285.

Wivstat, M. 1999. Nitrogen mineralization and crop uptake of N from decomposing ^{15}N labelled red clover and yellow sweet clover plant fractions of different age. *Plant and Soil*, **208**: 21-31.

Woodfield, D.R. 1999. Genetic improvements in New Zealand forage cultivars. *Proceedings of the New Zealand Grassland Association*, **61**: 3-7.

Woodfield, D.R. & Caradus, J.R. 1994. Genetic improvement in white clover representing six decades of plant breeding. *Crop Science*, **34**: 1205-1213.

CHAPTER 2

The Future Role of Forage Legumes in Mediterranean-climate Areas

L. Sulas[1]

ABSTRACT

In the areas of the world with a Mediterranean-type climate, forage and pasture legumes are a basic component of production systems. The Mediterranean basin is not only the centre of origin and primary diversity of almost all pasture and forage legumes, but also the current and future world source of their genetic resources. New varieties of legumes have been released in Mediterranean basin countries, and, in southern Australia, the presence of environmental, economic and biological constraints has led to a novel generation of "alternative" pasture legumes, with increasing attention paid to perennials.

As well as for traditional use, many pasture and forage legumes are being exploited for other purposes. Legumes can be specifically used in firebreak lines; in silvipastoral systems; in organic agriculture; as cover crops in cropping systems, vineyards and orchards; and for other agronomic and environmental purposes. Pharmaceutical and medicinal uses are also being explored.

In this chapter, new germplasm options for Mediterranean forage legumes and their future role are discussed in relation to constraints and recent perspectives, dealing with seed multiplication, valorization of legume germplasm, aspects of legume-*Rhizobium* symbiosis, forage quality and bio-active compounds.

INTRODUCTION

The Mediterranean basin, Cape Province of South Africa, California (USA), central Chile and southern Australia are the five areas of the world with Mediterranean-type climates (Aschmann,

[1]CNR – Istituto per il Sistema Produzione Animale in Ambiente Mediterraneo, Via E. De Nicola, 07100 Sassari, Italy.

1973). By far the largest is the Mediterranean basin, which includes 17 countries, plus Portugal, Iran and Iraq, with a total surface under a Mediterranean climate of 138 million hectares (Le Houérou, 1980); about half of this territory is located in Europe. In the Mediterranean basin and areas with homologous climate, agricultural activities and farming have traditionally been of great economic significance, and forage and pasture legumes represent a basic component of their production systems.

Due to the rich flora, the Mediterranean basin, with part of West Asia, is the centre of origin and of primary diversity of almost all pasture and forage legumes. Moreover, these areas represent the current and future world source of these species, whose genetic diversity for morphological and ecophysiological attributes is huge. This is documented by the several germplasm collections made in the recent past (Cocks, 1993) and still in progress.

The basic reasons for the forage legumes from the Mediterranean basin being breeding material of high value are that they derive from a wide range of different microclimates and pedologic situations. This ecological heterogeneity increases genetic variation over short distances (and consequently enhances collecting results). Another feature is the remarkable abiotic and biotic stress resistance characterizing natural legume populations, and in particular the adaptation of many forage legumes to grazing and high stocking rates.

To cope with the very high environmental variability, Mediterranean pasture and forage legumes have developed distinctive strategies to ensure adaptation and reproduction (Cocks, 1999); knowledge of these is fundamental to identifying and exploiting important attributes of these species – attributes not fully considered so far.

Moreover, an ancient heritage of human activity and social aspects (Papanastasis, 1998) has been very important in the natural selection of pasture legumes. This still exists in less developed countries, farming being still a primary activity, compared with Mediterranean Europe, where relatively few years of subsidy-based agricultural policy can easily modify or extirpate age-old rural traditions.

At the same time, legume germplasm collected from North African countries is much appreciated and used in selection programmes at world level, but it has produced very limited returns in the countries of origin, due to socio-economic, cultural and technical constraints that sometimes outweigh environmental limitations.

Spontaneous legumes, mainly annuals, are an important component of the ecosystem and grassland production. In addition, many forage legumes are currently grown in a complex mosaic of environmental conditions and farming systems dominated by small ruminants (Porqueddu and Sulas, 1998). Although, some perennial legumes have almost disappeared from the farming systems, there is renewed interest in them, such as sulla (*Hedysarum coronarium*).

Annual medics (*Medicago* spp.) and subterranean clover (*Trifolium subterraneum sensu lato*) are better known and grown as self-reseeding pasture species. From the Mediterranean basin, these species were introduced by European settlers to Mediterranean-type climatic areas of southern Australia, central Chile, California (USA), and, probably by Arab traders, to the Cape region of South Africa. Their introduction, naturalization and diffusion had a remarkable impact on the new environments, particularly in Australia (Cocks, 1999), where annual legumes have effectively contributed to sustaining and increasing cereal and animal production, being a basic component of ley and phase farming systems.

As a consequence, almost all commercial varieties available in the seed market have been developed in Australia, jointly with seed harvesting systems and special vacuum harvesters. Since the mid-1980s, more than 50 cultivars of annual legumes, mainly following germplasm collections, have been released in Australia for both domestic use and export as seed.

In California (USA), *Trifolium hirtum* and *T. subterraneum* arrived in 1930, either via Australia or directly from Mediterranean countries. These clovers were adapted to the annual rangelands of California and they had the potential to increase livestock production by improving forage quality and availability (Graves *et al.*, 2001). Later, they were also used as cover crops, and have spread to several south-eastern states of the USA.

In central Chile's Mediterranean-climate area, there are more than seven million hectares of pastures dominated by naturalized annual plants, originating in the Mediterranean basin and part of Eurasia, after the clearance of the indigenous sclerophyllus woodlands. Despite the proven potential in increasing pasture production, particularly on infertile soils, the utilization of local forage legumes has been little explored and developed (Ovalle *et al.*, 1999). Varieties of subterranean clover and annual medics are imported from Australia, and perennial legumes from USA and New Zealand.

In the winter rainfall subregions of South Africa, one of the main problems is the decrease of productivity and sustainability of the legume component, attributed to the use of varieties selected for other environments. In order to find an alternative to wheat monoculture, the identification of new productive forage legume systems is required. Recently, a collaborative research project to develop a selection and screening programme for naturalized annual medics has been started (M.T. Oberholzer, pers. comm.).

NEW GERMPLASM AND PROMISING LEGUME SPECIES FOR MEDITERRANEAN AREAS

New varieties released in Mediterranean basin countries

Commercial varieties of annual self-reseeding legumes imported from Australia have sometimes been unsuitable for pasture improvement in Mediterranean basin countries, due to the different management systems, low frost tolerance, and, in particular, to a genetic base too narrow to cope with the mosaic of existing situations. Equally, attempts to introduce Australian ley farming into the Mediterranean basin were unsuccessful, despite the annual legumes used having their origin exclusively in the same area.

Some failures in the introduction of Australian cultivars and a rising consciousness of the value of native legumes have stimulated genetic, morpho-physiological and agronomic studies aimed at the valorization of local germplasm. Several selection programmes have been carried out by various research institutions in the last 20 years.

In Spain (Gonzáles López, 1994), subterranean clover varieties have been selected, with five cultivars registered in 1985 (Orellana, Coria, Areces, Valmoreno, Gaitan), and two in 1989 (Zujaran and Cubillana). In Italy, four cultivars of subterranean clover (Antas, Campeda, Losa, Limbara) have recently been registered (Piano, Pecetti and Carroni, 1997) and a *Medicago polymorpha* L. cultivar, Anglona (Porqueddu, Loi and Fara, 1998). New cultivars of subterranean clover have also been selected in Portugal and France (P. Masson, pers. comm.) and also *M. polymorpha* cv. Mauguio (Prosperi, Ronfort and Genier, 1999). Some cultivars of annual medics were selected in Morocco (Abdelguerfi *et al.*, 2000). In Tunisia, two cultivars of *M. truncatula* and one cultivar of *Scorpiurus muricatus*, recommended for arid and semi-arid areas, have been released (Hassen *et al.*, 1998), while a cultivar of *T. subterraneum* is currently in process of registration, but all these cultivars have not been multiplied due to the lack of proper seed harvesters (A. Zoghlami, pers. comm.).

The new varieties proved to be superior to imported Australian ones, being specifically selected for the environmental and management conditions of the countries of origin. Unfortunately, the absence of a seed multiplication industry has prevented their wide diffusion, being available in the seed market only if multiplied in Australia. This great disadvantage has probably discouraged interest in selection and valorization activities for these species.

Species and varieties for the future: alternative annual pasture legumes for Mediterranean-type systems in Australia

The major difference between legume use in Australian ley farming and the grasslands of the Mediterranean basin is the uniformity of Australian systems, where ley farming is based on very few legume species, which are managed as monocultures. The changing pattern of agriculture makes the maintenance of a wide genetic base essential. For many forage species that become a priority, plant collecting missions are still necessary to provide germplasm for adaptation to economic, sociological and environmental changes (Francis, 1999). Ecogeographical study permits the prediction of areas and habitat that are likely to contain particular species, thus making the collection work easier.

In contrast, the collections of annual medics and subterranean clover available in Australia are considered more than adequate for the foreseeable future, but potentially valuable alternatives have been neglected. For example, the genus *Trifolium,* with more than two hundred species, of which only 15 are used commercially around the word, is just one source where new varieties could be found. Similarly, among the 30 annual medic species native to the Mediterranean basin, only seven have commercial cultivars (*Medicago littoralis, M. murex, M. polymorpha, M. rugosa, M. scutellata, M. tornata* and *M. truncatula*). Even if not all the annual medics have agronomic potential, their great diversity has been not fully evaluated and exploited within the genus. If work is not restricted to a particular species or genus, the potential for improvement is unlimited in the short term (Evans, 1996).

A priority list of promising legumes was drawn up by Francis (1999) (see Table 2.1), with emphasis on easily harvestable species, as an alternative to the traditional subterranean clover and annual medics and to promote a wider range of environmental adaptation. This list can be expected to change progressively as germplasm is evaluated and as new challenges arise.

Nevertheless, it is important to point out that species used currently will continue to be improved. In fact, subterranean clover is still considered a very important pasture legume in southern Australia, and it is grown on around 20 million hectares. This species, with a wide range of maturities, and with three subspecies adapted to different soil types, can be used successfully in both permanent pastures and crop rotations. Incremental improvements in subterranean clover will continue to be made as more rigorous selection procedures are applied, using both traditional breeding and modern biotechnology techniques (Evans, 1996).

By using traditional breeding methods, varieties with adaptation to specific conditions and pests can be developed. Better winter growth and tolerance to different management systems could be achieved. Such improvements are generally made from existing, well-adapted species.

Genetic engineering is another tool that can be used to introduce desirable characters into already successful cultivars. Resistance to herbicides has been incorporated into some varieties of *Trifolium subterraneum,* while resistance to viral diseases and root rots may be transferred.

Table 2.1 Pasture and forage legumes of which more germplasm is needed for different soil conditions and uses

Acid sandy soil	Fine-textured calcareous soils	Non-calcareous fine soils
Ornithopus sativus	*Trifolium purpureum*	*Trifolium pallidum*
Lupinus luteus	*T. lappaceum*	*T. vesiculosum*
Biserrula pelecinus	*T. alexandrinum*	*T. clypeatum*
Trifolium incarnatum	*T. dasyurum*	*T. glanduliferum*
T. cernuum	*T. leuchantum*	*T. michelianum*
	T. argutum	*T. squarrosum*
	Trigonella balansae	*T. spumosum*
	Hedysarum coronarium	*T. nigrescens*
	Onobrychis viciifolia	*T. hirtum*
Forages and hays	**Salinity/waterlogging**	**Agroforestry/shade places**
Lathyrus clymenum	*Melilotus indica*	*Trifolium clusii*
L. tingitanus	*M. officinalis*	*T. boisseri*
L. ochrus	*Lotus corniculatus*	
L. cicera	*L. sativus*	
Vicia villosa	*Trifolium resupinatum*	
V. sativa subsp. *macrocarpa*	*T. fragiferum*	
V. pannonica	*T. clusii*	
V. articulata	*T. physodes*	

SOURCE: Modified from Francis, 1999.

Annual medics are still key species in areas of southern Australia, but the older cultivars are very susceptible to aphids. Five medic cultivars (*Medicago truncatula* cvs. Caliph, Jester and Mogul; *M. littoralis* cv. Herald; *M. tornata* cv. Toreador), suitable for various purposes in the dryland regions, have recently been released (Korte *et al.*, 2001) as part of the National Annual Pasture Legumes Improvement Programme (NAPLIP). These cultivars are resistant to one or more aphids, and will produce more herbage and consequently fix more nitrogen in the soil than older, susceptible forms. Moreover, *M. truncatula* has emerged lately as a worldwide model system for functional genomics and biotechnology; in fact, this diploid species, with its relatively small genome, is very attractive for molecular genetics (Colebatch, Trevaskis and Udvardi, 2002).

However, much larger improvements in production and persistence can come from "alternative legumes" in ecological niches unsuitable for subterranean clover or annual medics (i.e. traditional pasture legumes). In fact, some alternative species can provide a close match, not only to ecological niches but also to different farming systems. Species that tolerate salinity and waterlogging can allow an increased stocking rate in areas normally considered wasteland, while releasing "better" land for cropping.

In southern Australia, the start of legume-based farming, using subterranean clover, annual medics and superphosphate fertilization, has been fundamental in supporting wheat-wool sheep rotations in the last 100 years. Today, this system is considered unsustainable as a result of environmental, economic and biological problems and constraints (Howieson, O'Hara and Carr, 2000). Problems include the development of secondary salinity and soil acidification, the continuously declining terms of trade, the decline in the price ratio for animal and plant products, the high cost of seed production and associated environmental problems, the low persistence of subterranean clover and annual medics in acid and sandy soils, the development of herbicide resistance in weeds, and the arrival of new insect pests and diseases. In addition, there are still wide agro-ecological niches without well-adapted pasture species.

These problems have led to a definition of a new ideotype of pasture annual legume, with relevant and desirable agronomic and ecological traits (Carr, 1997). The most important agronomic attributes are cheap and easy seed harvestability with conventional cereal harvesters; tolerance of acid soils; deep-rootedness to exploit water and nutrients better; good seed production; small seed size; an adequate level of hardseededness; and tolerance to grazing, diseases and pests. The current or future replacement of ley farming by phase farming, where cropping will be continuous for 4-6 years followed by a period of legume-dominated pastures for 2-3 years, implies additional roles for forage legumes. In particular, the forage legumes should also be able to compete against herbicide-tolerant weeds and to transpire deep water from the soil profile in order to mitigate the serious problem of rising water tables, which is causing losses of arable land to salinity in Australia, due to the salt dissolved in the groundwater. Moreover, the acidification of soil under annual pasture is a significant environmental problem in the wheat belt and permanent pastures. This has been attributed to the leaching of nitrate beyond the root zone of shallow-rooted annual legumes, where it cannot be effectively utilized by traditional pasture species.

A new generation of annual pasture legumes has been released in the Australian market since the establishment of the Centre for Legumes in Mediterranean Agriculture (CLIMA) in 1992, followed by the activity, still in progress, of NAPLIP. The first step was germplasm collection of new legumes species, looking at alternatives to subterranean clover and annual medics. Cultivars of several new species have been released, such as biserrula (*Biserrula pelecinus* cv. Casbah), French serradella (*Ornithopus sativus* cv. Cadiz), crimson clover (*Trifolium incarnatum*, cv. Caprera), gland clover (*T. glanduliferum* cv. Prima), arrowleaf clover (*T. vesiculosum*, cv. Cefalù) and improved varieties of balansa clover (*T. michelianum* cvs. Bolto and Frontier) and yellow serradella (*Ornithopus compressus* cvs. Charano and Santorini). In a very short time, they have had a strong positive impact on profitability and sustainability of southern Australian farming systems (Loi *et al.*, 2000).

For instance, on waterlogged grey clay soils, *T. michelianum* and *T. resupinatum* significantly outproduced the annual medics and subterranean clover previously recommended for the area (Evans and Snowball, 1993), while arrowleaf clover, a late-maturing annual with a deep tap root has a prolonged growth period. As these species have small seeds, which survive passage through the digestive tract of herbivores, they can be severely grazed during summer.

Biserrula pelecinus is typical of the alternative legumes; originating from the Mediterranean basin, it is a new pasture legume suitable for acid, sandy surfaced, duplex soils (Loi, Howieson and Carr, 2001), showing agronomic adaptation and highly desirable attributes. It is a prolific seed producer and the soft papery pods are easily threshed and processed by conventional harvesters. It also has a root system three times deeper than subterranean clover and serradella, a characteristic that may provide a longer period of green feed for grazing animals, due to its ability to utilize water and nutrients from deeper in the soil profile. In addition, it is extremely hard seeded, with only a small proportion (<2%) of seeds softening during the first summer.

Indeed, the "alternative legume" designation comes from a strictly Australian point of view, as some of these species cannot be considered new or alternative in Mediterranean basin countries, where *T. incarnatum* (Plate 2.1), for example, is a very common component of an annual mixture with oats. Varieties of arrowleaf clover were already released in the United States of America in the late 1950s, based on germplasm from Italy. Sulla is not only a native species but also a traditional legume crop. In fact, its attributes have been documented in several Italian papers during the last three hundred years, but it is a totally new species in Australia.

Plate 2.2 *Biserrula pelecinus* (top) and *Trifolium glanduliferum* cv. Prima (Gland clover) (bottom). Photographs by Dr Angelo Loi, Centre for Legumes in Mediterranean Agriculture (CLIMA), The University of Western Australia, Nedlands WA 6907, and Dr Brian Dear, Principal Research Scientist, NSW Agriculture, Wagga Wagga Agricultural Institute, Australia.

For the full valorization of forage legumes, in particular, in new cultivation environments, the role of associated rhizobia is fundamental. Some unsuccessful introductions of forage legumes in new Mediterranean-type environments were blamed on the lack of adequate microsymbionts, whose presence in fields by inoculation is strictly necessary. In southern Australia, efforts to improve the level of nitrogen fixation in pastures dominated by *Trifolium* and *Medicago* species have focused on the selection and introduction of better rhizobia. Improved rhizobia have been selected for high levels of nitrogen fixation, adaptation to specific soil conditions (such as acidity), and for good competitive ability (Hebb *et al.*, 1998). In areas where rhizobia are already present, their importance and potential have been traditionally overlooked. Nevertheless, the inoculation of Mediterranean forage legumes by introducing more effective strains for nitrogen fixation can enhance symbiotic effects (Sulas *et al.*, 1998).

Deriving maximum benefits from the development of new legume varieties requires concurrent detailed investigation of rhizobia ecology. Currently, all collecting of Mediterranean forage legumes germplasm also includes the isolation of the related rhizobia, because plant and microsymbiont have evolved together. This does not imply optimal nitrogen fixation, but it can be improved after the introduction of more competitive and efficient strains. An interesting experiment was carried out in Syria to study the performances of *M. aculeata* and *M. rigidula* accessions inoculated with different strains of *Rhizobium meliloti* (Sultan *et al.*, 2001). It demonstrated that efficient nodulation remains the only satisfactory alternative source of nitrogen that increases dry matter production, and may extend the vegetative growth period, particularly in rangelands where it is neither practical nor economical to add nitrogen fertilizer. In addition, some strains were more efficient than others under low winter temperatures, enhancing the growth and forage yield of the annual medics during this season.

In use, rhizobia are incorporated into a carrier, and commercial formulations for legume inoculation are produced in many countries (Date, 2001). Different forms are available (powdered, granular, liquid) according to the seeding technique. Unfortunately, some inoculants are still not available in many Mediterranean countries, and this constrains the use of more efficient strains.

Recent studies have found important and complex aspects requiring investigation in the future in order to achieve full efficiency from the legume-rhizobia symbiosis (Amarger, 2001; Thies, Holmes and Vachot, 2001). For example, where there are large, natural populations of rhizobia in the soil, the inoculation is invariably of little immediate effect, but, once introduced into soil, rhizobia can exchange genetic information with saprophytic or rhizosphere bacteria, generating in a just few years new populations of legume-nodulating rhizobia by lateral transfer of symbiotic genes to non-symbiotic resident bacteria (Sullivan and Ronson, 1998). Moreover, these ineffective populations of rhizobia could form a barrier to the introduction of more effective strains. For all these reasons, knowledge of the fate of the introduced strains in the soil environment is critical.

Given the constraints in the introduction of new inoculant strains, plant selection has been recently proposed as an alternative strategy for improving pasture legume symbiosis in the field. Thus, a plant that forms a highly effective association with all sources of soil rhizobia is desirable. However, in species such as *Medicago sativa*, *M. polymorpha*, and to a lesser extent *M. littoralis* and *T. michelianum*, significant improvements in nitrogen fixation appear possible (Ballard *et al.*, 2000). There is ample evidence that pasture legumes vary at both an inter- and intra-species level in their ability to form an effective symbiosis across a range of soil rhizobia occurrences. Following this approach, seed inoculation can be avoided with related practical implications for farmers.

FORAGE QUALITY PREDICTION AND IMPORTANCE OF BIO-ACTIVE COMPOUNDS

In Mediterranean pasture legumes, the quality of dry forage as standing hay and pods is very important. Their contribution to feed animals is important as the high crude protein contents complement the poor quality of stubble and dry crop residues in late spring and summer. This topic is still little investigated, despite the long duration of the dry season and the associated effects on legume persistence.

Forage quality in annual medics was reviewed by Porqueddu (2001), who suggested also primary selection criteria (e.g. leaf:stem ratio, stem thickness, chemical composition, digestibility and intake, palatability, and anti-nutritional factors) to improve feeding value in annual medics to be used in Mediterranean farming systems.

Due to wide genetic variation in annual legumes, much still unexplored, traditional methods of selection and breeding should generate important results in terms of improved forage quality. The production of transgenic plants does not seem to be a priority for the immediate future, except in subterranean clover. In fact, high-sulphur subterranean clover has been developed to increase wool growth, and work continues to obtain high-sulphur, rumen-degradable protein in the leaves.

Recent improvement in the prediction of forage quality in Mediterranean legumes

The major factor affecting legume nutritive value is the stage of growth at time of grazing or harvest, which is markedly influenced by season. An interesting method for quantifying the morphological stage of forage legumes is that of Kalu and Fick (1981), which proposes a numerical scale for individual stems and procedures for determining a mean stage weight (MSW) for the lucerne (*M. sativa*) canopy.

Interesting adaptations of this method are being applied to other legumes, both perennials and annuals. After definition of a specific phenological scale for sulla, its quantified morphological stage was evaluated as a predictor of pre-harvesting herbage quality in Italy (Borreani *et al.*, 1998). The MSW showed itself to be a good predictor of sulla organic matter digestibility. The use of the quantified morphological stage proved to be an easy pre-harvest tool to enhance yield and quality of sulla and to determine cutting or grazing schedules for more profitable management of the crop.

A similar methodology was used by Zu, Sheaffer and Barnes (1996), who developed a system for quantifying maturity in annual medics. Twelve Australian cultivars of six species of annual medics were evaluated at four locations, with determination of forage yield, quality and maturity. When forage crude protein (CP), neutral detergent fibre (NDF) and acid detergent fibre (ADF) concentrations were regressed against maturity, a significant positive correlation was observed between fibre concentration and maturity.

The equations based on MSW, suitably re-calibrated for homogeneous environmental areas, could be an easy and fast tool for researchers and farmers to assess digestibility of forage species.

In the future, this method could be expanded to other Mediterranean legumes, reducing the need for expensive chemical analysis, and with other advantages, including the possibility of comparing different legumes at the same morphological stage.

The role and potential of secondary plant compounds: the condensed tannins in forage legumes

Several groups of secondary plant compounds are present in forage legumes, being involved in the defence against herbivores, insects and pathogens, and in the regulation of symbiosis and allelopathy (Reed *et al.*, 2000). Some secondary plant compounds of forage legumes are associated with improved nutritive value and may have beneficial effects on animal health and welfare.

Condensed tannins (CTs) are phenolic compounds found in a number of important forage genera of Leguminosae, including *Coronilla, Hedysarum, Lespedeza, Lotus, Onobrychis* and *Trifolium* (McMahon *et al.*, 2000). They are found in cell walls or as intracellular inclusions in stems, bark, leaves, flowers or seeds. Tissue-specific localizations have been observed in a number of forage species. In *T. repens*, CTs are found only in the flowers; in lucerne only in the seed coat; while in sainfoin (*Onobrychis viciifolia*), CTs are present in all organs except the cotyledons and roots, with the highest concentration occurring in the leaves (Lees, Gruber and Suttil, 1995).

Condensed tannin (CT) can react by hydrogen bonding with plant proteins in the near neutral pH range to form CT-protein complexes, which are stable and insoluble at pH 3.5-7.0, but which dissociate and release protein at pH <3.5. Thus, CTs contained in forage plants can protect dietary proteins against degradation in the rumen and increase amino acid supply for absorption in the small intestine. CT-containing species do not cause bloat, while moderate CT concentrations (2-4% of the dry matter) also enhance forage nutritive value in grazing ruminants without depressing rumen fibre digestion or voluntary intake (Douglas *et al.*, 1995).

CTs increased liveweight gains in sheep grazing sulla relative to those on lucerne or perennial pasture (Niezen *et al.*, 1995), and in sheep consuming *L. corniculatus* compared with those receiving also polyethylene glycol (PEG) (Douglas *et al.*, 1995). Other effects of CT include reduced carcass fatness in sheep consuming *L. pedunculatus*, as compared with sheep feeding on lucerne, as well as increased milk production and wool growth in sheep grazing *L. corniculatus*, compared with sheep treated with PEG (Min *et al.*, 1998). It is important to define the concentration of forage CT that increases the efficiency of protein digestion and animal productivity in forage-fed ruminants, and to develop more ecologically sustainable systems of controlling some diseases under grazing.

CTs extracted from *Lotus* spp., sulla and sainfoin have anthelmintic-like activity against deer lungworm and gastrointestinal larvae, and grazing these forages may therefore be used as an alternative method for controlling internal parasites in deer (Molan *et al.*, 2000). The effect of CT extracted from *L. pedunculatus, L. corniculatus, Dorycnium pentaphylum, D. rectum, H. coronarium* and *O. viciifolia* was investigated on hatching of eggs and the development of the first stage larvae of the sheep nematode *Trichostrongilus colubriformis*, under *in vitro* conditions. The results suggested that CT may be able to break the life cycle of nematodes, and thus reduce the contamination of pastures with viable eggs, and help to decrease the need for anthelmintics; CT then becoming the principal and "natural" method of parasite control (Molan, Waghorn and McNabb, 2002).

There is a great interest in exploiting the bloat-preventing characteristics of CT to allow safe grazing of lucerne. As these CTs are produced only in the seed coat of lucerne, this could be achieved by altering the location of biosynthesis and storage from the seed coat to the leaves and stems or by the transfer of genes from another legume species; neither of these strategies has been entirely successful (McMahon *et al.*, 2000)

The current and future perspectives are more and more related to exploit bio-active compounds of phytotherapeutic interest to increase both animal welfare and performances. In addition, by the utilization of bio-active substances from the native flora or from forage crops, the use of chemicals in animals can be reduced, with consequently reduction in their residues in products such as meat and milk, thus increasing consumer food safety. The EU regulations for organic farming encourage this possibility, but knowledge on this topic is still restricted. A good approach is to start from the local veterinary ethnobotany, to avoid also the loss of this cultural heritage. For the future, new forage legumes from native flora should be identified and investigated, taking into account also the level of useful bio-active compounds in order to achieve additional benefits from the phytotherapeutic potential of the fodder germplasm.

Multiple purposes and new special uses of forage legumes in Mediterranean-climate areas

With increasing environmental concern and interest in sustainable agricultural systems, many forage legumes are being utilized for non-traditional use, or not exclusively for fodder production, stimulating renewed interest in them.

Legumes in silvipastoral systems and firebreak lines

Silvipastoral systems involving the coexistence of trees and pasture are a common land use practice well adapted to Mediterranean environments (Papanastasis, 1998). On partly wooded areas, agronomic improvements also include annual forage legume overseeding, with or without fertilization. Several varieties of subterranean clover proved to be suitable for overseeding with pine and eucalyptus forestation in France and Greece. These improvements supply high quality forage for domestic animals and wildlife, reducing the impact of climatic hazard on both forage productivity and grazing calendar organization. They also increase competition with shrubs and protect well developed trees.

The irrational use of fire can result in progressive land degradation in many areas of the Mediterranean basin, while grazing can reduce the fire hazard by preventing the build-up of herbaceous vegetation and palatable shrubs. In southern France, original management systems have been developed by combining grazing and firebreak lines, limiting the development of big fires (Etienne, 1996). The establishment of firebreak lines is thus an important tool in preventing summer fires. For this purpose, traditional and alternative forage legumes could be sown within silvipastoral areas, where the fire prevention effect derives from grazing the biomass within woody areas or in the firebreaks. Heavy grazing can reduce biomass at the beginning of the summer, thus minimizing the combustibility of the dry canopy and, consequently, the fire risk. In addition, the animals feed on protein-rich forage, such as annual medics, improving their diet and increasing their consumption of ligneous vegetation, generally less palatable, and thus facilitating better control of bushes in firebreak areas (Prosperi, Ronfort and Genier, 1999). Moreover, by using firebreak lines, other advantages can be achieved: reduced soil erosion, increased grazing carrying capacity, improved landscape and wildlife potential, multiple use of land, and diversified human activities. In addition, land abandonment can be reduced in more marginal areas.

Recent results obtained with the sowing of some annual legumes for firebreaks in Corsica (France), Tuscany and Sardinia (Italy) proved satisfactory (Antona *et al.*, in press).

Legumes for organic agriculture

In recent years, an exponential increase in the implementation of organic agriculture has been observed in Mediterranean areas of Europe, with about 60 000 farms and a million hectares of agricultural land in Italy, for example. EU Regulation no. 1804/99 was an additional stimulus for the conversion of extensive animal production systems into organic farming systems. As a consequence, the importance has increased of forage legumes in organic cultivation systems, in which the use of genetically modified organisms is forbidden (Younie and Hermansen, 2000). Moreover, the prohibition of meat-mill use, due to the risk of spread of bovine spongiform encephalopathy (BSE), has lead to an increase in the importance of plant protein for animal feeding. Because of these factors, and considering the high number of animals bred in Mediterranean-type areas, forage and grain legumes have a renewed and fundamental role in organic cropping and farming systems, related to their ability to fix nitrogen and produce plant proteins. In organic agriculture, biological nitrogen fixation is an essential and irreplaceable nitrogen source, the use of chemical fertilizers being forbidden. Equally, herbicides are not permitted, but developments in cultural methods of weed control have included the use of novel weed-suppressing cover crops (Bond and Grundy, 2001). Weed suppression achieved by including a legume cover crop in a cropping system becomes an important benefit. However, due to the peculiarity of Mediterranean organic agriculture (Guet, 1997), namely water scarcity, soil types and climatic conditions, for example, proper agronomic management strategies should be still developed. They could promote sustainable organic systems based on forage legumes, not related just to subsidies, and provide experimental data on this so far little investigated topic. Consequently, future research on legumes should be re-oriented to this sector. For the next 10-15 years, the increasing economic importance of the organic products market is expected to continue, reflecting rising demand from consumers.

Legume cover crops for crop systems

The use of legume cover crops for green manure is an agronomic practice that dates back more than two thousand years in the Mediterranean basin. In the last 50 years, due to the cheap availability of artificial fertilizers, this ancient practice has been neglected.

Production systems with legumes as a primary nitrogen source for subsequent non-legume crops use full-season green manure crops, and intersown legume and cover crops (Griffin, Liebman and Jemison, 2000). Each of these two options presents a significant management challenge, but cover crops, generally grown over the winter between harvest of one crop and the sowing of a subsequent crop, is a third option and can overcome some disadvantages associated with the first two options.

It is well documented that cover crops can supply sufficient nitrogen for production of the following grain or vegetable crop. Other than nitrogen supply, the use of cover crops can provide additional benefits, from their effects on soil physical and chemical properties, weeds, pests and diseases (Campiglia, 1999).

A cropping system for durum wheat, based on the replacement of fallow with a winter legume, has been proposed for the Italian environment (Campiglia *et al.*, 2000), as an alternative to the conventional system. Subterranean clover is established in the wheat crop as a living mulch; it is used subsequently as a cover crop and a green manure crop.

The development of sustainable agricultural strategies based on the use of cover crops is receiving increasing attention in both Mediterranean and temperate agro-ecosystems. Nevertheless, there are sometimes negative effects, such as extra water consumption, potential reduction in grain yield and additional problems in controlling pest and disease. So the success of a cover crop depends on achieving a positive balance between potential benefits and detrimental effects. Due to the substantial effects on several components of the agro-ecosystem, more research is needed. Currently, there is interest in annual pasture legumes for their use as cover crops in temperate regions in the USA with prolonged freezing temperatures (Fish *et al.*, 2001); when these species are sown in autumn and allowed to winter-kill, they may be able to provide the benefits of winter annuals, without reducing available water moisture and eliminating the need for chemical or mechanical control in spring.

Cover crops in vineyards and orchards

The use of annual self-reseeding legumes as cover crops in vineyards under Mediterranean climatic conditions can allow an alternative and low impact management of the soil compared with the traditional application of herbicides in the rows and tillage between rows (Porqueddu, Fiori and Nieddu, 2000). Moreover, a legume provides green cover when the erosion risk is high, while the dense mulch of dead herbage left in summer increases the soil bearing capacity for the passage of agricultural equipment for field operations (e.g. treatments, harvesting) and limits loss of water by evaporation from the soil. A substantial reduction in soil erosion was obtained in hilly vineyards of Italy and France, when comparing cover crop with traditional tillage treatments.

Self-reseeding cover crops can also reduce the costs associated with annual sowing (tillage, seed). In addition, they can successfully improve physical and chemical soil characteristics and limit nitrate leaching and herbicide pollution in intensive fruit production systems. Subterranean clover in acid soils and annual medics, mainly *Medicago polymorpha, M. rigidula* and *M. truncatula* in alkaline soils, have been tested as cover crops in Australia, California (USA), France, Italy and Portugal (Masson and Gintzburger, 2000) with no negative effects for grape production and for qualitative and organoleptic characteristics of wine. In vineyard production, properly managed cover crops can be used to stress the vines and to reduce their vigour in the spring. In addition, the high quality biomass of the legume can be grazed in winter by ewes, with undoubted advantages for both cover crop and farming management.

The possibility of using self-reseeding legume cover crops is based on both species and cultivar adaptation to the climatic and edaphic environments, and to the constraints of the tree system. A recent list of species and commercial varieties of annual legumes with potential for use as cover crops in vineyards, and available in the seed market, has been provided by Piano (1999).

The considerations for vineyards apply also for orchards and in particular for the olive groves that are traditionally widespread and economically important in all Mediterranean-basin countries.

Annual legumes are very common cover crops in orchards and vineyards in California (USA) (Graves and Christensen, 1995). Improvements in fruit quality is most specific to grapes, but cover crops have general benefits that are more widely applicable in a range of cropping systems. The ability of cover crops to suppress weeds between tree and vine rows is often overlooked. This is a major benefit of using cover crops; in fact cover crops can also be an effective weed control tool, particularly where herbicide resistant weeds are issues, breaking the life cycle of a particular weed species, or acting as a mulch or cover that suppresses weed growth. Another clear benefit from legume cover crops is that they can be used as habitat for predaceous arachnids and insects in biological control programmes for reducing pest populations on tree crops. Other specific reasons for planting cover crops are related to special problems with the orchard site. For example, growing conditions in orchards can be improved by reducing dust around the perimeter and during harvest operations. Nevertheless, some problems related to weed control requires the setting up of appropriate management techniques and choice of suitable species in order to maximize benefits.

Soil protection and rehabilitation and other special uses

Other possible uses of legumes include the environmental restoration and rehabilitation of degraded sites, such as quarries, mines, etc. For example, sulla was first introduced to New Zealand for the revegetation of sloping areas and sites degraded from soil erosion, with an additional function of landscape embellishment, due to its beautiful inflorescences. Annual medics are used in mixtures sown to green highway and railway slopes in the Mediterranean part of France (Prosperi, Ronfort and Genier, 1999). Stands of *M. scutellata* have been established in Montana, USA, to protect soil from wind erosion during the autumn windy season (Sims, 1998).

Field experiments in Italy (Chisci *et al.*, 1998), demonstrated the value of sulla for environmental protection and valorization in clay soil marginal areas, where, due to wheat monoculture, progressive soil degradation and land degeneration have been observed. Sulla canopy cover offers very good protection to the soil during the rainy season. In fact, its canopy dissipates storm kinetic energy, reducing detachment of soil particles by raindrop impact, increasing impedance to overland-flow, decreasing runoff velocity and elongating lag-time in water concentration on slopes. The results are increased water infiltration, decreased runoff and erosion, and reduced landslide incidence.

Some annual legumes are being evaluated for phytodepuration of wastewater as a complement to the traditional water depuration system (G. Rivoira, pers. comm.), while another potential use is based on the capacity of some legumes to phytodecontaminate metal in mine wastes (S. Bullitta, pers. comm.). These phytoremediation uses require the identification of both legume species and related root nodule bacteria, capable of colonizing soils contaminated with heavy metal, and the isolation of tolerant genotypes, in order to exploit their ability to remove, degrade or stabilize pollutants.

A novel use of legumes is to produce forage specifically for wildlife on farms where hunting is an important source of income. A singular situation exists in areas of USA, where annual legume research is now directed to "food plots" for deer or other game animals and birds (W. Ocumpaugh, pers. comm.). The aim is to improve the number and quality of the game animals, as the more animals there are, the more can be hunted, because there are people that will pay for hunting rights. Consequently, domestic livestock are being relegated to a role of strategic forage harvesters to manage the vegetation on a piece of property that is primarily managed for wildlife to be hunted for profit. In fact, the income potential can be much greater than it has been for decades from livestock-related enterprises. *M. polymorpha* cv. Armadillo has been recently released; it is pod-spiny and not friendly to sheep and goat management systems, but very useful for cattle and wildlife (Ocumpaugh, 1999).

In the near future, this less conventional use of legumes could have increased relevance in southern areas of Europe, where the integration of different rural activities is being promoted, while a decrease in economic importance of traditional activity will be unavoidable.

Sulla, like some other legumes, is an excellent plant traditionally used for the production of high quality honey, in combination with other important uses.

In the past, Arabs ate lucerne sprouts or the entire plant for its benefits to their health and strength. Today, lucerne is still considered a nutritive herb, rich in minerals and vitamins, and lucerne leaf tea is commercially available.

Seeds of *Vicia sativa* are traditionally eaten in some areas by Sardinians, while shrub legumes are still used for handcraft products (Pira and Camarda, in press); other species are used as vegetables (Bennett, Reid and Francis, 2000).

Some legumes can be used as ornamental plants, such as *Anthyllis tetraphylla* (S. Bullitta, pers. comm.). Novelty uses include the possibility of making a special wine from the flower-heads of *Trifolium repens* or *T. pratense,* and the development of 'lucky' four-leaved cultivars of *T. repens* as garden or pot plants (Frame, Charlton and Laidlaw, 1998).

A new use of lucerne as a biofuel for electricity generation has been suggested; the strategy is to separate whole plant hay into leaf and stem, with the stem gasified and the leaves processed into leaf meal for livestock feeding.

Insecticidal, pharmaceutical and medicinal uses

Due to the antinutritional effects of some secondary metabolites, genetic selection of forage legumes has traditionally aimed to obtain varieties with low contents of such chemicals.

The high contents of phyto-oestrogenic compounds in *Trifolium subterraneum* and *T. pratense,* which are detrimental for fertility in sheep, causing the well known "clover disease", are, in contrast, of interest for pharmaceutical use. Annual legumes are specifically grown for pharmaceutical purposes in Australia, being requested by the medical sector (Adlercreutz and Mazur, 1997) for prevention of diseases. Among secondary metabolites of legumes, there are several examples of useful compounds with insecticidal and pharmaceutical properties. *M. scutellata* seeds show a higher content of trypsin inhibitors than other *Medicago* species, both annual and perennial types (Ceciliani *et al.,* 1997). These substances have insecticidal effects and anticancer activity. *M. scutellata* is also potentially useful as a source of germplasm for improving resistance to insects in lucerne. Saponin content in alfalfa is considered detrimental or harmful from a nutritive point of view, yet saponin's toxicity has been demonstrated against insects, fungi and in blood tests (haemolytic activity); insecticidal properties and antileukaemic potential have been also investigated (Tava and Odoardi, 1996).

Genistein, biochan A and formononetin are isoflavones present in subterranean clover. Evidence seems to indicate that genistein and biochanin A have anticancer effects (Fung *et al.,* 1997); genistein inhibits the *in vitro* proliferation of tumour cells and the growth of human prostate cancer cells. For this reason the selection of accessions characterized by high contents of flavonoids is important in order to exploit these potential pharmacological activities. The factors and interactions responsible for high concentrations of these compounds should be also identified (Piano, Pecetti and Tava, 2001).

There is much interest in the health-promoting effects of isoflavones (Mazur *et al.,* 1998), particularly in soy. In fact, high consumption of soybean by Asian people has been associated with a lower incidence of breast cancer and prostate cancer compared to Western people (Stephens, 1997). Although *Trifolium pratense* forage is rich in formononetin, which is responsible for causing reproductive problems in sheep, pharmaceutical products use it, and *T. pratense* is considered one of the richest natural sources of isoflavones, containing all the most important isoflavones that are present in Asian, Mediterranean and Latin American diets. These products have been formulated to promote urinary function and prostate health naturally, without modifying the diet to include large amounts of legumes.

Another interesting example comes from *Trigonella foenun-graecum,* a legume widely grown for flavouring and as a food and forage crop around the Mediterranean basin. Interest in cultivating this annual legume in temperate regions has increased because the seeds contain a steroid, diosgenin. Diosgenin, extracted from its seeds is used by pharmaceuticals companies in manufacture of commercial steroids (such as progesterone), contraceptive steroids and corticosterone.

Taking into account the great biodiversity and potential of germplasm, these new uses should be further investigated in the future, looking at the evaluation of several bio-active compounds that are present in many forage legumes, but so far little studied or unknown.

Implications of alternative and new uses in forage legumes

Even if many legumes prove to be suitable for several uses, one problem is the absence or limited availability of cultivars specifically selected for those uses, and well adapted to the necessary utilization methods. This requires selection activities focusing specifically on these uses, specific characteristics being required for each use (Roggero and Porqueddu, 1999). For example, earliness, ability to survive without utilization, and high regeneration ability are required for vineyard cover crop. For firebreak lines, low sward height and high regeneration capacity are important. For sward to be grazed in the forest, shade tolerance and limited dead residues are needed. In fact, agroforestry is an example of a land use form requiring new legume species or varieties. To ensure an optimum soil cover year after year, it might be wise to establish populations based on a mixture of hard- and soft-seeded genotypes.

Sometimes, accessions rejected for forage purposes, due to low yield potential, could come in useful for other uses, for which phytomass production is not an essential trait, while a short life cycle may be an important advantage. At the same time, if the main use is for environmental purposes, without grazing, some aspects of forage quality are not very important. Accessions characterized by high contents of bio-active substances, making them unsuitable for animal feeding, can instead be interesting for pharmaceutical uses.

In southern Australia, the new roles for legumes will require the identification, introduction, selection and evaluation of new pasture species and genotypes suitable for the emerging phase farming systems. Nevertheless, the new attributes required might not necessarily be essential or useful for other Mediterranean-type areas.

It is to be expected that the evolution from productive to alternative and multiple uses of legume species differ between developed and developing countries. In the latter, the traditional productive role would remain fundamental, while new uses may not be immediately relevant, but could be very possible in the future. It is foreseeable that seed demand for new varieties adapted to specific non-productive uses will increase.

More than other forage species, legumes can better support the multi-functionality of European agriculture, are environmentally friendly and can produce safe and affordable high quality food in a competitive and sustainable manner. Consequently, current and future research on forage legumes should be re-oriented to satisfy these objectives.

According to Spangenberg *et al.* (2001), several characteristics exhibited by pasture legumes make them potential suitable candidates for "molecular farming". Advances in genetic manipulation technologies can provide transgenic legumes for non-forage uses. In the near future, it will be possible to exploit some forage plants as bioreactors for the production, among others, of industrial enzymes, pharmaceuticals, vaccines, antibodies and biodegradable plastics. The advent of plant biotechnology has created novel possibilities for future uses of forage legumes crops (Austin and Bingham, 1997).

CONCLUSIONS

In the Mediterranean basin countries, the full valorization and exploitation of forage legumes have not yet been achieved. There has been a sort of "fashion" of germplasm collecting, sometimes based more on external than internal initiatives, but a basic problem is that collection is only the first step in the legume germplasm valorization process. Initial characterization and evaluation is followed by the selection process, with subsequent development of new varieties, development of seed production and, finally, the use of the newly available cultivars by farmers. If each step is not followed through, the continuing activity of collecting is almost meaningless. Clearly, this is a complex process and requires proper planning, involving all the interested parties in order to achieve the indispensable coordination between researchers, seed companies, extension services and farmers. Regrettably, there is too often an absence of effective coordination and collaboration at regional, national and international levels at all stages of valorization.

Moreover, it is important to overcome the existing gap between research outputs and production system needs; for example, legume varieties for alternative and multiple purposes are not always available, even if requested.

Sometimes, a lack of adequate human, cultural or financial resources, or a combination, is a limiting factor, particularly in developing countries but, very often, the valorization of local germplasm is not considered a strategic activity, being in general ignored by the public institutions, as happens also in developed countries.

Stronger connections between countries of Mediterranean Europe and North Africa are desirable to develop common strategies, follow similar objectives and thereby use limited resources more efficiently. It is important to pay attention to the existing situations and peculiarities, because better improvements can be obtained from the development of existing production systems, rather than the wholesale transfer of foreign management systems (e.g. ley farming). Due to the natural richness of legume species, a useful strategy could be the optimization of suitable agronomic techniques and management to increase the contribution of native legumes in pastures as an alternative to the introduction of less-adapted exotic varieties.

In the future, research on pastures and forage legumes for Mediterranean-basin countries should be aimed at direct benefit from the valorization of the rich local germplasm, not only for current purposes, but also for potential uses yet to be discovered. It is therefore important not to underestimate all the potential uses of these species that could create economic returns. Research should follow original and more independent lines of development in each context to cope with local needs, which are very diversified within the Mediterranean-type areas. Only a multidisciplinary approach in research can provide effective results.

For the development of a seed industry, some technical problems must be solved, and particularly for the Mediterranean areas of the European Union. A basic change for the better would be to modify agricultural regulations that currently discourage use of forage legumes and favour other, less sustainable but heavily subsidized, crops.

Australian activity is a driving force in legume valorization at global level; new cultivars are being, and will continue to be for the foreseeable future, released and their seed marketed. As a consequence, legume seed is available in the whole Mediterranean-climate world, representing an effective opportunity, sometimes the only one, for many situations.

However, the undoubted advantage of being able to buy these varieties in seed markets should always be balanced by an adequate and satisfactory return for less developed countries that have freely shared their germplasm.

The need for precise legislation concerning local germplasm protection in the Mediterranean basin is supported by recent legislation and agreements on germplasm protection at national and international level: the Convention on Biological Diversity (CBD), adopted by UNCED in 1992; the International Undertaking on Plant Genetic Resources and the International Code of Conduct on Collection and Transfer of Plant Germplasm, adopted by FAO in 1983 and 1993, respectively; EU Council Regulation no. 1467/94/EU concerning conservation of genetic resources in agriculture; and a local bill proposal "Protection of Sardinian local plant genetic resources and regulations regarding collection activities for scientific, economic and commercial purposes" (Fois, Mura and Bullitta, 2000). These initiatives are also clear evidence of a commitment to sharing of all benefits deriving from genetic resources among donors, holders and users of germplasm, and of the involvement of local populations and institutions in the enjoyment of all benefits stemming from the collection, transfer and utilization of plant genetic resources from their region. In fact, the use of collections made by non-local people, initially acting for scientific purposes and often continuing their activities until the release of new commercial varieties, could be considered as reducing the economic potential of local resources.

At the same time, strict protection of plant genetic resources *per se* is senseless; it must be coupled with valorization programmes able to exploit actual or potential uses for scientific, economic, ecological and cultural purposes.

Joint action between countries sharing germplasm and countries that mainly benefit from it, for both internal use and seed export activities, could contribute to sustain the development of less favoured Mediterranean countries, For example, this could be achieved by the common and contextual developing of varieties selected for the needs of both donor and user countries. Other concrete aid could result from the sharing of human and technical resources, not only during the germplasm collection phase, but also in subsequent steps required for the full exploitation of pasture and legume species.

Finally, the potential of the application of biotechnology to forage legumes to improve forage quality, persistence and nitrogen fixation ability, and for the extraction of new products from forage, will be very important in the future. Nevertheless, biotechnology means, at the moment, mainly multinational corporations and not public research. For this reason, public research in biotechnology should be encouraged and supported concretely.

ACKNOWLEDGEMENTS

Special thanks are due to the author's colleagues, Professor Salvatore Caredda, Dr Antonello Franca, Dr Rosella Muresu and Dr Giannella Piluzza, and to Dr Angelo Loi, Dr Ricos Thanopulos, Dr Aziza Zoghlami and Professor William Ocumpaugh for information provided. The author also acknowledges researchers whose references are omitted due only to constraints of space.

REFERENCES

Abdelguerfi, A., Laouar, M., Tazi, M., Bounejmate, M. & Gaddes, N.E. 2000. Present et futur des pâturages et des légumineuses fourragères en région méditerranéenne: Cas du Nord de l'Afrique et de l'Ouest de l'Asie [Present and future of pastures and forage legumes in Mediterranean Region: Case of North Africa and West Asia]. *Cahiers Options Méditerranéennes*, **45**: 461-468.

Adlercreutz, H., & Mazur, W. 1997. Phyto-oestrogens and Western Diseases. *Annals of Medicine*, **29**(2): 95-102.

Amarger, N. 2001. Rhizobia in the field. *Advances in Agronomy*, **73**: 109-168.

Antar, A. & Christiansen, S. 1996. Harvesting of medic seed. Lecture notes, ICARDA.

Antona, M., Casanova, J.B., Franca, A., Seddaiu, G., Argenti, G., & Lombardi, P. In press. Les interventions pastorales au service de la prevention des incendies en Corse, en Toscane et en Sardaigne [Pastoral interventions for fire prevention in Corsica, Tuscany and Sardinia]. *In:* Proceedings of the Euro-Mediterranean Conference "The Future of the Green Mediterranean". Alghero, Italy, 1-2 June 2001. (In press).

Aschmann, H. 1973. Distribution and peculiarity of Mediterranean ecosystems. pp. 11-19, *in:* F. Di Castri and H.A. Mooney (eds). *Mediterranean type ecosystems*. Berlin: Springer-Verlag.

Austin, S. & Bingham, E.T. 1997. The potential use of transgenic alfalfa as a bioreactor for the production of industrial enzymes. pp. 409-424, *in:* B.D. McKersie and D.C.W. Brown (eds). *Biotechnology and improvement of forage legumes*. Wallingford, UK: CAB International.

Ballard, R.A., Charman, N., Craig, A.D. & Hughes, S.J. 2000. Symbiotic performance of pasture legumes with naturalised soil rhizobia. *Cahiers Options Méditerranéennes*, **45**: 315-320.

Bennett, S., Reid, R. & Francis, C. 2000. Minor and underutilised legumes. Vol. 1, pp. 357-386, *in:* N. Maxted and S. Bennett (eds). *Plant genetic resources of legumes in the Mediterranean*. Dordrecht, The Netherlands: Kluwer Academic.

Bond, W., & Grundy, A.C. 2001. Non-chemical weed management in organic farming systems. *Weed Research*, **41**: 383-405.

Borreani, G., Caredda, S., Peiretti, P.G., Roggero, P.P., Sargenti, P., Sulas, L. & Valente, M.E. 1998. Evolution of yield and quality of sulla in relation to the morphological stage of development. pp. 233-234, *in:* Proceedings of 5th ESA congress. Nitra, the Slovak Republic, 28 June-2 July 1998.

Campiglia, E. 1999. Colture di copertura utilizzate in agroecosistemi mediterranei. Nota 1: modificazione dell'ambiente colturale [Cover crops in Mediterranean agro-ecosystems. Note 1: Field environment modifications]. *Rivista di Agronomia*, **33**: 90-103.

Campiglia, E., Paolini, R., Caporali, F. & Mancinelli, R. 2000. Sistemi colturali alternativi per la produzione di frumento duro (*Triticum durum* Desf.). [Alternative cropping systems for durum wheat (*Triticum durum* Desf.)]. *Rivista di Agronomia*, **34**: 313-320.

Carr, S.J. [1997]. A pasture ideotype for farming systems on the acid soils of the drier Western Australian wheatbelt. *CLIMA*, in press.

Ceciliani, F., Tava, A., Iori, R., Mortarino, M., Odoardo, M. & Ronchi, S. 1997. A trypsin inhibitor from snail medic seeds active against pest proteases. *Phytochemistry*, **44**: 393-398.

Chisci, G.C., Bazzoffi, P., Pagliai, M. & Pellegrini, S. 1998. Sulla: a cover crop for protection and improvement of clayey soils. Proceedings of the Workshop on Sustainable tillage Systems, Auburn, AL. 8-10 July 1998.

Cocks, P.S. 1993. Legumes from the Mediterranean basin: a continuing source of agricultural wealth for southern Australia. *CLIMA, Technical Paper*, no. 1: 1-14.

Cocks, P.S. 1999. Reproductive strategies and genetic structure of wild and naturalised legume populations. pp. 20-31, *in:* S.J. Bennett and P.S. Cocks (eds). *Genetic Resources of Mediterranean Pasture and Forage Legumes*. Dordrecht, The Netherlands: Kluwer Academic.

Colebatch, G., Trevaskis, B. & Udvardi, M. 2002. Symbiotic nitrogen fixation research in the postgenomics era. *New Phytologist*, **153**: 37-42.

Date, R.A. 2001. Advances in inoculant technology: a brief review. *Australian Journal of Experimental Agriculture*, **41**: 321-325.

Douglas, J.B., Wang, Y., Waghorn, G.C., Barry, T.N., Foote, A.G. & Wilson, G.F. 1995. Liveweight gain and wool production of sheep grazing *Lotus corniculatus* and Lucerne (*Medicago sativa*). *New Zealand Journal of Agricultural Research*, **38**: 95-104.

Etienne, M. (ed). 1996. *Western European silvopastoral systems*. France: INRA. 276p.

Evans, P. 1996. Pasture varieties for the future. Proceedings of the Riverina Outlook Conference. Wagga Wagga, 15 August 1996.

Evans, P.M. & Snowball, R. 1993. Balansa and Persian clover lines outproduce registered cultivars, and subterranean clover and medics, in a 400 mm annual rainfall zone in Western Australia. pp. 53-56, *in:* Proceedings of the 7th Australian Agronomy Conference, Adelaide, South Australia, 19-24 september 1993.

Ewing, M., Poole, C., Skinner, P. & Bennett, A. 2001. Sulla and other forage species for southern Australia. *RIRDC Publication*, No. 01/41.

Fish, J.W., Hesterman, O.B., Sherstha, A., Kells, J.J., Harwood, R.R., Squire, J.M. & Sheaffer, C.C. 2001. Weed suppression by annual legume cover crops in no-tillage corn. *Agronomy Journal*, **93**: 319-325.

Fois, P., Mura, L. & Bullitta, S. 2000. Plant genetic resources protection in the Mediterranean basin: the case of Sardinian forage species. *Cahiers Options Méditerranéennes*, **45**: 112-115.

Frame, J., Charlton, J.F.L. & Laidlaw, A.S. (eds). 1998. *Temperate forage legumes*. Wallingford, UK: CAB International.

Francis, C.M. 1999. The need to collect new pasture and forage species. pp. 90-95, *in:* S.J. Bennett and P.S. Cocks (eds). *Genetic Resources of Mediterranean Pasture and Forage Legumes. Genetic Resources of Mediterranean Pasture and Forage Legumes.* Dordrecht, The Netherlands: Kluwer Academic.

Fung, M.C., Szeto, Y.Y., Leung, K.N., Wong-Leung, Y.L. & Mak, N.K. 1997. Effects of biochanin A on the growth and differentiation of myeloid leukemia WEHI-3B (JCS) cells. *Life Sciences,* **61**: 105-115.

Gonzáles López, F. 1994. Variedades Españolas de trébol subterráneo [Spanish varieties of subterranean clover]. *Junta de Extremadura, Consejería de Agricultura y Comercio.* 31.

Graves, W. & Christensen, P. 1995. Evaluation of *Medicago* spp. as self-reseeding cover crops in California vineyards. *Dryland Pasture, Forage & Range Network News* (ICARDA), **11**: 19-20.

Graves, W., Williams, W.A., Vaughn, C.E., Thomsen, C.D. & Jones, M.B. 2001. Australian varieties improve pasture in long-term annual legume trial. *California Agriculture,* **55**: 60-63.

Griffin, T., Liebman, M. & Jemison, J. 2000. Cover crops for sweet corn production in a short season environment. *Agronomy Journal,* **92**: 144-151.

Guet, G. 1997. *Agricoltura biologica mediterranea* [Mediterranean organic agriculture]. Bologna, Italy: Edagricole.

Hassen, H., Zoghlami, A., Sayari, R. & Kalbousi, R. 1998. Evaluation agronomique d'écotypes spontanés de *Medicago* et de *Scorpiurus* pour la sélection de cultivars adaptés aux régions arides et semi-arides. [Agronomic evaluation of *Medicago* and *Scorpiurus* populations for the selection of cultivars suited to arid and semi-arid regions]. *Annales de l'INRAT,* **72**: 211-235.

Hebb, D.M., Richardson, A.E., Reid, R. & Brockwell, J. 1998. PCR as an ecological tool to determine the establishment and persistence of *Rhizobium* strains introduced into the field as a seed inoculant. *Australian Journal of Agricultural Research,* **49**: 923-934.

Howie, J.H., Ballard, R.A, De Koning, C., Sandral, G. & Charman, N. 2001. *Trigonella balansae* – a new pasture legume for the alkaline soils of southern Australia? Proceedings of the 10th Australian Agronomy Conference, Hobart, 2001.

Howieson, J.G., O'Hara, G.W. & Carr, S.J. 2000. Changing roles for legumes in Mediterranean agriculture: developments from an Australian perspective. *Field Crops Research,* **65**: 107-122.

Kalu, B.A. & Fick, G.W. 1981. Quantifying morphological development of alfalfa for studies of herbage quality. *Crop Science,* **21**: 267-271.

Korte, C.J., Lim, P., Hill, N. & Latta, R. 2001. Evaluation of annual medics (*Medicago* spp.) in the Victorian Mallee. Proceedings of the 10th Australian Agronomy Conference, Hobart, 2001.

Lees, G.L., Gruber, M.Y. & Suttil, N.H. 1995. Condensed tannins in sainfoin. Occurrence and changes during leaf development. *Canadian Journal of Botany,* **73**: 1540-1547.

Le Houérou, H.N. 1980. L'impact de l'homme et de ses animaux sur la forét méditerranéenne [The impact of man and animals on the Mediterranean forest]. *Foret méditerranéenne,* **2**(1): 31-46; **2**(2): 155-174.

Lelievre, F., Porqueddu, C., Gènier, G. & Sulas, L. 1995. Seed production and harvest of annual burr medic (*Medicago polymorpha* L.): first results in south of France and Italy (Sardinia). pp. 314-317, *in:* Proceedings of 3rd International Herbage Seed Conference, Halle, Germany, 13-18 June 1995.

Loi, A., Howieson, J.G. & Carr, S.J. 2001. *Biserrula pelecinus* cv. Casbah. *Australian Journal of Experimental Agriculture,* **41**: 841-842.

Loi, A., Nutt, B.J., McRobb, R. & Ewing, M.A. 2000. Potential new alternative annual pasture legumes for Australian Mediterranean farming systems. *Cahiers Options Méditerranéennes,* **45**: 51-54.

Loi, A., Sandral, G.A., Gajda, K., McRobb, R. & Snowball, R. 2002. *Lotus ornithopodioides* L., a new pasture legume with potential for Mediterranean farming systems. pp. 144-146, *in:* Plant Breeding Conference, Perth, Western Australia, 15-20 September 2002.

Masson, P. & Gintzburger, G. 2000. Les légumineuses fourragères dans les systèmes de production méditerranéens: utilisations alternatives. [Forage legumes in the Mediterranean production systems: alternative uses]. *Cahiers Options Méditerranéennes,* **45**: 395-405.

Mazur, W.M., Duke, J.A., Wahala, K., Rasku, S. & Adlerceutz, H. 1998. Isoflavonoids and lignans in legumes: nutritional and health aspects in humans. *Journal of Nutritional Biochemistry,* **9**: 193-200.

McMahon, L.R., McAllister, T.A., Berg, B.P., Majak, W., Acharya, S.N., Popp, J.D., Coulman, B.E., Wang, Y. & Cheng, K.J. 2000. A review of the effects of forage condensed tannins on ruminal fermentation and bloat in grazing cattle. *Canadian Journal of Plant Science,* **80**: 469-485.

Min, B.R., Barry, T.N., McNabb, W.C. & Kemp, P.D. 1998. Effects of condensed tannins on the production of wool and on its processing characteristics in sheep grazing *Lotus corniculatus. Australian Journal of Agricultural Research,* **49**: 597-605.

Molan, A.L., Hoskin, S.O., Barry, T.N. & McNabb, W.C. 2000. Effect of condensed tannins extracted from four forages on the viability of the larvae of deer lungworms and gastrointestinal nematodes. *Veterinary Record,* **147**: 44-48.

Molan, A.L., Waghorn, G.C. & McNabb, W.C. 2002. Effect of condensed tannins on egg hatching and larval development of *Trichostrongylus colubriformis in vitro*. *Veterinary Record*, **150**: 65-69.

Niezen, J.H., Waghorn, T.S., Charleston, W.A.G. & Waghorn, G.C. 1995. Growth and gastro-intestinal nematode parasitism in lamb grazing either lucerne (*Medicago sativa*) or Sulla (*Hedysarum coronarium*) which contains condensed tannins. *Journal of Agriculture Science* (Cambridge), **125**: 281-289.

Nutt, B.J. & Loi, A. 1999. Harvestability of annual Mediterranean pasture legumes using conventional crop machinery. pp. 78-81, *in:* Proceedings 4th Intl Herbage Seed Conf., Perugia, Italy, 23-27 May 1999.

Ocumpaugh, W.R. 1999. Armadillo burr medic: a new legume variety for central and South Texas. *TFGC Advances*, **12**: 1-2.

Ovalle, C., Aronson, J., Del Pozo, A. & Avendaño, J. 1999. Restoration and rehabilitation of mixed espinales in central Chile: 10-year report and appraisal. *Arid Soil Research*, **13**: 369-381.

Papanastasis, V. 1998. Livestock grazing in Mediterranean ecosystems: an historical and policy perspective. pp. 3-5, *in:* Proceedings Int. Workshop on Ecological Basis of Livestock Grazing in Mediterranean Ecosystems. Thessaloniki, Greece, 23-25 October 1997.

Piano, E. 1999. Leguminose autoriseminanti nell'inerbimento del vigneto. [Self-reseeding legume cover crops for vineyards]. *L'Informatore Agrario*, **38**: 41-45.

Piano, E., Pecetti, L. & Carroni, A.M. 1997. Campeda, Limbara, Losa, Antas: le prime varietà italiane di trifoglio sotterraneo. [Campeda, Limbara, Losa, Antas: the first Italian varieties of subterranean clover]. *Sementi Elette*, **3-4**: 31-36.

Piano, E., Pecetti, L. & Tava, A. 2001. Variabilità in germoplasma di trifoglio sotterraneo per il contenuto in isoflavoni estrogenici di potenziale interesse farmacologico. [Variability for oestrogenic isoflavone contents of potential pharmaceutical interest in subterranean clover]. Atti del 33° convegno SIA, Agripolis, Padova, Italy, 21-23 September 1999.

Pira, G. & Camarda, I. In press. Ethnobotany and traditional learnings in the agropastoral world of Putifigari. Proceedings 3rd International Congress of Ethnobotany, Naples, Italy, 22-30 September 2001 (in press).

Porqueddu, C. 2001. Screening germplasm and varieties for forage quality: Constraints and potentials in annual medics. *Cahiers Options Méditerranéennes*, **45**: 89-98.

Porqueddu, C., Fiori, P.P. & Nieddu, S. 2000. Use of subterranean clover and burr medic as cover crops in vineyards. *Cahiers Options Méditerranéennes*, **45**: 445-448.

Porqueddu, C., Loi, A. & Fara, G. 1998. Anglona: la prima varietà italiana di medica annuale [Anglona: the first Italian variety of annual medic]. pp. 891-894, *in:* IV Congresso Nazionale Biodiversità: Germoplasma locale e sua valorizzazione. Alghero, Italy, 8-11 September 1998.

Porqueddu, C. & Sulas, L. 1998. Mediterranean Grassland systems. pp. 335-348. *in:* Proceedings 17th General Meeting of the EGF, Debrecen, Hungary, 18-21 May 1998.

Prosperi, J., Ronfort, J. & Genier, G. 1999. Constraints to the introduction of medics in French Mediterranean farming systems. pp. 154-161, *in:* Proceedings of the 13th EUCARPIA *Medicago* spp. Group Meeting. Perugia, Italy, 13-16 September 1999.

Reed, J.D., Krueger, C., Rodriguez, G. & Hanson, J. 2000. Secondary plant compounds and forage evaluation. pp. 433-448, *in:* D.I. Givens, E. Owen, R.F.E. Axford and H.M. Omed (eds). *Forage evaluation in ruminants nutrition*. Wallingford, UK: CAB International.

Roggero, P.P. & Porqueddu C. 1999. The agronomic purpose of evaluation. pp. 9-19, *in:* S.J. Bennett and P.S. Cocks (eds). *Genetic Resources of Mediterranean Pasture and Forage Legumes*. Dordrecht, The Netherlands: Kluwer Academic.

Sims, J.R. 1998. Snail medic for green manure, saline seep and wind erosion and grazing black medic with cattle in a medic pasture small grain rotation. *In:* Proceedings 36th North American Alfalfa Improvement Conference, Bozeman, Montana, 2-6 August 1998.

Spangenberg, G., Kalla, R., Lidgett, A., Sawbridge, T., Ong, E.K. & John, U. 2001. Transgenesis and genomics in molecular breeding of forage plants. Proceedings 10th Australian Agronomy Conference, Hobart, 28 January-1 February 2001.

Stephens, F.O. 1997. Phyto-estrogens and prostate cancer: possible preventive role. *Medical Journal of Australia*, **167**: 138-140.

Sulas, L., Fara, G. & Re, G.A. 2001. La raccolta del seme in mediche annuali e trifoglio sotterraneo [Seed harvesting of annual medics and subterranean clover]. *Sementi Elette*, **6**: 49-58.

Sulas, L., Re, G.A., Loi, A., & Howieson, J.G. 1998. The selection of optimal root-nodule bacteria inoculants increases the forage yield of *Hedysarum coronarium* (sulla). pp. 899-904, *in:* Proceedings 17th EGF Meeting, Debrecen, Hungary, 18-21 May 1998.

Sullivan, J.T. & Ronson, C.W. 1998. Evolution of rhizobia by acquisition of a 500-kb symbiosis island that integrates into a pre-tRNA gene. *Proceedings ot the National Academy of Sciences, USA*, **95**: 5145-5149.

Sultan, K., Gintzburger, G., Obaton, M., Robin, C., Touchane, H. & Guckert, A. 2001. Growth and nitrogen fixation of annual *Medicago-Rhizobium* associations during winter in Mediterranean region. *European Journal of Agronomy*, **15**: 221-229.

Tava, A. & Odoardi, M. 1996. Saponins from *Medicago* spp.: chemical characterization and biological activity against insects. pp. 97-109, *in:* G. Waller and K. Yamasaki (eds). *Saponins used in food and agriculture*. New York, NY: Plenum.

Thies, J.E., Holmes, E.M. & Vachot A. 2001. Application of molecular techniques to studies in *Rhizobium* ecology: a review. *Australian Journal of Experimental Agriculture*, **41**: 299-319.

Tuncks, B. 1987. Early developments in medic harvesting. *Agricultural Engineering* (Australia), **16**: 15-20.

Unkovich, M.J. & Pate, J.S. 2000. An appraisal of recent measurements of symbiotic N_2 fixation by annual legumes. *Field Crop Research*, **65**: 221-228.

Younie D. & Hermansen, J. 2000. The role of grassland in organic livestock farming. pp. 493-509, *in:* Proceedings of the General Meeting of the EGF, Aalborg, Denmark, 22-25 May 2000.

Zu, Y., Sheaffer, C.C. & Barnes, K. 1996. Forage yield and quality of six annual Medicago species in the North-Central USA. *Agronomy Journal*, **88**: 955-960.

CHAPTER 3

New Grasses and Legumes: Advances and Perspectives for the Tropical Zones of Latin America

L. Jank, C.B. do Valle[1] and P. de F. Carvalho[2]

ABSTRACT

As the global population is projected to increase by 40 percent by 2020, world meat demand is expected to rise globally by about 75 percent and by 160 percent in developing countries. The demand for new, more productive, better quality forages is potentially enormous. The neotropical savannah ecosystem, with 250 million hectares of well-drained lands, constitutes the largest undeveloped land resource area, and represents one of the world's most important potential future food sources. While several species of grasses and legumes have been introduced, tested and successfully commercialized in the lowland tropics of South and Central America during the past two decades, breeding is the key for future development of superior forages. The tropical world has still not profited from genetic manipulation of tropical forages through breeding, but possibilities are infinite. Breeding programmes are currently being developed, involving the most important species of forages, and the challenges and perspectives for future use are discussed.

INTRODUCTION

The world's population is increasing at an alarming rate. It is projected to rise from the present 6.3 billion to 7.54 billion by 2020, which is a 20 percent increase from current levels.

Concurrently, demands for food – cereals and meat – will increase, and mostly in developing countries, which are projected to increase their cereal demand by about 80 percent. Interestingly,

[1]Embrapa Beef Cattle Centre, C.P. 154, 79002-970 Campo Grande, MS, Brazil

[2]Federal University of Rio Grande do Sul, Agronomy Faculty, C.P. 776, 91501-970 Porto Alegre, RS, Brazil

meat demand in developing countries is expected to increase by a staggering 160 percent, and world meat demand will increase by about 75 percent (Pinstrup-Andersen and Pandya-Lorch, 1997). Under these circumstances, it seems most untimely to downsize pasture research conducted by international organizations, such as the International Centre for Tropical Agriculture (Centro Internacional de Agricultura Tropical, CIAT) in association with national institutions in tropical America and other developing regions of the world. Successful and sustainable savannah development has been, and should continue to be, research driven.

Forage species exploration, collecting missions and evaluation and selection of new introductions are the integral aspects of any pasture improvement programme in the tropical world. Several species of grasses and legumes previously unknown to agriculture have been introduced, tested and successfully commercialized in the lowland tropics of South and Central America during the past two decades. The variability in as yet untested "wild species" of grasses and legumes is far from being exhausted. It has been estimated that approximately 3 000 species have potential value as cultivated forages in the tropics (Ralph, 1982).

South American savannahs have a very high potential for pasture-based livestock industries. There are 250 million hectares of well-drained savannah grasslands in tropical America (Kerridge, 2001). The neotropical savannah ecosystem constitutes the largest area of undeveloped land resources, and represents one of the world's most important potential future food sources. Although these ranges support large populations of livestock, productivity is generally low, due to poor nutritive value of native pastures, especially during the dry season, where no more than 60 percent of the animal's maintenance requirements are provided. A fundamental approach in correcting nutritional deficiencies is through improved pastures based on new grasses, or on new tropical grass-legume mixtures. It has been demonstrated that, even on poor savannah soils, legume-based pastures are capable of doubling yields per animal and increasing yields per hectare by a factor of ten to fifteen compared to production from traditionally managed native savannahs (Lascano and Estrada, 1989; Toledo, 1985).

The integration of international markets and the search for better quality products at a lower cost are pressing farmers to intensify their production. Consequently, the great challenge is to make animal production competitive by offering cheaper and better quality products (Pereira *et al.*, 2001). The demand for new more productive, better quality forages is enormous. While forage breeding is still in its infancy (Hacker and Jank, 1998), challenges are great and possibilities seem unlimited. This chapter discusses the achievements of forage selection and breeding in the past thirty years, the present situation and perspectives for the future.

CONTEXT OF FORAGE SELECTION AND BREEDING PROGRAMMES

Selection and breeding of food crops date back to early times in agricultural history. In contrast, with the possible exception of lucerne (*Medicago sativa* L.), selection and breeding of forages, particularly in the case of tropical species, are still in their infancy. For many genera, especially of grasses, organized collection trips have never been made, sexual forms in most apomictic species have not been collected, and basic aspects have not been determined, such as reproductive behaviour, ploidy levels and chromosome behaviour, essential for deciding selection and breeding procedures to be used in a breeding programme.

In contrast, while breeding of most crops such as maize involves only one species of one genus, even if close relatives may be found in other genera, forage breeders need to tackle often more than one species among 60 or more legume, grass or shrub genera, and all the wild relatives of each species. Nevertheless, there are many more maize breeders in the world than there are forage breeders. In Brazil, the forage seed market value parallels that of hybrid maize, but forage breeders are very few and generally in public institutions (mostly in research centres, with very few in universities), whereas crop breeders abound, in both the public and private sectors. Consequently, new grasses and legumes are offered to farmers around the world, and particularly in Latin America, at only a comparatively slow pace.

The main international institution in Latin America responsible for the development of tropical pastures for the last 30 years has been CIAT. Through the Tropical Pasture Programme, created in 1971, CIAT played a major role in collecting, maintaining, exchanging and evaluating forage germplasm, especially legumes, in many tropical countries (Miles, 2001; Valle, 2001). The objectives of this programme were firstly, to develop the savannahs, expanding to forest margins, as in Peru and Costa Rica, and lastly to other regions, such as hillsides. Selection for adaptation to edaphic and climatic conditions has been the primary objective of pasture research initiated by CIAT in the Eastern Plains (Llanos) of Colombia. Through CIAT's programme and the International Tropical Pastures Evaluation Network (RIEPT), established with the participation of national institutions, the adaptations of species and genera were determined. Accessions within these species and genera have been evaluated and some cultivars have been released.

Plant breeding presumes the need to modify a variety whose deficiencies and qualities are known, and has as objectives increasing productivity, improving quality and extending environmental adaptability (Pereira *et al.*, 2001). To attain this, various factors need to be overcome in breeding forages for tropical conditions. There must be access to the germplasm, and because most grass species are from Africa, more knowledge is needed of the basic biology, genetics and variability of the species to be improved. Difficulties in manipulation of high ploidy levels must be overcome. There is also a need to domesticate species and adapt common breeding and experimental techniques to the specific requirements of forage species and shorten the process to the release of cultivars, due to the need to test them under grazing. A long-term commitment by the institution and the team involved is absolutely necessary in order to develop new cultivars of any forage species.

International cooperation is a requirement for efficient genetic resource management (Harlan, 1983; Valle, 2001). However, the success of such investment depends on government investments and international cooperation in the conservation of the specimens, otherwise all effort spent in the collecting and multiplication will be futile. The ease with which collections are lost is alarming, and it needs long-term efforts and investments for the preservation of large volumes of natural resources (Harlan, 1983). However, the prospect today is one of concern, as government restrictions have reached institutions such as the Commonwealth Scientific and Industrial Research Organization (CSIRO), depository of a large forage collection, and CIAT, also affecting its collection. Budget restrictions have cut back CIAT's staff and consequently its scope for action in Latin America (Valle, 2001). The Tropical Pastures Programme at CIAT had 22 scientists and staff in 1984, but only 4–5 in 1997 (Maass and Pengelly, 2001). At CSIRO, the number of scientists involved in pasture improvement dropped from 15 in 1985 to 3 in 1999. The Australian Tropical Forage Genetic Resource Centre (ATFGRC) of CSIRO had to send to CIAT and the International Livestock Research Institute (ILRI) storage around 15 000 accessions of the "less important" genera due to downsizing of its operations (Maass and Pengelly, 2001).

GERMPLASM AVAILABILITY IN GENE BANKS

The three main international forage gene banks in the world together hold around 30 000 accessions of the main forage species (Hanson and Maass, 1999). Collection of forages was prioritized in the 1970s and 1980s by CSIRO, CIAT, ILRI and the International Plant Genetic Resources Institute (IPGRI), with the collaboration of many national institutes (Schultze-Kraft, Williams and Koeghan, 1993; Hanson and Maass, 1999). The main focus, however, was tropical legumes, as can be seen by the number of accessions and species in each bank. At CIAT, in 1996, 89.2 percent of the specimens were legumes, and 9.8 percent were grasses (Torres, 1996). The main genera in the banks are 22 legume genera, 12 grasses and 14 shrubs, whose number of species and accessions are summarized in Table 3.1.

Table 3.1 Number of species (and accessions) of the main genera of legumes, grasses and shrubs in the three main international gene banks

	CIAT		CSIRO		ILRI	
Legumes	345	(15 980)	518	(9 210)	233	(4 550)
Grasses	105	(1 880)	252	(2 670)	139	(1 820)
Shrubs	66	(720)	154	(1 420)	183	(1 450)

Key: CIAT = Centro Internacional de Agricultura Tropical, Cali, Colombia; CSIRO = Commonwealth Scientific and Industrial Research Organization, Australia; ILRI = International Livestock Research Institute, Ethiopia.
Source: Adapted from Hanson and Maass (1999).

It is estimated that accessions in these gene banks are 30 percent duplicated and may be repeated in national gene banks (Hanson and Maass, 1999). Stored legume seeds may show variation in different gene banks after renewal because of cross-pollination in species originally thought to reproduce by autogamy (Penteado, Mira and Vega, 1996; Hanson and Maass, 1999). From 1986 to 1991, CIAT evaluated more than 1 000, 400, 270 and 430 grasses in Colombia, Brazil, Peru and Costa Rica, respectively, and 1 000, 1 100, 640 and 430 herbaceous legumes, respectively. The ecosystems were the llanos in Colombia, cerrados in Brazil, humid tropics in Peru and Costa Rica, and subhumid tropics in Costa Rica (Miles and Lapointe, 1992). Altogether, in Brazil, 343 *Brachiaria* and 42 *Paspalum* accessions were evaluated. Of the legumes, 393 *Centrosema*, 234 *Desmodium*, 37 *Arachis*, 47 *Pueraria*, 225 *Stylosanthes*, 215 *Calopogonium* and 15 *Cajanus* accessions and a few from other genera were evaluated (Pizarro and Carvalho, 1992).

Despite such large numbers, new forages have not been released proportionally. It is absolutely necessary that in most institutions where these evaluations have taken place, continuity be given through the complete process from evaluation to release. In other words, selected accessions need to be evaluated in regional trials, and under grazing, and then officially released, with an adequate supply of seed and a strong marketing strategy.

THE SAVANNAHS

The greatest extension of land in Latin America suitable for pasture exploitation is the savannahs, divided into well-drained and seasonally-flooded savannahs.

Well-drained savannahs

There are more than 260 million hectares of well-drained savannahs in South America, of which over 200 million hectares are situated in Brazil, in the region known as the cerrados; the rest is in Bolivia (14 million hectares), Colombia (17 million hectares), the Guyanas (4 million hectares) and Venezuela (28 million hectares) (Kerridge, 2001). The characteristics of the savannahs are the dominance of oxisols, which are deep, well-structured, acid soils (pH between 4.0 and 5.5, with high aluminium saturation), and well-defined wet and dry seasons extending from four to six months (Sano, Barcellos and Bezerra, 2000). The cerrado region of Brazil has suffered from the enormous advance of the agricultural frontier in recent decades. This biome has already lost over 40 percent of its native vegetation, and nowadays 20 percent is occupied by cultivated pastures (Sano, Barcellos and Bezerra, 2000), which are responsible for 55 percent of Brazilian beef production (Macedo, 2001). Therefore, the dramatic increases in human population experienced in this region (the population has quadrupled since 1960) imposed pressure for more intensive systems, which in turn demanded better outputs from the forage breeding and selection programmes.

Large-scale pasture improvement in the vast tropical savannahs of South America began in Brazil in the 1960s, and is continuing at an ever-increasing rate. The "cerrados revolution", a pasture-based livestock system, was developed on monospecific swards of introduced grasses. The initial species was *Brachiaria decumbens* Stapf cv. Basilisk. It is estimated that 40 million hectares of savannah have been sown to this cultivar. The initial impact was a dramatic 2- to 3-fold increase in animal production compared with the native grassland. Subsequently, two other species of the genus have been commercialized, *B. brizantha* (Hochst. ex A. Rich.) Stapf and *B. humidicola* (Rendle) Schweick. Approximately 50 percent of these areas have serious problems of resource degradation and marginal economic viability (Macedo, 2001).

Seasonally-flooded savannahs

Seasonally wet lands with a high water table constitute significant areas in the tropical American savannahs. It is estimated that there are 12 million hectares of hygromorphic soils in Brazilian cerrados (Goedert, 1986). Commercial forages, especially grass-legume associations, have not been available for this ecosystem until recently. *Brachiaria humidicola* is one of the species adapted to short periods of inundation and a high water table. This species was notorious for being too aggressive to associate with leguminous forage species, but *Desmodium ovalifolium* Guill. & Perr., *D. heterophyllum* (Willd.) DC. and *Arachis pintoi* Krapov. & W.C. Gregory were found to be tolerant to wetland situations and compatible with *B. humidicola* in the high rainfall isohyperthermic savannahs in the Colombian llanos.

Paspalum atratum Swallen, a native of seasonally wet soils in Brazil, attracted considerable research, and, more recently, commercial interest for seasonally flooded lands. The apomictic accession BRA-009610 has been released in several countries, and is known by six different cultivar names: Suerte (Florida), Pojuca (Brazil), Cambá-FCA (Argentina), Ubon Paspalum (Thailand), Hi-gane (Australia), and Terenos (the Philippines). The long growth season, high nutrient content and dry matter and seed yields of this accession in seasonally wet situations are of economic significance. This species can provide an adequate diet for the maintenance of adult cattle by filling the "nutritional gap" that regularly occurs in tropical American savannahs.

TROPICAL FORAGE GRASSES

Origin

Africa is the world's main centre of origin and of variation in the Gramineae (Clayton, 1983), but collections in the country are expensive and access to collection sites difficult. Collecting expeditions in the last decades have evolved from small expeditions where all that was found was collected, to well planned expeditions organized for collecting specific species or genera.

Very few forage grasses are native to South America, the two most important being *Paspalum* spp. and *Axonopus* spp. *Paspalum* is widely distributed from temperate to tropical regions of the Western Hemisphere, and is abundant in central and southern Brazil, eastern Bolivia, Paraguay, northern Argentina and Uruguay (Quarín, Valls and Urbani, 1997).

Reproductive behaviour

At least 60 percent of the tropical forage grass species reproduce by apomixis (Hacker and Jank, 1998), and these include *B. brizantha*, *B. decumbens*, *B. humidicola*, several species of *Paspalum* spp., *Panicum maximum* Jacq., *Cenchrus ciliaris* L., *Melinis minutiflora* P. Beauv., and *Hyparrhenia rufa* (Nees) Stapf. Species like *Andropogon gayanus* Kunth, *B. ruziziensis* R. Germ. & Evrard, *Pennisetum purpureum* Schumach., *Paspalum notatum* Flügge, *Setaria sphacelata* (Schumach.) M.B. Moss ex Stapf & C.E. Hubb., *Chloris gayana* Kunth, *Cynodon dactylon* (L.) Pers. and *C. nlemfuensis* Vanderyst reproduce by cross-pollination. It is noteworthy that, within the same genus, apomixis and cross-pollination may be found in different species (Valle and Glienke, 1991; Burson, 1997).

Breeding of apomictic species depends on the availability of totally or highly sexual plants, which are found in the centre of origin of the species. Apomixis is asexual reproduction by means of seeds, and in *P. maximum* and *Brachiaria* is the result of apospory followed by parthenogenesis (Savidan, 1982; Valle and Savidan, 1996). In apospory, meiosis occurs, but all four megaspores abort, and a somatic cell of the nucellus develops into an egg, which is, therefore, 2n and of the maternal type. Subsequently, fertilization of the egg cell fails and through parthenogenesis the 2n egg cell develops and forms the embryo. The sperm does, however, fertilize the polar nuclei, which become 3n. The ratio embryo/endosperm 2n/3n confers the fertility to the seeds (Savidan, 1982).

Pennisetum purpureum is an alogamous and protogynic species and breeding may rely on germplasm of closely related species, such as *P. glaucum* (L.) R. Br. and *P. squamulatum* Fresen. (Pereira *et al.*, 2001). The cross with *P. glaucum* yields sterile triploids that, if doubled, restores fertility to the chromosomes, and thus seed-propagated cultivars may be obtained (Dujardin and Hanna, 1985). The species *P. squamulatum* is apomictic and apomixis has already been transferred to *P. purpureum* (Dujardin and Hanna, 1985). The advantage of apomixis in the hybrid progeny is that hybrid superiority found in an individual may be immediately fixed through apomixis and the individual released as a cultivar.

Germplasm availability

The tropical world has benefited in the past forty years from collection trips organized exclusively for the collection of only one genus or species in the botanical centre of origin of the species in Africa. The two forages thus collected were *P. maximum* and *Brachiaria* spp. In the centre of origin in Latin America, extensive collection of *Paspalum* spp. has taken place over the past thirty years.

The French and the Japanese collected *P. maximum* germplasm in the 1960s and 1970s in Kenya and Tanzania (Combes and Pernès, 1970; Hojito and Horibata, 1982). Both groups found sexual plants. After 1984, a sub-set of the French germplasm was sent to Brazil (Brazilian Corporation for Agricultural Research (Embrapa) Beef Cattle Centre), Colombia (CIAT), Mexico (Colegio de Postgraduados) and Cuba (Estación Experimental de Pastos y Forrajes "Indio Hatuey"). Only Brazil received the sexual forms. The French and Japanese germplasm can now be considered representative of the natural variability of the species and further collections may not yield more genetic diversity (Savidan *et al.*, 1989).

The genus *Brachiaria* was collected in eastern Africa by CIAT and ILRI, with the collaboration of six eastern African countries (Keller-Grein, Maass and Hanson, 1996), totalling about 800 accessions of 17 different species. Sexual forms have been discovered in *B. decumbens, B. brizantha* and *B. dictyoneura* (Fig. & De Not.) Stapf (Valle, Savidan and Jank, 1989; Valle and Glienke, 1991). The more drought-tolerant *B. mutica* (Forssk.) Stapf, found in West Africa, still needs to be extensively collected (Keller-Grein, Maass and Hanson, 1996). The germplasm of *Brachiaria* has been shared with CIAT (687 accessions), ILRI (520 accessions) and Brazil (Embrapa – 446 accessions). In Brazil, the collection was sent to the Beef Cattle Centre in Mato Grosso do Sul State, the Savannahs Centre in the capital Brasilia, and the Humid Tropical Centre in Pará State.

There is little information about the evolution and centres of diversification of *P. purpureum* and there are no records of collecting expeditions to the centres of diversity, or the introduction of germplasm with wide genetic variability (Pereira *et al.*, 2001). The large ILRI collection assembled throughout the years contains accessions of *P. purpureum* and its hybrids with *P. glaucum*. This collection may contain duplicates and is still not representative of the natural distribution of the species (Wouw, Hanson and Luethi, 1999).

Various collection trips have been organized for the genus *Paspalum*. Brazil is a rich source of species and ecotypes. Valls (1987) reviewed the distribution of the species and the Embrapa Genetic Resource Centre maintains a collection of about 1 600 accessions (J.F.M. Valls, pers. comm).

TROPICAL FORAGE LEGUMES

Origin

Several important forage legume genera and species have originated in tropical America, such as *Stylosanthes, Arachis, Calopogonium, Centrosema, Leucaena, Macroptilium* and *Galactia striata* (Jacq.) Urb. (Argel and Pizarro, 1992; Pizarro and Carvalho, 1992; Valls and Pizarro, 1994). The region between Mexico and the Caribbean to the north of Argentina is considered the greatest centre of origin and diversity of tropical legumes with forage potential in the world (Paterson, 1987).

Legumes originating in other regions include *Pueraria* and *D. ovalifolium* from east Asia and the south Pacific; alyce clover (*Alysicarpus*) from India; and *Lotononis bainesii* Baker, *Lablab purpureus* (L.) Sweet and *Neonotonia wightii* (Graham ex Wight & Arn.) Lackey from Africa (Harlan, 1983).

Reproductive behaviour

The reproductive structure of the main tropical legume species is still not very well known. Even though autogamy prevails, recent studies by Maass and Torres (1998) using flower colour as a marker to assess the presence of outcrossing, have shown 4 percent of outcrossing in *D. heterocarpon*

(L.) DC., 13 percent in *G. striata* and *Chamaecrista rotundifolia* (Pers.) Greene, 18 percent in *Centrosema virginianum* (L.) Benth. and 23 percent in *C. gyroides*. In *C. virginianum*, some accessions only reproduced by autogamy, but others presented as much as 89 percent outcrossing. In *Stylosanthes*, outcrossing has been found in the rate of 13.8 percent in *S. guianensis* (Aubl.) Sw. (Miles, 1985), and 19.5 percent in *S. capitata* Vogel (Miles, 1983). According to Maass and Torres (1998), as more data become available, more legumes previously thought to be autogamous should show significant rates of outcrossing.

Germplasm availability

Embrapa Natural Resources Centre, with the cooperation of various national and international institutes, has substantially covered Brazilian territory in the past thirty years in the search for tropical forage legumes. *Arachis pintoi* and *Stylosanthes guianensis* are amongst Brazilian native legume species with the highest potential for plant breeding. Much effort has been invested in *Arachis* collection, and there is ample variability and the possibility of increasing variability through intraspecific hybridization using stoloniferous male parents (Valls *et al.*, 2001). However, there are no breeding programmes yet.

Stylosanthes is the genus with the most history of breeding, but the use of this legume commercially is still very limited (Miles and Lascano, 1997). Eight cultivars have been released in Tropical America and 12 in Australia directly from the germplasm collection, but many of them have not been adopted (Miles and Lascano, 1997). The problem with *Stylosanthes* for the breeder is not the lack of germplasm, but how to rationalize so much variability to select adequate criteria and evaluation methods in the selection of progenitors.

Other countries have also been extensively covered in the search for native legumes, including Paraguay (Hacker, Glatzle and Vanni, 1996), Cuba (Albert *et al.*, 2000) and Venezuela (Faría-Mármol, 1995).

TROPICAL FORAGE SHRUBS

Origin

Many of the most useful forage shrub legume species originate in the area from Mexico and Central America to South America, such as *Albizia*, *Leucaena*, *Cratylia argentea* (Desv.) O. Kuntze and *Gliricidia sepium* (Jacq.) Kunth. & Walp. (Queiroz *et al.*, 1997; Francisco and Hernandez, 1998; Shelton, 2001). *G. sepium* is the most used multipurpose tree. Other tropical shrubs and trees, such as *Acacia*, originate in North Africa and the Near East, and *Prosopis africana* in eastern Africa (Shelton, 2001).

Reproductive behaviour

Legume shrubs and trees reproduce predominantly by allogamy, but some species reproduce by mixed systems and autogamy. Reproduction studies in *C. argentea* and *C. mollis* Mart. ex Benth. have shown that they were both self compatible, and performed well under forced autogamy and allogamy (Queiroz *et al.*, 1997). The first forage species that left their centres of origin were usually represented by a few seeds of a few plants, so of very restricted genetic diversity, as in the case of *G. sepium*, *Calliandra calothyrsus* Meisn., and *L. leucocephala* (Lam.) de Wit. Because the last-named species is a self-fertile polyploid, it is composed of identical genetic material (Shelton, 2001).

Most forage tree legume improvement programmes lack an accurate classification and documentation of the species, so important taxa are frequently omitted. Also, natural hybridization occurs between the different taxa when brought to introduction banks, and hybrid seed are spread without the knowledge of the breeder (Shelton, 2001).

Germplasm availability

Leucaena has always aroused interest in many tropical countries and many collecting expeditions were organized. The University of Hawaii and CSIRO assembled collections of *Leucaena* that include many different species. Later, the Oxford Forestry Institute collected *Leucaena* in its native range. This is considered the most complete collection in terms of taxa (Stewart and Dunsdon, 1998). Although many taxa were found to be promising as a fodder crop, the well-known *L. leucocephala* subsp. *glabrata* was still superior due to its good palatability (Stewart and Dunsdon, 1998). Hybridization may be possible with other species, such as *L. pallida*, resistant to psyllid (Castillo *et al.*, 1997).

NEW GRASSES FOR LATIN AMERICA

There is not one record that shows how many forage grasses have been released throughout the Latin American countries in the past decades, but probably not more than thirty, in contrast to Australia, where 113 cultivars have been released since 1961 (Pengelly and Staples, 1997). Many grasses have been released in South America in recent years and are cited in the literature: *A. gayanus* cv. Carimagua-1, *B. dictyoneura* (=*B. humidicola*) cv. Llanero, and *B. brizantha* cv. La Libertad among others. These are excellent grasses that are being used in various countries to some extent.

The genera *Panicum* and *Brachiaria*, in general, show outstanding performance in many countries when compared with other forage grasses. These genera were found to be the most suited for various surface types in Colombia (Parra and Gómez-Carabalí, 2000) and are responsible for 90 percent of the grass seed sales in Brazil (Andrade, 2001).

The evaluation of the French *P. maximum* collection in Brazil has led to the release of many cultivars, and the expectation of many future releases (Savidan *et al.*, 1989; Jank *et al.*, 1994; Jank *et al.*, 1997). This species is a high yielding, good nutritive quality forage adapted to medium to high fertility soil and regions with over 600 mm rainfall. Results of the germplasm evaluation showed that, compared with the standard cv. Colonião, 37 percent of the 426 accessions produced higher leaf dry matter yield, 42 percent produced higher leaf per cent and 70 percent showed better regrowth following cuts (Jank, 1995). Twenty-five morphologically diverse accessions were selected for evaluation in seven states, covering distinct ecosystems of the country (Savidan, Jank and Costa, 1990). Accessions with the broadest adaptation and others with regional adaptation were selected (Jank *et al.*, 1993), and the best seven accessions were compared under grazing (Euclides, Macedo and Oliveira, 1995) (Plate 3.1) and then for animal performance (Euclides *et al.*, 2000). These evaluations resulted in the release of cultivar Tanzânia-1 in 1990 (Institut de Recherches Scientifique (IRD) T58), cv. Mombaça (IRD K190A) in 1993, and cv. Massai (IRD T21) in 2001. The first two accessions total 10 percent of the seed market in the country (Andrade, 2001) and are being exported to 26 other Latin American countries. Cv. Tanzânia-1 is of medium size, 80 percent more productive than the traditional cv. Colonião, and more easily managed in comparison

Plate 3.1 *Panicum maximum* cultivar evaluation under grazing in Brazil. [Photograph by Dr Stephen Reynolds].

with other cultivars of the same species. Cv. Mombaça is 130 percent more productive than Colonião, and used in intensive production systems with rotational pasture management. Cv. Massai, in contrast, is very short, maximum 90 cm height, with thin leaves and is used for pasture diversification and grazing of cattle, horses, sheep and goats (Jank, 1995; Euclides *et al*., 2000; Jank *et al*., 2001).

The evaluations of the French collection in Mexico and Cuba are also leading to the selection of promising accessions (Valles *et al*., 1995; Segui and Mendoza, 1999). In Mexico 13 accessions were considered promising (Valles *et al*., 1995) and in Cuba, 64 percent of the accessions showed adaptability to seasonally wet acid soils (pH 5.36) with low fertility, of which 14 showed genetic potential for these stress conditions (Segui and Mendoza, 1999).

The evaluation of the *Brachiaria* collection in Brazil showed that 22 selected accessions produced 10 percent more leaf dry matter yield than the commercial cultivar cv. Marandu, and 68 percent more than the average of the collection (Valle *et al*., 1999). Based on these results, 21 accessions were selected for evaluation in regional trials to assess ecotype × environment interaction, and eight were selected for grazing trials (Euclides *et al*., 2001). Four are currently being evaluated in a regional grazing trial and one was released as *B. brizantha* cv. Xaraés in 2002 (Valle, 2002). This cultivar has shown faster regrowth compared with cv. Marandu, resulting in 35 percent more animal productivity per hectare in the first year of grazing. Even though it does not display antibiosis in the same way as cv. Marandu, no insect damage has yet been observed. The evaluation in distinct ecosystems in Brazil has led to the selection of seven accessions for the seasonal evergreen forest ecosystem (Simão Neto, Dias Filho and Serrão, 1995) and ten accessions in the northeast region (Moreno, Neto and Santana, 1995). The evaluation at CIAT has led to the release of cv. Pasto Toledo in Costa Rica. It is a high yielding accession selected at CIAT, with nutritive values similar to other *B. brizantha* accessions (Argel, 2000).

In the search for grasses adapted to seasonally-wet soils, *P. atratum* attracted considerable research and commercial interest. Agronomic characterization of the species and selection of productive ecotypes was carried out at the Embrapa Cerrados Centre in Brazil, in the 1980s. Subsequently, superior accessions were evaluated under grazing in association with *A. pintoi* at the Embrapa Beef Cattle Centre. As the result of agronomic evaluation of wetland grasses, three accessions of *P. atratum* of the Section Plicatula have been selected, BRA 009610, BRA 003913 and BRA 009661 (Grof *et al.*, 1989).

Concurrently, a large number of accessions of *Paspalum* spp. were evaluated in various regions of Brazil. At the Embrapa Cerrados Centre, in Brasilia, 42 accessions were evaluated (Valls, Pizarro and Carvalho, 1993), which presented more constant seed production and better flowering synchronization than most tropical grasses (Pizarro, 2001). Later, 84 new accessions were evaluated (Pizarro, 2001).

In southeast Brazil, 58 accessions (including accession BRA 009610) out of 215 evaluated, were selected for dry matter production in the first year similar to or higher than *B. decumbens* and *A. gayanus* cv. Baeti. These preliminary agronomic investigations clearly indicated the superiority of late-flowering accessions of *P. atratum*, and especially BRA 009610 (Batista and Godoy, 2000).

This accession – BRA 009610 – was collected on a low-humic gley soil in a seasonally wet area with a high water table in Mato Grosso do Sul State, Brazil. On seasonally wet, and acidic, soils, *P. atratum* was the best grass in Florida. It produced high dry matter yield, averaging more than 20 tonne ha^{-1} on a low humic gley soil, as well as on a grey podzolic soil in Thailand. Both soils were acid (pH 4.6), low in N, P and K, and seasonally wet. Liveweight gains on a *P. atratum* BRA 009610 – *Arachis pintoi* BRA 031143 association ranged from 580 kg ha^{-1} to 800 kg ha^{-1} in Brazil (Barcellos, Pizarro and Costa, 1997), and from 460 to 860 kg ha^{-1} in Florida (Kalmbacher *et al.*, 1997).

A newcomer to the suite of tropical grasses was *Andropogon gayanus* var. *bisquamulatus* Hack. The accession CIAT 621 was officially released by the national institutions in Colombia and Brazil as cvs. Carimagua-1 and Planaltina, respectively, in 1981/82. Adoption was the greatest in the cerrados of Brazil, where about 3 million hectares were established.

The *A. gayanus* improvement project, at CIAT had as one aim the production of a uniformly late-flowering and vigorous cultivar. With the polycross technique, plants with the desired traits were selected in large paddocks and clonally established in a polycross nursery. Late-flowering, vigorous genotypes were selected by recurrent selection from two successful generations reproduced by sexual seed. Progenies of these late-flowering genotypes were compounded in synthetic line II and exposed to grazing in a grass-legume association, with *C. acutifolium* CIAT 5568 as the companion legume. A final selection for desirable plant types was carried out under grazing. These selections included two late-flowering distinct growth forms, a short and a tall one, which were compounded separately to form synthetic line III. The chief advantages of these synthetic varieties of andropogon are ease of management through grazing and control of excessive stemmy herbage during the active period of growth in the wet season. They have not been commercialized, but the experimental evidence clearly indicates the feasibility of production of improved cultivars of this species well adapted to savannah conditions.

In Brazil, an *A. gayanus* breeding programme has recently been developed. The objectives were to improve slow establishment and stand uniformity, the main setback of cv. Planaltina. After many cycles of selection, cv. Baeti was released. It is similar to cv. Planaltina, but more productive, more uniform and establishes easily from seed, thus improving weed competition (Batista and Godoy, 1995). To our knowledge, however, these breeding programmes are not being further developed.

Two very successful breeding programmes are not located in Latin America, but the products of these programmes have had a large spread through Latin American countries. These are the bermudagrass (*Cynodon dactylon*) and bahiagrass (*Paspalum notatum*) programmes conducted by G. Burton and colleagues in Georgia, USA. Bermudagrass is a pan-tropical species and very variable (Hacker and Jank, 1998). Breeding has been underway since the 1960s and many cultivars have been released (Hacker and Jank, 1998). The last release so far has been cv. Tifton 85, a hybrid between an accession of *C. dactylon* and Tifton 68 (*C. nelmfuensis*), and is the most productive of all bermudagrasses released. Its production and digestibility are 26 percent and 11 percent higher, respectively, than Coastal bermudagrass (Burton, Gates and Hill, 1993). In Brazil, in a savannah ecosystem with a six-month dry period, Tifton 85 was compared to *B. brizantha* cv. Marandu and *P. maximum* cv. Mombaça, the two forages that occupy over 50 percent of the Brazilian pastures. Tifton 85 had a higher production and carrying capacity in the rainy season, associated with protein levels up to 16.7 percent and *in vitro* organic matter digestibility (IVOMD) of up to 79.3 percent, resulting in 7 percent and 14 percent higher average daily gain than cv. Marandu and cv. Mombaça, respectively, and 20 and 25 percent higher production per area than the two forages, respectively (L.R.L. S´Thiago, pers. comm.).

Paspalum notatum cv. Pensacola is a low production natural introduction in the USA. After nine cycles of recurrent restricted phenotypic selection, Tifton 9 was released, and presented greater seedling vigour, palatability and 47 percent higher yield (Burton, 1989). However, this yield increase was partly due to a lower photosynthetic partitioning to roots when compared with cv. Pensacola (Pedreira and Brown, 1996).

NEW LEGUMES FOR LATIN AMERICA

Some of the excellent legumes released in recent years include *C. acutifolium* Benth. cv. Vichada, *S. guianensis* cv. Capica, and *A. pintoi* cv. Amarillo. In the search for tropical legumes for use in associations with grasses, the most promising was found to be *A. pintoi* (Argel and Pizarro, 1992). It is justly considered the ideotype of tropical legumes, that is actually comparable to white clover of the temperate regions (B. Grof, pers. comm.). High forage quality, tolerance to heavy grazing, stoloniferous growth habit, compatibility with aggressive C_4 grasses (*Brachiaria* spp., *Cynodon* sp., *P. atratum*), a subterranean seed bank, high seed production and regeneration from self-sown seed are its chief attributes. In many respects, it performs as well as white clover or even better. It tolerates some soil acidity, short periods of saturated soil conditions, does not cause bloat, and retains its leaves in the six-month dry season in Brazil (Pizarro, 2001).

The evaluation of *Arachis* accessions in South America by CIAT began in 1976, and so far more than 50 evaluation experiments have been conducted. These experiments have been carried out in the llanos of Colombia, and in the Brazilian cerrados, involving seasonally flooded land, well-drained savannahs and the humid tropics (Pizarro, 2001). The species *A. pintoi* showed promise in various ecosystems, on all sites evaluated, and in Central America and the Caribbean.

On low fertility acidic soils in Colombia, annual liveweight gains in associations including *A. pintoi* ranged from 160 to 200 kg head^{-1} and from 250 to 600 kg ha^{-1}. The inclusion of *A. pintoi* in grass pastures resulted in increased liveweight gains of 20 to 200 percent, and milk yield increases of 17 to 20 percent, compared with sole grass pastures (Lascano, 1993). On a soil of volcanic origin on the Atlantic coast of Costa Rica, *A. pintoi* of the same origin (CIAT 17434, cv. Maní Forrajero Perenne) in association with *B. brizantha* cv. Marandu produced daily gains from 452 to 551 g in grazing cattle at a stocking rate of 3 head ha^{-1}. The highest annualized mean liveweight gain was 937 kg, obtained on the association grazed by 6 head ha^{-1}. The authors (Hernandez *et al.*, 1995) concluded that *A. pintoi* is the best herbaceous legume currently available for the humid tropics of Costa Rica. In Brazil, in association with *B. dictyoneura*, the selected and released cv. Belmonte yielded 558 g head^{-1} day^{-1} initially and 568 kg ha^{-1} year^{-1} after four years grazing by cattle (Pereira, 2003).

Many cultivars of *S. guianensis* have been released, of which cv. Mineirão is the most popular in Brazil, retaining green colour in the dry season and being anthracnose resistant (Andrade and Karia, 2000). However, its spread is limited by low seed production. *Stylosanthes capitata* is a good seed producer, but is very susceptible to the anthracnose fungus (*Colletotrichum gloeosporioides* Penz.), which devastated commercial lines of *Stylosanthes* in Brazil in the early 1970s (Grof, Fernandes and Fernandes, 2001a). Of 296 accessions of *S. capitata* evaluated in Brasilia, Brazil, 25 were resistant to anthracnose, but in Campo Grande, only 4 were resistant (Grof, Fernandes and Fernandes, 2001a). Venezuelan *S. capitata* accessions, in contrast, showed high levels of resistance in Brazil, but are prostrate, have small leaves and inflorescences and lack vigour. Brazilian ecotypes are erect or semi-erect, have large leaves, and produce 21 to 35 percent more dry matter (Grof, Fernandes and Fernandes, 2001a). Of all *Stylosanthes* species, *S. macrocephala* M.B. Ferreira & S. Costa accessions are the most resistant to the fungus, but they are not very good seed producers. Recently, a new technique for breeding for anthracnose resistance in *Stylosanthes* spp. was developed at the Embrapa Beef Cattle Centre. This technique involves a mass hybridization scheme to produce a high degree of genetic diversity and good quantities of hybrid seed. After many cycles of breeding among anthracnose-resistant Brazilian and Venezuelan *S. capitata*, and among *S. macrocephala* accessions, the cultivar Campo Grande was released in 2000 (Grof, Fernandes and Fernandes, 2001a). This cultivar is a physical mixture of the two species, *S. capitata* and *S. macrocephala*, in the proportion 80:20 by weight, respectively, formed from 17 and 6 accessions, respectively, of each (Grof, Fernandes and Fernandes, 2001a). Breeding for this cultivar also involved improvement of seed production, so that it could be machine harvested.

Sixty-eight accessions of three species of *Centrosema* were evaluated in the cerrados of Brazil, but only *C. brasilianum* showed promise for this region (Pizarro and Carvalho, 1996). In Venezuela, four out of ten accessions of *C. pubescens* evaluated were selected for further studies in the state of Zulia, based on forage and seed production (Faría-Mármol, 1995).

Desmodium spp. are used in the recovery of soils in Colombia, as a cover crop under African palm tree plantations in Colombia, and for grazing in Venezuela in pastures on acid soils, despite it having a very low palatability (Schultze-Kraft and Schmidt, 1998). However, all 133 accessions evaluated in Brazil had weak establishment and were susceptible to drought and nematodes (Pizarro and Carvalho, 1996). Some factors that limit the adoption of *D. ovalifolium* are lack of knowledge of its management, and low animal intake (Schultze-Kraft and Schmidt, 1998). One cultivar (cv. Itabela), however, was officially released for the cacao production zone of the Atlantic coast of Brazil. This accession (CIAT 350) has disease and nematode problems in the llanos, but other resistant accessions have been identified in Brazil, as well as in the llanos and in the

Philippines. One type was outstanding in seed production (400 kg ha^{-1} clean seed). The species received insufficient publicity in spite of the fact that it is adapted to acid soils, is compatible with *B. humidicola* and tolerates saturated soil conditions (B. Grof, pers. comm.). For legumes, such as *D. ovalifolium,* that contain anti-quality components, such as tannins, it is important to evaluate acceptability by ruminants early in the selection process. A core collection of 18 accessions were evaluated in 30 m^2 plots in a cafeteria-type trial, under grazing, in the Colombian coffee zone, and in the Colombian Amazon zone. This type of experiment was found a useful tool to adjust and confirm laboratory data and was recommended for use as a standard step in germplasm selection projects (Schmidt *et al.*, 2001).

Calopogonium mucunoides Desv. is an important forage legume used in Brazil as it associates well with *B. decumbens* (Seiffert and Zimmer, 1988). It tolerates drought, is used as a cover crop, and variously as forage or hay at critical periods of the year (Pizarro, Ramos and Carvalho, 1996). Despite being perennial, it behaves as an annual or biennial, and its persistence in the pasture depends largely on its natural recovery (Seiffert and Zimmer, 1988; Pizarro, Ramos and Carvalho, 1996). The evaluation of a recently introduced germplasm line may result in the release of new options for the diversification of associated pastures in the country. A comparison of 215 accessions showed a variation of 32 to 54 percent in *in vitro* dry matter digestibility (IVDMD), 1 to 4 percent in production and 0 to 450 kg ha^{-1} in seed yields. Seventeen accessions were selected, of which two were outstanding because of high leaf retention in the dry season (Pizarro and Carvalho, 1997).

In Brazil, promising accessions have been selected from germplasm of *Macroptilium* spp., *Neonotonia wightii* and *Glycine* spp. (Pizarro, 2001). As with other important material, the evaluation of these species in this environment should have continuity.

NEW SHRUBS FOR LATIN AMERICA

New uses for legume trees and shrubs have been appearing. Trees are being studied as alternative sources of feed to enhance production and quality of forages in countries that need such alternative sources. The addition of foliage of *Bauhinia purpurea* DC. ex Walp. to guinea grass (*P. maximum*) increased the quantity of forage and kept the quality acceptable (Hernández, 2000). The tree-grass association restored the grass nutrient export and increased the pasture's productivity (Hernández, 2000).

CIAT coordinated an evaluation programme in the Latin American tropics. Eleven accessions of *Cratylia* were evaluated in Colombia, Costa Rica, Mexico, Guatemala, Brazil and Peru, in locations with various vegetation and with precipitation varying from 997 to 4000 mm. The eleven accessions have shown good adaptation over this range of climates and soils with high Al concentrations (Argel and Lascano, 1998). *Cratylia argentea* showed high foliar retention and high regrowth in the dry season. It has a deep root development (up to 2 m), which favours growth in the dry season and in poor and acid soils, and is currently being evaluated in dual-purpose livestock exploitations in subhumid areas of Central America by the TROPILECHE (Tropical milk) Consortium, coordinated by CIAT and ILRI (Argel and Lascano, 1998; Lascano *et al.*, 2002). The purpose of this consortium is to increase milk and beef production in dual-purpose systems on smallholder-farms of tropical Latin America through improved legume-based technologies. The target areas contemplated by this Consortium are the dry hillside areas in Costa Rica and the forest margins in Peru, representative of the dry and humid tropics, respectively.

The shrubs *D. velutinum* and *Flemingia macrophylla* (Willd.) Kuntze ex Merr., are promising in South America because of their tolerance to acid soils, although the quality of the latter is not good (Kexian *et al.*, 1998). The genera *Acacia, Prosopis, Calliandra* and *Erythrina* do not have high quality, but *Leucaena, Gliricidia, Sesbania* and *Chamaecytisus* have much variability and are of higher quality (Osuji *et al.*, 1997). Thirty accessions of 14 genera and 16 species of shrub legumes from CIAT were evaluated in sites of tropical rain forest and seasonal semi-evergreen in Cochabamba, Bolivia, under annual precipitation exceeding than 3200 mm·year^{-1} (Vallejos and Cardona, 1995). Considering forage yield, nutritive quality and yield distribution, from seven to nine accessions showed promise at each site, but three accessions of *F. macrophylla*, one of *Calliandra* sp., and one of *C. gyroides* were promising at all sites.

In another experiment, out of 48 accessions of 12 genera of shrub legumes evaluated for adaptation to different surface types in the Valley of Cauca, Colombia, only *Leucaena* and *Cratylia* showed outstanding performance (Parra and Gómez-Carabalí, 2000). Thirty-eight accessions of 22 taxa and five interspecific hybrids of *Leucaena* were evaluated in Honduras for their potential value as a tropical fodder crop (Stewart and Dunsdon, 1998). Three taxa had higher crude protein and digestibility scores than *L. leucocephala*, but when yield and palatability were included, the latter was superior.

The use of forage tree legumes is limited by the low nutritional quality of many species, due to anti-quality compounds, especially tannins. The challenge of releasing a new variety of shrub legume is that the known varieties are frequently not improved, so are of inferior quality and susceptible to diseases and insects. Because the farmer only knows these cultivars, he rejects the new releases. Also, once the new variety is released into a region, it may hybridize with the inferior forms and lose its advantage because of loss of purity. Because farmers save seed of only a few trees, inbreeding depression may occur and the next crop will have inferior performance (Shelton, 2001).

CHALLENGES AND PERSPECTIVES

Researchers worldwide are studying the most important genera and species of tropical forage grasses and legumes. From the data presented above, it is clear that much has already been evaluated in South and Central America. It seems that with the exception of tropical African grasses, large germplasm sources are available for selection and breeding, and there is germplasm that shows great promise in many distinct ecosystems. Many accessions have been indicated as promising in the past years, but somehow there seems to be a gap between the indication of these accessions and their final utilization and release to the public. It is absolutely imperative that promising accessions be evaluated under grazing before or during the release process to aid management at a farm level, and, very importantly, to convince farmers of the superiority and the indication of the specific cultivar. Grazing experiments are very expensive, and that may be the reason for the lack of continuity observed in many selection programmes.

Breeding is the key for future development of superior forages. As the selection programmes based on large germplasm sources suffer from continuity problems, resulting in the irregular release of superior accessions, breeding is the next step in the generation of productive high quality adapted forages. The tropical world has still not profited from the genetic manipulation of tropical forages through breeding, but possibilities are infinite.

Breeding programmes are being developed, at least involving the most important species of forages. Much work is yet to be done. Breeding of *P. maximum* has been underway at Embrapa

Beef Cattle Centre in Brazil since the early 1990s. The programme deals with the transference of the stolon characteristic to productive and adapted accessions (Savidan *et al.*, 1989; Jank, 1995) and increase in productivity and general adaptability (Jank *et al.*, 2001). Six percent of the collection of 426 accessions present stolons, a desirable trait, since most ecotypes are tall growing and caespitose. Good soil cover is especially important in high rainfall areas and in recently cleared land where weed encroachment becomes limiting. These accessions are short and provide good soil cover, but have low year-round production, and do not persist well under frost conditions or long periods of drought. Breeding involves sexual × apomictic crosses; sexual plants being tall and productive. Embrapa Beef Cattle Centre and the Instituto Agronômico do Paraná (IAPAR) will release another productive accession in 2003, as cv. Millenium. In November 2002, another national network was established, to evaluate 22 selected accessions and hybrids obtained from breeding in five regions in Brazil. New cultivars should arise from selection among these superior accessions under grazing (L. Jank, pers. comm.).

Particularly promising are plant breeding programmes within the genus *Brachiaria*, involving the recombination of genes from sexual plants of *B. ruziziensis* (high forage quality and determined flowering cycle) as the female parent with those of apomictic *B. brizantha* (source of resistance to spittlebug (Homoptera: Cercopidae)) and *B. decumbens* (source of vigour and adaptation to acid, infertile soils) as male parents. The progeny of these crosses are segregating for apomixis and sexuality, thus the F_1 apomicts are ready to move directly into selection and evaluation programmes while the sexual plants are available for further crossing (Valle, Macedo and Calixto, 2000). Breeding has been underway at CIAT and at the Embrapa Beef Cattle Centre since 1988. In 1999, three promising hybrids were developed and selected for further evaluation in grazing trials outside CIAT (Miles, 1999). Two of them have shown moderate resistance to spittlebug, which should limit its dispersal through Brazil. The third one was found to be highly resistant to this insect, but has shown limitations in seed production (Miles, 1999). This hybrid has been registered as cv. Mulato in Brazil. Nevertheless, they are very promising for regions of Central America and Colombia, where the area planted to *Brachiaria* is not so extensive.

The breeding goals for *Brachiaria* in Brazil also involve the development of improved *B. humidicola* for better quality while maintaining the advantage of this species, which is the adaptation to periodically waterlogged soils (Pereira *et al.*, 2001). Breeding in *Brachiaria* was limited until recently due to little knowledge concerning species taxonomy, mode of reproduction, breeding behaviour, and chromosome numbers. This information is now becoming available. Breeding also had to use the sexual *B. ruziziensis* as a bridge for improvement of the species *B. decumbens* and *B. brizantha* (Miles and Valle, 1996). However, with the discovery of sexual plants within these species and the duplication of their chromosome numbers, the need of a bridge of different species, which commonly implies chromosome pairing problems, is no longer necessary. An obstacle to using some of the treated accessions, however, is the very late flowering exhibited, thus isolating these genotypes (Valle, unpublished). These ongoing research programmes should result in a number of new options becoming available in the very near future.

Promising accessions of *Brachiaria* are being currently evaluated under grazing in various localities in Colombia and Brazil. In 2001, a network of grazing trials was established in five localities in Brazil, to evaluate the potential of four *Brachiaria* accessions selected from regional trials. Each locality is evaluating two accessions and a common standard, so that each accession will be evaluated in at least two localities (C.B. do Valle, pers. comm.). In Colombia, two hybrids (CIAT 36030 and 36061), one accession of *B. brizantha* (CIAT 26318) and two of *B. humidicola* (CIAT 26159 and 26427), selected from the combined analysis in regional trials, have been established in 3-ha plots to compare animal performance. Other grazing trials should be implemented when funded (Miles and Franco, 1999).

In the Amazon tropical humid forest environment in the Amazonian Piedmont, on an oxisol, 60 percent of the pastures are *Brachiaria,* but at various levels of degradation. To find alternative *Brachiaria* for this region, 24 accessions and hybrids of the CIAT collection were evaluated. One accession of *B. brizantha* showed spittlebug tolerance and high palatability, and should advance to grazing trials (Gomez *et al.*, 2000).

Pennisetum purpureum is very productive and requires good soil conditions. Despite this, some accessions show promise in poor, acid soils characteristic of most soils in Brazil, with low N availability. Ten of the 48 accessions evaluated in Minas Gerais showed promise for use under low N availability conditions and should be further evaluated (Xavier *et al.*, 1998). Breeding is underway at the Embrapa Dairy Cattle Centre and involves *P. purpureum* with the species *P. glaucum* (L.) R. Br. and *P. squamulatum* Fresen (A.V. Pereira, pers. comm.). The cross within the species *P. purpureum* yields superior elephant grass cultivars, which are multiplied vegetatively as clones. One such hybrid has recently been released as cv. Pioneiro. The cross between *P. purpureum* and *P. glaucum* yields sterile triploid progenies that, when the chromosome number is doubled by colchicine, yields fertile hexaploid progenies. The breeding of this type is in an advanced stage and a population reproducing by seeds will start evaluation under grazing in 2003, and should be released within two years (A.V. Pereira, pers. comm.). The advantage of this new cultivar is seed production rather than vegetative propagation, facilitating field establishment and the adoption of the cultivar.

Within the genus *Paspalum*, accession *P. regnellii* Mez BRA-000159 shows good promise. This species tolerates low temperatures better than *P. atratum.* It originates from higher latitudes in South Brazil and is a sexual type. Both of these *Paspalum* species are compatible with the stoloniferous *A. pintoi* and in trials the associations produced high quality grazing with a stable legume content. In the sixth year under intermittent grazing of one week of grazing and three weeks of rest, the legume content of the *P. atratum-A. pintoi* BRA 15598 mixture was 27.0 percent, and the *P. regnellii-A. pintoi* association contained 29.3 percent legume on a dry weight basis. Other species are also promising for wet soils. In an evaluation of *Paspalum* spp. in seasonally waterlogged soils in Quilichao, Colombia, three accessions (*P. atratum*, *P. guenoarum* Arechav. and *P. plicatulum* Michx.) were selected for further testing under grazing in association with legumes (Peters *et al.*, 2001).

Within the genus *Stylosanthes*, very successful composite hybrids have been obtained from the crosses between *S. guianensis* (Aubl.) Sw. var. *vulgaris* × var. *pauciflora*. These hybrids exhibit high genetic diversity in comparison with pure-line cultivars, which was found to confer resistance to the fungus anthracnose (Grof, Fernandes and Fernandes, 2001b). The stability of the disease resistance, yield attributes and adaptation to a wide range of ecological conditions were tested in multilocational trials in four regions in Brazil (located from longitude 6°9′ S to 20°27′ S), and also in China and Australia (Grof, Fernandes and Fernandes, 2001b). At least one of the hybrids should be released to farmers in Brazil in the next few years.

A very interesting experiment was established in 2001, to select promising accessions of *Stylosanthes* under grazing. Five accessions selected at Embrapa Beef Cattle for their productivity, adaptation and disease resistance were each planted in three plots of 200 m^2 in each of three, 1-ha pastures of *B. brizantha* cv. Marandu, *B. decumbens* cv. Basilisk and *Andropogon gayanus* cv. Planaltina. These pastures were replicated. The management is continuous, with a flexible stocking rate. After one year of grazing, two of the legume accessions are showing much promise in association with all three grasses (M. Pereira, pers. comm.).

The genus *Centrosema* is very plastic in terms of its adaptation and presents known qualities for animal production (Machado, 1999). In Cuba, 32 accessions were evaluated in 20-m^2 plots with four replications, associated with *A. gayanus* CIAT-621 in mob-grazing conditions (Machado, 1999). Three accessions were outstanding in production and legume proportion throughout the trial. Seed is being multiplied so that further studies can be carried out with these accessions (Machado, 1999). In Brazil, eight accessions of *C. brasilianum* were selected and should be evaluated in regional trials and associated with grasses under grazing (Pizarro, 2001).

Another interesting study is being developed on farms in eastern Venezuela. Many genera of legumes exhibit amphicarpy in some of the species. Amphicarpy is the production of both aboveground and belowground seeds, and has been observed in the genera *Centrosema, Clitoria* and *Macroptilium,* among others. It provides the possibility of increasing persistence under grazing due to the production of belowground seed. Five species of *Centrosema* have been shown to exhibit amphicarpy (Schultze-Kraft, Schmidt and Hohn, 1997), and three accessions of *C. rotundifolium* Mart. ex Benth. will be evaluated on farms in eastern Venezuela in association with grasses to assess their increased persistence due to amphicarpy (Rodríguez, Schultze-Kraft and González, 2001).

Eleven accessions of *A. pintoi* were evaluated in association with *B. dictyoneura* in Colombia. Although quality characteristics, consumption and persistence of the accessions were similar to those of cv. Mani Forrajero Perenne, their establishment was faster and forage production higher (Castillo, 2001). In Brazil, seven accessions of *A. pintoi* were selected and evaluated in association with an accession of *P. maritimum*. After two years, one accession of *A. pintoi* occupied 60 percent of the association, presented good vigour and was free of diseases. This accession showed high rate of establishment, persistence, aggressiveness and needs to be further evaluated on slopes and in other ecosystems (Pizarro, Ramos and Carvalho, 1997). The cost of seed production is often mentioned as the major drawback to wider use of the species. It is worth pointing out that a mechanical harvester capable of harvesting and cleaning this seed in one operation is technically feasible, and would reduce the cost of seed and hence establishment (B. Grof, pers. comm.). Seed production in *A. pintoi* accessions is high, with from 979 to 6 133 kg ha^{-1} having been obtained on three sites (Argel and Pizarro, 1992).

Leucaena breeding was developed in the 1980s, involving the species *L. leucocephala, L. diversifolia* (Schltdl.) Benth. and *L. shannnoni* Donn. Sm., with the aim of selecting genotypes tolerant to acid soils with low concentrations of Ca and high concentrations of Al, typical of the cerrados soils in Brazil (Hutton, 1985). The hybrids were evaluated in regional trials in the country, and the best ones advanced through generations with selection, so that lines that presented excellent development in acid soils were obtained. One such line (*L. leucocephala* × *L. diversifolia*), tolerant to high Al concentrations, is being evaluated under grazing at the Embrapa Cerrados and will be released in the next two years (C.T. Karia, pers. comm.).

CONCLUDING REMARKS

To ensure diversity of pasture systems, further collecting is needed to obtain genera not yet thoroughly collected in Africa, including *Hyparrhenia, Melinis, Setaria, Andropogon, Hemarthria, Chloris,* and *Cenchrus,* with more directed collections of *Brachiaria*. The species *B. mutica* has not yet been collected and new sources of sexuality in *B. brizantha* must be sought. Sexual forms within apomictic *Hyparrhenia rufa* and *Melinis minutiflora* must be collected.

A few species of forages have not received much deserved attention, such as *Pueraria phaseoloides* (Roxb.) Benth., which presents much variability. The commercial cultivar produces little seed in grazed swards; consequently it is vulnerable to overgrazing, fire and mismanagement. Since it is not stoloniferous, free-seeding strains are preferable. Such accessions exist in CIAT's collection, but are not being explored.

Most universities around the world have maize breeders in their genetics and plant improvement departments, but only very few, and especially in the United States of America, actually have forage breeders. This results in most plant breeding courses and research being biased towards crops of allogamous or autogamous reproduction, whereas forages or apomixis becomes the theme of only a few lectures or special topics in research. Much is yet to be done, and studies concerning apomixis and the apomictic grasses must not be omitted from the genetics courses, especially since currently many attempts are being made to transfer the apomixis gene(s) to important cultivated crops, including rice, maize and wheat.

A further challenge, and perhaps one of the most difficult to undertake, is to convince farmers, agronomists and extensionists that diversity is a must at the farm level. In Brazil, 80 percent of the cultivated pastures (80 million hectares) are planted to *Brachiaria* spp. and *B. brizantha* cv. Marandu alone is estimated to cover 40 million almost contiguous hectares. Considering that this grass is apomictic and therefore with no variation in its constitution, and that pasture grasses are perennial, the risk for these production systems is unbelievable. The consequences of such genetically uniform monocropping start to show up when news of pasture losses becomes a daily event. Certainly, the ease of establishment, high animal production and ease of management have led this grass to become so popular in pasture systems. It is of great concern that forage specialists endorse the rapid spread and adoption of a commercial cultivar of *B. decumbens* in the countries of the East, such as the Philippines, and Australia. However, one can understand that to avoid the tropical world's grassland monoculture, other good forages must be selected, bred and made available to farmers.

New techniques, such as the use of molecular markers, tissue culture and assisted selection using Quantitative Trait Loci (QTLs), will be very useful in the future for the selection of new improved cultivars (Valle, 2001). Significant progress has been made in the past fifteen years in the establishment of methodologies required for molecular breeding of forage plants. The application of molecular methodologies in forage plant improvement will increase the efficiency of phenotype-based selection, and plant genomics will be important for accelerating the application of biotechnology to forage and grassland agriculture (Hancock & Ulyatt, 2001; Spangenberg *et al.*, 2001).

Because of the unfortunate downsizing of pasture research programmes conducted by well-recognized international and national institutions, including CIAT, CSIRO and the Department of Primary Industries (DPI), it is of paramount importance that more progressive and large national programmes covering a range of agro-ecosystems, such as Embrapa, in Brazil, assume a leading role in this area of research. A number of important advances in pasture and livestock research have been made in the last decade that are expected to have a major impact on the productivity, persistence and sustainability of pasture-based livestock systems. Further progress must be research driven.

ACKNOWLEDGEMENTS

The authors would like to express their gratitude to Dr Bela Grof for the many suggestions and valuable contributions.

REFERENCES

Albert, A., Morán, Y., Valdespino, A., & Alvarez, O. 2000. Nota técnica: coleccion e identificacion de leguminosas de interes para la ganaderia em Topes de Collantes [Technical note: Collection and identification of legumes of interest to cattle rearing in Topes de Collantes]. *Pastos y forrajes*, **23**: 123–127.

Andrade, R.P. de. 2001. Pasture seed production in Brazil. pp. 129–132, *in*: *Proceedings of the XIX International Grassland Congress*, São Pedro, São Paulo, Brazil, 11–21 February 2001.

Andrade, R.P. de & Karia, C.T. 2000. Uso de *Stylosanthes* em pastagens no Brasil [Use of *Stylosanthes* in Brazilian pastures]. pp. 273–309, *in*: Anais do Simpósio de Forragicultura e Pastagens: Temas em evidência Proceedings of the Symposium of Forrages and Pastures: Themes in evidence]. Lavras, Minas Gerais, Brazil, 1–3 July 2000.

Argel, P.J. 2000. Pasto Toledo, nuevo cultivar para zonas tropicales de América [Toledo pasture, new cultivar for tropical zones of America]. *Pasturas Tropicales*, **22**(3): 38–39.

Argel, P.J. & Lascano, C.E. 1998. *Cratylia argentea* (Desvaux) O. Kuntze: Una nueva leguminosa arbustiva para suelos ácidos en zonas subhúmedas tropicales [*Cratylia argentea* (Desvaux) O. Kuntze: A new shrub legume for acid soils in subhumid tropical zones]. *Pasturas Tropicales*, **20**(1): 37–43.

Argel, P.J. & Pizarro, E.A. 1992. Germplasm case study: *Arachis pintoi*, pp.57–73. *In*: *Pastures for the tropical lowlands, in: CIAT's Contribution*. Cali, Colombia. 283p.

Barcellos, A.O., Pizarro, E.A. & Costa, N.L. 1997. Agronomic evaluation of novel germplasm under grazing: *Arachis pintoi* BRA-031143 and *Paspalum atratum* BRA-009610. pp. 47–48, *in*: *Proceedings of the XVIII International Grassland Congress*, Winnipeg, Manitoba, and Saskatoon, Saskatchewan, Canada, 8–19 June 1997.

Batista, L.A.R. & Godoy, R. 1995. BAETÍ – Embrapa 23, uma nova cultivar do capim andropogon (*Andropogon gayanus* Kunth.) [BAETÍ – Embrapa 23, a new cultivar of andropogon grass (*Andropogon gayanus* Kunth.)]. *Revista Brasileira de Zootecnia*, **24**(2): 204–213.

Batista, L.A.R. & Godoy, R. 2000. Caracterização preliminar e seleção de germoplasma do gênero *Paspalum* para produção de forragem [Preliminary characterization and selection of germplasm of the genus *Paspalum* for hay production]. *Revista Brasileira de Zootecnia*, **29**(1): 23–32.

Burson, B.L. 1997. Apomixis and sexuality in some Paspalum species. *Crop-Science*, **37**: 4, 1347–1351.

Burton, G.W. 1989. Registration of 'Tifton 9' Pensacola bahiagrass. *Crop Science*, **29**: 1326.

Burton, G.W., Gates, R.N. & Hill, G.M. 1993. Registration of 'Tifton 85' bermudagrass. *Crop Science*, **33**: 644.

Castillo, A.R. 2001. Potencial produtivo de ecotipos de *Arachis pintoi* en Piedemonte de los Llanos Orientales de Colombia [Productive potential of ecotypes of *Arachis pintoi* in the Piedmonte of the Eastern Llanos of Colombia]. *Pasturas Tropicales*, **23**(1): 19–24.

Castillo, A.C., Cuyugan, O.C., Fogarty, S. & Shelton, H.M. 1997. Growth, psyllid resistance and forage quality of *Leucaena leucocephala*, *L. pallida*, *L. diversifolia* and the F_1 hybrid of *L. leucocephala* × *L. pallida*. *Tropical Grasslands*, **31**: 188–200.

Clayton, W.D. 1983. Tropical grasses. pp. 39–46, *in:* J.G. McIvor and R.A. Bray (eds). *Genetic resources of forage plants*. East Melbourne, Australia: Commonwealth Scientific and Industrial Research Organization (CSIRO). 321p.

Combes, D. & Pernès, J. 1970. Variations dans le nombres chromosomiques du *Panicum maximum* Jacq. en relation avec le mode de reproduction [Variation in the number of chromosomes in *Panicum maximum* Jacq. in relation to mode of reproduction]. *Comptes Rendues Academie des Science Paris, Sér. D.*, **270**: 782–785.

Dujardin, M. & Hanna, W.W. 1985. Cytology and reproductive behavior of pearl millet-napiergrass hexaploids × *Pennisetum squamulatum* trispecific hybrids. *Journal of Heredity*, **76**(5): 382–384.

Euclides, V.P.B., Macedo, M.C.M. & Oliveira, M.P. de. 1995. Avaliação de ecotipos de *Panicum maximum* sob pastejo em pequenas parcelas [Evaluation of ecotypes of *Panicum maximum* in small plots]. pp. 97–99, *in:* XXXII Anais da Reunião Anual da Sociedade Brasileira de Zootecnia [Proceedings of the 32nd Annual Meeting of the Brazilian Society for Animal Science]. Brasilia, DF, Brazil, 17–21 July 1995.

Euclides, V.P.B., Macedo, M.C.M., Valério, J.R. & Bono, J.A.M. 2000. Cultivar Massai (*Panicum maximum*) uma nova opção forrageira: características de adaptação e produtividade [Massai cultivar (*Panicum maximum*) a new forage option: adaptation and productivity characteristics]. ID No. 397, *in:* XXXVII Anais da Reunião Anual da Sociedade Brasileira de Zootecnia [Proceedings of the 37th Annual Meeting of the Brazilian Society for Animal Science]. Viçosa, Brazil. CD-ROM.

Euclides, V.P.B., Valle, C.B. do, Macedo, M.C.M. & Oliveira, M.P. de. 2001. Evaluation of *Brachiaria brizantha* ecotypes under grazing in small plots. pp. 535–536, *in:* Proceedings of the XIX International Grassland Congress, Piracicaba, Brazil. , São Pedro, São Paulo, Brazil, 11–21 February 2001.

Faría-Mármol, J. 1995. Potencial forrajero de accesiones venezolanas de *Centrosema pubescens* [Forage potential of Venezuelan accessions of *Centrosema pubescens*]. *Pasturas Topicales,* **17**(1): 14–17.

Francisco, G. & Hernández, I. 1998. *Gliricidia Sepium* (Jacq.) Kunth. y Walp., arbol multiproposito para una ganaderia sostenible [*Gliricidia Sepium* (Jacq.) Kunth. y Walp., multipurpose tree for sustainable cattle rearing]. *Pastos y Forrajes,* **21**: 191–212.

Goedert, W.J. (ed). 1986. *Solos dos Cerrados* [Cerrados soils]. São Paulo: Nobel. 422p.

Gomez, M.M., Velásquez, J.E., Miles, J.W. & Rayo, F.T. 2000. Adaptación de *Brachiaria* en el Piedemonte amazónico colombiano [Adaptation of *Brachiaria* in the Colombian Amazonian Piedmont]. *Pasturas Tropicales,* **22**(1): 19–32.

Grof, B., Andrade, R.P.de, Souza, M.A. & Valls, J.F.M. 1989. Selection of *Paspalum* spp. adapted to seasonally flooded várzea lands in Central Brazil. pp. 291–292, *in*: Proceedings of the XVI International Grassland Congress, Nice, France, 4–11 October 1989.

Grof, B., Fernandes, C.D. & Fernandes, A.T.F. 2001a. A novel technique to produce polygenic resistance to anthracnose in *Stylosanthes capitata*. p. 525, *in*: Proceedings of the XIX International Grassland Congress, São Pedro, São Paulo, Brazil, 11–21 February 2001.

Grof, B., Fernandes, C.D. & Fernandes, A.T.F. 2001b. New *Stylosanthes guianensis* for tropical grasslands. pp. 526–7, *in*: Proceedings of the XIX International Grassland Congress, São Pedro, São Paulo, Brazil, 11–21 February 2001.

Hacker, J.B., Glatzle, A. & Vanni, R. 1996. Paraguay – a potential source of new pasture legumes for the subtropics. *Tropical Grasslands,* **30**: 273–281.

Hacker, J.B. & Jank, L. 1998. Breeding tropical and subtropical forage plants. pp. 49–71, *in*: J.H. Cherney and D.J.R. Cherney (eds). *Grass for Dairy Cattle.* New York/Wallingford: CABI Publishing.

Hancock, K.R. & Ulyatt, M.J. 2001. Opportunities in molecular biology: enhancement of the nutritional value of forages. pp. 627-632, *in*: Proceedings of the XIX International Grassland Congress, São Pedro, São Paulo, Brazil, 11–21 February 2001.

Hanson, J. & Maass, B.L. 1999. Conservation of tropical forage genetic resources. pp. 31–36, *in*: Proceedings of the XVIII International Grassland Congress, Winnipeg, Manitoba, and Saskatoon, Saskatchewan, Canada, 8–19 June 1997. Docs. V3. Sessions. lii_031.pdf CD-ROM.

Harlan, J.R. 1983. The scope for collection and improvement of forage plants. pp. 3–14, *in*:, J.G. McIvor and R.A. Bray (eds). *Genetic resources of forage plants.* East Melbourne, Australia: Commonwealth Scientific and Industrial Research Organization (CSIRO).

Hernández, M. 2000. Efecto de la adicion de follaje de *Bauhinia purpurea* em la produccion y calidad de *Panicum maximum* [Effect of the addition of foliage of *Bauhinia purpurea* in the production and quality of *Panicum maximum*]. *Pastos y forrajes,* **23**: 141–147.

Hernández, M., Argel, P.J., Ibrahim, M.A. & 't Mannetje, L. 1995. Pasture production, diet selection and liveweight gains of cattle grazing *Brachiaria brizantha* with and without *Arachis pintoi* at two stocking rates in the Atlantic Zone of Costa Rica. *Tropical Grasslands,* **29**: 134–141.

Hojito, S. & Horibata, T. 1982. Plant exploration, collection and introduction from Africa. *In*: Nekken Shiryo 58. TARC. Japan. 1-120.

Hutton, E.M. 1985. Acid-tolerant hybrids in the tree legume *Leucaena*. pp. 199–201, *in:* Proceedings of the XV International Grassland Congress, Kyoto, Japan. 24-31 August 1985.

Jank, L. 1995. Melhoramento e seleção de variedades de *Panicum maximum* [Breeding and selection of *Panicum maximum* varieties]. pp. 21–58, *in*: Simpósio sobre Manejo da Pastagem, 12., Piracicaba. [Proceedings of the 12th Symposium on Pasture Management]. Piracicaba, Sao Paulo, Brazil, 5–7 September 1995.

Jank, L., Calixto, S., Costa, J.C.G., Savidan, Y.H. & Curvo, J.B.E. 1997. Catalog of the characterization and evaluation of *Panicum maximum* germplasm: morphological description and agronomical performance. EMBRAPA-CNPGC, Documentos 68. 53p.

Jank, L., Costa, J.C.G., Savidan, Y.H. & Valle, C.B. do. 1993. New *Panicum maximum* cultivars for diverse ecosystems in Brazil. 1: 509–511, *in*: Proceedings of the XVll International Grassland Congress, Palmerston North, New Zealand, 8–21 February 1993.

Jank, L., Savidan, Y.H., Souza, M.T. de & Costa, J.C.G. 1994. Avaliação do germoplasma de *Panicum maximum* introduzido da África: 1. Produção forrageira [Evaluation of *Panicum maximum* germplasm introduced from Africa: 1. Forage production]. *Revista da Sociedade Brasileira de Zootecnia,* **23**(3): 433–440.

Jank, L., Valle, C.B. do., Carvalho, J. de. & Calixto, S. 2001. Evaluation of guineagrass (*Panicum maximum* Jacq.) hybrids in Brazil. pp. 498–499, *in*: Proceedings of the XIX International Grassland Congress, São Pedro, São Paulo, Brazil, 11–21 February 2001.

Kalmbacher, R.S., Brown, W.F., Colvin, D.L., Kretschmer, A.E. Jr. & Martin, F.G. 1997. Suerte atra paspalum. Its management and utilization. *Florida Agricultural Experiment Station Circular*, S-397.

Keller-Grein, G., Amézquita, M.C., Lema, G. & Franco, L.H. 1993. Multilocational testing of grasses and legumes in the humid tropics of South America. pp. 217–219, *in*: Proceedings of the XVll International Grassland Congress, Palmerston North, New Zealand, 8–21 February 1993.

Keller-Grein, G., Maass, B.L. & Hanson, J. 1996. Natural variation in *Brachiaria* and existing germplasm collections. pp. 16–42, *in*: J.W. Miles, B.L. Maass &, C.B. do Valle (eds). Brachiaria*: Biology, agronomy, and improvement*. CIAT Publication, N° 259. Cali:CIAT.

Kerridge, P.C. 2001. A historical perspective of agropastoral system research in the savannahs of South America. pp. 1–17, *in:* T. Kanno and M.C.M. Macedo (eds). JIRCAS/Embrapa Gado de Corte International Joint Workshop on Agropastoral System in South America. Campo Grande, Mato Grosso do Sul, Brazil, 7-8 December 1999. *JIRCAS Working Report*, No. 19.

Kexian, Y., Lascano, C.E., Kerridge, P.C. & Avila, P. 1998. The effect of three tropical shrub legumes on the intake and acceptability by small ruminants. *Pasturas Tropicales,* **20**: 31–35.

Lascano, C.E. 1993. Nutritive value and animal production of forage *Arachis*. pp. 109–121, *in*: P.C. Kerridge and B. Hardy (eds). *Biology and agronomy of forage* Arachis. CIAT publication, No. 240.

Lascano, C.E. & Estrada, J. 1989. Long-term productivity of legume-based and pure grass pastures in the Eastern Plains of Colombia. pp. 1179–1180, *in*: Proceedings of the XVI International Grassland Congress, Nice, France, 4–11 October 1989.

Lascano, C.E., Holmann, F., Romero, F., Hidalgo, C. & Argel, P. 2002. Advances in the utilization of legume-based feeding systems for milk production in Sub-humid Tropical regions. pp. 43–59, *in*: Anais de Palestras da XXXIX Reunião Anual da Soceidade Brasileira de Zootecnia [Proceedings of the Invited Papers of the 39th Annual Meeting of the Brazilian Animal Science Society]. Recife, Pernambuco, Brazil, 29 July–1 August 2002.

Liu, C.J., Musial, J.M. & Thomas, B.D. 1999. Genetic relationships among *Stylosanthes* species revealed by RFLP and STS analyses. *Theoretical and Applied Genetics,* **99**: 1179–1186.

Maass, B.L. & Pengelly, B.C. 2001. Tropical forage genetic resources - Will any be left for future generations? pp. 541–542, *in:* Proceedings of the XIX International Grassland Congress, São Pedro, São Paulo, Brazil, 11–21 February 2001.

Maass, B.L. & Torres, A.M. 1998. Off-types indicate natural outcrossing on five tropical forage legumes in Colombia. *Tropical Grasslands,* **32**: 124–130.

Macedo, M.C. 2001. Sustainability of pasture production in the savannahs of Tropical America. pp. 391–399, *in*: Proceedings of the XVIII International Grassland Congress, Winnipeg, Manitoba, and Saskatoon, Saskatchewan, Canada, 8–19 June 1997.

Machado, R. 1999. Seleccion de Germoplasma de *Centrosema* spp. bajo condiciones de pastoreo simulado [Selection of germplasm of *Centrosema* spp. under simulated grazing conditions]. *Pastos y forrajes,* **22**: 307–316.

Miles, J.W. 1983. Natural outcrossing in *Stylosanthes capitata*. *Tropical Grasslands,* **17**: 114–117.

Miles, J.W. 1985. Evaluation of potential genetic marker traits and estimation of outcrossing rate in *Stylosanthes guianensis*. *Australian Journal of Agricultural Research,* **36**: 259–265.

Miles, J.W. 1999. Nuevos híbridos de *Brachiaria* [New hybrids of *Brachiaria*]. *Pasturas Tropicales,* **21**(2): 78–80.

Miles, J.W. 2001. Achievements and perspectives in the breeding of tropical grasses and legumes. pp. 509–515, *in*: Proceedings of the XIX International Grassland Congress. São Pedro, São Paulo, Brazil, 11–21 February 2001.

Miles, J. & Franco, L.H. 1999. The Brachiaria Evaluation Network in Colombia. pp. 141–143, *in:* Tropical grasses and legumes: Optimizing genetic diversity for multipurpose use (Project IP-5). Annual Report 1999. CIAT, Cali, Colombia.

Miles, J.W. & Lapointe, S.L. 1992. Regional germplasm evaluation: A portfolio of germplasm options for the major ecosystems of tropical America, pp. 9–28, *in*: Pastures for the tropical lowlands – CIAT's Contribution. Cali, Colombia. 283p.

Miles, J.W. & Lascano, C.E. 1997. Status of *Stylosanthes* development in other countries. I. Stylosanthes development and utilization in South America. *Tropical Grasslands,* **31**(5): 430–434.

Miles, J.W. & Valle, C.B. do. 1996. Manipulation of apomixis in *Brachiaria* breeding. pp. 164–177, *in*: J.W. Miles, B.L. Maass and, C.B. do Valle (eds). Brachiaria: *Biology, Agronomy, and Improvement*. Centro Internacional de Agricultura Tropical - CIAT/Empresa Brasileira de Pesquisa Agropecuária - EMBRAPA . CIAT Publication No.259.

Moreno, M.A.R., Neto, J.C. & Santana, J.C. 1995. Evaluación preliminar de dos especies de *Brachiaria* en Itabela, Bahia, Brasil [Preliminary evaluation of two species of *Brachiaria* in Itabela, Bahia, Brazil]. *Pasturas Tropicales*, **17**(1): 31–33.

Osuji, P.O., Odenyo, A.A., Acamovic, T., Stewart, C.S. & Topps, J.H. 1997. The role of legume forages as supplements to low quality roughages – ILRI experience. Selected papers from an international conference on *Evaluation of forages for ruminants in the tropics*, Harare, Zimbabwe, 28 August–1 September 1995. *Animal Feed Science and Technology*, **69**: 1–3; 27–38.

Parra, F.A. & Gómez-Carabalí, A. 2000. Introducción y elaluación de especies herbáceas y arbustivas forrajeras en zonas de ladera de Cauca y Valle del Cauca, Colombia [Introduction and evaluation of herbaceous and shrub forages in the zones of arable land in the Cauca Valley, Colombia]. *Pasturas Tropicales*, **22**(2): 54–62.

Paterson, R.Y. 1987. Genetic resources of tropical pasture. pp. 21–31, *in*: J.E. Moore, K.H. Quesenberry and M. Michaud (eds). *Forage livestock research needs for the Caribbean basin*. Gainesville, Florida: University of Florida.

Pedreira, C.G.S. & Brown, R.H. 1996. Yield of selected and unselected bahiagrass populations at two cutting heights. *Crop Science*, **36**: 134–137.

Pengelly, B.C. & Staples, I.B. 1997. Opportunities for the development of new tropical forage cultivars. *Tropical Grasslands*, **31**: 344–349.

Penteado, M.I.de O., Mira, L.E. de & Vega, M.P.de la. 1996. Genetic resources of *Centrosema* spp.: genetic changes associated to the handling of an active collection. *Genetic Resources and Crop Evolution*, **43**: 85–90.

Pereira, A.V., Valle, C.B. do, Ferreira, R. de P. & Miles, J.W. 2001. Melhoramento de forrageiras tropicais [Breeding of tropical forages]. Capítulo 18, pp. 549–601, *In*: L.L. Nass, A.C.C. Valois, I.S de Melo, M.C. Inglis-Valadares, (eds). *Recursos Genéticos & Melhoramento - Plantas*. Fundação MT, Rondonópolis.

Pereira, J.M. 2003. Amendoim forrageiro cv. Belmonte [Forage peanut cv. Belmonte]. (see: http://www.ceplac.gov.br/amendoim_forrageiro.htm)

Peters, M., Hincapié, B., Avila, P. & Lascano, C.E. 2001. Evaluation of *Paspalum* spp. with adaptation to poorly drained soils in the tropical Americas. pp. 530–531, *in*: Proceedings of the XIX International Grassland Congress, São Pedro, São Paulo, Brazil, 11–21 February 2001.

Pinstrup-Andersen, P. & Pandya-Lorch, R. 1997. A vision of the future world food production and implications for the environment and grasslands. Opening speaker. pp. 11–16, *in*: Proceedings XVIII International Grassland Congress, Canada, Winnipeg, Manitoba, and Saskatoon, Saskatchewan, Canada, 8–19 June 1997.

Pizarro, E.A. 2001. Novel grasses and legumes germplasm: advances and perspectives for tropical zones. pp. 93–10, *in*: Proceedings of the XIX International Grassland Congress, São Pedro, São Paulo, Brazil, 11–21 February 2001.

Pizarro, E.A. & Carvalho, M.A. 1992. Cerrado: Introduccion y evaluacion agronomica de forrajeras tropicales [Introduction and agronomic evaluation of tropical forages]. pp. 1–68, *in*: Red. Internacional de Pastos Tropicales, 1a. Reunion Sabanas, 23 a 26 noviembre, 1992, Brasilia, Brasil. 686p.

Pizarro, E.A. & Carvalho, M.A. 1996. Introducción y evaluación de leguminosas forrajeras en el Cerrado brasileño: *Centrosema* spp. y *Desmodium* spp. [Introduction and evaluation of forage legumes in the Brazilian Cerrados: *Centrosema* spp. y *Desmodium* spp]. *Pasturas Tropicales*, **18**(2): 14–18.

Pizarro, E.A. & Carvalho, M.A. 1997. Evaluation of a collection of *Calopogonium mucunoides* Desv. for the Cerrado ecosystem, Brazil. *Journal of Applied Seed Production*, **15**: 17–21.

Pizarro, E.A., Ramos, A.K.B. & Carvalho, M.A. 1996. Potencial forrajero y producción de semillas de accesiones de *Calopogonium mucunoides* preselecionadas en el Cerrado brasileño [Forage potential and seed production of preselected accessions of *Calopogonium mucunoides* in the Brazilian Cerrados]. *Pasturas Tropicales*, **18**(2): 9–13.

Pizarro, E.A., Ramos, A.K.B. & Carvalho, M.A. 1997. Produccíon y persistencia de siete accesiones de *Arachis pintoi* asociados com *Paspalum maritimum* en el Cerrado brasileño [Production and persistence of seven accessions of *Arachis pintoi* associated with *Paspalum maritimum* in the Brazilian Cerrados]. *Pasturas Tropicales*, **19**(2): 40–44.

Quarín, C. L., Valls, J.F.M. & Urbani, M.H. 1997. Cytological and reproductive behaviour of *Paspalum atratum* – promising forage grass for the tropics. *Tropical Grasslands*, **31**: 114–116.

Queiroz, L.P. de, Silva, M.M. da, Ramos, A.K.B. & Pizarro, E.A. 1997. Reproduction studies in *Cratylia argentea* (Desv.) O. Kuntze and *Cratylia mollis* Mart. ex Benth. (Leguminosae-Papilionoideae). *Pasturas Tropicales*, **19**: 3; 20–23.

Ralph, W. 1982. Introducing new forage plants to the tropics. *Rural Research in CSIRO*, **114**: 14–19.

Rodríguez, I., Schultze-Kraft, R. & González, S. 2001. Evaluation of *Centrosema rotundifolium* for sand-soil savannahs in Eastern Venezuela. pp. 565–566, *in*: Proceedings of the XIX International Grassland Congress, São Pedro, São Paulo, Brazil, 11–21 February 2001.

Sano, E.E., Barcellos, A.O. & Bezerra, H.S. 2000. Assessing the spatial distribution of cultivated pastures in the Brazilian savannah. *Pasturas Tropicales*, **22**(3): 2–15.

Savidan, Y.H. 1982. Nature et hérédité de l'apomixie chez *Panicum maximum* Jacq. [Nature and heredity of apomixis in *Panicum maximum* Jacq]. *ORSTOM Travaux et Documentos*, **153**. 159p.

Savidan, Y.H., Jank, L. & Costa, J.C.G. 1990. Registro de 25 acessos selecionados de *Panicum maximum* [Register of 25 selected accessions of *Panicum maximum*]. *Embrapa Gado de Corte, Documentos*, **44**. 68p.

Savidan, Y.H., Jank, L., Costa, J.C.G. & Valle, C.B. do. 1989. Breeding *Panicum maximum* in Brazil: 1. Genetic resources, modes of reproduction and breeding procedures. *Euphytica*, **41**: 107-112.

Schmidt, A., Schultze-Kraft, R., Maass, B.L. & Lascano, C. 2001. The use of cafeteria trials for the selection of *Desmodium ovalifolium* genotypes. pp. 570–571, *in*: Proceedings of the XIX International Grassland Congress, São Pedro, São Paulo, Brazil, 11–21 February 2001.

Schultze-Kraft, R. & Schmidt, A. 1998. Segundo taller de Trabajo de proyecto *Desmodium ovalifolium* [Second work meeting of the project *Desmodium ovalifolium*]. *Pasturas Tropicales*, **20**(1): 44–46.

Schultze-Kraft, R., Schmidt, A. & Hohn, H. 1997. Amphicarpic legumes for tropical pasture persistence. ID No. 191: 1-13-14. *In:* Proceedings of the XVIII International Grassland Congress, Winnipeg, Manitoba, and Saskatoon, Saskatchewan, Canada, 8–19 June 1997.

Schultze-Kraft, R., Williams, W.M. & Keoghan, J.M. 1993. Searching for new germplasm for the year 2000 and beyond. pp. 1: 181–188, *in:* Proceedings of the XVII International Grassland Congress, Palmerston North, New Zealand, 8–21 February 1993.

Segui, E. & Mendoza, F. 1999. Posibilidades del *Panicum maximum* para suelos acidos [Possibilities of *Panicum maximum* for acid soils]. *Pastos y forrajes*, **22**: 221–229.

Seiffert, N.F. & Zimmer, A.H. 1988. Contribución de *Calopogonium mucunoides* al contenido de nitrógeno en pasturas de *Brachiaria decumbens* [Contribution of *Calopogonium mucunoides* to the nitrogen content in *Brachiaria decumbens* pastures]. *Pasturas Tropicales*, **10**(3): 8–13.

Shelton, H.M. 2001. Advances in forage legumes: shrub legumes. pp. 549–556, *in:* Proceedings of the XIX International Grassland Congress, São Pedro, São Paulo, Brazil, 11–21 February 2001.

Simão Neto, M., Dias Filho, M.B. & Serrão, E.A.S. 1995. Avaliação da adaptação de acessos de *Brachiaria* para a Amazônia Oriental do Brasil [Evaluation of the adaptation of *Brachiaria* accessions in the Eastern Amazonia, Brazil]. *Pasturas Tropicales*, **17**(1): 9–13.

Spangenberg, G., Kalla, R., Lidgett, A., Sawbridge, T., Ong, E.K. & John, U. 2001. Transgenesis and genomics in molecular breeding of forage plants. pp. 615-625, *in*: Proceedings of the XIX International Grassland Congress, São Pedro, São Paulo, Brazil, 11–21 February 2001.

Stewart, J.L. & Dunsdon, A.J. 1998. Evaluación preliminar de la calidad potencial como forraje de rango de especies de *Leucaena* [Preliminary evaluation of the forage quality potential of a range of *Leucaena* species]. *Pasturas Tropicales*, **20**(3): 36–50.

Toledo, J.M. 1985. Pasture development for cattle production in the major ecosystems of the tropical American lowlands. pp. 74–81, *in*: Proceedings of the XV International Grassland Congress, Kyoto, Japan, 24–31 August 1985.

Torres, G.A.M. 1996. Un herbario de referencia para la colección de germoplasma de forrajes tropicales conservada por el Centro Internacional de Agricultura Tropical [A reference herbarium for the collection of germplasm of tropical forages maintained by CIAT (International Centre of Tropical Agriculture)]. *Pasturas Tropicales*, **18**(3): 71–74.

Valle, C.B. do. 2001. Genetic resources for tropical areas: Achievements and perspectives. pp. 477–482, *in*: Proceedings of the XIX International Grassland Congress, São Pedro, São Paulo, Brazil, 11–21 February 2001.

Valle, C.B. do. 2002. Pastagens - Xaraés: nova alternativa para a *Brachiaria brizantha* cv. Marandu [Pastures - Xaraés: a new alternative for *Brachiaria brizantha* cv. Marandu]. *A Granja*, **58**(642): 50–51.

Valle, C.B. do & Glienke, C. 1991. New sexual accessions in *Brachiaria*. *Apomixis Newsletter*, **3**: 11–13.

Valle, C.B. do, Glienke, C. & Leguizamon, G.O.C. 1995. Inheritance of apomixis. p.14, *in*: *Harnessing apomixis: A new frontier in Plant Science*. Proceedings of the International Conference. College Station, Texas, USA, 23–27 September 1995.

Valle, C.B. do, Macedo, M.C.M. & Calixto, S. 2000. Avaliação agronômica de híbridos de *Brachiaria* [Agronomic evaluation of *Brachiaria* hybrids]. *In:* Anais da XXXVII Reunião Anual da Sociedade Brasileira de Zootecnia [Proceedings of the 37th Annual Meeting of te Brazilian Society for Animal Science]. Viçosa, Brazil, 24–27 July 2000. CD Rom, ID No. 383.

Valle, C.B. do & Savidan, Y.H. 1996. Genetics, cytogenetics, and reproductive biology of *Brachiaria*. pp. 147–163, *in:* J.W. Miles, B.L. Maass and C.B. do Valle (eds). Brachiaria*: Biology, Agronomy, and Improvement*. Centro Internacional de Agricultura Tropical - CIAT/Empresa Brasileira de Pesquisa Agropecuária - EMBRAPA. CIAT Publication No. 259.

There is evidence of indigenous use of unripe *Leucaena* pods and seeds for human consumption in the Tehuacan Valley in Mexico, dating back to 6800 BC. It seems that *Leucaena* cultivation may have begun about 2 000 years ago (Hughes, 1998a,b) and continues to be cultivated for human consumption in Mexico today, but rarely for forage.

Current interest in MPTs is principally for forage. In developing countries, an increasingly affluent urban population is consuming more meat, and demanding higher quality meat. Smallholders are responding to this new and profitable opportunity with feeding strategies designed to fatten animals for slaughter at a younger age. This requires either high quality locally grown feeds or use of expensive concentrates. Farmers are finding that forage tree legumes meet their needs, enabling them to achieve increasing levels of profitability.

In Australia, large-scale leucaena (*Leucaena leucocephala*) plantings are occurring because farmers appreciate that leucaena–grass systems are both sustainable and highly productive. This production system allows them to produce cattle for high value domestic markets, and for export to east and southeast Asia (Larsen *et al.*, 1998; Shelton and Jones, 1995).

In Indonesia and the Philippines, numerous small-scale plantings of MPTs (boundary, fence line, border and small grove plantings) have occurred because farmers need the multiple benefits of fuelwood, control of animal movement, conservation of agricultural land, and supply of forage.

The purpose of this paper is to review the important issues for research and development (R&D) in forage tree legumes in tropical and subtropical farming systems.

DIVERSITY, QUALITY AND AVAILABILITY OF PLANTING MATERIAL

Lack of diversity in original introductions

In most cases, the movement of tree legume germplasm around the world began more than a century ago. Original seed introduced into an area was usually harvested from a few readily accessible trees. This meant that the early developments were based on unimproved, inferior and often seedy varieties that contained little genetic diversity.

The movement of small amounts of seed of *Leucaena leucocephala* subspecies *leucocephala* from Mexico to Southeast Asia in the 1600s is the most celebrated example of this phenomenon. This species is a highly self-fertile polyploid, so that further movement of the introduced variety was of almost identical genetic material. The spread of readily harvestable seed from country to country, region to region, organization to organization and farmer to farmer has resulted in over-reliance on an extremely narrow genetic base (Hughes, 1998a,b). This original leucaena, now known as the "common weedy type", has invaded disturbed sites, and become a weed in many countries. We now appreciate a much greater diversity in *Leucaena* from a range of ecological zones, some of which germplasm has significant potential value for agriculture, with much lower weed risk (Mullen *et al.*, 1998b).

Similarly, Dutton (1992) reported that most seed of *Prosopis* planted around the world was of unknown origin and from a narrow genetic base. The thorny *Prosopis* shrubs, widespread in Africa and India, came from introductions of inferior germplasm, and this has led to a poor appreciation of the genus. Research trials have shown that there is superior germplasm for different rainfall zones and soil types, and information on this new material needs to be disseminated.

The introduction of *Gliricidia sepium* from Trinidad to Sri Lanka was reportedly based on seed from one tree (Stewart, Allison and Simons, 1996). They suggested that the genetic diversity in many introduced populations will not be sufficient to ensure long-term stability.

There are other examples. Only two seed samples of *Calliandra calothyrsus* were first introduced into Indonesia (Java) from Guatemala in 1936 to provide shade for coffee. After 1974, seed was further spread by forest rangers for fuelwood use, and the species now covers more than 30 000 ha in Java alone (Kartasubrata 1996).

There is an added problem, as selection of best-performing varieties by farmers has narrowed diversity. In Flores (Indonesia), a wide diversity of tree legumes was grown in farming systems in the 1960s. Species such as *Acacia, Albizia, Calliandra, Cassia, Gliricidia, Pterocarpus, Sterculia* and *Tamarindus* were all grown in various mixed farming systems. With intensification and commercialization, there was greater reliance on a few species, notably leucaena (Djogo, Siregar and Gutteridge, 1995). The arrival of the leucaena psyllid in this region in the mid-1980s was particularly devastating.

The current recommendation for selection of seed from a native range is to obtain seed from at least 25, and preferably 50, trees, with sufficient distance between them (50 m) to minimize the likelihood of co-ancestry (Allison and Simons 1996). This simple approach for capture of local diversity was not appreciated when the first introductions were made.

Obtaining high quality germplasm

It is clear that many farmers are using inferior planting material and that overcoming this limitation will not be a simple matter. Cromwell, Brodie and Southern (1996), in surveys of farmers using MPTs in Honduras, Malawi and Sri Lanka, found that the quality and reliability of MPT germplasm supply was limited. Purchase from formal sources was often expensive, and seed was often obtained locally, as it was cheaper and more accessible. For this reason, germplasm was often:

- of unknown genetic quality,
- collected and distributed with weak protocols,
- selected on timber criteria,
- distributed with no knowledge or understanding of provenance quality, provenance origins, or the importance of genetic diversity, and
- not reliably available.

Other workers have found that the market is not discerning. Much of the demand for *Gliricidia* has been met with seed of inferior quality with no premium paid for quality (Simons, 1996). A major and serious limitation for *Prosopis* spp., which is restricting adoption, is lack of availability of seed of well documented provenances or improved varieties. There is also no source of clones of the best vegetatively propagated material (Dutton, 1992).

This has not been the case with leucaena, as high quality seed is marketed under species, cultivar or provenance name.

There are two major difficulties when introducing new varieties to areas where inferior varieties are already well established.

- Farmers may not be prepared to purchase seed of improved varieties, e.g. of *Gliricidia*, as they consider it a low value crop (Simons, 1996). They can use existing material at no cost. New germplasm would need to be markedly superior. Although seed of the best provenances of *Gliricidia* is now becoming available from seed orchards around the world, much of the current demand for *Gliricidia* is being satisfied by inferior unlabelled material, and it may be difficult to promote superior provenances, such as Retalhuleu (Simons, 1996). Simons (1996) suggested that new varieties will need to be at least 30% better in terms of woody and leaf biomass to interest farmers.

- When new material is introduced into existing areas, there may be hybridization, loss of purity and therefore loss of advantage. This will be less of a problem for vegetative propagation and self-pollinated seed. There may also be inbreeding depression if farmers collect seed from just a few trees for propagation, such as from those that seed prolifically.

However, other experience is that farmers will plant new varieties if they recognize the benefits. For instance, farmers in Batangas Province in the Philippines were immediately enthusiastic about the new F_1 hybrid KX2 (*L. leucocephala* K636 × *L. pallida* K748) (A. Castillo, pers. comm.). They appreciated its improved growth potential and its resistance to psyllids. In contrast, K636 (cv. Tarramba) may not be accepted as quickly as it is similar to the giant leucaenas already present in the Philippines. A similar positive response by farmers to the introduction of hybrid *Leucaena* was observed in northern Viet Nam (Plate 4.1).

Plate 4.1 Feeding hybrid KX2 leucaena to goats in North Viet Nam. [Photograph: H.M. Shelton]

In order to introduce new varieties, it may therefore be important to:

- ensure that the variety has clear benefits, such as insect or disease resistance, or greatly improved productivity;
- create an appropriate local name for the new variety;
- provide seed or planting material of high quality; and
- ensure that seed is readily available through traditional channels at a reasonable price, and promote farmer-level (smallholder) seed production schemes to provide income for farmers, and thereby local availability of seed.

THE PRINCIPAL SPECIES AND THEIR AGRO-ECOLOGICAL RANGE

Principal species

The literature lists several hundred species of leguminous trees with potential for forage (Le Houérou, 1980; Attah-Krah, Sumberg and Reynolds, 1986). Most have not been investigated and few are in current use to any significant extent. Of the 5 000 known nitrogen-fixing woody species, Brewbaker (1986) suggested that only about 80 leguminous tree and shrub species might have potential multipurpose agroforestry roles, including fodder, in tropical farming systems. Roshetko *et al.* (1996) listed 46 species suitable for fodder, but many fewer have found significant use in world animal production systems.

Twenty species and the primary references are given in Table 4.2. There are additional species which have forage potential, but which are not in significant use in world farming systems. Within each species, there is genetic variation that can be exploited. In this review, only those species that are in significant use for forage have been chosen as the principal MPTs.

Species that have potential, but are not yet in significant use, include the leucaena hybrids (e.g. *Leucaena pallida* × *L. leucocephala*), *L. collinsii* and *L. trichandra;* the last-named species is suited to the high altitude tropics (Shelton *et al.*, 1998).

Table 4.2 Most-used tree legume species for forage purposes (primary references in parentheses)

Higher quality species		Lower quality species	
Albizia lebbeck	(Lowry, 1989)	*Acacia aneura*	(Beale, 1994)[(1)]
Chamaecytisus palmensis	(Snook, 1982)	*Acacia nilotica*	(Audru *et al.*, 1992)
Cratylia argentea	(Argel and Lascano, 1998)	*Acacia tortilis*	(Wickens *et al.*, 1995)[(1)]
Desmodium rensonii	(Djogo, Sregar and Gutteridge, 1995)	*Albizia chinensis*	(Zabala, 1997)
Desmanthus virgatus	(Gutteridge, 1994a)	*Albizia saman*	(Roshetko, 1995)
Gliricidia sepium	(Stewart, Allison and Simons, 1996)	*Calliandra calothyrsus*	(Evans, 1996)
Leucaena leucocephala	(Shelton *et al.*, 1998)	*Erythrina* spp.	(Westley and Powell, 1993)
Leucaena diversifolia	(Shelton *et al.*, 1998)	*Faidherbia albida*	(Wickens *et al.*, 1995)[(1)]
Sesbania grandiflora	(Gutteridge and Rekib, 1995)	*Flemingia macrophylla*	(Gutteridge, 1994a)
Sesbania sesban	(Gutteridge and Rekib, 1995)	*Prosopis juliflora*	(Dutton, 1992)

Note: (1) Principal application is in indigenous semi-subsistence systems.

Agro-ecological range

Agroforesters require forage tree legumes adapted to a wide range of environments. Species are needed for climates ranging from the humid tropics of some Pacific Island countries, the seasonally dry tropics of Southeast Asia, the cooler high altitude tropics of eastern Africa, and the arid zones of the Sahel and North Africa. Similarly, adaptation to a great range of edaphic conditions is sought, from the alkaline vertisiols of the brigalow soils in Central Queensland to the very acid, high-aluminium, grasslands of South America.

A summary of the general ecological adaptation of key species is given in Table 4.3. As for other parameters, there is considerable diversity both within and between species for all genera. Multivariate genotype × environment trials are required to identify this variation. This has been completed for the *Leucaena* genus only (Mullen *et al.*, 1998b), and showed that there was specific adaptation for a range of environments, and that the suitability of a particular environment for *Leucaena* could be predicted by a growth model (Figure 4.1).

Table 4.3 Tolerance of some fodder tree legumes to various environments

Species	Acid soils (pH<5.5)	Cool temp. (15–25°C)	Low rainfall (<500 mm)	Medium rainfall (500 –1000 mm)	High rainfall (>1000 mm)	Poor drainage	High salinity
Acacia aneura	T	NT	T	NT	NT	NT	NT
Acacia angustissima	T	NT	NT	NT	T	NT	NT
Acacia nilotica	NT	NT	NT	NT	NT	T	T
Acacia tortilis	NT	T	T	T	NT	NT	NT
Albizia chinensis	T	T	NT	T	T	NT	NT
Albizia lebbeck	T	NT	T	NT	T	NT	T
Albizia saman	T	NT	NT	NT	T	T	NT
Calliandra calothrysus	T	T	NT	T	T	NT	NT
Chamaecytisus palmensis	NT	NT	NT	NT	NT	NT	NT
Cratylia agentea	T	NT	NT	NT	T	NT	NT
Desmodium rensonii	T	NT	NT	T	T	NT	NT
Desmodium virgatus	NT	NT	NT	NT	T	NT	NT
Erthrina spp.	T	NT	NT	NT	T	NT	NT
Faidherbia albida	NT	NT	T	NT	NT	T	NT
Flemingia macrophylla	T	NT	NT	NT	T	T	NT
Gliricidia sepium	T	T	NT	T	T	NT	NT
Leucaena diversifolia	NT	T	NT	T	T	NT	NT
Leucaena KX2 hybrid	NT	NT	NT	T	T	NT	NT
Leucaena leucocephala	NT	T	NT	T	T	NT	NT
Leucaena pallida	NT	T	NT	T	T	NT	NT
Leucaena trichandra	NT	NT	NT	NT	T	NT	NT
Prosopis juliflora	NT	NT	T	NT	NT	NT	NT
Sesbania grandiflora	NT	T	NT	NT	T	T	T
Sesbania sesban	NT	NT	NT	NT	T	T	T

Notes: T = tolerant; NT = not tolerant
Sources: Roshetko *et al.*, 1996; Shelton, 1994a.

There is a wide range of ecological adaptation among tree legumes (Plates 4.2 and 4.3), although there is no single species suited to the entire range of conditions. Managers must select carefully to ensure successful growth of tree species in their environment. The topic of environmental adaptation is treated in detail elsewhere (Shelton, 1994a). The vexed question of suitable leguminous trees for acid soils continues to be an important research topic for many workers (Evans and Szott, 1995). Several species, such as *Cratylia argentia*, *Desmodium velutinum* and *Flemingia macrophylla*, are valued in South America because of their acid soil tolerance, but need to be more thoroughly tested for nutritional quality for ruminant feeding (Kexian *et al.*, 1998). No tolerance of severely acid soils (pH <5.0, with high aluminium saturation) was found in *Leucaena* (Mullen *et al.*, 1998a).

Taxonomic confusion

Accurate classification and documentation of genus and species relationships within important plant groups are a prerequisite for effective plant improvement programmes. Accurate naming,

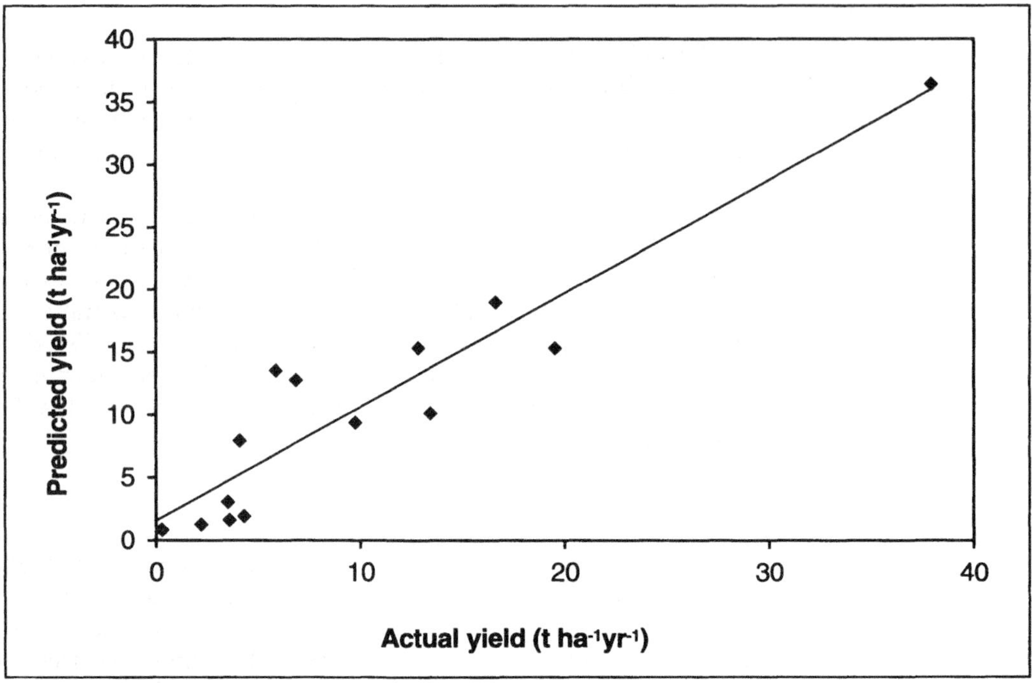

Figure 4.1 Growth model for *Leucaena*.
Yield = 21.4 FI + 43.1 T_{min} + 38.4 AI − 16.5 T_{max} − 0.42 R − 22.0 PI − 494 (r^2 = 0.68).
Where: FI = Fertility index, T_{min} = minimum temperature, T_{max} = maximum temperature, AI = Acidity index, R = Rainfall (mm), PI= Psyllid index.
Source: After Mullen, 2001.

and knowledge of the diversity available, influences the direction of plant evaluation (Hughes, 1998b). Without this, key taxa may be omitted, and uncertainty is often the main result. This has occurred with most forage tree legume plant improvement programmes.

After a long period of disagreement, *Leucaena* has finally been described in detail, agronomically and nutritionally (Shelton *et al.*, 1998), and 22 species have been named (Hughes, 1998a,b), compared with only 17 species in an earlier report (Hughes and Harris 1995). This is not the case with all-important genera. Only a small proportion of the total diversity in *Albizia* has been surveyed, and there is considerable unmapped potential in this genus (Hughes and Pottinger, 1997). There is incomplete understanding and confusion regarding the taxonomy, genetics and ecology of *Prosopis* (Dutton, 1992). Hybridization, intra-specific polymorphism and heterogeneity make it very difficult to identify some *Prosopis* spp. (Ibrahim, 1992).

Another source of taxonomic confusion arises when natural hybridization occurs among taxa otherwise normally well separated in their native range, but that are brought together in evaluation programmes. Hybrid seed is unknowingly collected and spread. Examples of this phenomenon have been documented in *Leucaena* (Hughes, 1998a), *Erythrina* (Neill, 1993) and *Gliricidia* (Lavin, 1996). In Papua New Guinea, where *Leucaena leucocephala* and *L. diversifolia* were introduced separately, a vigorous inter-specific hybrid has appeared spontaneously. It is highly favoured and known locally by the incorrect name "*L. mexicana*".

Plate 4.2 *Gliricidia sepium* growing as a live fence in Bali, Indonesia. [Photograph: H.M. Shelton]

Plate 4.3 *Sesbania grandiflora* along rice paddy bunds in Lombok, Indonesia [Photograph: H.M. Shelton]

Further, our ability to describe variation in plant material has greatly improved, and this may warrant a re-assessment of some genera. Traditionally, morphological and agronomic traits have been used to characterize patterns of diversity in plants. It is now known that these represent only a small proportion of the genome. Such traits are influenced by environmental factors, thus limiting their use for description of genetic relationships and variability. Molecular approaches, such as the use of isozymes and other genetic markers, which may be more independent of environmental influence, are likely to provide a more powerful method to gauge species relationships and origins (Dawson and Chamberlain, 1996). Macqueen (1996) confirmed that studies of molecular data, polyploidy and hybridization research, rather than morphological work, were needed to understand the complex patterns of variation in *Calliandra*.

Unfortunately, in the present economic environment, it is difficult to find support for taxonomic studies, yet such activity underpins all plant improvement programmes, and ultimately influences the quality of new varieties made available for farmers.

ANIMAL PRODUCTIVITY

Productivity and nutritive value

High nutritive value for livestock is an essential prerequisite for successful adoption of forage species. Without high quality, commercially-oriented farmers may not achieve the level of animal response necessary to justify their investment. At the same time, in areas where feed resources are grossly inadequate, or other uses are equally important, farmers may accept a more modest contribution from browse, especially if the species is indigenous and does not require specific introduction and management, as is the case with the *Acacia* communities of the Sahel, North Africa and the Near East (Wickens *et al*., 1995).

New varieties that demonstrate good agronomic and ecological adaptation must therefore be assessed for nutritive value (Dicko and Sikena, 1992). Plants that grow well but contribute little to livestock production are of little value as forage species. For instance, the species *Cratylia argentia*, *Desmodium velutinum* and *Flemingia macrophylla* are valued in South America for their acid-soil tolerance, but *Flemingia macrophylla* has low intake, high condensed tannin content, low digestibility and poor protein quality (Kexian *et al*., 1998).

The most important measure of forage quality is intake of digestible dry matter, and ultimately the production of animal product. Whilst this is known for well researched species such as *Leucaena leucocephala* (Middleton *et al*., 1995), *Sesbania sesban* (Gutteridge, 1994b), *Calliandra calothrysus* (Shelton *et al*., 1996) and *Gliricidia sepium* (Stewart, 1996), there is much less information on other species. Much of the data available are chemical composition only, and therefore of limited value. Apart from between-laboratory variation in methodology, the presence of anti-nutritive compounds, such as tannins, in many tree legumes can have a large effect on their *in vitro* dry matter digestibility and an even larger effect on *in vitro* nitrogen digestibility (Jones and Palmer, 2000). El Hassan *et al.* (2000) confirm that chemical analysis alone will be of limited value for predicting nutritive value of tree legumes, as anti-microbial components affect outcomes. The concepts of nutritive value of tree legumes are described in detail by Norton (1994a,b,c).

Once established, forage legume systems can deliver both sustainable and high productivity. In developed countries, such as Australia, the oldest leucaena–grass systems have been grazed for over 30 years and remain productive. In Central Queensland, liveweight gains of greater than 1 kg day^{-1} are regularly achieved in cattle during the growing season on leucaena–buffel grass pastures (Larsen *et al*., 1998) (Plate 4.4).

Plate 4.4 Aerial view of buffel grass–leucaena pastures in Central Queensland, Australia [Photograph: S. Burle]

Similar results are reported from southern Texas in the United States of America (Dale Elrod, pers. comm.). This greatly exceeds that obtained from herbaceous grass–legume pastures in similar environments. Galgal (in press) showed that maximum liveweight gain occurred with a leucaena intake (cv. Tarramba K636) of around 1.6 kg day^{-1}, equivalent to 0.5 percent of body weight of 300-kg steers.

Other advantages of the grazed leucaena–grass system compared to herbaceous grass–legume pastures in Australia include:

- Relative ease of maintaining a suitable grass–legume balance as the two components are separated spatially.
- High levels of nitrogen cycling, which promotes strong grass growth, or ingress of desirable grasses, between leucaena rows. An exception occurs when nitrogen fixation (Dart, 1994; Dommergues, 1992; Lesueur, Date and Mullen, 1998) and cycling are inhibited by nutrient deficiencies, particularly those that are crucial to the biological nitrogen fixation process, such as phosphorus, sulphur and molybdenum (Burle, 2002).
- Insulation of farms against the worst effects of drought, provided conservative stocking rates are employed. This arises due to the deep-rooted character of leucaena, enabling it to produce high protein sprouts during dry periods. Graziers regularly report modest liveweight gains and good condition of cattle grazing droughted leucaena paddocks due to the ingestion of small amounts of leaf sprouting from the defoliated woody stems.
- Control of runoff during heavy precipitation and excellent infiltration, thus minimizing soil erosion.

In developing countries, tree legume fodders are commonly used as protein supplements to improve the productivity of ruminants fed lower quality grasses and crop residues. Feeding healthy tree legume foliage to ruminants increases the supply of microbial protein and undigested plant protein, and other essential nutrients, and has the capacity to increase intake and therefore productivity of animals. Norton (1994a,b,c) reported that the amount of legume fodder needed varied with the quality of the basal diet, the quality of the supplement, and the level of animal production expected. He concluded that the benefits of supplementation with tree legumes were an increase in total intake, often higher intake of the basal diet, and overall improved diet digestibility. Stewart, Allison and Simons (1996) indicated that animal performance of ruminants on *G. sepium*-supplemented diets was maximized at around 30 percent of the diet. Drying tree legume foliage has been reported to improve its nutritive value, perhaps due to increased by-pass protein, reduced anti-nutritive factors or improved palatability (Norton, 1994b).

An interesting method for feeding tree legume fodder in the Philippines was described by Elaine Lanting (pers. comm.). The method ("Supak system") is predominantly practised in Batangas Province for the purpose of fattening cattle to slaughter weight more quickly, and thereby gain access to higher value markets. Chopped leucaena, concentrates like rice bran or copra meal, salt and molasses are mixed together in a pail or similar container, the mixture is prepared into a slurry and then force-fed using a bamboo tube. The forced feeding method enhances voluntary feed intake of the animal, particularly of the fibrous basal feed.

In the drier regions of the world, such as the arid and semi-arid zones of the Sahel and North Africa, tree legumes, principally *Acacia* spp., provide a high proportion of total herbage intake, and most of the protein intake, for livestock. This proportion increases during dry periods (Baumer, 1992).

Another example is the use of *Acacia nilotica* subspecies *tomentosa* in northwest and southwest Djibouti, known locally as Kassal-to, which occurs naturally on silty clay soils in intermittently flooded valleys (Audru *et al.*, 1992). It is planted in gardens throughout the country and is appreciated for its tolerance of drought and high levels of salinity. In the Madgoul Valley, *A. nilotica* is a highly valued forage source for milking goats, but also for fat-tailed sheep and dromedaries. The optimum density of trees is 100 plants ha^{-1}. The branches are lopped and leaves, pods and flowers are collected and fed. The foliage is much sought after by small ruminants and camels, even though the leaves and especially the pods are rich in tannins. Flowers are especially palatable and are reserved for young goats and young lambs. Camels eat whole pods, while goats eat green pods but do not relish the seeds.

The pods (80 to 100 kg per adult tree) are used to tan the hides of goats to be made into water containers. The pods, ground in water, are also used medicinally for treatment of diabetes and ulcers. The Government of Djibouti, with foreign assistance, wishes to develop livestock production by transferring this indigenous system to the vast alluvial plains that are flooded twice each year.

Toxicity

Unfortunately, tree legumes can contain an array of secondary compounds that can negatively affect nutritive value or even lead to toxicities in ruminant and non-ruminant animals. They appear to have no functional role, although they may impart ecological advantage by limiting or preventing damage from insects, fungi, bacteria, protozoa or grazing animals. Common effects are reduced growth rates, reduced nitrogen and dry matter digestibility, and decreased absorption of specific nutrients. The range of compounds present in tree legumes as reported in the literature

effective, the most practical, the most cost-effective, and the most ecologically sound approach is the use of resistant varieties. We still do not understand the mechanisms of resistance in *Leucaena* necessary to develop effective screening programmes (Mullen *et al.*, 1998a).

It is vital that more detailed information is assembled on the diseases and pests affecting the cultivation and productivity of tree legumes. Country and regional surveys are needed to describe the location and extent of problems. The existing networks are an appropriate way to gain information on disease and insect problems currently experienced (Lenné and Boa, 1994). Catalogues and manuals illustrating the main insect and disease species are required to assist field workers, not only with identification but also with formulation of control measures. Preparation of quarantine guidelines to ensure the safe movement of seed to limit the spread of pests is another priority.

Unfortunately, there is often a lack of specialist expertise to address these problems.

CONSERVATION ISSUES

The development and improvement of tree legumes for farm use is contingent on the availability of germplasm from the centres of origin of species to underpin improvement programmes. It is therefore imperative that the native ranges of tree legumes are protected from exploitation and overuse. Some examples of the current status for selected species are given below.

According to Wickens *et al.* (1995), the former *Acacia* communities in the Sahel, North Africa and the Near East have deteriorated almost beyond recovery. This has been due principally to excessive demand for fuelwood, but also to overgrazing, and demand for more agricultural land, all driven by increasing population pressures. The result is almost irreversible. Rehabilitation of these areas will be very slow where desertification and soil movement has occurred, as there is little soil seed reserve. They stress the need for low-cost, participatory approaches, emphasizing preventative rather than remedial measures.

Stands of *Faidherbia albida* in Wadi Aribo in western Sudan are endangered due to indiscriminate lopping for browse by nomads for their camels (Wickens *et al.*, 1995).

Due to its colonizing nature, *G. sepium* is not under threat at the species level, but certain important provenances, such as Retalhuleu in Guatemala, are under serious threat from human encroachment and river erosion (Stewart, Allison and Simons, 1996).

Most *Albizia* species are severely depleted in their native range in Mexico and Central America. Most species remain abundant in only a few areas. Promotion of greater use of the species would assist with their *in situ* conservation (Hughes and Pottinger, 1997).

In *Leucaena*, the majority of species are of no conservation concern. However, three species, *L. matudae*, *L. magnifica* and *L. involucrata*, are rare and threatened. According to Hughes (1998a), there are less than 400 known individual plants of *L. magnifica*.

Prosopis africana is seriously threatened in the semi-arid lowlands of West Africa in Burkina Faso, Niger, Mali and Senegal. ICRAF has organized seed collections of this species to capture the genetic diversity before invaluable genetic resources are lost (Tchoundjeu, Weber and Guarino, 1998). Patterns of genetic diversity in populations of *Calliandra calothyrsus* were examined using isozyme analysis and their conservation status was reported by Chamberlain (1998).

There are various methods used to conserve genetic resources. Hughes (1998a) discusses the merits of *in situ* (maintenance of natural population), *ex situ* (e.g. germplasm banks and botanic gardens) and *circa situm* (maintenance while in agricultural use, e.g. as hedgerow) conservation. *In vitro* techniques for conservation and multiplication of germplasm, and elimination of disease, have been applied to the conservation of *Leucaena leucocephala*, *Erythrina brucei* and *Sesbania sesban* by Ruredzo and Hanson (1988). Perhaps a combination of all approaches may be necessary. As with taxonomic studies, it is now exceedingly difficult to obtain financial support for conservation of undomesticated genetic resources in their native range.

THE DEBATE OVER EXOTIC VERSUS NATIVE SPECIES

"Too often in extension work, a few exotic species have been strongly promoted without any attention being given to the rich indigenous flora and local knowledge of it" (Bekele-Tesemma, Birnie and Tengnäs, 1993).

In recent years there has been increasing interest in indigenous species as an alternative to introducing exotic species, with debate concerning the appropriateness of introducing exotic species into indigenous ecosystems. There are many reasons for this trend:

- Farming communities have very detailed knowledge of the use and value of indigenous species, and often this has been neither documented, assessed nor verified (B. Calub, pers. comm.; Schrempp, Tato and Hurni, 1992).
- There are ecological advantages in using a diversity of indigenous species, compared to a monoculture of exotics.
- There is concern about preserving and conserving indigenous germplasm.
- A reduced emphasis on promotion of exotic species and greater *in situ* use of local tree diversity may reduce the risk of unwanted weed invasion and genetic pollution through hybridization (Hughes, 1994).

There is no simple answer to this debate and decisions have to be made on merit, based on the important issues. Combined use of native and exotic species may be an option in some instances. Lundgren (1992) reported that in the Rift Valley of Kenya, farmers have long practised traditional forms of agroforestry using indigenous species combined with some use of exotics (eucalypts, cypress and grevillea) for poles, timber and fuel. Pastoral communities manage the scattered indigenous trees so that they play an essential role in meeting subsistence requirements.

In the Kenyan highlands, with greatly increased population pressure, intensively managed exotic leguminous trees for fodder and green manure production are common. In these areas, there is greater, not reduced use of tree species. The result is a forest mosaic. The traditional species are not disappearing; they are being combined with a variety of new species, to fill complementary roles (Lundgren, 1992).

Indigenous forage tree species have generally been used for subsistence feeding rather than in commercial systems. Wickens *et al.* (1995) describe how fuelwood and grazing were the principal uses of the former *Acacia* communities in the Sahel, North Africa and the Near East.

Often, exotic species are more vigorous and produce higher yields than indigenous species, and this is their great attraction. This was the case in Malawi, where *L. leucocephala*, *Cassia spectabilis* and *Gliricidia sepium* have been promoted over the indigenous *Faidherbia albida*, which is slow growing (Cromwell, Brodie and Southern, 1996). There are many regions where exotic species have made invaluable contributions. It has been estimated that 150 to 200 million people use gliricidia worldwide, the majority of whom live outside its native range (Simons, 1996). *Leucaena* is now naturalized in the Philippines where it is the principal source of tree fodder and of fuelwood. This species underpins a sustainable, highly productive beef cattle production system in northern Australia (Middleton *et al.*, 1995).

In India, fast growing, multipurpose exotic tree species were introduced alongside the relatively slow growing *Acacia nilotica* (an indigenous tree) to enhance biomass production. However, competition reduces growth of the indigenous tree. Careful planning and thoughtful species selection was recommended before implementation of exotic large-scale afforestation programmes (Neelam-Bhatnagar, Bhandar and Promila Kapoor, 1993).

Significantly, most commercial use of forage from tree legumes has been with exotic species. Indigenous species appear to have been confined to subsistence feeding systems, perhaps due to their lower productivity and lower quality. The communal management regimes employed in traditional systems place few limits on use, and this has led to exploitation (Wickens *et al.*, 1995), and difficulties in management.

Sometimes, indigenous species are better adapted to difficult soils. In Costa Rica, native leguminous species had more potential for reforestation and agroforestry on acid soils high in aluminium and manganese than exotic species (Tilki and Fisher, 1998). In contrast, in the mountainous area of Minas Gerais, Brazil, where acid infertile soils predominate, the exotic species *Acacia mangium* and *A. auriculiformis* achieved faster growth than indigenous species when introduced into an existing *Brachiaria decumbens* pasture (Carvalho, 1997).

Exotic species can have significant effects on associated ecosystem species. In Hawai'i (USA), four species of native birds rarely feed on the fruits of the exotic nitrogen-fixing tree *Myrica faya* that is invading Hawaii Volcanoes National Park. However, five species of exotic birds were seen ingesting the fruit (Woodward *et al.*, 1990). In South Africa, invasive exotic plants, such as *Acacia longifolia* and *A. mearnsii*, were detrimental to native, ground-living, invertebrate fauna. There was no significant effect on species richness and diversity, but there was a different assemblage of species associated with exotic compared with indigenous vegetation. Management should therefore be sensitive to the needs of the ecosystem to ensure conservation of desirable species when native vegetation is supplanted by exotics (Samways, Caldwell and Osborn, 1996).

Over the last three decades, there has been movement of plant material around the world on an unprecedented scale, with few restrictions covering movement. Hughes (1994) advocated a more cautious approach to species introduction and a more thorough assessment of the advantages and limitations of native and exotic species to lessen the risks of introduction of a weed.

Many native plants are incompletely studied. Some species are only now undergoing preliminary domestication and are still harvested by the traditional gathering activities associated with wild species. A case in point is *Acacia albida* (syn *Faidherbia albida*), which is now the focus of international collaborative efforts to extend its versatility of utilization (Nouaille, 1992). There is large variability in performance of individual trees as little plant improvement has occurred, and little is known of the silviculture of the species (Cromwell, Brodie and Southern, 1996).

It is likely that the most appropriate path through this minefield is judicious use of both native and exotic species. For instance, in tropical humid Africa, research may continue on species such as *Leucaena*, *Gliricidia* and *Sesbania*, but emphasis may gradually shift to local species, as adoption may be more rapid and widespread. Schrempp *et al.* (1992) (cited in Schrempp, Tato and Hurni, 1992) noted from their work in the eastern highlands of Ethiopia that preferred species in fields were indigenous species such as *A. albida* (syn. *Faidherbia albida*), while preferred species off-field were fast growing exotics such as *A. mearnsii*, *A. saligna*, *Eucalyptus* spp. and *P. procera*.

WEEDINESS

A number of introduced tree legumes have become serious weed pests. Given the large number of introductions to many new environments, this is not surprising. Weediness of introduced exotic trees has generally occurred when:

- the purpose for the introduction has failed, or has resulted in only partial use of trees;
- seedlings and trees are protected from grazing by thorns, or by low palatability;
- trees have abundant, precocious seed production;
- seeds are only partially digested by ruminant grazers, and viable seeds are spread in faeces;
- seed is spread on the hoofs of animals, or transported by flood waters;
- seeds are long-lived in the soil;
- young plants grow and colonize rapidly, and tolerate drought, grazing and fire;
- trees are long-lived;
- there are disturbed areas nearby suitable for invasion; or
- there is unpredictable growth as trees perform beyond expectations away from natural predators, or in new climatic, edaphic or management environments.

These conditions typify in part a number of introduction events. Thus *Acacia nilotica* was introduced to provide shade and fodder for sheep in western Queensland, but now infests 6 million hectares of *Astrebla* grasslands (Carter, 1994).

Over the past 80–100 years, mesquite (*Prosopis* spp.) has become an aggressive invader of desert grasslands in the southwestern United States of America (Ibrahim, 1992) due to interference in the natural ecological balance by man and his activities. Strategies for control and management of this problem are still not available. Grazing livestock and reduced occurrence of fire were key factors in the increase in density of mesquite.

A case study of mesquite showing weedy tendencies in Wadi Hadramout (2.2 million hectares) in Yemen was reported by E.A. Elsiddig (unpublished report to FAO). *Prosopis cineraria* and *P. juliflora* were introduced to Seiyun Town Park in the Wadi in the mid-1960s. In 30 years, *P. juliflora* (mesquite) has spread from just a few trees and now occurs extensively throughout the Wadi. *P. cineraria* was also planted around Seiyun in 1980, but has not spread to the same extent, and is less vigorous.

Mesquite was spread via ruminants, which ingest seeds and excrete them in faeces. Manure is a common source of fertilizer for agriculture (*P. cineraria* is not spread in this way, as seeds appear to be digested by the animals). With a high content of hard seed, passage through the digestive system of ruminants acts to scarify mesquite seed.

Mesquite was found to be well adapted to environments with high moisture, such as drainage lines, irrigation canals, bridge inlets and outlets, and irrigated farmlands. It has invaded disturbed land but has been less successful on land where native vegetation is still vigorous.

Farmers are not unanimous in condemning mesquite as it has a number of beneficial features, such as:

- use as building poles, as *P. juliflora* is less susceptible to borer attack than other *Prosopis* species;
- honey made from mesquite is of excellent quality;
- withstands severe drought;
- provides shade and forage for livestock, particularly during the dry season;
- protects wadi banks against erosion;

- supplies fuel for rural poor, and makes excellent charcoal; and
- can be used as a live hedge to control stock movements.

However, the disadvantages of mesquite are substantial, and include:

- blocks flow of water along waterways, resulting in flooding;
- competes with palm trees, eventually killing them;
- only the pods are palatable to livestock, and not the leaves. This means that large quantities of pods are consumed and excreted in faeces;
- very high pod production occurs, up to 3.34 kg per tree, ensuring rapid spread, with seedlings germinating at very high densities, forming thickets; and
- seeds can be transported to favourable sites long distances away by goats and sheep, and by water.

Mesquite is difficult to control. Burning is effective with young trees, and management also plays a role. Thus it is important not to spread manure that contains a lot of seeds. This is a difficult option due to the importance of the manure as fertilizer. Composting the manure prior to spreading kills the seeds, and mechanical uprooting of plants can also be effective.

In conclusion, the spread of mesquite in Wadi Hadramout has been rapid and caused many negative effects. Although there have been benefits as well, it appears that these may not outweigh the negative effects.

The original movement of *Leucaena leucocephala* subspecies *leucocephala* around the world has lead to this inferior but seedy variety becoming a weed in many tropical environments (Hughes, 1994).

Species may also become a weed in their own environment. *Albizia tomentosa* is a weed in disturbed areas in Mexico (Hughes and Pottinger, 1997) and *Acacia aneura* is often weedy in southwest Queensland (Australia), when poorly managed (Beale, 1994).

The question of weed risk raises many difficult questions, with few easy answers. Some suggest that only indigenous species should be considered in agroforestry programmes. But this is an unrealistic constraint on the development of competitive farming systems, and indeed on the environment. Forage tree legumes are already major contributors to the sustainability of our farming systems. They have the potential to become even more important in our livestock industries, thus enhancing the quality of life of rural communities. However, it will be imperative to pursue our objectives with environmental responsibility. The key is to carefully evaluate the level of risk, reject high-risk introductions, and carefully manage introductions to minimize the chances of weed outbreak.

When introducing new species to an environment, it may be necessary to first:

- review susceptibility of seedlings and trees to grazing (thorns, toxins and anti-palatability will reduce animal access);
- assess seed production and longevity;
- assess seed dispersal mechanisms;
- study climatic and soil characteristics to predict areas potentially susceptible to invasion;
- study insect predators and pathogens in the native range of species;
- evaluate methods of control, e.g. pests, fire, chemical and mechanical methods;
- after introduction, install long-term monitoring and rapid reaction systems; and
- ensure that farmers have support to make full beneficial use of MPT introductions.

Other approaches should also be used. For example, to minimize the weed risk status of *Leucaena*, it will be necessary to (i) educate farmers to achieve good grazing management to minimize seed production and to prevent seedling growth; (ii) spray or slash isolated seedlings;

(iii) introduce less seedy varieties, e.g. the KX2 hybrid; (iv) develop sterile hybrids that will eliminate risk; and (v) utilize biological control methods, such as the leucaena bruchid beetle (*Acanthosceloides macrophthalmus*), which greatly reduces the amount of viable seed produced. A number of these strategies can be combined to reduce the risk of this species spreading as a weed.

Perhaps the most important step is to ensure that the rural community adopting the new species has the tools to make full use of the MPT. There are many examples of apparent weediness occurring because villagers, unaware of the many uses of new plants, have abandoned the original plantings.

Nevertheless, tree legumes should not be introduced where risk is high, or where nearby disturbed vegetation might be ecologically threatened.

INTEGRATING FORAGE TREE LEGUMES INTO FARMING SYSTEMS, AND FARMER UPTAKE

Examples of successful adoption of exotic and indigenous tree legumes for multipurpose uses, including forage, are numerous. Outstanding examples are *Leucaena leucocephala* in Australia (Middleton *et al*., 1995) (Plate 4.4) and Asia (Moog, Bezkorowajnyj and Nitis, 1998), *Gliricidia sepium* in Southeast Asia (Stewart, 1996), *Sesbania grandiflora* in Indonesia (Gutteridge, 1994b), *Calliandra calothyrsus* in Indonesia (Palmer, Macqueen and Gutteridge, 1994), and *Acacia* spp. in Africa (Wickens *et al*., 1995).

Nevertheless, despite high levels of promotion, farmer uptake has been lower than anticipated. Recent attempts to achieve adoption of complex agroforestry packages, such as alley cropping, have been only partially successful due to unrecognized failings in approach (Gutteridge, 1998). Difficulties in achieving high levels of adoption for *Leucaena* are reported for Africa (Dzowela *et al*., 1998), South America (Argel, Lascano and Ramirez, 1998) and Asia (Moog, Bezkorowajnyj and Nitis, 1998).

Although the forage bank concept has been shown to be feasible in tropical Africa, the rate of adoption has been low due to socio-economic constraints, such as land tenure insecurity and lack of infrastructure support (Cromwell, Brodie and Southern, 1996).

The value of alley cropping is hotly debated in Africa (Cromwell, Brodie and Southern, 1996). Research groups have become cautious about the sustainable benefits of the system and, in particular, the value of the mulch in terms of increased crop yields on farm. Part of the problem has been identified by Grist, Menz and Nelson (1999). They found that while alley cropping of *Gliricidia* in *Imperata* grasslands can increase soil fertility, farmers were likely to incur a loss in the first year of development, and that it would take about four years to begin making a profit. Alley cropping systems may be more successful when prunings are used for livestock feeding rather than for fertility improvement alone. Gutteridge (1998) argued that alley cropping was less successful in subhumid environments where competition for moisture reduced yields of inter-planted crops.

There have been many reasons advanced to explain success or lower than anticipated levels of adoption (Smith, 1992; Cromwell, Brodie and Southern, 1996; Larsen *et al*., 1998). Simple innovations such as a new variety overcoming a key problem, may be adopted with relatively little intervention. However, complex innovation, involving a new farming system, generally requires sustained, high-profile intervention. Shelton *et al*. (2000) reviewed a number of case studies and found that several factors were common to successful adoption, and these are considered below.

Technical

- Technical constraints must be resolved promptly to avoid farmers and extension workers becoming discouraged and losing interest, e.g. appropriate methods of establishment (Shelton, 1994b).
- Technical information needs to flow frequently, accurately and in a variety of appropriate formats (field visits, manuals, videos, newsletters, discussion groups) to farmers.
- A range of MPT species may need to be available to meet the diverse needs of farmers, their environments and farming systems. New germplasm needs to be markedly superior to existing material, as is the case with the KX2 hybrid *Leucaena* being introduced into the Philippines, Viet Nam and Indonesia.
- The best planting material should be available to farmers. This will require education of both extension agents and farmers to ensure adequate farmer knowledge concerning suppliers and varieties available.

Socio-economic

- Farmers, local leaders and groups, and government all need to be closely involved in the process and there needs to be frequent contact among all players. All need to feel some ownership and all need to be respected for their contribution to the innovation. The importance of communication, training, extension and research networks needs to be stressed (Dutton, 1992).
- Innovation needs to have positive commercial outcomes for individual farmers, as well as environmental outcomes. Cook, Grut and Christoffersen (1989) stressed the importance of understanding the economics of agroforestry systems from the farmer's point of view, as well as from the broader perspective of benefits to society. Project implementation should take into account local markets and opportunities for off-farm employment offered by tree products, as well as the opportunity costs perceived by farmers in making adoption decisions. A full cost–benefit analysis of new agroforestry systems is essential. It is unlikely that farmers will adopt new MPT systems based on environmental benefit only.

Other

- Successful tree legume-based interventions in Southeast Asia have commonly involved long-term, top-down extension methodologies (Shelton *et al.*, 2000). The need for institutional direction and long-term commitment may be necessary due to the complexity of many of the tree legume systems being promoted. However, successful adoption of tree legume interventions through a process of on-farm and participatory research has also been a consistent theme.
- Perhaps the most important elements of successful adoption are the time, enthusiasm and long-term commitment of the farmers, researchers and extension agents involved. Successful innovation needs champions to ensure continuity of interest and support over an extended period (often >10 years, and sometimes up to 30 years).

One thing is certain: without increased adoption, and more explicit demonstration of the relevance and benefits of forage tree legumes, the good will and support of funding and donor agencies will be limited.

CONCLUSIONS

The principal commercial use and future potential of multipurpose tree legumes is forage for livestock production. There is great potential to expand MPT use for both smallholder and large-scale ranch production systems. The main conclusions and R&D recommendations are:

- While there are significant examples of adoption and use of MPT species worldwide, there remains excellent opportunity to increase the use of this valuable plant group using participatory research and extension approaches.
- Only a few MPT species are currently in commercial use by farmers (about 20), but they cover a wide range of ecological environments. Many existing MPT plantings worldwide are from a narrow germplasm base, and access to the highest quality varieties needs to be improved. Apart from improved distribution networks, further study of taxonomy and nomenclature of important genera is required. There is opportunity for mixed use of both native and exotic MPTs for both environmental and production objectives.
- The range of MPTs that will be valuable as forage is restricted by low palatability and low nutritive value of many species. This is primarily due to their high content of anti-nutritive compounds such as condensed tannins.
- The conservation status of some MPTs is threatened due to overexploitation in their native range, and there is potential for weediness of both new and existing introductions, which needs to be carefully assessed.

REFERENCES

Allison, G.E. & Simons, A.J. 1996. Propagation and husbandry. pp. 49–71, *in:* J.L. Stewart, G.E. Allison and A.J. Simons (eds). Gliricidia sepium – *Genetic resources for farmers*. Tropical Forestry Paper, No. 33. Oxford Forestry Institute.

Anttila, L.S., Alakoski Johansson, G.M. & Johansson, S.G. 1994. Browse preference of Orma livestock and chemical composition of *Prosopis juliflora* and nine indigenous woody species in Bura, Eastern Kenya. Forestry in irrigation schemes with special reference to Kenya. *East African Agricultural and Forestry Journal*, 58. Special Issue: 83–90.

Argel, P.J. & Lascano, C.E. 1998. *Cratylia argentea* (Desvaux) O. Kuntze: una nueva leguminosa arbustiva para suelos acidos en zonas subhumedas tropicales. *Pasturas Tropicales*, **20**: 37–43.

Argel, P.J., Lascano, C.E. & Ramirez, L. 1998. Leucaena in Latin American farming systems: Challenges for development. pp. 319–323, *in:* Shelton *et al.*, 1998, q.v.

Attah-Krah, A.N., Sumberg, J.E. & Reynolds, L. 1986. Leguminous fodder trees in farming systems – an overview of research at the humid zone programme of ILCA in Southwestern Nigeria. pp. 307–329, *in:* I. Haque, S. Jutzi and P.J. Weate (eds). *Potentials of forage legumes in farming systems of sub-Saharan Africa*. Addis Ababa: ILCA.

Audru, J., Labonne, M., Guerin, H. & Bilha, A. 1992. *Acacia nilotica*: a traditional forage species among the Afar of Djibouti. pp. 277–293, *in:* Speedy and Pugliese, 1992, q.v.

Baumer, M. 1992. Trees as browse to support animal production. pp. 1-10, *in:* Speedy and Pugliese, 1992, q.v.

Beale, I.F. 1994. Management of mulga (*Acacia aneura*) scrublands in southeast Queensland. pp. 325–337, *in:* R.C. Gutteridge and H.M. Shelton (eds). *Forage Tree Legumes in Tropical Agriculture*. Wallingford, UK: CAB International.

Bekele-Tesemma, A., Birnie, A. & Tengnäs, B. 1993. *Useful trees and shrubs for Ethiopia*. Regional Soil Conservation Unit, Swedish International Development Authority (SIDA). 474p.

Boa, E.R. & Lenné, J.M. 1995. Diseases and pests of *Leucaena*. pp. 129–134, *in:* Shelton, Piggin and Brewbaker, 1995, q.v.

Boa, E.R. & Lenné, J.M. 1996. Diseases and insect pests. pp. 73-76, *in:* J.L. Stewart, G.E. Allison and A.J. Simons (eds). Gliricidia sepium – *Genetic resources for farmers*. Tropical Forestry Paper, No. 33. Oxford Forestry Institute.

Brewbaker, J.L. 1986. Nitrogen-fixing trees for fodder and browse in Africa. pp. 55–70, *in:* B.T. Kang and L. Reynolds (eds). *Alley farming in the humid and subhumid tropics*. Proceedings of a workshop held at Ibadan, Nigeria, 10–14 March 1986. IDRC, Ottawa.

Burle, S.T.M. 2002. Nitrogen accumulation and cycling in *Leucaena/Brachiaria decumbens* pastures on an acid infertile soil at Mt. Cotton, southeast Queensland, Australia. M.Agr.St Thesis, University of Queensland. 115p.

Carter, J.O. 1994. *Acacia nilotica* – a tree legume out of control. pp. 338–351, *in:* R.C. Gutteridge and H.M. Shelton (eds). *Forage Tree Legumes in Tropical Agriculture*. Wallingford, UK: CAB International.

Carvalho, M.M. 1997. Asociaciones de pasturas con arboles en la region centro sur del Brasil. *Agroforesteria en las Americas*, **4**: 5–8.

Chamberlain, J.R. 1998. Isozyme variation in *Calliandra calothyrsus* (Leguminosae): its implications for species delimitation and conservation. *American Journal of Botany*, **85**: 37–47.

Cook, C.C., Grut, M. & Christoffersen, L.E. 1989. Agroforestry in Sub-Saharan Africa: a farmer's perspective. *World Bank Technical Paper*. No. 112. 99p.

Cromwell, E., Brodie, A. & Southern, A. 1996. Germplasm for multipurpose trees: access and utility in small-farm communities. Case studies from Honduras, Malawi and Sri Lanka. Overseas Development Institute. 93p.

Dalzell, S.A. 1999. Genotype and environmental effects on proanthocyanidins in the *Leucaena* genus. PhD Thesis. The University of Queensland. 262p.

Dalzell, S.A., Stewart, J.L., Tolera, A. & McNeill, D.M. 1998. Chemical composition of leucaena and implications for forage quality. pp. 227-246, *in:* Shelton *et al.*, 1998, q.v.

Dart, P.J. 1994. Microbial symbioses of tree and shrub legumes. pp. 143–157, *in:* Gutteridge and Shelton, 1994, q.v.

Dawson, I.K. & Chamberlain, J.R. 1996. Molecular analysis of genetic variation. pp. 77–91, *in:* J.L. Stewart, G.E. Allison and A.J. Simons (eds). Gliricidia sepium – *Genetic resources for farmers*. Tropical Forestry Paper, No.33. Oxford Forestry Institute.

Dicko, M.S. & Sikena, L.K. 1992. Fodder trees and shrubs in range and farming systems in dry tropical Africa. pp. 27–41, *in:* A. Speedy and P.-L. Pugliese (eds). *Legume trees and other fodder trees as protein sources for livestock*. Proceedings of an FAO Expert Consultation. Malaysian Agricultural Research and Development Institute (MARDI), Kuala Lumpur. FAO, Rome.

Djogo, A.P.Y., Siregar, M.E. & Gutteridge, R.C. 1995. Opportunities and limitations in other MPT genera. pp. 39–43, , *in:* Shelton, Piggin and Brewbaker, 1995, q.v.

D'Mello, J.P.F. 1992. Nutritional potentialities of fodder trees and fodder shrubs as protein sources in monograstric feeding. pp. 115–127, *in:* Speedy and Pugliese, 1992, q.v.

Dommergues, Y. 1992. Maximising nitrogen fixation in Prosopis. pp. 207–218, *in:* R.W. Dutton, M. Powell and R.J. Ridley (eds). Prosopis *species – Aspects of their value, research and development*. Proceedings of Prosopis Symposium, held by CORD, University of Durham, UK, 27–31 July 1992.

Dutton, R.W. 1992. A research and development strategy for Prosopis. pp. 3–27, *in:* R.W. Dutton, M. Powell and R.J. Ridley (eds). Prosopis *species – Aspects of their value, research and development*. Proceedings of Prosopis Symposium, held by CORD, University of Durham, UK, 27–31 July 1992.

Dzowela, B.H., Wandera, P.F., Were, J. & Mohammed-Saleem, M.A. 1998. *Leucaena* in smallholder farming systems in Africa: Challenges for development. pp. 311–318, *in:* Shelton *et al.*, 1998, q.v.

El Hassan, S.M., Kassi, A. Lahlou, Newbold, C.J. & Wallace, R.J. 2000. Chemical composition and degradation characteristics of some African multipurpose trees. *Animal Feed Science and Technology*, 86: 27–37.

Evans, D.O. 1996. International Workshop on the Genus *Calliandra*. Forest, Farm and Community Tree Research Reports, Special Issue, 1996. Winrock International. 268p.

Evans, D.O. & Szott, L. 1995. Nitrogen fixing trees for acid soils. Nitrogen Fixing Tree Research Reports (Special Issue). Winrock International and NFTA, Morrilton, Arkansas, USA. 328p.

Faint, M.A., McNeill, D.M., Stewart, J.L., Castillo, A.C., Acasio, R.N. & Lynch, J.J. 1998. Palatability of *Leucaena* to ruminants. pp. 215-226, *in:* Shelton *et al.*, 1998, q.v.

Galgal, K.K. In press. Forage and animal production from new *Leucaena* accessions. PhD Thesis, University of Queensland. In Press.

Grist, P., Menz, K. & Nelson, R. 1999. Multipurpose trees as improved fallow: An economic assessment. *International Tree Crops Journal*, **10**: 19–36.

Gutteridge, R.C. 1994a. Other species of multipurpose forage tree legumes. pp. 97-108, *in:* Gutteridge and Shelton, 1994, q.v.

Gutteridge, R.C. 1994b. The perennial *Sesbania* species. pp. 49-64, *in:* Gutteridge and Shelton, 1994, q.v.

Gutteridge, R.C. 1998. Leucaena in alley cropping systems: Challenges for development. pp. 337-341, *in:* Shelton *et al.*, 1998, q.v.

Gutteridge, R.C. & Rekib. 1995. Opportunities and limitations in *Sesbania*. pp. 24–28, *in:* Shelton, Piggin and Brewbaker, 1995, q.v.

Gutteridge, R.C. & Shelton, H.M. (eds). 1994. *Forage Tree Legumes in Tropical Agriculture*. Wallingford, UK: CAB International.

Hughes, C.E. 1994. Risks of species introductions in tropical forestry. *Commonwealth Forestry Review*, **73**: 243–252, 272–273.

Hughes, C.E. 1998a. Leucaena. *A genetic resources handbook*. Oxford Forestry Institute, *Tropical Forestry Papers*, No. 37. 274p.

Hughes, C.E. 1998b. Taxonomy of Leucaena. pp. 27–38, *in:* Shelton *et al.*, 1998, q.v.

Hughes, C.E & Harris, S.A. 1995. Systematics of Leucaena: Recent findings and implications for breeding and conservation. pp. 54–65. *in:* Shelton, Piggin and Brewbaker, 1995, q.v.

Hughes, C.E. & Pottinger, A.J. 1997. *Albizia* species from Mexico and Central America. pp. 57–65, *in:* N.Q. Zabala (ed). *International Workshop on* Albizia *and* Paraserianthes *Species*. Proceedings of a workshop held in Bislig, Suriago del Sur, the Philippines, 1994. Farm, Forestry and Community Tree Research Reports, Special Issue. Winrock International.

Ibrahim, K.M. 1992. *Prosopis* species in the south-western United States, their utilization and research. pp. 83–115, *in:* R.W. Dutton, M. Powell and R.J. Ridley (eds). Prosopis *species - Aspects of their value, research and development*. Proceedings of Prosopis symposium held by CORD, University of Durham, UK, 27–31 July 1992.

Jones, R.J. & Palmer, B. 2000. *In vitro* digestion studies using ^{14}C labelled polyethylene glycol (PEG) 4000: comparison of six tanniniferous shrub legumes and the grass *Panicum maximum*. *Animal Feed Science and Technology*, 85: 215–221.

Kaitho, R.J. 1997. Nutritive value of browses as protein supplements(s) to poor quality roughages. PhD Thesis. Wageningen Agricultural University. 189p.

Kaitho, R.J., Umunna, N.N., Nsahlai, I.V., Tamminga, S., van Bruchem, J. & Hanson, J. 1997. Palatability of wilted and dried multipurpose tree species fed to sheep and goats. pp. 59–73, *in:* R.J. Kaitho. Nutritive value of browses as protein supplements to poor quality roughages. PhD Thesis. Wageningen Agricultural University.

Kaitho, R.J., Umunna, N.N., Nsahlai, I.V., Tamminga, S., van Bruchem, J., Hanson, J. & van de Wouw, M. 1996. Palatability of multipurpose tree species: effect of species and length of study on intake and relative palatability by sheep. pp. 47–58, *in:* R.J. Kaitho. Nutritive value of browses as protein supplements(s) to poor quality roughages. PhD Thesis. Wageningen Agricultural University.

Karachi, M., Shirima, D. & Lema, N. 1997. Evaluation of 15 leguminous trees and shrubs for forage and wood production in Tanzania. *Agroforestry Systems*, **37**: 253–263.

Kartasubrata, J. 1996. Culture and uses of *Calliandra calothyrsus* in Indonesia. pp. 101–107, *in:* D.O. Evans (ed). International Workshop on the Genus *Calliandra*. Forest, Farm and Community Tree Research Reports, Special Issue, 1996. Winrock International.

Kexian, Yi, Lascano, C.E., Kerridge, P.C. & Avilia, P. 1998. The effect of three tropical shrub legumes on intake and acceptability by small ruminants. *Pasturas Tropicales*, **20**: 31–35.

Larsen, P.H., Middleton, C.H., Bolam, M.J. & Chamberlain, J. 1998. Leucaena in large-scale grazing systems: Challenges for development. pp. 324–330, *in:* Shelton *et al.*, 1998, q.v.

Lavin, M. 1996. Taxonomy. pp. 3–17, *in:* J.L. Stewart, G.E. Allison and A.J. Simons (eds). Gliricidia sepium – *Genetic resources for farmers*. Tropical Forestry Paper, No.33. Oxford Forestry Institute.

Le Houérou, H.N. 1980. Chemical composition and nutritive value of browse in tropical West Africa. pp. 261–289, *in:* H.N. Le Houérou (ed). *Browse in Africa, the current state of knowledge*. Addis Ababa: ILCA.

Lenné, J.M. 1992. Diseases of multi-purpose woody legumes in the tropics: a review. *Nitrogen Fixing Tree Research Reports*, **10**: 13–29.

Lenné, J.M. & Boa, E.R. 1994. Diseases of tree legumes. pp. 292–308, *in:* Gutteridge and Shelton, 1994, q.v.

Lesueur, D., Date, R.A. & Mullen, B.F. 1998. Rhizobium specificity in Leucaena. pp. 86-95, *in:* Shelton *et al.*, 1998, q.v.

Lowry, J.B. 1989. Agronomy and forage quality of *Albizia lebbeck* in the semi-arid tropics. *Tropical Grasslands*, **23**: 84–91.

Lundgren, D. 1992. Foreword. pp. 9–10, *in: A selection of useful trees and shrubs for Kenya*. International Centre for Research in Agroforestry (ICRAF).

Macqueen, D. 1996. *Calliandra* taxonomy and distribution, with particular reference to the series Racemosae. pp. 1–17, *in:* D.O. Evans (ed). *International Workshop on the Genus* Calliandra. Forest, Farm and Community Tree Research Reports, Special Issue, 1996. Winrock International.

McNeill, D.M., Osborne, N., Komolong, M.K. & Nankervis, D. 1998. Condensed tannins in the genus *Leucaena* and their nutritional significance for ruminants. pp. 205–214, *in:* Shelton *et al.*, 1998, q.v.

Middleton, C.H., Jones, R.J., Shelton, H.M., Petty, S.R. & Wildin, J.H. 1995. Leucaena in northern Australia. pp. 214–221, *in:* Shelton, Piggin and Brewbaker, 1995, q.v.

Moog, F.A., Bezkorowajnyj, P. & Nitis, I.M. 1998. Leucaena in smallholder farming systems in Asia: Challenges for development. pp. 303–310, *in:* Shelton *et al.*, 1998, q.v.

Mullen, B.F. 2001. Agronomic adaptation to environmental challenges in the genus *Leucaena*. PhD Thesis. The University of Queensland. 288p.

Mullen, B.F., Gabunada, F., Shelton, H.M., Stur, W.W. & Napompeth, B. 1998a. Psyllid resistance in *Leucaena*. pp. 51-60, *in:* Shelton *et al.*, 1998, q.v.

Mullen, B.F., Shelton, H.M., Basford, K.E., Castillo, A.C., Bino, B., Victorio, E.E., Acasio, R.N., Tarabu, J., Komolong, M.K., Galgal, K.K., Khoa, L.V., Co, H.X., Wandera, F.P., Ibrahim, T., Clem, R.L., Jones, R.J., Middleton, C.H., Bolam,

M.J.M., Gabunada, F., Stur, W.W., Horne, P.M., Utachak, K., & Khanh, T.T. 1998b. Agronomic adaptation to environmental challenges in the genus *Leucaena*. pp. 39–50, *in:* Shelton *et al*., 1998, q.v.

Murphy, S.T. 1990. Pests of *Sesbania* species and *Sesbania* species as pests: A review. pp. 123–130, *in:* B. Macklin and D.O. Evans (eds). *Perennial* Sesbania *Species in Agroforest Systems*. Proceedings of a Workshop, Nairobi, Kenya. Nitrogen Fixing tree Association, Hawaii.

Neelam-Bhatnagar, Bhandar, D.C., & Promila Kapoor. 1993. Competition in the early establishment phases of an even-aged mixed plantation of *Leucaena leucocephala* and *Acacia nilotica*. *Forest Ecology and Management*, **57**: 213–231.

Neill, D.A. 1993. The genus *Erythrina*: taxonomy, distribution and ecological differentiation. pp. 15–27, *in:* Westley and Powell, 1993, q.v.

Norton, B.W. 1994a. The nutritive value of tree legumes. pp. 177–191, *in:* Gutteridge and Shelton, 1994, q.v.

Norton, B.W. 1994b. Tree legumes as dietary supplements for ruminants. pp. 192–201, *in:* Gutteridge and Shelton, 1994, q.v.

Norton, B.W. 1994c. Anti-nutritive and toxic factors in forage tree legumes. pp. 202–215, *in:* Gutteridge and Shelton, 1994, q.v.

Nouaille, C. 1992. At the frontiers of domestication. *Biofutur*, **111**: 43–46.

Nygren, A. 1993. Traditional uses and cultural significance of three *Erythrina* species among rural population of Tuís District, Turrialba, Costa Rica. pp. 62–67, *in:* Westley and Powell, 1993, q.v.

Osuji, P.O., Odenyo, A.A., Acamovic, T., Stewart, C.S. & Topps, J.H. 1997. The role of legume forages as supplements to low quality roughages – ILRI experience. Selected papers from an international conference on Evaluation of forages for ruminants in the tropics, Harare, Zimbabwe, 28 August–1 September 1995. *Animal Feed Science and Technology*, **69**: 1–3, 27–38.

Palmer, B. Macqueen, D.J & Gutteridge, R.C. 1994. *Calliandra calothyrsus* – a mutltipurpose tree legume for humid locations. pp. 65–74, *in:* Gutteridge and Shelton, 1994, q.v.

Roshetko, J.M. 1995. *Albizia saman*: pasture improvement, shade, timber and more. NFT Highlights, NFTA 95-02. Forest, Farm, and Community Tree Network (FACT Net), Winrock International.

Roshetko, J.M., Dagar, J.C., Puri, S., Khandale, D.Y., Takawale, P.S., Bheemiah, G. & Basak, N.C. 1996. Selecting species of nitrogen fixing trees. pp. 19-24, *in:* J.M. Roshetko and R.C. Gutteridge (eds). *Nitrogen fixing trees for fodder production – A field Manual*. Morrilton, AR: Winrock International.

Ruredzo, T.J. & Hanson, J. 1988. The use of *in vitro* culture in germplasm management. *ILCA Germplasm Newsletter*, **18**: 4–9.

Samways, M.J., Caldwell, P.M. & Osborn, R. 1996. Ground-living invertebrate assemblages in native, planted and invasive vegetation in South Africa. *Agriculture, Ecosystems and Environment*, **59**: 19–32.

Schrempp, B., Tato, K. & Hurni, H. 1992. Non-conflicting multipurpose tree integration: a case study in the Harerge Highlands, eastern Ethiopia. pp. 109–117, *in: Soil Conservation for Survival*. A selection of papers presented at the 6th International Soil Conservation Organisation, held in Ethiopia and Kenya.

Shelton, H.M. 1994a. Environmental adaptation of forage tree legumes. pp. 120–131, *in:* Gutteridge and Shelton, 1994, q.v.

Shelton, H.M. 1994b. Establishment of forage tree legumes. pp. 132–142, *in:* Gutteridge and Shelton, 1994, q.v.

Shelton, H.M. & Jones, R.J. 1995. Opportunities and limitations in leucaena. pp. 16–23, *in:* Shelton, Piggin and Brewbaker, 1995, q.v.

Shelton, H.M., Piggin, C.M. & Brewbaker, J.L. (eds). 1995. *Leucaena – Opportunities and Limitations*. Proceedings of workshop held in Bogor, Indonesia. 24–29 January 1994. *ACIAR Proceedings*, No. 57.

Shelton, H.M., Norton, B.W., Mullen, B.F., Gutteridge, R.C. & Dart, P.J. 1996. Utilization and nutritive value of *Calliandra calothrysus* for forage: a review of research at the University of Queensland. pp. 210–221, *in:* D.O. Evans (ed). *International Workshop on the Genus* Calliandra. Forest, Farm and Community Tree Research Reports, Special Issue, 1996. Winrock International.

Shelton, H.M., Gutteridge, R.C., Mullen, B.F. & Bray, R.A. (eds). 1998. *Leucaena – Adaptation, Quality and Farming Systems*. Proceedings of workshop held in Hanoi, Viet Nam. *ACIAR Proceedings*, No. 86. 358p.

Shelton, H.M., Piggin, C.M., Acacio, R., Castillo, A., Mullen, B.F., Rika, I.K., Nulik, J. & Gutteridge, R.C. 2000. Case studies of locally successful forage tree systems. pp. 120–131, *in:*, W.W. Stur, P.M. Horne, J.B. Hacker and, P.C. Kerridge (eds). *Working with farmers: The Key to adoption of forage technologies*. Proceedings of Workshop held in Cagayan de Oro City, The Philippines, 12–15 October 1999. *ACIAR Proceedings*, No. 95.

Simons, A.J. 1996. Seed orchards and breeding. pp. 119–125, *in:* J.L. Stewart, G.E. Allison and A.J. Simons (eds). Gliricidia sepium – *Genetic resources for farmers*. Tropical Forestry Paper, No.33. Oxford Forestry Institute.

Smith, O.B. 1992. Alley farming and protein banks for tropical Africa. pp. 245–256, *in:* Speedy and Pugliese, 1992, q.v.

Snook, L.C. 1982. Tagasaste (tree lucerne) *Chamaecytisus palmensis*. A shrub with high potential as a productive fodder crop. *Journal of Australian Institute of Agricultural Science*, **48**: 209–213.

Speedy, A. & Pugliese, P.-L. (eds). 1992. *Legume trees and other fodder trees as protein sources for livestock.* Proceedings of the FAO Expert Consultation held at the Malaysian Agricultural Research and Development Institute (MARDI) in Kuala Lumpur, Malaysia. 14-18 October 1991. FAO, Rome.

Stewart, J.L. 1996. Utilization. pp. 33-48, *in:* J.L. Stewart, G.E. Allison and A.J. Simons (eds). Gliricidia sepium – *Genetic resources for farmers.* Tropical Forestry Paper, No.33. Oxford Forestry Institute.

Stewart, J.L., Allison, G.E. & Simons, A.J. (eds). 1996. Gliricidia sepium – *Genetic resources for farmers.* Tropical Forestry Papers, No. 33. Oxford Forestry Institute. 125p.

Tchoundjeu, Z., Weber, J. & Guarino, L. 1998. Germplasm collections of endangered agroforestry tree species: the case of *Prosopis africana* in the semi-arid lowlands of West Africa. *Agroforestry Systems,* **39**: 91–100.

Tilki, F. & Fisher, R.F. 1998. Tropical leguminous species for acid soils: studies on plant form and growth in Costa Rica. *Forest Ecology and Management,* **108**: 175–192.

Walter, G.H. & Parry, W.H. 1994. Environmental adaptation of forage tree legumes. pp. 309–321, *in:* Gutteridge and Shelton, 1994, q.v.

Westley, S.B. & Powell, M.H. 1993. *Erythrina* in the New and Old Worlds. Nitrogen Fixing tree Research Reports, Special Issue. 358p.

Wickens, G.E., Sief El Din, A.G., Sita, G. & Nahal, I. 1995. Role of *Acacia* species in the rural economy of dry Africa and the Near East. *FAO Conservation Guide,* No. 27. 56p.

Woodward, A. & Reed, J.D. 1997. Nitrogen metabolism of sheep and goats consuming *Acacia brevispica* and *Sesbania sesban. Journal of Animal Science,* **75**: 1130–1139.

Woodward, S.A., Vitousek, P.M., Matson, K., Hughes, F., Benvenuto, K. & Matson, P.A. 1990. Use of the exotic tree *Myrica faya* by native and exotic birds in Hawai'i Volcanoes National Park. *Pacific Science,* **44**: 88–93.

Zabala, N.Q. 1997. *Albizia* and *Paraserianthes* species. Forest, Farm, and Community Tree Research Reports, Special Issue. Winrock International and NFTA, Morrilton, Arkansas, USA. 164p.

Section II
FORAGE CONSERVATION

CHAPTER 5

Silage: A Global Perspective

R.J. Wilkins[1]

ABSTRACT

The adoption of silage-making is considered in relation to the availability of suitable techniques to accomplish efficient preservation of silage. Limitations are noted and alternative approaches are outlined. The current and prospective roles for silage in different regions are discussed.

The essentials for efficient silage-making are to rapidly achieve and successfully maintain anaerobic conditions, and to lower water activity sufficiently to prevent degradation. Practical approaches for achieving these objectives were not available in Europe until the late 1960s. Several factors combined to give reliable technology: the ready availability of sheet polythene for effective sealing; the production of forage harvesters capable of rapid harvesting and efficient chopping; and the ability to direct the course of fermentation through wilting or the use of easily applied and effective additives. This technology was rapidly adopted on farms.

However, the efficiency of the system is still limited by reduction in feeding value and losses of nutrients during ensiling, and by the costs and resource inputs required for making silage. In all regions, it will be necessary to constrain costs of silage-making, particularly by the full and efficient use of machinery, and to use appropriate technology to maintain feeding value and to reduce losses. In temperate areas, it is possible that the use of grass silage in particular may lessen with increasing competition from grazed grass and from alternative feeds. In tropical and subtropical countries, suitable techniques for making silages from forages and by-products are emerging, and it is probable that the importance of silage will increase, particularly in areas where the markets for milk and milk products are well developed.

[1]Institute of Grassland and Environmental Research, North Wyke, Okehampton, UK, and University of Plymouth, Seale-Hayne Faculty, Newton Abbot, UK

INTRODUCTION

This chapter outlines the development and adoption of silage-making, and considers the principles for successful conservation of forage by ensiling, and alternative approaches that could be followed to achieve success. The final section discusses the current and prospective situation for silage in different regions.

Silage has been made for approaching 4 000 years. McDonald, Henderson and Heron (1991) detail biblical references to silos and silage, and refer to old paintings suggesting that ensiling was used to preserve forages in 1000 to 1500 BC, and that fodder was stored in silos in Carthage about 1200 BC. There are sporadic references to silage being made in Europe through to the early nineteenth century, and then a spate of interest from the middle of the 1900s in northern and central Europe and in North America. Silage was much advocated as a method for forage conservation that, in contrast to haymaking, is relatively independent of weather conditions and enables the preservation of high-quality forage. Many farms made small quantities of silage, but it was only a minor contributor to the winter feeding of ruminants. In the United Kingdom, for example, some 100 to 150 thousand tonnes of silage were made in the late 1880s, and this would have contributed 2 or 3 percent of winter feed requirements (Brassley, 1996). Furthermore, despite the theoretical attractions of silage-making, the production of silage in the United Kingdom barely doubled from 1880 to 1940.

Despite substantial research (A.I. Virtanen from Finland was awarded the Nobel Prize for his work on silage in the 1920s and 1930s), much advisory effort and innovation by farmers, silage did not make a major impact in Europe until the last 40 years. There is now more than 50 million tonnes of silage made in the United Kingdom annually, and the quantity of grass dry matter (DM) conserved as silage exceeds that conserved as hay in practically all west European countries (Wilkinson, 1996). The adoption in North America was more rapid, particularly with the use of maize as a silage crop. Silage has, though, yet to make a major impact in animal production in tropical and subtropical regions and Wilkins, Syrjälä-Qvist and Bolsen (1999) estimated that, on a world basis, silage supplied only 2 percent of the nutrients consumed by ruminant livestock.

PRINCIPLES OF SILAGE-MAKING

Conservation as silage relies on the storage of forage in anaerobic conditions. Under these conditions, acids, particularly lactic acid, are produced by fermentation, thus lowering pH to a level sufficient to inhibit the development of putrefactive micro-organisms. If conditions are not met, these micro-organisms survive to break down lactic acid and amino acids to give butyric acid, ammonia, amines and other degradation products. If a sufficiently low pH is reached, the silage will be stable for many years, provided that anaerobic conditions are maintained. The principal putrefactive micro-organisms are clostridial bacteria. They are sensitive to water activity (a_w), a measure of relative vapour pressure determined by moisture content and the molecules and ions in plant juice, rather than pH per se (McDonald, Henderson and Heron, 1991: 111), so that the critical pH required for stability is lowest for the wettest materials (Table 5.1).

Table 5.1 Critical water activities (a_w) and pH values for silages

DM (g kg^{-1})	Water activity	pH required for stability
150	0.985	4.10
200	0.980	4.20
250	0.975	4.35
300	0.971	4.45
350	0.966	4.60
400	0.961	4.75
450	0.956	4.85
500	0.952	5.00

Source: Weissbach, 1996.

Changes in pH during ensiling are illustrated in Figure 5.1 for silages with a stable lactic fermentation and for silages that did not reach the critical pH value required for the prevention of the proliferation of clostridia. In the latter case, after an initial fall in pH, values rise as lactic acid is degraded to acetic and butyric acids and there is progressive breakdown of protein to ammonia. Typical compositions for grass prior to ensiling and for the resulting silages are given in Table 5.2.

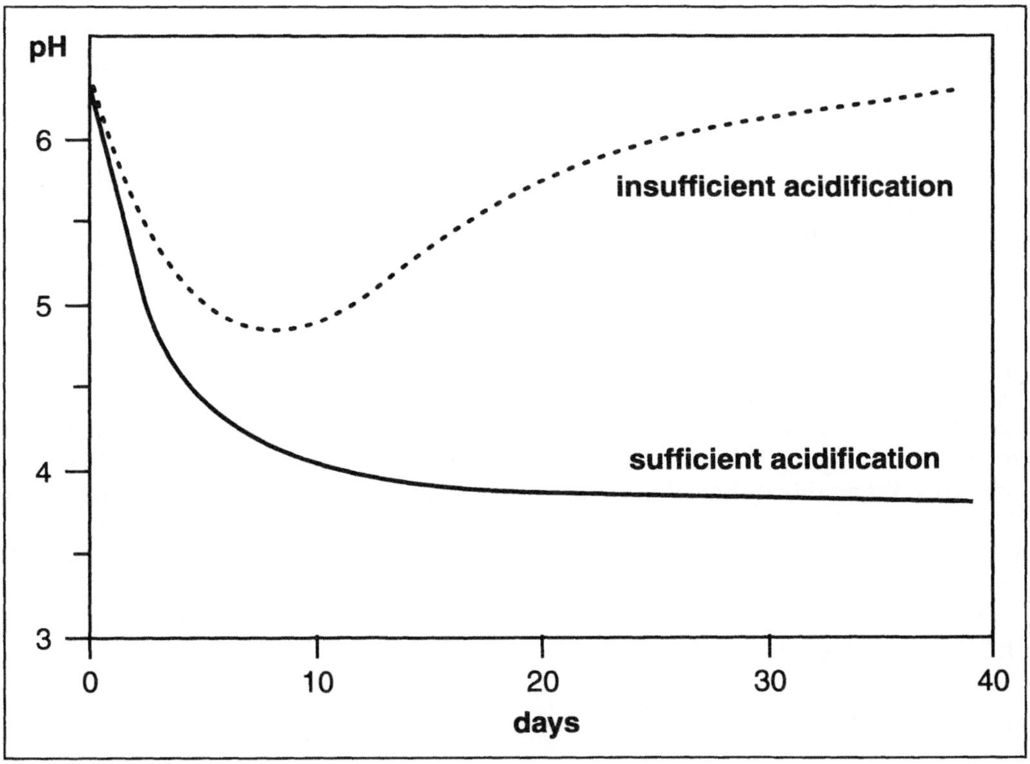

Figure 5.1 Course of pH changes in silages differing in extent of acidification.
Source: Weissbach, 1996.

Table 5.2 Typical composition of grasses prior to ensiling and after undergoing either a clostridial or a lactic fermentation

	Direct cut grass		Wilted grass	
	Unensiled	After clostridial fermentation	Unensiled	After lactic fermentation
DM (g kg^{-1} fresh matter)	180	200	300	290
pH	6.0	5.0	6.0	4.3
WSC (g kg^{-1} fresh matter)	20	0	30	7
Composition of DM (g kg^{-1})				
Water-soluble carbohydrates	110	0	100	25
Neutral detergent fibre	500	550	510	525
Crude protein	150	160	155	160
Lactic acid		25		55
Acetic acid		30		15
Butyric acid		20		0
Composition of N (g kg^{-1} N)				
True protein	800	300	700	400
Ammonia		220		80

Note: The nutritive value of silages that have undergone a clostridial fermentation will be reduced, associated with extensive protein breakdown giving high ammonia contents and the high contents of acetic and butyric acids.
Key: DM = dry matter; WSC = water-soluble carbohydrate.

The essentials for silage-making are (i) rapidly achieving and successfully maintaining anaerobic conditions by effective sealing; and (ii) lowering water activity sufficiently to prevent degradation.

Silage-making will not be successfully adopted until these conditions can be achieved on the farm in a practical and economic manner. The availability of techniques to achieve these objectives did not occur in Western Europe until the late 1960s. Critical factors that needed to be satisfied before rapid adoption took place were (i) the ready availability of sheets of polythene, which, with protection, provided rapid and effective sealing; (ii) progress in mechanization, with development of forage harvesters capable of harvesting rapidly, chopping efficiently and initiating good conditions for air exclusion and rapid fermentation; and (iii) knowledge of fermentation control and the ability to direct fermentation through either wilting or the use of easily-applied acid additives. When this technical package was available, the production of silage in the United Kingdom increased by at least 20 percent per year for a number of years. A further stimulus to adoption occurred in the 1980s, with the development of big-bale techniques. These techniques made a major contribution in encouraging the adoption of silage-making on small farms, but are now widely used on farms of all sizes in the United Kingdom, contributing at least 20 percent of the total silage production.

LIMITATIONS TO SILAGE

While this technology has made a major impact in Europe and North America, it still has limitations, and research and development has been directed to these problems. These concern (i) reduction in feeding value; (ii) losses during ensiling; and (iii) costs of silage-making.

Reduction in feeding value

Ensiling may result in reductions in feed intake and in the efficiency of protein utilization compared with the unensiled material. Reductions in feed intake are most severe in silages that have undergone a clostridial fermentation and are characterized by high contents of ammonia-N. Barry *et al.* (1978) demonstrated that the limitation to intake was associated principally with decarboxylation rather than deamination of amino acids. However, when clostridial fermentation is prevented by extensive fermentation and acid production, intake may also be reduced. Thus addition of lactic acid to silage will reduce intake (McLeod, Wilkins and Raymond, 1970), although not as severely as when clostridial fermentation occurs. In order to maximize intake it appears necessary to start with a crop of high intake potential (usually this means high digestibility), restrict protein breakdown and restrict acid production through fermentation. This is difficult to achieve, although these conditions may occur with silages of very high DM content and where additives are used to severely restrict fermentation, such as formic acid applied at a rate of above 5 litres per tonne. The prediction of intake of silage-based diets is discussed by Huhtanen (2002). The intake of grass silages can be predicted from near infrared spectra (NIRS) (Gordon *et al.*, 1998) or from digestibility (g kg^{-1} DM), total fermentation acid content (g kg^{-1} DM) and ammonia N (g kg^{-1} N) (Huhtanen *et al.*, 2002). The Silage Intake Index of Huhtanen *et al.* (2002) is calculated as follows:

$$\text{Silage Intake Index} = 100 + 0.151^{*}\,(\text{D-value} - 690) - 0.000531^{*}\,((\text{fermentation acids})^2 - 6400) - 4.765^{*}(\log_n \text{ammonia-N} - \log_n 50)$$

In addition to reducing feed intake, ensiling generally results in a lowering of the efficiency of protein utilization. This results from substantial breakdown of protein to amino acids and simpler N compounds and reductions in water-soluble carbohydrate (WSC) content through fermentation to acids. Thus, when silage is fed, there is rapid production of ammonia in the rumen, but shortage of readily available carbohydrate limits the synthesis of microbial protein. Losses of N in urine are generally higher than with unensiled forages and the total supply of amino acids to the animal is reduced. However, the provision of readily-available carbohydrate as starch or WSC in feed supplements leads to more efficient N capture in the rumen and a marked improvement in microbial protein production. Actions taken to reduce the fermentation of WSC and the extent of protein breakdown during ensiling will also improve the utilization of silage N. Thus alterations to ensiling technique to improve feed intake will also increase the utilization of N in silage. Another approach is to seek to reduce the rate of protein breakdown in the silo and in the rumen. There is now evidence that polyphenol oxidase in red clover restricts protein breakdown during ensiling (Jones, 1993). Protein breakdown will also be restricted in tannin-rich forages (Albrecht and Muck, 1991) and when either tannins or formaldehyde are used as additives. While protein breakdown in the rumen will be reduced, there is a risk that some of the protein will also be rendered indigestible in the intestines and be lost in faeces.

In contrast to the lowering of intake and protein utilization that results from ensiling, there are only small changes in digestibility, with small increases in fibre digestibility often compensating for the higher proportion of cell walls in silage (Wilkins, 1988). However wilting prior to ensiling will reduce digestibility by about 1 percent per day in the field, and a clostridial fermentation will reduce digestibility by up to 4 percent. The digestibility of silages is thus most strongly influenced by the digestibility of the crop at harvest. Provided that clostridial fermentation is prevented, the digestibility at harvest is the most important factor determining the level of production that can be achieved with silage. Within forage species there is a positive relationship between digestibility and feed intake, but there is variation between plant species in this relationship, with levels of intake at a given digestibility being higher in silages made from legumes than from grasses, as shown in Figure 5.2 (Paul, Auerbach and Schild, 2002).

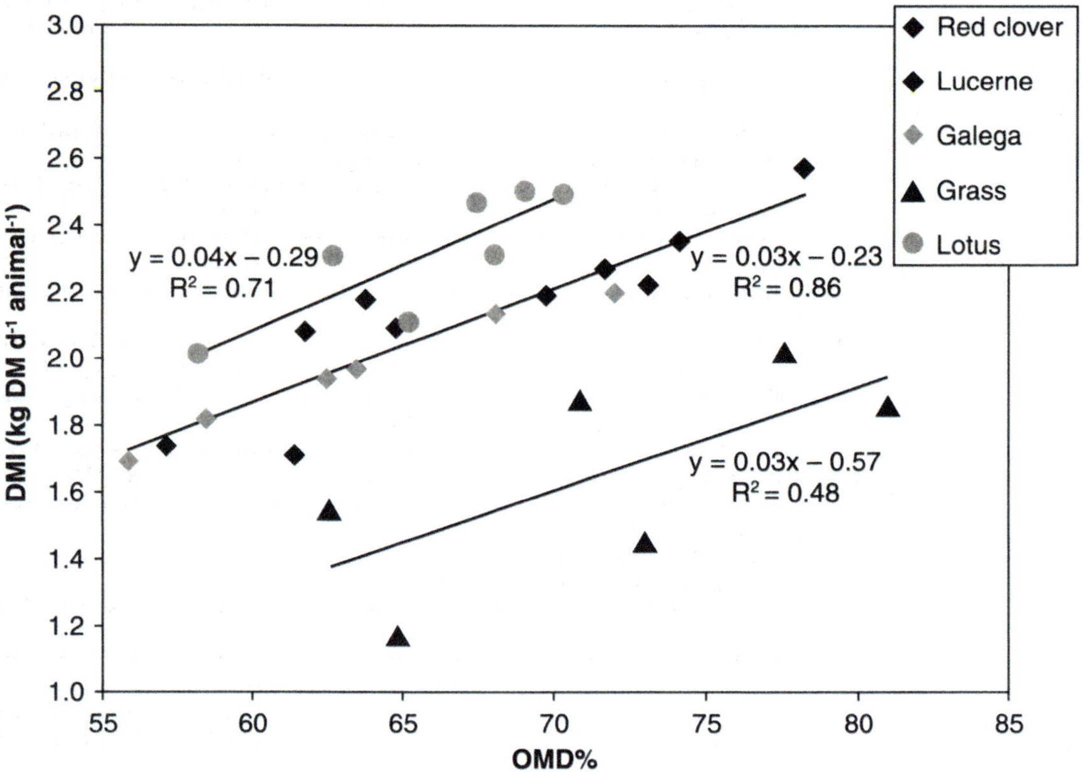

Figure 5.2 Dry matter intake by sheep as a function of organic matter digestibility *in vivo* for silages prepared from legumes and grass.

Source: Adapted from Paul, Auerbach and Schild, 2002.

Losses during ensiling

The magnitude and nature of losses during silage-making and feeding will have a major influence on both the costs of the silage fed and on its feeding value. The range of DM losses may be from only 5 percent with a lactic fermentation and completely anaerobic conditions, to 100 percent when small silos are made without sealing, where there will be very high gaseous losses and the remaining material is inedible. A high proportion of the loss is of material such as WSC, of high feed value, so that as losses increase, the feeding value of the remaining silage is generally reduced.

Field losses

Losses in the field are greatest in systems involving long periods of field wilting. They arise principally from plant respiration after cutting and from mechanical losses.

Figure 5.3 illustrates the relationship between respiration loss and DM content at different temperatures. In European conditions, a usual figure would be around 2 percent of the DM lost per day in the field, but losses in high temperatures and poor drying conditions may reach 5 percent per day.

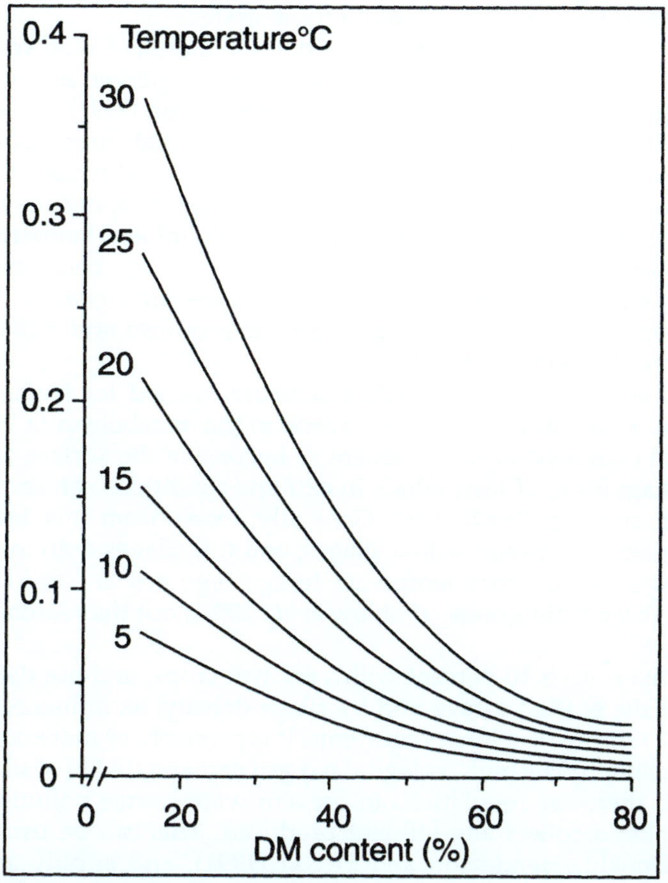

Figure 5.3 Respiration losses from cut grass in the field.
Source: Honig, 1980.

Mechanical loss of forage that is not picked up during harvesting increases with the degree of fragmentation that occurs during field operations and the efficiency of pick up. Honig (1980) found that the DM loss increased from 0.9 percent to 2.9 percent for each mechanical treatment as grass dried from 25 percent to 70 percent DM; losses with legumes are likely to be at least twice these figures because of the leaves drying faster than stems and high leaf fragility.

There are also possible losses of soluble components through leaching, and indirect losses, because lengthy field wilting periods will delay regrowth, as discussed by Wilkins (1988).

In-silo losses

The main sources of loss are from fermentation, oxidation and effluent.

Losses associated with anaerobic fermentation are small. A totally homolactic fermentation will convert WSC to lactic acid with no loss of DM and a loss of only 1 percent of the gross energy from the WSC. With clostridial fermentation of lactate, the losses of DM and gross energy are 51 percent and 18 percent respectively (McDonald, Henderson and Heron, 1991). When expressed on the basis of total forage DM, these losses are reduced because less than 20 percent of forage components are generally used as fermentation substrates.

Losses through oxidation are much more serious. Oxidation of WSC resulting from air trapped in the forage at ensiling is much less important than the prevention of air ingress during the storage and feeding periods. McDonald, Henderson and Heron (1991) calculated that the entrapped air caused a loss of only 0.3 percent of the DM and a temperature rise of around 2°C. Continued ingress of air was shown by Honig (1969) in Germany to result in losses at the rate of 2 percent per month of storage. These losses would represent a higher proportion of the DM in small rather than large silos, and would also increase with increased ambient temperature. In addition to the gaseous losses, oxidation will result in the remaining forage being broken down to form a structureless and inedible mass. In less extreme conditions, air ingress will result in the growth in the surface layers of the silo of moulds giving toxic products and pathogenic bacteria such as *Listeria monocytogenes* (Wilkinson, 1999).

A final stage of aerobic loss occurs when silos are opened for feeding. Air penetration will stimulate activity of yeasts and bacteria and result in the metabolism of WSC, fermentation acids and even structural carbohydrates. The extent of heating in the surface zone of the silo gives an indication of the magnitude of loss, which in extreme conditions can reach 30 percent over a 10-day period (Honig and Woolford, 1980). Generally, losses from this source are increased with (i) slow feed-out rates; (ii) heavily wilted silages; and (iii) silages with low contents of acetic and butyric acids. Silages made from temperate forage legumes are much less prone to aerobic deterioration than those from grasses (Pahlow *et al.*, 2002), but the reason for this high stability is not known.

Effluent losses may reach 10 percent with very wet crops, and are determined principally by the DM content of the ensiled forage and by silage density, as influenced by silo depth (Figure 5.4). While effluent represents a loss of nutrients, it is probably of more consequence in relation to the risk of water pollution. The high biological oxygen demand (BOD) of silage effluent is notorious. In many countries, there are regulations to prevent watercourse pollution, and silos have to be equipped with tanks to collect any effluent produced. This can be used as an animal feed, as discussed by McDonald, Henderson and Heron (1991), and supply substantial quantities of nutrients, with the DM containing about 250 g CP kg, with substantial quantities of sugar or acids. It is, however, normally applied back to farmland. One tonne of effluent of 6 percent DM will supply as fertilizer about 2.5 kg of N, 1 kg of P and 2 kg of K. Care must be taken not to apply effluent in hot dry conditions as this may cause scorch of grass.

System losses

Where silages are made under laboratory conditions, with complete control of air ingress, in-silo losses of DM rarely exceed 10 percent. However, in research carried out at a farm scale, but with careful management, total system losses generally approach 20 percent. Field losses are higher with high-DM systems, but in-silo losses fall with increase in crop DM content (see Figure 5.5). Overall, losses are generally somewhat lower in systems involving wilting to 30 to 40 percent than when wetter crops are ensiled, as shown in Table 5.3, based on results from a coordinated experiment carried out in eight European countries.

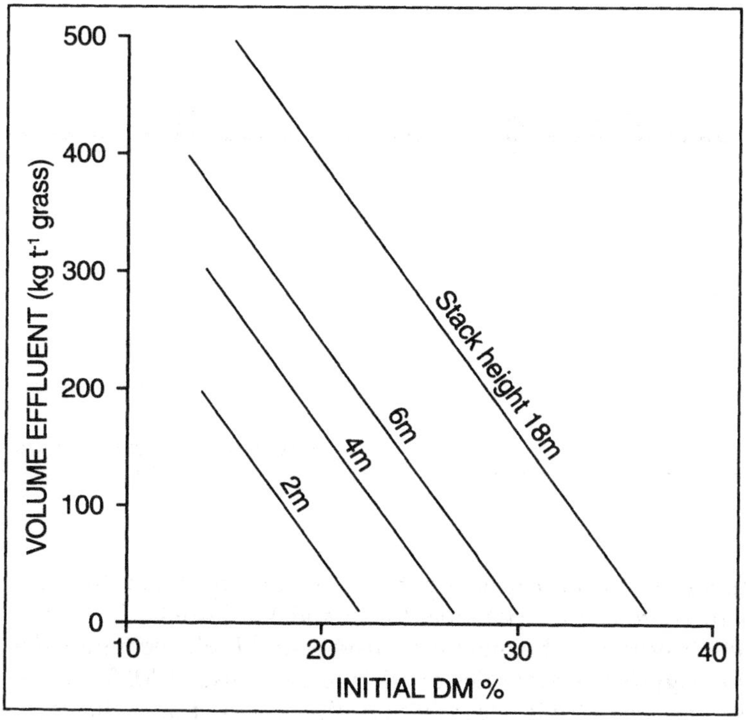

Figure 5.4 Effect of initial dry matter content and silo height on effluent production. Source: Peters and Weissbach, 1980.

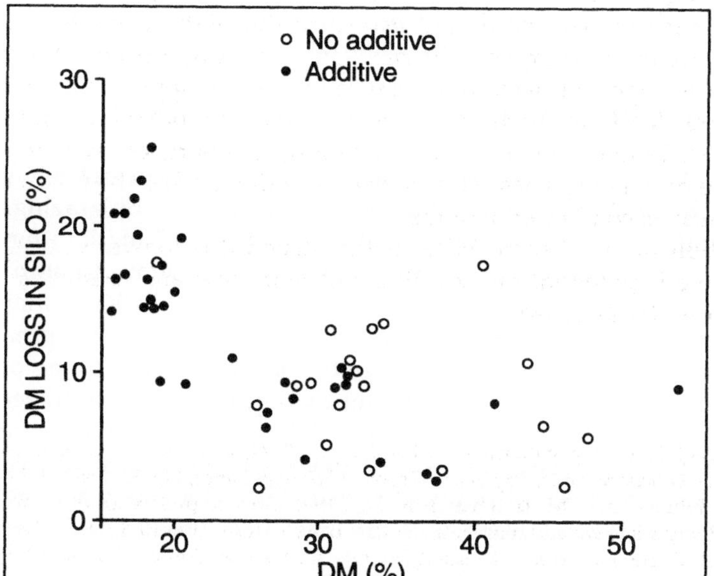

Figure 5.5 In-silo dry matter losses in relation to dry matter content of ensiled forages. Source: Zimmer and Wilkins, 1984.

Table 5.3 Losses of dry matter (percentage) from wilted and unwilted silages in Eurowilt experiments

	Unwilted silage(1)	Wilted silage(2)
Field	2.5	8.0
Effluent	3.2	0
Other in-silo losses	12.9	8.5
Total	18.6	16.5

Notes: (1) Well preserved with formic acid as additive. DM content ca 20 percent. (2) Wilted to 30 to 35 percent DM, with up to 2 days in the field.
Source: Zimmer and Wilkins, 1984.

It is difficult to assess losses accurately at the farm level, and in studies in the United Kingdom, loss levels are generally assumed to be in the range of 20 to 30 percent. However, Peel, Matkin and Huckle (1988) found that metabolizable energy (ME) losses at the level of the whole farm varied from 27 percent to 45 percent in studies in which they directly measured the quantity and composition of the harvested forage and the contribution of the silage to the ME supplied to the farm animals.

Costs and resource requirements

With the great variation that exists in silage-making methods, it is not surprising that there is also considerable variation in the costs of producing silage DM or nutrients. Some general points can be made. Almost always the costs of supplying nutrients as silage are higher than those for forage that is either grazed or cut and fed fresh (zero grazing). With grazing, there are no harvesting or storage costs and with zero grazing there is no storage cost, while with silage-making there are both harvesting and storage costs. Thus, in general, silage should only be considered in circumstances where the use of fresh forage is not possible. In the United Kingdom and in Finland, the costs of supplying nutrients are some 60 percent higher with silage than with grazing (Wilkins, Syrjälä-Qvist and Bolsen, 1999). There are some exceptions to this generalization, particularly if ensiling permits the utilization of feeds that cannot be grazed or fed fresh over an extended period. For instance, some crop by-products may be available for a short season and at low or zero price to the system. Also silages may be produced at relatively low cost from crops such as maize, which give high yields, but are not adapted to use by grazing.

Figure 5.6 illustrates costs of silage production in the United Kingdom for low-DM and high-DM systems. Total costs varied relatively little between the systems. In both cases, about half of the total cost arose from land and the production of the grass, while harvesting costs were around one third of the total. Clearly, there are substantial effects on total cost if production costs can be reduced. Doyle and Topp (2002) calculated a reduction in cost per tonne of DM of around 10 percent through the use of forage legumes rather than grass with N fertilizer. Actions to reduce system losses will also reduce total cost per tonne of silage fed. Machine (or labour) use efficiency is extremely important, as discussed by Rotz (2001). There must be a balanced set of equipment that is used for a large number of hours per year. There has been a large increase in use of contractors for silage-making in Europe, with the machinery used more fully than when operated on a single farm. The shared use of big-balers for handling straw and hay as well as silage will reduce machinery costs for silage.

The higher costs of silage compared with grazing are also reflected in higher requirement for support energy for silage. White, Wilkinson and Wilkins (1983), for instance calculated that the

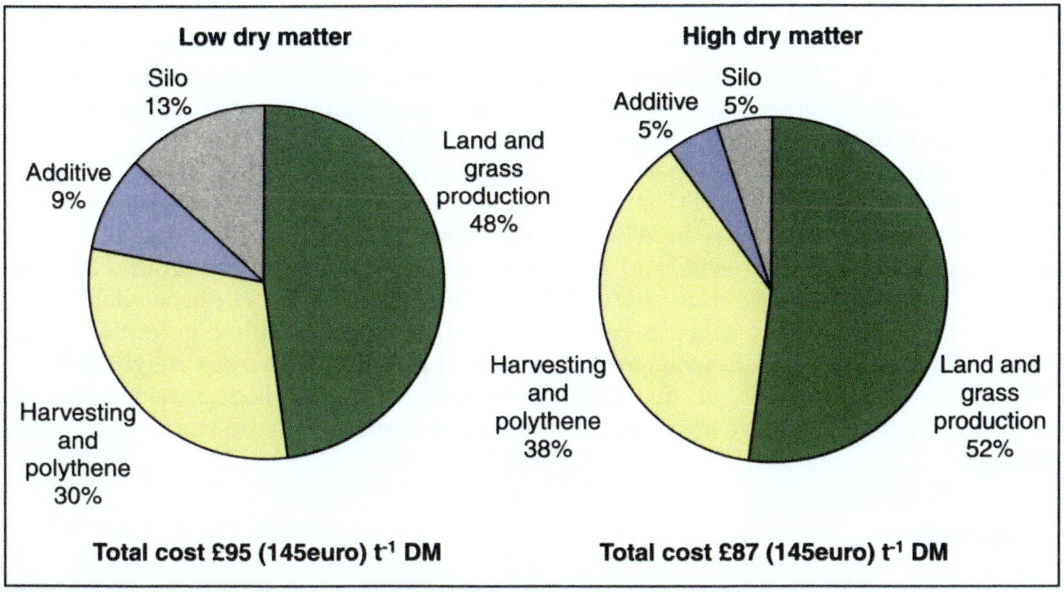

Figure 5.6 Cost of production of low and high dry matter silage in 3-cut system in Northern Ireland, UK. Source: Adapted from Mayne, Ferris and Keady, 1998.

support energy required per unit of ME produced with perennial ryegrass was 50 percent higher for direct-cut silage than for grazing, while for alfalfa (lucerne) (with no requirement for fertilizer N), wilted silage required twice as much support energy as did zero-grazing.

There has, perhaps, been insufficient consideration of the full costs and the trade offs between feeding value, prevention of losses and the cost of resource inputs. In small silos, the risk of very high losses of DM and of nutrients is great. It is probable that there will be good financial rewards for the costs incurred in polythene and the protection of the polythene in order to limit losses and improve preservation.

THE DIFFERENT STAGES OF SILAGE-MAKING

Crops for silage

The choice of crop (or by-product) obviously has a large impact on silage-making and the way that that silage can be used as a feed. Crops vary in their nutrient composition at harvest and also in their suitability for fermentation. Composition of materials for ensiling in relation to digestibility (or ME content) and crude protein (CP) content are generally well known, as is the importance of growth stage at harvest in relation to these characteristics. Ensiling generally results in only small changes in digestibility and CP content.

There are, though, large differences between different crops in terms of DM and WSC contents, and buffering capacity (BC), all factors that have major effects on silage fermentation and on silage feeding value. These factors have been integrated into a single expression, the Fermentability Coefficient (FC), calculated by Pahlow and Weissbach (1999) as:

$$FC = DM(\%) + (8\ WSC \div BC)$$

An FC value of above 45 indicates good fermentability and that satisfactory preservation should be achieved without additives. This has been shown to be widely applicable to temperate grasses and legumes. There is evidence, though, that the FC required may be higher in materials with very low nitrate content than in materials with higher nitrate content, because of the anti-clostridial effects of nitrate (Kaiser, Weiss and Polip, 2002).

Generally the temperate grasses, particularly ryegrass, and whole-crop cereals have high contents of WSC and low BC, giving an FC above 45, provided they have a DM percentage above 25. Tropical grasses are much lower in WSC and thus have lower FCs and ensilability. Temperate forage legumes are all low in WSC and high in BC and require wilting to around 35 percent to achieve an FC of 45 (Pahlow *et al.*, 2002). When information on BC is not available, the WSC content in the fresh matter gives a reasonable indication of ensiling potential. Clostridial fermentation only rarely occurs when WSC is at least 3 percent of the fresh weight (Wilkinson *et al.*, 1981). However, practically all materials can be satisfactorily ensiled, provided that they are either heavily wilted or receive applications of effective additive and are stored under anaerobic conditions.

Field operations

Crops may be cut and transported to the silo in one operation (direct cutting), or cut and then transported from the field after a period of minutes through to several days. There may be subsequent field operations during this wilting period.

Direct cutting with flail harvesters was the predominant method used in northern Europe for grasses in the middle of the twentieth century, but this has now been largely superseded by systems involving field wilting. The main drivers for change were the requirements to reduce effluent production and the opportunities for improved fermentation and feeding value with wilted silages. It is unlikely that there will be any reversion to direct cutting with these crops. Direct cutting is, though, the normal approach for maize and other whole-crop cereals. These crops have sufficiently high DM content at cutting to minimize effluent production and provide satisfactory fermentation without additives.

There is, though, ample evidence that leaving cut grass in the field during poor weather conditions will result in high losses and reductions in feeding value (Zimmer and Wilkins, 1984). Silage-making systems must be designed to reduce the risks of this occurring. This can be done by (i) adjusting cutting plans according to the weather forecast (this is not always possible, particularly if contractors are used); (ii) taking steps to increase the drying rate by, for instance, using a mower-conditioner and spreading the cut crop to increase interception of solar radiation (note that this approach increases risk of loss if rain falls on the spread crop); and (iii) being prepared to harvest at a lower DM content than planned, but using an additive, possibly at increased rates.

The crop may be chopped in the field using a precision-chop harvester or into the silo using a stationary chopper. Big-bale systems have generally not involved chopping, although balers equipped with a chopping mechanism are now available. Chopping results in (i) improved consolidation in the silo and increased silage density; and (ii) more rapid release of cell contents and more rapid fermentation in the silo. In some situations this will give improvements in fermentation and in feeding value, particularly with wet forages. Chopping, though, is not an essential part of silage-making, as has been demonstrated by the success of big-bale systems, both in research and in practice. However, when the forage is not chopped, the DM content required for good preservation is increased and there may be some feed intake reduction.

Additives

There has been an immense amount of research on silage additives in recent years, but most silage worldwide is made without additives. There is, however, clear evidence that effective additives, when applied to crops with low FC, will result in some reduction in DM losses and a substantial improvement in feed intake and animal performance (see Wilkins, 1988; Muck and Shinners, 2001). The evidence is less clear cut when additives are used with crops with high FC, but, on average, effective additives will give a small improvement in animal performance (Table 5.4).

Table 5.4 Effects of bacterial inocula and formic acid on intake and animal performance relative to no additive silage at 100 – average results from experiments published in International Silage Conferences and in review of Northern Ireland experiments by Mayne and Steen (1993)

		Inocula	Formic acid
Intake		107	110
Liveweight gain (cattle)	(g animal^{-1} day^{-1})	106	110
Milk fat and protein yield	(kg cow^{-1} day^{-1})	104	106

Source: Based on Wilkins, 1996.

Generally, the main purpose of using an additive is to prevent clostridial fermentation by (i) direct provision of an acid to lower pH, as with organic and inorganic acids; (ii) the addition of a source of readily available carbohydrate, such as molasses, to increase the quantity of fermentable substrate; (iii) the addition of amylolytic or cellulolytic enzymes to generate more fermentable substrate from crop components; (iv) the addition of compounds with specific effects against clostridia, such as sodium nitrite; or (v) the addition of bacteria to improve the efficiency of conversion of WSC to lactic acid. All of these approaches may be successful, but there is now a long and successful track record with the use of formic acid in a wide range of situations, and with the use of inocula of lactic acid bacteria in all but the most demanding of situations. With bacterial inocula, it is important to have applications of at least 10^6 organisms per gram of forage in order for the inoculum to dominate the epiphytic flora and ensure quantitative conversion of WSC to lactic acid. At present, freeze-dried inocula may be used or a culture may be grown on the farm in an approach analogous to home brewing, as described by Merry, Jones and Theodorou (2000) in the "Live System" method. There remains a need for WSC for fermentation, and with materials of very low FC, the use of an inoculum alone will not ensure good fermentation, as shown for wet alfalfa and galega (*Galega orientalis*) by Pahlow *et al.* (2002). There will, though, be future possibilities for developing inocula with the capability of fermenting a wider range of crop substrates.

The decision on whether or not to use an additive will be influenced by the crops to be ensiled, the suitability of weather conditions for wilting, the cost and availability of additives and effective application systems in relation to the returns from animal products, and the importance of ensuring effective preservation and high feeding value. While in Europe and North America most attention in recent years has been devoted to acids (and their salts) and inocula, in tropical and subtropical regions there is a strong case for considering whether crops and by-products with high content of fermentable carbohydrate and high DM content can be added to forages with low FC at the time of ensiling.

In addition to the use of additives to prevent clostridial fermentation, additives have been used for other purposes. Aerobic deterioration after the silo is opened is a substantial problem, particularly with dry silages and those with low volatile acid contents. There has been some success with additives based on propionic acid or biological additives that will generate increased quantities of acetic or propionic acids in the silo (Dreihuis, Oude Elferink and Van Wikselaar, 2001). Effluent production can be restricted by the addition of dry materials such as straw or grain at ensiling, but large quantities are required to have a marked effect on effluent production. Attempts have been made to apply additives that will increase digestibility by attacking structural carbohydrates or the bonding of carbohydrates and lignin, but there has, as yet, been only limited success (see Wilkins, 1996). The addition of cellulolytic enzymes will have little effect on digestibility as their action in the silo is on carbohydrates, which are readily digested in the rumen anyway. Urea (as a source of ammonia) and sodium hydroxide may give some increase in digestibility with low quality forages, but there are difficulties in developing techniques appropriate for on-farm use.

The silo and sealing

Silos vary from hermetically sealed tower silos up to 20 m high, through bunker silos containing up to 20 000 tonne of silage, to wrapped big-bales (see Plate 5.1) containing 500 kg of silage and small bags with only 10 kg of silage. Many laboratory studies are carried out in test tubes with only 100 g of forage. In terms of effects on losses and preservation, silage management is more important than the type of silo used, with the prevention of air ingress throughout the storage period being the key to achieving low levels of loss. Table 5.5, taken from a review of world literature on grass and grass-clover silage by Zimmer (1980), indicates average levels of in-silo loss for different silage systems and silo types.

Plate 5.1 Plastic-wrapped big silage bales. [Photograph by Dr Stephen Reynolds]

Table 5.5 Effect of silo type on dry matter losses (percentage) (excluding field losses).

		DM loss (%)
Direct-cut silages (<20% DM)	Stacks or pits	27
	Bunkers or clamps	16
	Towers	20
Wilted silages	Bunkers or clamps (25–30% DM)	16
	Towers (30% DM)	12
	Towers (>30% DM)	8
	Gas-tight towers (>35% DM)	8

Note: In-silo losses from wilted crops ensiled in big-bales will be in the range of 5 to 10 percent, provided the polythene is not broken during storage. However, damage to the polythene may result in complete loss of the bale.
Source: From a review of world literature by Zimmer, 1980.

The ready availability of polythene sheeting has had a tremendous impact on the ability to make silages with low levels of loss. It is, however, important to ensure that the polythene is protected from damage during the storage period. For bunker or clamp silos, this can be done by weighting with soil, straw bales or a complete network of tyres. The storage of wrapped bales is aided by the provision of a net or cover on top of the bales, and the use of double or triple bags in small bag systems is important. As noted earlier, the smaller the size of the silo, the more critical it is to achieve good sealing and protection. The North American approach of making extremely large silos, but without covering, has survived only because the ratio of surface area to mass is low in those circumstances, but the magnitude of losses has probably not been fully appreciated. Bolsen *et al.* (1993) showed typical losses of 40–80 percent in the top 1 m of an uncovered bunker.

It is likely that a wide range of silo types will continue to be used. In many European countries, the design requirements for new bunker silos are extremely demanding in order to prevent risk of effluent loss from the silo system, even if it is intended that only dry crops be ensiled. This has a big impact on the capital costs of silage systems, and is likely to further encourage the adoption of wrapped bale systems, with their lower capital costs. The costs for both the purchase and the disposal of polythene sheeting are likely to become increasingly important issues. Practical approaches for filling silos and for silo management are described well in Raymond and Waltham (1996), and big-bale techniques are described by Lingvall (1995).

SILAGE IN DIFFERENT REGIONS

This section discusses the current and prospective situation for silage-making in four regions of the world: the Nordic countries of northern Europe; northwest Europe; south and east Asia; and Australia and New Zealand. Much of the material for the first three regions is taken from Wilkins, Syrjälä-Qvist and Bolsen (1999), while the section on Australia relies heavily on material supplied by Dr Alan Kaiser, including Kaiser, Jacobs and Davies (1996), Kaiser and Evans (1997), and Kaiser and Piltz (2002).

Nordic Europe

This region is characterized by a short growing season and a long winter. Grass has traditionally been the main forage used, and conserved grass forms the basis of winter feeding. Silage-making increased considerably from the 1960s and is the predominant method of conservation. Additives based on formic acid have been very widely used, with practically all silage in Finland and Norway made with an additive. Traditionally, silages were direct cut in Finland and Norway, but wilted in the Baltic countries of Estonia, Latvia and Lithuania. Wilting systems have increased in recent years throughout the region, with the main objective of reducing effluent production.

It is anticipated that silage will remain the major conserved forage and that total production will at least be maintained. However, grass will experience more competition from other forages as farmers strive to reduce costs of production and risks to the environment. The climate is not suited to maize production, but there is increasing interest throughout the region in using whole crop cereals (principally barley and wheat) for silage. These crops can produce yields of over 10 tonne DM ha^{-1} in a single cut. Yield is reliable, as the crops are spring sown and not at risk from winter damage, and they can be harvested with DM contents above 30 percent, with little risk of effluent production. Although intake levels of rations that include whole cereals may be quite high, values for digestibility and CP content are below those that can be achieved with grass silage.

There is also interest in greater use of forage legumes for silage. Well-preserved forage legumes will give higher intakes and levels of milk production than grass silage (Bertilsson, Dewhurst and Tuori, 2002) (Table 5.6) and result in increased profitability (Doyle and Topp, 2002). There is, however, a greater risk of winter damage with most legumes than with grass. In view of the high CP content of legumes, it is likely that they will provide a good complement for whole-crop cereals in feeding dairy and beef cattle. Red clover is the legume that is used most widely, but there is increased use of alfalfa and white clover in southern parts of the region. Galega is used to a small extent in Finland and in the Baltic countries, but has high potential because of high yield potential and extreme persistence (Halling *et al.*, 2002; Kravale, Adamovic and Adamavicha, 2002).

Table 5.6 Feed intake and milk production for cows fed silages made from grass, red clover and white clover with 8 kg of concentrates per cow per day (means from experiments in Finland, Sweden and United Kingdom)

		Grass	Red clover	White clover
Silage intake	(kg DM day^{-1})	12.6	13.6	13.2
Milk production	(kg day^{-1})	28.1	29.7	32.1

Source: Bertilsson, Dewhurst and Tuori, 2002.

The trend for increased adoption of high-DM systems is likely to continue and it is possible that there will be reduced reliance on acid additives. Big-bale techniques have made a major impact, particularly on the recently established small farms in the Baltic countries.

Northwest Europe

This region has high annual grass production, but limited herbage growth during winter. The predominant herbages are grasses and grass–white clover mixtures. Swards are grazed during the growing season and silage is the main method of grass conservation. There has been a large increase in the importance of maize for silage, particularly in the Netherlands and France.

There was a large increase in silage-making in this region from 1970 to 1990, with silage supplanting hay as the main method of conservation, with an increase in the total quantity of feed conserved. In the United Kingdom, this increase occurred largely at the expense of grazing, and Wilkins (1992) calculated that the proportion of grass that was conserved rather than grazed increased from 28 percent to 47 percent over that period.

The position of grass silage in this region has been increasingly challenged in recent years because of trends for reduction in the price of animal products and in the price of grain, increases in the yields and reliability of maize production, and increases in the costs of silage harvesting and storage. As noted earlier, the costs of supplying nutrients in silage are higher than with grazing. There has already been a reduction in the quantity of grass silage made, falling by 10 percent from the peak figures of ten years ago. Many farmers, particularly in areas to the west of the region, with high rainfall and long growing seasons, are seeking to increase the length of the grazing season and reduce the quantities of silage required in their systems (see Rook and Penning, 2000).

It is probable that much of the silage required would be produced as a by-product from grazing systems, with silage being made mainly in the early summer when grass growth exceeds animal requirements. The silage made will be leafy and of high quality as it will be important to harvest at a young stage to ensure rapid regrowth and continuity of supply of high quality grass for grazing. If there is a requirement for extra silage in the system, it may well be supplied from maize or forage legumes, rather than from grass managed specifically for silage production.

It will be important to constrain the costs of silage production, because of increased competition from cereal grain and other feeds. Reductions in machinery costs and silo costs will be particularly important. High capital costs (to prevent effluent pollution) may limit the building of new bunker silos and increase the proportion of silage made in wrapped bales. Increased concerns for environmental protection will continue to encourage the use of high-DM systems and actions for the recovery and recycling of polythene.

South and east Asia

In contrast to the regions in Europe discussed above, there is no tradition of silage-making in south and east Asia, and silage currently makes only a very minor contribution to animal feeding. However, one of the major limitations to livestock production in the dry-humid and humid tropics is the uneven supply of pasture forage during the year. In many systems, the surplus remains as standing hay to be consumed (or wasted) during the dry season. The material is of very low quality and is used inefficiently. It has been suggested for at least the last 50 years that ensiling may have an important part to play in these systems, but there has been little adoption in practice. It is possible that the right combination of appropriate techniques and economics has now come together to encourage greater use of silage in this region.

Problems that have limited adoption included (i) the high concentration of cell walls and low contents of WSC in tropical grasses and legumes; (ii) low levels of available mechanization; (iii) limited availability and high cost of materials for effective sealing of silos; (iv) small farm size and ready availability of labour for daily harvesting and green feeding of animals; (v) low price of animal products; and (vi) lack of familiarity with silage-making amongst farmers and their advisers.

There has been a recent increased interest in silage-making amongst both scientists and farmers in this region. There were several papers from this region in the proceedings of the recent electronic conference on *Silage making in the tropics with particular emphasis on smallholders* (FAO, 2000), and possibilities were assessed and described by Chin (2002).

It was acknowledged that the suitability for fermentation of tropical grasses is lower than that of temperate grasses, but, with good technique, satisfactory preservation can be achieved. There are also situations in which silage may be made from carbohydrate-rich crops such as maize or sorghum, either alone or in mixture with grass. Bolsen and Faylon (1994) and Machin (2000) have pointed out the large range of crop by-products that may be available at low cost and are appropriate for use directly for silage or that can be added to improve the ensilability and nutritive value of grasses. By-products of high nutritive value include molasses, cereal brans, citrus pulp and other fruit processing residues, while materials with lower feeding value, such as sugar-cane bagasse and oil-palm fronds, may also make a contribution to overall feed supply.

The problem of limited mechanization remains in many areas, but hand labour is becoming less available. Increases in the prices of animal products and, particularly, the growing commercial markets for milk and milk products, have provided an effective demand for higher quality feeds and increased the ability of producers to pay for milk production inputs.

It is important to chop tropical forages in order to achieve successful fermentation and storage. This may be done with a stationary chopper at the silo. Programmes are currently in hand to encourage farmers to form cooperatives and share the equipment needed to make silage from surplus forage. There has also been increase in the availability of appropriate silos for both small and larger farms. Nakamanee (2000) surveyed the attitude of farmers in Thailand to alternative types of silos. There was diversity in preference, with 38 percent preferring bunker silos, 31 percent plastic buckets, 23 percent small polythene bags and 8 percent large polythene bags (containing about 800 kg of forage). As noted earlier, it is particularly important that small silos are sealed very effectively to avoid total losses. The high ambient temperatures in this region add to the need for good sealing. These temperatures also exacerbate the problem of aerobic deterioration of silage when silos are open, particularly if there are few animals and feed-out rates are slow.

Polythene sheeting or polythene bags are now more readily available at relatively low price. The small-bag approach to silage-making described by Nakamanee (1997) and Lane (2000) is particularly interesting. Forage is cut (normally by hand), chopped and filled into a bag containing 10–40 kg of forage; this is then compressed and sealed and put inside a second bag and sealed. The bags need to be protected from pests, birds and vermin during storage. Each bag will provide silage for only a few cow-days, thus reducing risks of aerobic deterioration during feed-out. Because of risks of damage to polythene by vermin, Chin (2002) advocated the use of plastic drums of about 100 kg capacity as an alternative to bags. Old drums are currently readily available and may be re-used many times. The small-silo approach may find a significant role in smallholder dairy systems, while systems based on bunkers or large wrapped bales will be important for larger operations.

Wilkins, Syrjälä-Qvist and Bolsen (1999) and Chin (2002) both concluded that there would be some increase in silage-making in this region, but that care must be taken to develop appropriate technologies and to look at ways in which silage can best complement other available feeds. Silage from special forage crops is likely to be one of the most expensive sources of nutrients (other than grain) and so the prevention of waste and the achievement of good preservation and high quality are particularly important.

Australia and New Zealand

Although silage and silage-making techniques are well known, at least in the southern parts of Australia and New Zealand, silage has not been a significant feature of animal production systems.

There is much variation climatically between different parts of Australia, with temperate, Mediterranean, subtropical, tropical and arid areas. Herbage production patterns are often affected more by availability of water than by variation in temperature. The continent is also characterized by large variation in rainfall between years. In these circumstances, consideration has been given to forage conservation to provide a drought reserve to guard against inter-annual variation in herbage production. However, the opportunity costs of this strategy are high, and forage conservation is now considered principally within the annual cycle of forage production and utilization, particularly for provision of feed during shortage periods and for control of grass growth and quality during periods of rapid growth.

Calculations from Kaiser, Jacobs and Davies (1996) indicate that in the early 1990s hay and silage provided only some 5 percent of the feed consumed by ruminants in Australia, with only 6 percent of the conserved forage being in the form of silage. Silage-making has, however, considerably increased in recent years, from 0.6 million tonnes in 1991 to 2.8 million tonnes in 1999 (Stubbs, 2000). This increase has been concentrated mostly in the dairy sector, and to a more limited extent for feedlot beef production. It has occurred during a period of increase in cow numbers and increase in milk price. This has provided economic conditions suitable for the adoption of silage-making, while, particularly in southern Australia, wrapped big-bales have provided a suitable technique. Silage-making has been possible with minimal additional capital investment in machinery and silos. Weather conditions are generally suited to field wilting and risks of loss through aerobic deterioration during feed-out are less with wrapped bales than when high-DM material is ensiled in clamps. In southern Australia, most silage is made from grass-based swards, but the major crops for silage in Queensland are maize, sorghum and millets. These species are less suited to ensiling in wrapped bales and are generally direct cut and ensiled in bunkers. Cowan (2000) projects further increases in the use of silage in typical Queensland dairy farms (Figure 5.7).

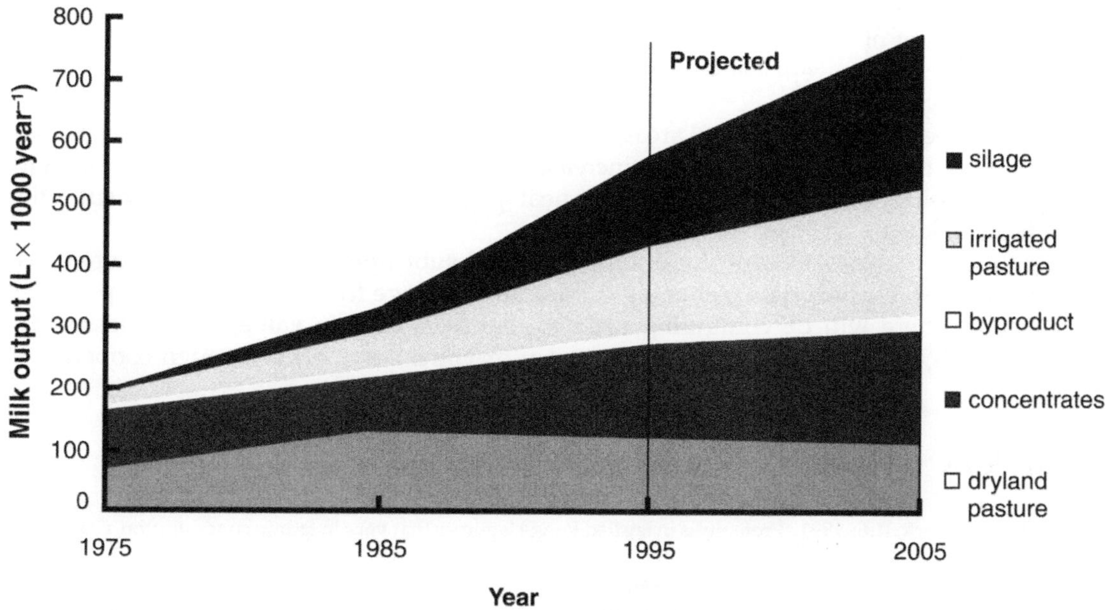

Figure 5.7 Past estimates and projections of milk and feed use on a typical Queensland, Australia, dairy farm. Source: Cowan, 2000.

Conserved forage has traditionally had only a minor role in systems in New Zealand, because of the long grass-growing season and the concentration of milk production during this period of growth. However, there is now increased emphasis on extending the season of milk production and also increasing dairying in parts of the country (south of South Island) with a longer winter. These developments, together with increases in the world prices for milk products, have led to a substantial increase in silage-making. Some of this silage is made from surplus herbage during periods of particularly rapid grass growth, but there is increased interest and practical adoption of forages grown specially for silage, with maize being used in North Island and whole-crop barley, wheat and triticale in South Island. Silages are generally fed-out to stock at pasture and may contribute during periods of slow growth, in both winter and summer. The widespread use of big-bales has been a particularly important feature of increased adoption of silage-making, particularly silage made from grass swards.

There are some parallels in the developments in Australia and New Zealand in the last decade with those in northwest Europe in the 1970s, with technology and economics combining to produce practical systems that have been adopted rapidly on farms. However, it has already been noted that the costs and energy requirements for provision of nutrients in silage are generally considerably higher than for grazing. It will be important to remain fully aware of the costs and efficiency of the complete system and to endeavour to ensure maximum possible utilization of the lowest cost feed resources. It is probable, however, that there will be considerable further increase in silage-making in Australia and New Zealand in the next decade.

CONCLUSIONS

Silage will continue to have a substantial role in animal production systems in temperate areas of the world to supply nutrients to housed animals during periods of low crop growth. There will be a continuing need for efficient use of good technology and inputs to constrain the costs of silage production in view of competition from grazed grass and grain as feed resources. It is probable that there will be some replacement of grass silage by silages made from maize, whole crop cereals and forage legumes.

In some temperate and subtropical areas, where silage currently supplies only a small proportion of the nutrients, there will probably be an increase in silage-making, encouraged by improvements in silage technology, increases in milk and meat prices and increases in the consumption of animal food products.

The potential of silage to contribute to tropical and subtropical animal production is, as yet, largely unrealized. Appropriate technologies for making silage from forages and by-products are emerging, but there will be continuing need to maintain feeding value, reduce losses during storage through effective sealing, and to constrain machine costs, often through cooperative use of machinery.

REFERENCES

Albrecht, K.A. & Muck, R.E. 1991. Proteolysis in ensiled forage legumes that vary in tannin concentration. *Crop Science*, **31**: 464–469.

Barry, T.N., Mundell, D.C., Wilkins, R.J. & Beever, D.E. 1978. The influence of formic acid and formaldehyde additives and type of harvesting machine on the utilization of nitrogen in lucerne silages. 2. Changes in amino-acid composition during ensiling and their influence on nutritive value. *Journal of Agricultural Science*, **91**: 717–725.

Bertilsson, J., Dewhurst, R.J. & Tuori, M. 2002. Effects of legume silages on feed intake, milk production and nitrogen efficiency. *Landbauforschung Voelkenrode*, **234**: 39–45.

Bolsen, K.K., Dickerson, J.T., Brent, B.E., Somon, R.N., Dalke, B.C, Lin, C.J. & Boyer, J.E. 1993. Rate and extent of spoilage in horizontal silos. *Journal of Dairy Science*, **76**: 2940–2962.

Bolsen, K.K. & Faylon, P.S. 1994. *Silage technology for cattle feeding: a training manual.* Manila, the Philippines: Pioneer Development Foundation.

Brassley, P. 1996. Silage in Britain 1880-1990: the delayed adoption of an innovation. *Agricultural History Review*, **44**: 61–82.

Chin, F.Y. 2002. Ensilaging of tropical forages with particular reference to South East Asian systems. pp. 22–36. *Proceedings of the XIII International Silage Conference.* Auchincruive, Scotland, UK, 11–13 September 2002.

Cowan, T. 2000. Use of ensiled forages in large-scale animal production systems. pp. 31–39, *in:* FAO, 2000, q.v.

Doyle, C.J. & Topp, C.F.E. 2002. An economic assessment of the potential for increasing the use of forage legumes in north European livestock systems. *Landbauforschung Voelkenrode*, **234**: 75–85.

Dreihuis, F., Oude Elferink, S.J.W.H. & Van Wikselaar, P.G. 2001. Fermentation characteristics and aerobic stability of grass silage inoculated with *Lactobacillus buchneri*, with or without homofermentative lactic acid bacteria. *Grass and Forage Science*, **56**: 330–343.

FAO. 2000. Silage making in the tropics with particular emphasis on smallholders. Edited by L. 't Mannetje. *FAO Plant Production and Protection Paper*, No. 161. 180p.

Gordon, F.J., Cooper, K.M., Park, R.S. & Steen, R.W.J. 1998. The prediction of silage intake potential and organic matter digestibility of grass silages by near infrared spectroscopy analysis of undried samples. *Animal Feed Science and Technology*, **70**: 339–351.

Halling, M.A., Hopkins, A., Nissinen, O., Paul, C., Tuori, M. & Soelter, U. 2002. Forage legumes – productivity and composition. *Landbauforschung Voelkenrode*, **234**: 5–14.

Honig, H. 1969. Der Einfluss unterschiedlich anarober Lagerungsbedingungen auf den Gaerverlauf. pp. 173–181, *in: Proceedings of the Third General Meeting of the European Grassland Federation.* Braunschweig, Germany, June 1969.

Honig, H. 1980. Mechanical and respiration losses during pre-wilting of grass. pp. 201–204, *in: British Grassland Society Occasional Symposium*, No. 11.

Honig, H. & Woolford, M.K. 1980. Changes in silage on exposure to air. pp. 76–87, *in: British Grassland Society Occasional Symposium*, No. 11.

Huhtanen, P. 2002. New developments in the prediction of intake of silage-based diets. pp. 236–251, *in: Proceedings of the XIII International Silage Conference.* Auchincruive, Scotland, UK, 11–13 September 2002.

Huhtanen, P., Khalili, H., Nousiainen, J.I., Rinne, M., Jaakkola, S., Heikkila, T. & Nousiainen, J. 2002. Prediction of the relative intake potential of grass silage by dairy cows. *Livestock Production Science*, **73**: 111–130.

Jones, B.A. 1993. Inhibition of proteolysis by red clover. pp. 106–107, *in: Proceedings of the X International Silage Conference.* Dublin, Ireland, September 1993.

Kaiser, A.G. & Evans, M.J. 1997. Forage conservation on Australian Dairy Farms. Animal Industries Report No. 3. NSW Agriculture, Orange, Australia.

Kaiser, A.G., Jacobs, J.L. & Davies, B.L. 1996. Pasture and forage crop conservation – hay and silage. In: J. Rowe and N. Cossins (eds). *A users guide to drought feeding alternatives.* Armidale, Australia: University of New England.

Kaiser, A.G. & Piltz, J.W. 2002. Silage production from tropical forages in Australia. pp. 48–61, *in: Proceedings of the XIII International Silage Conference.* Auchincruive, Scotland, UK, 11–13 September 2002.

Kaiser, E., Weiss, K. & Polip, I. 2002. A new concept for estimation of the ensiling potential of forages. pp. 344–358, *in: Proceedings of the XIII International Silage Conference.* Auchincruive, Scotland, UK, 11–13 September 2002.

Kravale, D., Adamovic, A. & Adamavicha, O. 2002. Forage legume and legume-grass swards productivity and silage quality in Latvia. *Grassland Science in Europe*, **7**: 434–435.

Lane, I.R. 2000. Little bag silage. pp. 79–83, *in:* FAO, 2000, q.v.

Lingvall, P. 1995. *The balewrapping handbook.* Smaalandsstenar, Sweden: Trioplast AB. 51p.

Machin, D.H. 2000. The potential use of tropical silage for livestock production, with special reference to smallholders. pp. 71–78, *in:* FAO, 2000, q.v.

Mayne, C.S., Ferris, C.P. & Keady, T.W.J. 1998. Is high quality grass silage cost effective for winter milk production? Agricultural Research Institute of Northern Ireland, Hillsborough. *Occasional Publication*, No. 26.

Mayne, C.S. & Steen, R.W.J. 1993. A review of animal production responses to formic acid and inoculant treatment of grass silage in trials at the Agricultural Research Institute of Northern Ireland. pp. 178–179, *in: Proceedings of the X International Silage Conference.* Dublin, Ireland, September 1993.

McDonald, P., Henderson, A.R. & Heron, S.J.E. 1991. *The biochemistry of silage.* Marlow, England: Chalcombe Publications. 340p.

McLeod, D.S., Wilkins, R.J. & Raymond, W.F. 1970. The voluntary intake by sheep and cattle of silages differing in free-acid content. *Journal of Agricultural Science*, **75**: 311–319.

Merry, R.J., Jones, R. & Theodorou, M.K. 2000. The conservation of grass. pp. 196–228, *in:* A. Hopkins (ed). *Grass: its production and utilization*. Oxford, England: Blackwell Science.

Muck, R.E. & Shinners, K.J. 2001. Conserved forage (silage and hay): progress and priorities. pp. 753–762, *in: Proceedings of the XIX International Grassland Congress*. Sao Pedro, Brazil, 11–21 February 2001.

Nakamanee, G. 1997. Dairy farming in Thailand. *South East Asia Feed Resources Research and Development Network*, **6**: 3–4.

Nakamanee, G. 2000. Successful smallholder silage production: a case study from northeast Thailand. pp. 15–16, *in:* FAO, 2000, q.v.

Pahlow, G., Rammer, C., Slottner, D. & Tuori, M. 2002. Ensiling of legumes. *Landbauforschung Voelkenrode*, **234**: 27–30.

Pahlow, G. & Weissbach, F. 1999. New aspects of evaluation and application of silage additives. *Landbauforschung Voelkenrode*, **206**: 141–158.

Paul, C., Auerbach, H. & Schild, G.-J. 2002. Intake of legume silages by sheep. *Landbauforschung Voelkenrode*, **234**: 55–61.

Peel, S., Matkin, E.A. & Huckle, C.A. 1988. Herbage growth and utilized output from grassland on dairy farms in southwest England: case studies on five farms, 1982 and 1983. II. Herbage utilization. *Grass and Forage Science*, **43**: 71–78.

Peters, G. & Weissbach, F. 1980. Factors influencing the formation of effluent in forage ensiling. pp. 1333–1335, *in: Proceedings of the XIII International Grassland Congress*. Leipzig, Germany, 18–27 May 1977.

Raymond, F. & Waltham, R. 1996. *Forage conservation and feeding*. Ipswich, England: Farming Press.

Rook, A.J. & Penning, P.D. (eds). 2000. Grazing management. *Proceedings British Grassland Society Occasional Symposium*, No. 34. 270p.

Rotz, C.A. 2001. Mechanization: planning and selection of equipment. pp. 763–768, *in: Proceedings of the XIX International Grassland Congress*. Sao Pedro, Brazil, 11–21 February 2001.

Stubbs, A.K. 2000. *Atlas of the Australian fodder industry: outline of production and trade*. Rural Industries Research and Development Corporation Publication No. 00/122, Kingston, ACT, Australia.

Weissbach, F. 1996. New developments in crop conservation. pp. 11–25, *in: Proceedings of the XI International Silage Conference*. Aberystwyth, Wales, UK, 8–11 September 1996.

White, D.J., Wilkinson, J.M. & Wilkins, R.J. 1983. Support energy use in animal production from grassland. pp. 33–42, *in: British Grassland Society Occasional Symposium*, No. 14.

Wilkins, R.J. 1988. The preservation of forages. pp. 231–255, *in:* E.R. Ørskov (ed). *Feed science*, Amsterdam, the Netherlands: Elsevier Science Publishers.

Wilkins, R.J. 1992. Grasslands for the future. *Third Research Conference, British Grassland Society*. Greenmount College, Northern Ireland, 2–4 September 1992.

Wilkins, R.J. 1996. Silage production and utilisation – perspectives from the silage research conferences. pp. 1–10, *in: Proceedings of the XI International Silage Conference*. Aberystwyth, Wales, UK, 8–11 September 1996.

Wilkins, R.J., Syrjälä-Qvist, L. & Bolsen, K.K. 1999. The future role of silage in sustainable animal production. pp. 23–40, *in: Proceedings of the XII International Silage Conference*. Uppsala, Sweden, 5–7 July 1999.

Wilkinson, J.M. 1996. *Silage in Europe: a survey of 33 countries*. Lincoln, England: Chalcombe Publications.

Wilkinson, J.M. 1999. Silage and health. pp. 67–81, *in: Proceedings of the XII International Silage Conference*. Uppsala, Sweden, 5–7 July 1999.

Wilkinson, J.M., Chapman, P.F., Wilkins, R.J. & Wilson, R.F. 1981. Interrelationships between pattern of fermentation during ensilage and initial crop composition. pp. 631–634, *in: Proceedings of the XIV International Grassland Congress*. Lexington, USA, 15–24 June 1981.

Zimmer, E. 1980. Efficient silage systems. pp. 186–197, *in: British Grassland Society Occasional Symposium*, No. 11.

Zimmer, E. & Wilkins, R.J. 1984. Efficiency of silage systems: a comparison between unwilted and wilted silages. *Landbauforschung Voelkenrode*, **69**: 1–88.

CHAPTER 6

Prospects for Haying Systems Around the World

D. McCartney[1]

ABSTRACT

Hay harvesting as a means of conserving forages can be traced back over 2000 years. In the past century, hay harvesting systems have developed from harvesting by hand to ultra-modern round and square bales, densified bales and pelleting systems. With the current move towards larger farming units, hay harvesting equipment has become larger and more expensive. However, in some parts of the world, hay is still harvested by hand. Extensive research on the development and evaluation of hay harvesting equipment occurred in the 1970s and 1980s in North America and Europe. Today, there is very little active hay harvesting research. In those areas of the world where weather permits, hay harvesting ranges from the traditional hand methods to the fully mechanized large round and square bale systems. If large volumes of feed are required, it is usually harvested as silage. With all the technical developments over the years, spoilage due to the weather is still the major limitation to forage preservation.

INTRODUCTION

In most parts of the world, forage conservation is a key element for productive and efficient ruminant livestock farms (Muck and Shinners, 2001). Harvesting, storage and feeding losses have always been the greatest problems in hay production. It has been estimated that at least 25 percent of the forage crop can be lost between harvesting and feeding (Rasmussen, 1969). These losses occur both in the field and during storage. In hay production, the forage crop is dried so that it is essentially biologically inactive, with respect to both plant enzyme activity and microbial spoilage (Muck and Shinners, 2001). Field and storage loss is dependent on weather conditions during the drying period and on the moisture content of the forage at time of storage (Thorlacius and

[1]Agriculture and Agri-Food Canada Research Centre, Lacombe, Alberta, Canada.

Robertson, 1984). Much effort has been expended in the development of hay harvesting systems to reduce these losses. The purpose of the review is to highlight and summarize research on haying systems technology and provide a view of future research and development needs of the industry around the world.

HARVESTING SYSTEMS

There are many options avalable to farmers around the world for harvesting and storing forage crops. In some parts of the world, hay is cut and harvested by hand, while in other areas mechanical harvesting and storage occurs. The type of forage harvesting system is dependent upon economic viability of the farm, the climate of the region and the end use of the forage crop. Small, square bales are preferred in some areas for ease of handling. If the forage is to be used in large quantities or transported any distance, big round and square bales are the most efficient. Weather is still the major influence on determining the type of forage harvesting system. In areas where it is difficult to have sufficient drying time for hay harvesting, silage is the preferred method for forage conservation.

There has been a lot of development and evaluation research in North America and Europe on hay harvesting systems. However, with all this advancement, research shows that losses in quantity and quality still occur between harvest and feeding. These losses can be excessive, and are often a result of wet weather or mechanical damage to the forage.

Several authors have reviewed hay harvesting systems throughout the world. Rotz (2001) has presented a review of the various options in forage harvesting equipment in North America. McLennan (1998) has reviewed Australian haying systems, and Augsburger and Methal (1993) have summarized haying systems technology in Uruguay. There are many forage harvesting equipment manufacturers around the world and computer programs have been developed to assist the forage producer in their selection of equipment (Rotz, 2001). The Prairie Agriculture Machinery Institute (PAMI, 2001) in Canada has tested the efficiency and performance of most North American and European hay harvesting equipment. This institute is one of the major centres in North America that tests farm equipment.

Drying the forage crop

Successful storage of hay requires field drying to a moisture content of approximately 20 percent (Robertson, 1983). The rate of drying is dependent on weather conditions, temperature, humidity, soil moisture, solar radiation and wind speed. The drying behaviour of the forage is also affected by plant species, stage of growth, leaf-to-stem ratio and the structure and volume of the swath or windrow. The rate of moisture loss follows an exponential pattern, such that a drop in moisture content from 30 percent to 20 percent may take as much time as dropping from 80 percent at cutting to 30 percent (Wilkinson, 1981).

Plant respiration continues after cutting until moisture levels of 30 to 40 percent are reached, and can result in dry matter loss of non-structured carbohydrates, ranging from 2 to 16 percent. The loss of nitrogen due to respiration is approximately 2.5 percent of the initial level. Proteolysis proceeds rapidly from the time of cutting until terminated by either the attainment of a high dry matter content or a low pH value (Robertson, 1983). Research in North America and Australia has estimated that leaf loss can be as high as 62 percent and dry matter loss as high as 37 percent in rain-damaged field-cured alfalfa hay. Other research has estimated losses ranging from 4 percent per day to as high as 80 percent per day for harvesting hay under poor drying conditions (Thorlacius and Robertson, 1984). Miller *et al.* (1967) found that storage losses caused by spoilage and heating increased at moisture levels above 20 to 25 percent. When animals were fed this hay, animal gain was reduced but dry matter intake was not affected. As outlined above, the issue of nutrient and dry matter loss has still not been completely solved and still needs research attention in the future.

Cutting methods

The major mowing options around the world are the cutterbar or sicklebar mower, and the rotary disk mower. The cutterbar mower uses reciprocating knives, while the rotary disk mower has knives rotating at high speed. The cutter bar mower has been used worldwide for many years and is a very reliable, low cost method of mowing. It can be pulled by animal power or a low-horsepower tractor. The major disadvantage is limited mowing capacity, as field speed is limited by the cutting capacity of the reciprocating knives. These knives are more likely to clog in heavy, wet or lodged forage. This limited capacity has led to the development and widespread use of rotary mowers (Rotz, 2001; Savoie, 1990) (Plate 6.1). With this equipment, forage is cut with knives attached to disks on a horizontal plane (disk, drum) or on a vertical plane (flail), rotating at a relatively high speed. This type of mowing offers virtually unlimited cutting capacity, as field speed is often limited only by the operator's ability to control the machine. The purchase cost is about 30 percent more per unit of cutting width, and maintenance and repair costs may be

Plate 6.1 Hay harvesting using a rotary disk mower.

greater, particularly as the machine ages. Disk mowers require about four times as much power to operate than do cutter bar mowers (Rotz and Muhtar, 1992). A larger tractor is therefore required and more fuel is consumed per hour of use, but with faster field speeds; thus less labour and tractor time is required (Rotz, 2001). Rotary mowers exert a certain amount of suction because of their high rotational speed. This can bring small amounts of dust or soil into the forage. The flail mower coarsely chops the forage stems in several pieces whereas the disk and drum mowers cut the stems at a single point at the base of the plant. Flail chopping increases not only the field drying rate but also increases field losses, as the short pieces are lost in the subsequent handling operations (Savoie, 1990).

Time of day of cutting

In the western United States, Fisher, Mayland and Burns (2002) have shown that shifting hay mowing from early in the day to late in the day was effective in increasing forage preference by sheep, goats and cattle. Plants accumulate sugars during the day and this results in a diurnal cycling of forage sugars and overall quality. By cutting forages during the late afternoon, extra sugar is captured, thus increasing the feeding value of the forage. However, hay producers must evaluate the need for extra drying time from late morning to late afternoon against the need for the higher quality forage.

Methods of mowing and drying

Most mowing machines used in forage production today include mechanical conditioning to help speed field drying of the crop. Conditioning modifies the plant structure to improve drying rate without causing leaf loss. This process eliminates the need for raking or turning the hay windrow for drying. Mower conditioners have deflecting baffles to adjust the swath width, as it is advantageous to leave as wide a swath as possible in order to increase drying rates (Savoie, 1990). A wide variety of conditioning devices is available, but the most common options are crimpers, crushing rollers, plastic brushes or flails. Some designs are more aggressive, which speeds drying but damages the crop, thus increasing field losses by disassociating leaves and other small forage particles high in nutritive content. Losses using mower-conditioners can vary between 1 and 5 percent of dry matter (DM) yield, with losses being similar for the two major mower types (Rotz, 2001). Roller design does not have much effect on loss. Adjustment of roll pressure and clearance has more effect than the pattern or material used in the rolls. The difference in loss between a well-adjusted mower-conditioner and a similar mower without a conditioner is 1 to 2 percent of crop yield. Intermeshing rubber or plastic rollers work best for conditioning alfalfa, but little difference has been found in drying rates when the common roll devices are used with proper adjustment (Shinners, Koegel and Straub, 1991; Rotz and Sprott, 1984).

Flail mower conditioners use rotating steel or plastic tines to abrade and break forage stems. These machines were developed in Europe for use in grass forage crops. When used on alfalfa and other legumes, this type of mower conditioner can lead to inferior drying and to a greater leaf loss, depending upon machine adjustment. In general, roll conditioning devices are recommended for alfalfa and legumes, and flail mower conditioners are recommended for grass-based forage (Rotz, 2001).

Robertson (1983) found that the mower conditioner produced the fastest drying time, but this faster drying treatment usually re-wetted more readily, so that dew or rain eliminated any advantage. However, the advantage of the mower conditioner could mean the difference between being able to bale in the evening versus the risk of exposing the hay to rain the following day. When this hay was fed to lambs, the cutting methods that decreased raking and conditioning of the hay increased dry matter digestibility and intake of digestible dry matter.

Conventional hay harvesting requires about 2 to 4 days under good climatic conditions. However, little can be done to protect dry or partly cured hay from rain damage. Once the plant cell has died, rainwater can dissolve its contents and leach them from the hay, the most digestible contents going first. In the case of light rain, the loss of nutrients may be minor, but heavy rain leaches most of the nutrients and also compacts the windrows.

Swath drying

The top of the forage swath dries more rapidly than the bottom. Mechanically turning the swath can speed the drying process by moving the wetter material to the upper surface, where it dries more quickly as more of the crop is exposed to the radiant solar energy and drying air. There are three methods used to move hay swaths: tedding, swath inversion, and raking.

Hay tedders use long rotating fingers to stir, spread and fluff the hay. The tedder increases the area over which solar radiation reaches the swath. Tedding can improve the drying rate by 28 to 58 percent on the day it is applied. However, there can be mechanical loss of the leaves due to the high speed of fingers fluffing the crop. Savoie (1990) found field losses of 4 to 8 percent when he tedded alfalfa (*Medicago sativa*), but losses of less than 2 percent with timothy (*Phleum pratense*). Tedding twice reduced the drying time by a full day. If the producer only uses the tedder once, then Savoie recommends doing it on the first day right after mowing. Other studies have found that alfalfa is not suited for handling by the tedder as the losses can range between 11 and 15 percent. Rotz and Savoie (1991) found that the loss caused by tedding in alfalfa was greater than the average rain loss. The increased machinery, fuel and labour costs may only justify routine use of tedding on grass crops in wet climates (Rotz, 2001).

Windrow inverting machines gently lift the swath and turn it over so the original bottom, wet layer is moved to the top. The windrow inverter is not as effective in reducing drying time as the tedder, but losses in alfalfa or grass hay were much lower after inversion than with the tedder (Savoie, 1990). In alfalfa haying systems, the inverter would be a more suitable machine to manipulate alfalfa swaths, but there is not much advantage for grass. Rotz and Savoie (1991) recommend minimal use of the tedder or the inverter in alfalfa as the added costs of the operation would probably exceed the benefit received.

Raking is used to turn or roll the windrows together to speed drying and for easier pick-up by the baler. Raking hay in the morning of the day of baling can reduce the field curing time by a few hours, thus allowing an earlier start at baling. However, raking at moisture levels of less than 40 percent can result in leaf losses of 10 to 25 percent (Friesen, 1980). In a light crop, raking losses will be considerably higher because of inadequate forage volume.

Rotary windrowers can cause greater losses compared to the more standard side-delivery rake. The rotary windrower uses rotating tines to sweep hay into a windrow, while side delivery rakes provide a rolling and wrapping action of plants in the swath (Rotz, 2001).

Recently, super conditioning or maceration equipment has been developed to aid in this drying process (Krutz, Holt and Miller, 1979; Savoie, 1990). The freshly mown hay is crushed through a series of serrated rollers and compressed into a thin mat. The stems are split longitudinally, allowing the forages to dry very quickly since the intracellular water is freed after the grinding process. Evaporation is faster and more efficient than after conventional roller conditioning. Studies have shown that intensive maceration of forage can improve both the intake and efficiency of forage fed to dairy cows in early lactation (Savoie, 1990). This might be caused by the macerated forage having a greater proportion of ruptured surface area compared to the conventional hay and the rumen microbes thus having faster access to the cell nutrients (Savoie, 1996). Too intense a maceration or conditioning will result in a large proportion of fine particles broken off during the conditioning and these particles would then be lost in the baling process. The maceration conditioning machines are now available in Canada, but acceptance by the manufacturing and hay industry has been very slow (PAMI, 2001).

A large proportion of the world's forage production is moving to arid climates in the western United States and in Australia (Shinners, 1997). Here, hay producers have fewer weather-related problems because of the low annual rainfall and relative humidity. Much of the forage production in these areas requires the use of irrigation. With the high daily temperatures and low relative humidities, short field curing times would be expected. However, typical field drying times are four to six days (Orloff, 1990). The long field drying periods occur because the hay producers use thicker windrows in order to reduce bleaching caused by exposure to the sun, and this increases drying time (Shinners, Koegel and Straub, 1991; Orloff, 1990). This bleaching removes green forage pigments, which downgrades its visual quality. There is no evidence that bleached forage has significantly different nutritional quality compared with "green" hay, but customers still want green hay, particularly in the dairy, equine or export markets. Further research and marketing strategies will be required in the future to alleviate this concern.

Dry matter content

Farmers harvesting hay need an accurate and rapid method of measuring dry matter content, an accurate weather forecast and a means of evaluating various harvesting options on a given day. Moisture can be measured with electronic moisture meters, which are based on conductance or capacitance of the crop. They are less accurate than the traditional laboratory moisture testers, which use heat or the domestic microwave oven method (Muck and Shinners, 2001; Pitt, 1993; PAMI, 2001). In some areas, weather forecasts are able to provide up-to-date information for the producer at hay harvesting time. In the future, such forecasts will need to cover wider geographical areas in order to provide more hay producers with information on expected drying time for hay harvesting. In addition, a rapid, automatic method of determining moisture content of the swathed hay will be required at the baler so that the operator can determine if the hay can be baled as-is or needs treatment with a chemical or biological hay preservative.

Hay balers

Small rectangular bales have been most popular over the past fifty years, but large round and rectangular bales are becoming the predominant hay package form. Hay harvesting equipment has also been developed to produce low-density stacks, but these constitute a relatively small portion of the hay produced.

The traditional plunger-type small baler makes a rectangular bale about 36 by 46 cm in cross-section, with lengths up to 130 cm (Plate 6.2). The bale weighs from 15 to 35 kg, but traditionally required manual labour for stacking and feeding. As a means of reducing labour, a mechanical bale thrower can be mounted at the exit of the baler. This throws the bales into a trailing hay

Plate 6.2 Hay baler – small square bales.

wagon. At the storage site, bales then have to be manually stacked. Other options are automatic, self-propelled bale wagons that mechanically lift bales dropped in the field and stack them on the bale wagon. These stacks can then be moved and placed in the storage area.

Long hay mechanical stacking wagons have been used as another labour saving haying system (Coates *et al.*, 1978). Models with a capacity ranging from 1 to 10 tonnes have been available. The non-compaction model partially shredded or chopped the forage and blew it into the stacking container. Packing occurred from the force of the material being blown into the container and by the vibration of the wagon as it travelled over the fields. The compaction wagons used a mechanical force to lower the roof to compress the entire stack.

The quality of the hay following storage was consistently higher for the compaction stacks compared with the non-compaction stacks, but both these types of hay harvesting equipment have now been replaced by the more efficient round and large square baling systems.

Large round balers were introduced in the early 1970s. These balers produce bales 90 to 180 cm in diameter and 120 to 160 cm in length, with a mass of 200 to 900 kg (Plates 6.3 and 6.4). The European round baler produces a narrower bale compared with the North American bales. Tractor-mounted loaders or automatic bale loading wagons can transport bales to the storage site. As power requirement is dependent on the baler design, large round balers require about a 35 to 55 kW tractor, which is higher than small square balers (Rotz and Muhtar, 1992; PAMI, 2001).

There are two types of large round balers (Beacom, 1991). The hard-core baler has a spring-loaded adjustable bale chamber that rolls the hay under continuous pressure from start to finish. The soft-core baler has a fixed bale chamber and the hay does not roll until it fills the chamber

Plate 6.3 Hay baler – large round bales.

and exerts pressure on the rollers. The incoming hay is continually wrapped around the hay mass, which folds the centre into a "star" shape. The centre is less dense than the perimeter, thus providing better aeration after baling. The soft-core baler requires twice as much fuel and a much larger tractor than the hard-core baler. High moisture, soft-core bales harvested at 23 percent moisture and stored outside were superior to hard-core bales at the same moisture level in terms of field loss and liveweight gain per hectare (Beacom, 1991). When baled at this higher moisture level, losses were significantly greater with the hard-core baler than with the soft-core baler. However there was no significant difference in losses attributable to the type of baler when the hay was baled at lower moisture level.

Dry matter losses during hay baling are considerably higher with the large round baler compared with the traditional small square baler. These losses can be as high as 11 percent for the pick-up losses and 17 percent for the bale chamber losses (Robertson, 1983). Pickup losses are high when the machine is pulled at a faster speed than the rotating speed of the pickup device. Chamber loss of high quality leaf material is highest in dried material, and thus reduces the nutritive content of the remaining forage. These chamber losses can be minimized by conditioning the forage crop, maintaining the fastest ground speed possible while bailing and by increasing the width of cut and reducing the baler speed in light crops. Excessive loss occurs at low feeding rates of hay into the baler, and by rolling the bale too many times in the chamber (Robertson, 1983; Rotz, 2001).

Round balers use either twine or plastic net wrapping to tie the bales. The baler must stop forward motion as the bale is rotated with the twine. Depending upon the number of wraps of twine, this can take one to two minutes. The plastic net wraps only require one to two rotations,

Plate 6.4 Round bales.

and its use greatly reduces the time required, to about 30 seconds per bale (Beacom, 1991). The net wrap is much more expensive than twine, but this is balanced out when considering the hourly cost of running the baler and the reduced time required to wrap or tie the bale. The plastic net wrap does not shed the rain, but some research has shown improved forage quality compared to twine-wrapped bales. In the future, engineering design research will be required to speed up the wrapping process to make round baling more efficient.

The large, rectangular, high-density bales have become more popular, particularly for hay needing to be transported long distances (Muck and Shinners, 2001). These bales have a height and width of 60 to 130 cm, a length of 120 to 250 cm and a mass of 200 to 1 000 kg (Plate 6.5).

Special heavy-duty equipment is needed for lifting, transporting and feeding these large bales. The balers are capable of harvesting twice the amount of hay per hour compared with the small square balers, but require more power. The big market for this type of baler is in the large hay- or alfalfa-growing areas of the world where the forage is transported to distant markets. In the future, hay equipment manufacturers anticipate that more of the major hay producing areas around the world will convert to the more efficient large square bales.

Hay storage losses

Safe storage of hay requires that the moisture content of the harvested forage be below 20 percent. Microbial activity, heating and moulding will occur in hay with a moisture content of over 20 percent. Hay will heat to a higher temperature when the moisture range is between 30 to 40

Plate 6.5 Large rectangular bales.

percent, and if oxygen is present, the wet hay can reach a temperature where spontaneous combustion can occur and cause major hay fires. In many parts of the world it is extremely difficult to dry hay in the field to below 20 percent due to possible rainfall or high humidity. In these cases, silage, use of hay additives or auxiliary hay drying systems become an option.

There has been a lot of research effort devoted to designing systems that would dry forage with ambient air or heated air, either in the field or at a processing plant. Alfalfa dehydration plants around the world use this process, but the energy costs are very high. If high quality small square bales are required, excess moisture can be removed in a hay-drying barn by placing the bales on a perforated floor and blowing ambient or heated air through the bales. The problem with this system is the low capacity, high labour requirement and possibly high local cost of electricity (Elsässer, 1990; Jan *et al.*, 1980).

The Schwarting hay-drying tower has been available to farmers in Germany for many years. The hay tower system incorporated a huge vertical hay stack that was hollow and cylindrical. Drying was accomplished by blowing unheated air into the hollow centre and through the stack. A movable cover prevented the air from exiting the cylindrical hole at the top of the stack. The hay tower was filled with chopped hay using a blower and levelled with a rotating arm. Unloading was accomplished by reversing the direction of the leveller and dropping the hay down the centre hole to a conveyor below. Jan *et al.* (1980) evaluated the Schwarting hay tower in western Canada, using smooth brome (*Bromus inermis*) + alfalfa hay, field dried to 40 percent. The chopped hay was dried and later fed to steer calves in a feeding trial and compared with smooth brome + alfalfa silage, small square bales and large round bales. The daily gains, feed efficiency, and gain per hectare were highest for both the silage and the artificially dried hay from the tower, followed by the hay baled as small square bales and the large round bales. The high capital cost and limited capacity of the hay drying tower system precluded further acceptance of this system in North America (Jan *et al.*, 1980).

Bacterial and fungal growth

Various researchers around the world have evaluated methods of preventing microbial growth in stored hay. Storage losses are directly proportional to microbial growth and heating, which depends on the moisture content of the hay, the size and density of the bale, the drying rate of the hay and the microbial population in the hay. If baling is done at 20 to 30 percent moisture as a means of preventing leaf loss, microbial activity can still exist in the forage after baling. Hay does not become stable until the bales reach a moisture content of about 12 percent and the humidity is below 65 percent. In these conditions, most fungi will not grow (Mahanna, 1997).

Fungal growth can occur on the standing forage plant, the wilted crop or during storage. The type of fungal growth is strongly influenced by the microenvironment in the field, swath and bale. Wittenberg (1997) has provided an excellent overview of the role of microbial activity in changing the nutritive values in hay production systems. Wittenberg (1997) observed that the bulk of moisture loss from hay stored at 24 to 35 percent moisture occurred between days 4 to 14 of storage, a period when stack temperatures were the highest. It is in the initial stages of storage that the temperature rise associated with aerobic respiration initiates micro-organism growth. Forage acid detergent fibre (ADF) and neutral detergent fibre (NDF) concentrations increased as a result of the enzymatic, microbial and chemical processes occurring during hay harvest and storage.

Forage analysis for glucosamine, a constituent of chitin (found in cell walls of spores and mycelium of fungi), has been used as a means to estimate the total fungal biomass in forage (Wittenberg *et al.*, 1989; Roberts *et al.*, 1987). Fungal biomass concentrations measured in stored forage represent the total fungal biomass accumulated from the time the plant is growing, to the time spent in storage. Glucosamine values can be as low as 0.9 g kg^{-1} DM under ideal growing, harvest and storage conditions, with levels as high as 10 g kg^{-1} DM for severely moulded hay (Wittenberg, 1994). Generally, the highest accumulation of fungal biomass is associated with fungal growth on cut forage lying in the field. However, it should be assumed that the biomass contributes to the N and carbohydrate fractions of hays. Availability of these fungal fractions as protein and energy sources for ruminants consuming them is not known.

Many environmental factors are involved in the extent of fungal growth and species succession in hay following cutting, including water and nutrient availability, temperature, pH, gaseous composition and interactions with other micro-organisms. Variation in the amount of moisture that micro-organisms can use for growth becomes critical. This is dependent on the moisture content of the forage in the windrows and in storage.

The availability of substrate and its effects on fungal growth are not as clearly defined. Although leaves have a higher nutrient density than stems, alfalfa and grass stems had a greater spoilage potential than leaves, with fungal growth initiating at breaks or internodes along the stem (Wittenberg, 1997).

Fungal growth can occur at a wide range of temperatures. Most fungi found in stored plants thrive in the range from 10° to 40°C, with 25° to 35°C being optimal. Hay produced under conditions where temperatures do not exceed this optimal range will have a diverse range of fungal species. If the temperatures are in excess of 40°C, non-thermotolerant species will disappear. Stack temperatures can readily reach 40°C when forage moisture content at stacking is in the range of 18° to 25 percent. High ambient temperatures and humidity, and restricted air movement due to bale density or stacking design will contribute to an increase in peak temperature.

The pH remains near neutral in hay that has undergone moulding (Wittenberg, 1997), unless storage moisture content exceeds 40 percent, when the pH value may rise to 7.0 or more. Most fungi grow over a wide pH range, but will compete poorly with bacteria at pH 7.0 or above, when there is moisture available.

Wittenberg (1997) found little research to determine oxygen levels in stored hay. It is likely that poor air exchange could result in lowered O_2 and elevated CO_2 levels when plant or microbial respiration occurs. Although the fungi causing spoilage in hay are considered obligate aerobes, many are capable of growth at low concentrations of O_2.

The process of cutting and conditioning the plants results in the introduction of soil- and airborne microbes to the plant material. Breton and Zwaenepoel (1991) and other European studies with grasses have identified soil-borne species present on forage during wilting and baling. Wilting trials have indicated that precipitation events can result in 10 to 100-fold increases in the forage bacterial populations, with little change in the fungal counts, although fungal biomass yield increased (Wittenberg, 1995).

Extent of fungal growth is often difficult to assess visually. A heavy dew or precipitation can double the fungal biomass on the forage. Fungal species that become dominant during storage of hay that has undergone heating thrive at low available moisture and higher temperatures. This determines the dominant species (Wittenberg, 1997).

Wittenberg (1997) and others have tried to quantify the adverse effects of fungal biomass in hay by studying its effects on hay palatability, intake and digestion. Presence of fungal biomass in hays having similar nutrient profiles reduced hay palatability (Undi and Wittenberg, 1996) for young heifers, but did not adversely affect intake in growing steers (Bossuyt, Wittenberg and Crow, 1996) when fed as the sole dietary energy source. The fibre fraction of hay with elevated levels of fungal biomass also appeared to be more digestible compared with less mouldy hay of a similar nutrient profile. This could be due to the action of mycelia penetrating the plant cuticle and causing splits or fractures in the xylem and lignified vascular bundles.

Many of the field and storage fungi identified are potential mycotoxin producers. Wittenberg (1997) found few field trials in the literature that included an assessment for mycotoxins in forage dried and stored as hay. Some fungi demonstrate very low or no toxin production at low available moisture levels, even though active growth was still occurring (Corry, 1977).

Hay preservatives

Chemical and biological preservatives offer the hay producer several options in humid climates. Preservatives allow forage to be baled at a higher moisture content, thus reducing field drying time and the chances of rain damage. Baling hay at 25 percent moisture reduced field curing time by about a day (Rotz and Muck, 1994). If the hay could be baled and stored at a moisture level higher than 20 percent, mechanical separation losses during baling would be reduced (Shinners *et al.* 1996; Mir *et al.*, 1995; Friesen, 1978; Thorlacius and Robertson, 1984). There has been considerable research around the world directed towards the development of hay preservatives designed to inhibit microbial activity during storage. Wittenberg (1997) has also reviewed these alternatives.

Alkaline solutions can be used as hay preservatives. By adding monovalent alkaline metal ions to a carbonate solution, it causes a dissociation of the carboxyl groups in the cutin matrix of the plant cuticle. As a result, the plant is unable to prevent water movement from its interior to exterior. Solutions made with potassium carbonate (K_2CO_3) have received the greatest amount of attention in field trials. Alkaline solutions (0.18 M or higher) are effective for legumes, but not for grasses. Most commonly, K_2CO_3 solution (0.18 to 0.20 M) is applied at 2 to 3 percent of forage weight at the mower cutters, with the majority of the solution directed toward plant stems.

Combining mechanical conditioners and K_2CO_3 has an additive effect on drying rates. Crump (1985) found that a combination of mechanical and chemical conditioning was usually more effective than mechanical conditioning alone.

Potassium carbonate has been combined with sodium carbonate to reduce costs without sacrificing effectiveness; current thinking is that the ratio should not be less than 1:1 (Wittenberg, 1997). Potassium carbonate has been combined with methyl esters and emulsifying agents in an effort to further improve forage drying rates, but response to these additions is varied. Other forms in which the alkaline metals have been tested include potassium hydroxide or sodium silicate. However, chemical drying agents or desiccants that cause disruption of the plant cuticle can cause increased plant susceptibility to fungal invasion during storage (Kraynyk and Wittenberg, 1989) and microbial attack in the rumen (Hong, Broderick and Walgenbach, 1983).

Organic phosphates have demonstrated an ability to increase forage drying rates and reduce cell respiration rates. However, they have not been tested under field conditions and could present toxicity problems for livestock. Compounds in this category include tri-n-butyl phosphate and a diphosphate ester.

Two other types of preservatives commonly used in hay production are fungicides, which kill fungi, and fungistats, which inhibit fungal growth. Categories of hay preservatives, described on the basis of mode of action, include direct acidifiers, specific antimicrobial agents, bacterial inoculants and nutrients. Most preservatives are applied at the time of baling as either a granular or liquid product, although some products, such as anhydrous ammonia, may be applied shortly after stacking. Desirable criteria for hay preservatives include the ability to prevent fungal invasion in hay baled and stored above defined safe moisture levels. The preservative needs to inhibit the production of the fungal end products, namely spores and mycotoxins that have direct adverse effects on hay handlers and livestock. The products need to be cost effective, easy and safe to handle and have no negative impact on the animals (Wittenberg, 1997).

When adequate levels are applied, organic acids are effective in preventing fungal invasion of moist hay during storage. Propionic, acetic, isobutyric and formic acids, at levels of 0.5 to 2.5 percent of wet forage weight, significantly reduced dry matter losses and on average increased crude protein. Most commercial acid-based preservatives contain propionic acid or a propionic-acetic acid mixture. These hay preservatives have been shown to reduce mould growth, and reduce heating in the short term but not in the longer term (Rotz, 2001). High-moisture hay treated with an organic acid preservative maintained higher moisture content during storage, which supported microbial growth at a lower rate. These additives significantly reduced dry matter losses and on average increased crude protein available at feeding by more than 100 kg ha^{-1} (Robertson, 1983). Economic benefit from applying organic acid preservatives at the typical rates is only justified when it is used to avoid damage from impending rain (Rotz, 2001).

The corrosive nature and the high vaporization losses of the organic acid products cause application problems. Neutralized propionates, including ammonium, calcium or sodium salts of propionic acid or buffer, make the product less corrosive and less volatile. Laboratory results indicate that neutralized acids are less effective in preventing fungal invasion of moist hay, as they do not volatilize during field application. The recommended application rates are similar to that of the unneutralized acid (Wittenberg, 1997).

Anhydrous ammonia is an effective hay preservative (Koegel *et al.*, 1992; Mir *et al.*, 1991; Robertson, 1983; Thorlacius and Robertson, 1984). Ammonia preservatives are fungistats and include anhydrous ammonia, and urea. Anhydrous ammonia is the most commonly used and requires the hay to be covered in plastic at the time of and after application. Anhydrous ammonia is applied at a rate of 2 percent of forage weight for hay at 25 to 30 percent moisture (Thorlacius and Robertson, 1984). Ammonia will inhibit most yeast and mould growth in baled hay; however, some bacteria and heat-resistant fungi can proliferate in this preservation system. The chemical reaction associated with ammoniation can generate enough heat to cause an undesirable browning of the forage.

Mir *et al.* (1991) found that ammoniation raised the crude protein content of large round baled brome–alfalfa and alfalfa hay harvested at less than 20 percent and at 30 percent moisture, compared with non-ammoniated field-cured hay at less than 20 percent moisture. The ammoniated hay was free of mould even after 14 weeks of storage. No significant improvement in animal performance was found due to the hay type, moisture content or ammoniation. It was concluded that ammoniation was effective in preserving the quality of the high-moisture hay, which would have otherwise spoiled.

Baron and Mathison (1990) found that baling large round bales above 25 percent moisture with or with out a preservative in excellent or poor weather conditions resulted in less nutrients for feeding than did baling between 18 and 21 percent moisture, due to the loss of DM and nutrients in storage. Baling at less than 18 percent under rain-free conditions or after rainfall resulted in a decrease in yield of nutrients. If rain was imminent, Baron and Mathison (1990) recommended that hay be baled at moisture levels slightly above those considered safe for storage. They found that neutralized propionate applied at a rate of 1.25 to 1.50 percent (wet weight basis) improved the nutrient yield, but was not an economic option unless the hay could not be harvested at all due to wet weather.

Ammonia can be toxic to ruminant animals when it is applied to high quality legume hay at rates greater than 3 percent on a wet weight basis (Rotz, 2001). Ammoniated hay on rare occasions has been associated with the development of a nervous disorder in livestock consuming high quantities of the treated forage, especially alfalfa hay. The factors contributing to this condition, commonly referred to as crazy cow syndrome or bovine bonkers, are not known (Mir *et al.*, 1991). In contrast, urea can be applied as a granular product at the time of baling or stacking, at a rate of 40 kg $tonne^{-1}$, in order to reduce mould and browning in hay (Mahanna, 1997). Caution should be exercised in feeding the 4 percent urea-treated hay by diluting it to 50 percent of the total forage intake by mixing it with untreated hay.

Bacterial inoculants

Most microbial hay inoculants marketed today were initially developed to aid in the fermentation of silage. The effectiveness of these bacteria in baled hay is questionable, as they are facultative anaerobes that prefer anaerobic and moist conditions. Mir *et al.* (1995) found no advantage in using *Lactobacillus plantarum* on alfalfa round bale forage at 18 percent moisture content. Baron and Greer (1988) found that live bacterial culture, *Bacillus subtilis* or 12 percent lactic acid extract, were not effective at any moisture level for use in hay harvesting and storage. Selected strains of

Bacillus pumulus have been shown to be effective on alfalfa hay baled in small square bales of up to 25 percent moisture, large round bales at 20 percent moisture, and large square bales up to 18 percent moisture (Mahanna, 1997). The current use of bacterial inoculants is relatively small compared with their use in silage production systems, and little development research is being done around the world. Thus, additional research will be required in the future if bacterial preservatives are to become more effective.

Other types of hay additives include sugars, protein, other nutrient supplements and crude enzyme preparations. Antioxidants such as butylated hydroxy anisole, ascorbic acid, propyl gallate and butylated hydroxy toluene are sometimes included in hay additives to maintain the green coloration of hay and to enhance the effect of organic acids (Wittenberg, 1997). Common salt (NaCl) has been used on wet hay, and in sufficient concentrations it will absorb free water on the surface of the hay, thus, inhibiting microbial growth. There are, however, few controlled research studies to evaluate rates, concentrations and palatability problems of adding NaCl to hay (Lacefield, 1987).

Storage systems

Ideally, all forms of hay should be stored under cover to prevent rain damage. This is especially true for loose hay and small square bales. Martin (1980) found hay stored inside at normal moisture levels below 20 percent could still have a 5 to 10 percent dry matter loss. Other studies have found that hay stored inside can lose as much as 1 percent of dry matter for each percentage unit of moisture. Hay baled at 20 percent moisture will probably lose 5 to 8 percent of its dry weight by the time it dries to 12 percent moisture (Mahanna, 1997).

Losses in round bales stored outside on the ground are about three times greater than bales stored inside. These losses can be even greater, especially in high rainfall areas of the world. Mahanna (1997) reviewed studies that found 40 percent storage and feeding losses for large round alfalfa bales stored outside with no cover. The potential storage losses will justify storing round-bale hay under cover in high rainfall areas. In areas of low rainfall, round bales should be stored end to end on well-drained ground (Beacom, 1991). Much of the weather damage derives from bales being in contact with the moist ground. This loss can be greatly reduced by storing the bales on crushed rock, raised tyres or wooden rails, which provided a small air space underneath the bales. The bales should be placed end to end in long rows with an air space of at least 3 to 4 cm around each bale to permit good air movement. If the bales touch each, other spoilage caused by rain will occur at those points. If space is limited, bales can be stacked in threes: two bales vertically on top of one another, and the top bale horizontal – the top bale sheds the rain off the two bottom bales. For short-term storage, round bales can be stacked in a pyramid in rows of 4-3-2-1, but rain can still penetrate this configuration.

OTHER HAY HARVESTING SYSTEMS

Dehydrated alfalfa pellets

Canada, France, Spain and Italy are the main exporters of dehydrated alfalfa pellets. Commonly referred to as "dehy", these dehydrated alfalfa pellets are processed from alfalfa cut in the late bud to early flowering stage and contain at least 18 percent crude protein. The alfalfa forage is partly field dried to minimize the effects of weathering yet ensure high quality. The chopped material is then dehydrated using natural gas-fuelled dryers at an air temperature of 110° to 120°C. The rapid heating and drying process preserves the nutrients in the alfalfa and reduces the solubility of the protein. This heating process changes approximately 60 percent of the protein into the bypass form, which leads to more efficient use and digestion of the protein in ruminants.

The chopped, dried alfalfa is then ground into a meal prior to pelleting. The texture of the meal determines the size of the pellet produced. An antioxidant is added to preserve the beta-carotene and other vitamins, and the material is made into a 6–9 mm pellet.

Sun-cured alfalfa pellets are available in 6 mm and 7 mm sizes and are produced from alfalfa that has been naturally dried in the field at the late bud to early bloom stage. After field drying, it is moved to the plant for final drying at a lower temperature than dehy, and pelleted. Sun-cured alfalfa contains about 17 percent protein and has less bypass protein.

The United States of America and Canada are the two main exporters of alfalfa-hay cubes. Regular cubes are 33 mm on each side and 50 to 100 mm long. The forage is coarsely ground for cubes and, because of longer fibre length and a larger particle size, it can be used as the only forage source for dairy cows.

Cubes are easier to handle, transport and blend in a ration compared with traditional hay. Feeding cubes increases forage intake and there is less feed wastage and spoilage. The product is also uniform in quality and has a high nutrient content.

Densified hay

With the increasing demand for fibre for livestock feeding in Japan, Republic of Korea, Taiwan (Province of China), United Kingdom, some Near Eastern and Asian countries, a large densified-hay market has developed. Australia, Canada and the United States of America are the main exporters of the product. Despite the shipping advantages of alfalfa pellets and cubes to markets in Japan, Taiwan (Province of China) and Republic of Korea, densified or double compressed long hay is now the product of choice because long hay is a better source of effective fibre for dairy cattle.

The bales are double compressed to increase density and reduce volume for container shipping. For top grade or price for the export market, densified hay must have no rain damage, must be bright green and free of weeds. The hay must also be free of *Agropyron* grasses (i.e. wheat grasses) in order to prevent the spread of the Hessian fly (*Mayetiola destructor* (Say)) or its eggs. The Hessian fly causes great damage to the rice crop and currently Japan, Republic of Korea and Taiwan (Province of China) are free of the fly. Since the Hessian fly also lives in *Agropyron* grass species, these grasses, together with any damp hay, must be hand separated and discarded from the hay before it is re-baled as densified or double compressed bales. These densified or double compressed bales are compressed to 34.5 MPa and are about 55 × 65 × 100 cm in size and weigh 40 to 60 kg. Timothy is the main forage exported as double compressed or densified hay from Canada, while Australia specializes in oat hay. The United States of America exports various fibre-based hay and straw products to these countries in the densified form. With this process, densified hay processors are able to put about twice the amount of compressed hay into a shipping container compared with conventional baling methods.

UTILIZING HARVESTED FORAGE CROPS

Considerable research has been done on the feeding value of harvested forage crops. The stage of maturity of the forage is the most important factor influencing feeding value. Quality decreases as the plant matures, while dry matter yield increases to the early bloom stage in grasses and legumes. The relative importance of quality and yield will depend on the kind of livestock operation involved and the availability of other feeds. The application of nitrogen fertilizer will also increase the protein content and yield per hectare of the forage crops (McCartney, Bittman and Nuttall, in press).

Hay is recognized to be nutritionally superior to silage with respect to protein quality for the high producing ruminant animal (Petit and Tremblay, 1992). Forage protein value is assessed as the amount of amino acid available for absorption in the animal's small intestine. These amino acids may be derived directly from the dietary protein being offered or may be derived from microbial protein synthesized in the rumen. Forage harvest and storage systems that increase the proportion of forage protein that is undegradable in the rumen without increasing the unavailable protein will result in greater concentrations of dietary protein available for absorption in the small intestine (Wittenberg, 1997).

Von Keyserlingk *et al.* (1996) showed that the soluble nitrogen (N) concentrations of alfalfa and grass hays range from 25 to 63 percent of total N. The fractional rate of crude protein degradation for the degradable N in the hays can range from 3.6 to 17.1 percent per hour. Thus, hay can be a source of escaped protein in ruminant animals. Hay harvested and stored at moisture levels that result in heating will also affect the protein quality at feeding.

The maximum amount of forage voluntarily consumed by an animal will depend on the nutritional quality of the forage, the physical form of the forage or its bulkiness, the palatability and freedom from spoilage or weeds.

Grinding hay through progressively smaller screens in a hammer mill or farm-type tub grinder mixer will increase the density of the product and results in higher intake by the ruminant (Beacom, 1991). Beacom (1991) has shown that, on average, grinding through a 13-mm screen increased the density of hay by 23 percent compared with processing through a 25-mm screen. Intake and rate of gain were increased by reducing the bulk by grinding. The grinding of poor quality hay had a more beneficial effect than processing high quality hay. Large farming operations have used large-scale tub grinders using a large screen size of 51 mm, as their capacity was greatly reduced with the smaller screens. Economics and labour determines the use of grinding as a hay processing method.

Large round bales are also fed in metal feeders that allow 12 to 18 cows access to the forage. Round bales are also rolled out and fed directly on the ground. These methods of winter-feeding are widely practiced due to low labour and machinery inputs. There has been lots of research on storage losses of round-baled hay, but there has been little work done to evaluate feeding losses according to the feeding method, even though some studies suggest that feeding losses may reach 20 to 30 percent. The cone- and ring-type feeders, where cows were able to keep their heads inside the perimeter of the feeder while eating, have been found to be the most effective (Buskirk, 1998).

Bale shredders are also manufactured for chopping or shredding and distributing round-bale hay for feeding (PAMI, 2001). These machines work well for spreading round-bale hay onto the ground and, if fed in proper amounts, there is very little wastage.

FORAGE HARVESTING SYSTEMS AROUND THE WORLD

Haying systems around the world have been influenced by or adapted from both North American and European technology and research. Farmers and hay growers around the world use small square balers, large round balers and have an increased interest in the large square balers. The adoption of the more expensive large square balers comes from the need to speed up baling time, reduce labour and increase efficiency in transport over long distances.

In the northern and southern temperate climatic zones, haying systems are used, although in many areas silage production is most prevalent, especially where large volumes of forage are harvested. Silage is the preferred method of forage conservation where the weather does not permit easy hay drying. In high-rainfall tropical zones, where year-round grazing prevails, silage is the system of choice.

Methods of harvesting, storing and transporting hay vary with the needs and economics in each region of the world. It ranges from the simple, small, hand-harvested haystack, to large mechanized operations.

The area used for hay production around the world has decreased consistently as the production of silage has become more favourable. However, hay is still the main system for conserving grass on small-sized farms. The larger and more intensive farms rely on silage, with most using conventional chopped, and a few using baled silage. The major problems for hay production everywhere are suitable field drying conditions and scarcity of labour.

There is some increase in interest in hay production in the context of grassland managed for environmental objectives. On some European and North American farms with environmentally sensitive areas, farmers receive payment for participating in schemes where hay is cut at maturity in order to help increase bird and botanical diversity. This helps ground-nesting birds and the return of grass seed to the soil. The adoption of these methods has been high, but hay quality can be poor.

North American and European manufacturers produce a wide variety of hay harvesting equipment. This equipment is available in many countries around the world. The European round balers are usually smaller in width than the North American types. There is a trend for larger dairy farms in Europe to use the larger, more efficient harvesting machinery, but there still exists a need for smaller-scale machinery for farmers on smaller farms. In some areas of Europe, where hard cheese is produced, silage cannot be fed, as there are laws against additives or feeds that could alter the taste and quality of the cheese. As a result, the majority of the stored feeds will be loose or baled hay.

Hay drying is still a concern in some areas, and, in Finland, round bale or batch dryers are becoming more common. Hay is dried in these units for 6 to 10 days using circulating fans and ambient air.

In Eastern Europe and the former Soviet Union, hay production ranges from traditional loose or long hay, requiring considerable manual labour, to large round or square balers. Limited amounts of European hay production equipment are available, but some old equipment can be found. However, villagers with one or two animals will harvest small quantities of hay as they did centuries ago, with scythe or sickle, forks and horse-drawn hay wagons.

Asia

Much of the hay fed to livestock in Japan is imported from Australia, Canada and the United States of America. The hay that is produced in Japan is packaged as small square bales or large round bales, but hay drying conditions remains the limiting factor.

Japan is probably the country where round-bale wrapped silage became more popular than any other country in the world. It is estimated that nearly half of the forage produced in Japan is harvested as round-bale wrapped silage.

In Indonesia, Republic of Korea, Taiwan (Province of China) and Thailand, limited amounts of hay are grown and harvested, and ranges from hand harvesting of roadside hay to mechanical harvesting in larger fields.

The loose haystack system is the most common haying system in the People's Republic of China. The majority of the farms are very small and not well mechanized, and labour is readily available. Forage is harvested manually with a sickle or scythe, and hauled to the farmyard, where it is allowed to dry, and later stacked for storage.

On some of the larger state farms in China, old style tractor-driven reciprocating mowers are used. A large tractor (55 hp [≈40 kW]) can pull 5 old-style mower-rake sets with a capacity of 10-15 ha day^{-1}. Hay is also gathered by buck or dump rakes, or small square-bale systems may be used (on an estimated 10 percent of the farms). Round balers have not been used to any extent due to the lack of larger tractors and funds to purchase the equipment. On a few large state and privately owned farms with access to credit, modern haying equipment is used. This hay is produced in northeast China (Hulun Bieir League or Heilongjiang) and transported to dairy farms in the south – Beijing, Tianjin and Shanghai, for example.

Natural hay fields in Inner Mongolia have decreased sharply in recent years due to overgrazing. It is expected that there will be an increased use of hay harvesting technology and the adoption of more productive forage species in the future as there develops a greater demand for good quality forage to feed the increasing livestock numbers.

In northwestern China, there is a large amount of land that could be used for hay production if irrigation were available, but only 3 percent of the nation's hay comes from here. In this area, 75 percent of hay is still harvested by hand, hauled by donkey or horse carts, and hand stacked in the homestead. In northeastern China, the grasslands are harvested for hay by hand, with only 5 percent of the harvesting being done with machinery.

India has by far the largest number of cattle (270 million) of any country in the world. Hay production is very diverse. Most people have a few cattle and the owners bring some fresh green forage every day to their cattle. Hay is cut by hand, tied into bundles by hand, stacked in the field for drying and then brought home. In wet regions it is stored under the shed or stacked outside with a protective layer of straw.

In northern Punjab, the majority of the hay is put up by hand and brought to market in trucks or bullock carts, but some alfalfa is grown for commercial dairy operations. Most of the high quality forages are produced under contract from dairy operations, or sold for feeding livestock in urban centres. There is increased interest in intercropping trees with perennial grass for hay on marginal lands.

In the colder and higher-rainfall areas of Afghanistan and Northern Pakistan, grazing and hay harvesting is carried out on the mountainsides. Harvesting of forage is done by hand, dried in the sun and stored in big piles, which may be plastered over with mud to keep them dry during the winter. This stored hay is used to supplement the winter grazing of dairy and work animals, as cattle are grazed year round, and hay is often not needed. Similar practices occur in colder and northern areas of India and all of Nepal.

Near East

In Egypt, many crops are grown under irrigation in the Nile Delta. Farms average about 0.4 ha in size, and fresh berseem clover (*Trifolium alexandrinum*) is cut by hand with a sickle on a daily basis and fed to the farmer's livestock. With a few exceptions, some forage is harvested by machine and fed green with straw because of the high moisture content. There are hay projects outside the Nile valley that are able to quickly dry and harvest for the dairy industry.

Israel uses winter cereal crops for hay production. The production of winter legumes for hay is less popular, due to low yields and high field losses. Hay is also produced using modern equipment in areas where the forage needs to be transported long distances. Climate conditions in the subtropical areas of the country are favourable for hay production, as only a few days of drying in the sun are required. In the hot region of the Jordan Valley, irrigated alfalfa is harvested for hay up to 8 to 10 times during the summer, and twice during winter. On some government-owned state farms in Turkey, round and square balers are used. In the villages where livestock are not of high value, very little input is provided and hay is harvested manually.

In Iran, irrigated alfalfa is the main hay crop, grown for the large commercial dairy farms, similar to the California systems. On these larger farms, modern equipment is used, while on smaller farms hay is harvested and stacked by hand.

On the dairy farms and camel stations in the United Arab Emirates, irrigated alfalfa and Rhodes grass (*Chloris gayana*) are baled and transported as large round or square bales. There are large sites where land has been reclaimed from the desert and irrigated by subsurface water for alfalfa production. In Saudi Arabia, due to pending shortages of irrigation water, alfalfa hay acreages are decreasing. As aquifers are being drained for irrigation, salt water from the Persian Gulf is seeping into formations, resulting in irrigation water becoming salty. In the future, Saudi Arabia will become more dependent on imported hay products. The dairies sometimes prefer Australian oat hay, which has extremely good palatability. However, this product has to compete with the cheaper but currently heavily subsidized European big hay bales.

Africa

Hay production in South Africa is largely limited to the wetter parts of the country, where weeping love grass (*Eragrostis curvula*) is the dominant hay pasture species. In the past, small square bales were widely used, but there is an increase in use of large round bales. In the remainder of Africa, most cattle are grazed year around and there is limited hay or silage production.

Australia and New Zealand

Due to the high cost, very few Australian livestock farmers, with the exception of dairy farmers, conserve forage. It is more profitable to buy hay when needed for drought supplementation, as it is not profitable to routinely feed hay to sheep or cattle. Most hay in Australia is sun-cured, and some alfalfa hay producers use hay preservatives. Hay is baled using large square bales and large round balers. Australia has a well-developed hay harvesting technology information system for their producers. There is also a large export market for double compressed oat hay going to Japan and Republic of Korea.

New Zealand livestock systems rely on grazing pasture year round. Hay is made to control surplus pasture growth and preserve forage quality over the whole farm. The majority of hay is large round bales, but a significant quantity of hay is made in small square bales.

South America

Haying systems in South America, such as in Argentina, Brazil, Chile and Uruguay, are very similar to those of North America. Hay production is highly mechanized, using equipment that is common throughout North America and Europe. Although square bales are still used by many farmers, large round bales predominate among medium- and large-scale farmers. Alfalfa is by far the most common hay, and mixed pasture hay and grass-only hay are much less common.

WHAT ABOUT THE FUTURE?

The methodology and physiology of harvesting and storing hay is in most cases universal around the world. Research findings in general as reviewed in this paper can be used in any hay producing area. Similarly, hay harvesting equipment that is manufactured by most companies can be used domestically as well as exported to different countries. Major advances in equipment design and harvesting research were achieved in the past 30 years, but this trend has slowed down and at present there is very little research on haying systems occurring in any of the major hay growing countries due to lack of research funds. The major question that still has not been answered involves drying the crop faster by shortening the time between cutting and baling in order to reduce or eliminate the effects of weather. Hay producers around the world still face this problem. Being able to produce and store hay of high nutritional quality is still the biggest issue. With the increased need for high quality dairy hay as a fibre source, increased efforts in research and management are still required to reduce these quality losses.

In the future, more accurate weather forecasting data will be required at hay harvest time. In part of North America, a daily drying index is available for producers to use in order to decide when hay should be cut and baled. This system needs to be expanded, and the information could be used to develop computer simulation models to help hay producers to identify hay harvesting options as a means of reducing losses. As it develop, this process should include information on possible dry matter and nutritional losses during wilting and harvesting caused by plant respiration and precipitation, and include the various machinery and storage losses. This would aid the large-scale hay producer in purchasing equipment (Rotz, 2001) and identifying the critical points that need to be addressed to reduce the nutritional losses.

Rapid moisture determination is possible through the use of electronic metering systems, but multiple samples must be tested in order to obtain more accurate moisture readings throughout the field (PAMI, 2001). Many hay equipment manufacturing companies are in the process of developing on-the-move moisture metering and yield monitoring systems. These units will be hooked into the satellite-based Global Positioning System (GPS) and local geographical information systems (GIS) for on-farm use. The baler operator will then be able to monitor hay yield and the moisture content as each bale is being made. If the moisture content is too high in a certain spot in the field then they can leave it for another time or they can add a metered amount of hay preservative to the individual bale.

In many countries, farms are consolidating and become larger. Most hay harvesting equipment manufacturers are seeing a marketing trend towards bigger and more efficient equipment. The large round bale will continue to be the main haying system in the future because of its high productivity at low cost. Large square bales will become increasingly popular because they offer very high productivity, density and shipping efficiency. Small square bales will be of less importance due to high labour costs, but will remain as a specialty packaging system for small livestock producers, equestrian owners and exporters. Some areas of the world will continue to harvest hay by hand where quality has not yet become the major issue. In other areas, antiquated haying equipment will continue to be used. The lack of available financial resources has prevented the adoption of modern haying systems technology on these small-scale farms. There continues to be a great need for agricultural extension education on sound hay harvesting and pasture management in the developing world. The basic hay harvesting and storage technology can be adapted to meet the need for higher quality hay to satisfy the increasing demand for livestock products. As economies of scale continue to grow in the developing world, it is anticipated that the current small-scale equipment will meet their needs and little additional technical development will be required.

Haying machinery in general has not been developed for regions with short daily drying times combined with frequent showers. Evidently, this makes the present windrow-to-bale system perform very poorly on a quality basis, as most hay becomes weathered before baling. Alternatives that would allow for more sun drying with gentler raking systems are also required as a means of obtaining better quality hay. Further engineering research is required to enhance total hay harvesting systems.

Although the round baler is still the most economical, it still requires the operator to stop to tie the bale. This is a major productivity disadvantage and equipment manufactures need to address this issue in order to speed the harvesting process.

There are still handling problems with all forms of hay bales. Farmers need to automate their systems for feeding hay, as silage-feeding systems are much easier to automate. This is one of the main reasons that hay production will decrease. If forage needs to be transported over long distances, hay is the product of choice because of its high dry matter content. In large-scale operations and for transport markets, because of the high capital cost of large baling equipment, there is emerging an increased dependency on large-scale contractors owning and operating the expensive equipment.

Long-fibre forage exports will grow as countries around the Pacific Rim increase the proportion of meat and dairy products in their diets. This market will require consistent high quality products. Improved systems and quality standards will need to be developed for high-density bales, which are used to improve shipping efficiency. Currently the technology is available to produce double compacted bales, but the hand sorting of the unwanted plant material could be enhanced with the development of visual electronic scanners for identification and removal of damp or unwanted material prior to compaction. There is also a need to develop a baling system for producing high-density bales (i.e. 300 kg m^{-3}) in the field.

If high quality hay is to be produced on a consistent basis, improved hay inoculants and other types of hay preservatives will be required to provide more consistent and beneficial effects on animal performance. These products must be cost effective, environmentally friendly and safe, and easy to apply. The present extent of adoption of hay inoculants is relatively small due to cost and lack of consistent results.

In many parts of the world there is concern about pathogens, microbial toxins and insects being transmitted through hay. Research is needed to develop rapid systems for detecting and controlling the growth of these pathogens, toxins and insects in hay in a safe and economical way.

An increasing proportion of the world's hay supply can be expected to be grown in arid climates, under irrigation, as producers in these areas will have a greater control over growing, drying and harvesting forage, with less risk of weather damage. In some cases, re-hydration at baling will be required to lengthen the harvesting opportunity and enhance the baling process by reducing stem and leaf loss due to breakage. In other areas, there is still concern about bleaching or discoloration of hay. The market place wants green hay, and therefore, to reduce bleaching, producers tend to leave hay drying in large swaths for a long time, and this can lead to possible rain damage. There is very little scientific evidence that bleaching from the sun reduces nutritional quality. This question still needs to be addressed in these specific regions.

It is doubtful that systems for forage artificially dried before or after baling will develop, as such systems have low capacity, high capital cost and high fuel cost. Forage value will have to increase substantially before such systems become widely adopted.

In conclusion, the hay harvesting industry is still looking for ways to reduce losses in forage quality and quantity. Methods must be developed to shorten the time between cutting and baling, and reduce weathering losses. Improved bale handling and feeding systems will be required to reduce additional forage losses. Silage systems will become the way of the future in many regions of the world, but hay harvesting systems will continue to provide a long-fibre forage product to meet specialized needs in the global livestock industry.

ACKNOWLEDGEMENTS

The author would like to thank Sigrid Lohmann and Adele Depalme for their writing and editorial assistance, and Natalie Kuehn and Roberta Galenzoski for typing skills in the development of this review paper.

REFERENCES

Augsburger, H. & Methal, M. 1993. Henificacion [Haymaking]. *Instituto Nacional de Investigacion Agropecuaria Boletin,* No. 27.

Baron, V.S. & Greer, G.G. 1988. Comparison of six commercial hay preservatives under simulated storage conditions. *Canadian Journal of Animal Science,* **68**: 1195–1207.

Baron, V.S. & Mathison, G.W. 1990. Yield, quality and preservation of moist hay subjected to rain-free and weathered conditions. *Canadian Journal of Animal Science,* **70**: 611–612.

Beacom, S.E. 1991. Forage Crops in the Aspen Parkland of Western Canada, Harvesting. Canada Communications Group, Publishing Ottawa, Canada.

Bossuyt, C.V., Wittenberg, C.M. & Crow, C. 1996. Effect of fungal biomass in alfalfa hay on intake and total tract digestion in growing beef calves. *Journal of Animal Science,* **74**: 1336–1342.

Breton, A. & Zwaenepoel, P. 1991. Succession of moist hay microflora during storage. *Canadian Journal of Microbiology,* **37**: 248–251.

Buskirk, D. 1998. Evaluation of hay loss from large round bale feeders. Department of Animal Science, Michigan State University.

Coates, W.E., Robertson, J.A., Feldman, M. & Thorlacius, S.O. 1978. Observations on some mechanical hay stacking methods and procedures. *Canadian Journal of Agricultural Engineering,* **20**: 91–95.

Corry, J.E.L. 1977. Relationships of water activity to fungal growth. pp. 45–82, *in:* L.R. Beuchat (ed). *Food and beverage mycology.* 2nd ed. New York, NY: Van Nostrand Reinhold.

Crump, S.V. 1985. Potassium carbonate and forage drying. M.Sc. thesis, University of Alberta, Edmonton, Canada.

Elsässer, M. 1990. Drying of Forage Crops. The current practice, future possibilities and research needs. pp. 86–105, *in: Forage Conservation. Towards 2000*. Landbauforschung Volkenrode, Sonderheft 123. Braunscweig, Germany.

Fisher, D.S., Mayland, H.F. & Burns, J.C. 2002. Variation in ruminant preference for alfalfa hays cut at sunup and sundown. *Crop Science*, **42**: 231–237.

Friesen, O. 1978. Evaluation of hay and forage harvesting methods. *In:* G.R. Quick (ed). *Grain and Forage Harvesting*. St. Joseph, MI: ASAE.

Friesen, O.H. 1980. Economics of forage equipment and feeding systems. Alberta Cattlemen's Short Course, Banff, Alberta, Canada.

Hong, B.J., Broderick, G.A. & Walgenbach, R.P. 1983. Effect of chemical conditioning of alfalfa on drying rate and nutrient digestion in ruminants. *Journal of Dairy Science*, **71**: 1851–1859.

Jan, E.Z., Robertson, J.A., Feldman, M. & Lievers, K.W. 1980. Drying and storage of chopped hay with hay tower. Canadian Society Agricultural Engineering Proceedings. 80–220 Aug Edmonton, Alberta, Canada.

Koegel, R.G., Straub, R.J., Shinners, K.J., Broderick, G.A. & Mertens, D.R. 1992. An overview of physical treatments of lucerne performed at Madison, Wisconsin, for improving properties. *Journal of Agricultural Engineering Research*, **52**: 183–191.

Kraynyk, P.I. & Wittenberg, K.M. 1989. The effect of potassium carbonate, a drying agent, on storage of alfalfa and on the performance of growing lambs. pp. 272–276, *in: Proceedings of the Forage and Grassland Conference*. Guelph, Canada, June 1989.

Krutz, G.W., Holt, D.A. & Miller, D. 1979. For fast drying of forage crops. *Agricultural Engineering*, **60**(8): 16–17.

Lacefield, G.D. 1987. Why salt hay? *Hoard's Dairyman*, 25 May 1987: 463.

Mahanna, B. 1997. Hay additive review. "Where we've been; where we're going." Inoculant and Forages; Pioneer Hi-Bred International, Inc. Johnston, Iowa.

Martin, N.P. 1980. Harvesting and Storage of Quality Hay. p. 177, *in: Proceedings of the American Forage and Grassland Council Conference*. Louisville, Kentucky, February 1980.

McLennan, N. 1998. Producing, storing and improving native pasture hay. Agency for Food and Fibre Science. Queensland, Australia.

McCartney, D.H., Bittman, S. & Nuttall, W.F. [2004]. The influence of harvest management and applied N on seasonal yield, protein concentration and N uptake of pasture and hay type grasses in northeastern Saskatchewan. *Canadian Journal of Plant Science*, in press.

Miller, L.G., Clayton, D.C., Nelson, L.F. & Hoehne, O.E. 1967. Nutritive value of hay baled at various moisture contents. *Journal of Animal Science*, **26**: 1367–1373.

Mir, Z., Jan, E.Z., Robertson, J.A., McCartney, D.H. & Mir, P.S. 1991. Effects of ammonization of brome-alfalfa and alfalfa hay, stored as large round bales, on preservation and feed quality. *Canadian Journal of Animal Science*, **71**: 755–765.

Mir, Z., Jan, E.Z., Robertson, J.A., Mir, P.S. & McCartney, D.H. 1995. Effects of microbial inoculant and moisture content on preservation and quality of round-baled alfalfa. *Canadian Journal of Animal Science*, **75**: 15–23.

Muck, R.E. & Shinners, K.J. 2001. Conserved forage (silage and hay): progress and priorities. pp. 753–762, *in: Proceeding XIX International Grassland Congress*. São Pedro, Brazil, February 2001.

Orloff, S.B. 1990. Altering the windrow width to hasten curing. *In:* Proceedings of the 1990 California Alfalfa Symposia. Davis, CA: Department of Agronomy and Range Science, University of California, Davis.

PAMI [Prairie Agriculture Machinery Institute]. 2001. Annual Report. Humboldt, Saskatchewan, Canada.

Petit, H.X. & Tremblay, G.E. 1992. *In situ* degradability of fresh grass and grass conserved under different harvesting methods. *Journal of Dairy Plant Science*, **63**: 913–925.

Pitt, R.E. 1993. Forage Moisture Determination NRAES-59. Ithaca, NY. Northeast Regional Agricultural Engineering Service.

Rasmussen, K. 1969. The target for 1980. *In:* K.F. Nielsen (ed). *Canadian Forage Crops Symposium*. Modern Press. Saskatoon, Saskatchewan. 460p.

Roberts, C.A., Moore, K.J., Grafils, D.W., Kirby, H.W. & Walgenbach, R.P. 1987. Chitin as an estimate of mold in hay. *Crop Science*, **27**: 783–785.

Robertson, J.A. 1983. Influence of harvesting and conservation practices on forage quality. *Canadian Journal of Animal Science*, **63**: 913–925.

Rotz, C.A. 2001. Mechanization: planning and selection of equipment. pp. 763–768, *in: Proceeding XIX International Grassland Congress*, São Pedro, Brazil, February 2001.

Rotz, C.A. & Muck, R.E. 1994. Changes in forage quality during harvest and storage. pp. 828–868, *in:* G.C. Fahey (ed). *Forage quality, evaluation, and utilization*. Madison, WI: American Society of Agronomy, Inc.

Rotz, C.A. & Muhtar, H.A. 1992. Rotary power requirements for harvesting and handling equipment. *Applied Engineering in Agriculture,* **8**: 751–757.

Rotz, C.A. & Savoie, P. 1991. Economics of swath manipulation during field curing of alfalfa. *Applied Engineering in Agriculture,* **7**: 316–323.

Rotz, C.A. & Sprott, D.J. 1984. Drying rates, losses and fuel requirements for mowing and conditioning alfalfa. *Transactions of the ASAE,* **27**: 715–720.

Savoie, P. 1990. Forage handling systems. *In: Proceedings of the Field Crops Expo 90.* Agriculture and AgriFood Canada, Sainte Foy, Quebec, Canada.

Savoie, P. 1996. New developments in harvesting and preprocessing of forage. Canadian Dehydrators Association Conference and Trade Show, Saskatoon, Saskatchewan, Canada.

Shinners, K.J. 1997. Equipment for forage harvesting: what to look for and how to optimize. pp. 69-84, *in:* Silage: Field to Feedbunk, NRAES-99. Ithaca, NY: Northeast Regional Agricultural Engineering Service.

Shinners, K.J., Koegel, R.G. & Straub R.J. 1991. Leaf loss and drying rate as affected by conditioning roll type. *Applied Engineering in Agriculture,* **7**: 46–49.

Shinners, K.J., Straub, R.J., Huhnke, R.L. & Undersander, D.J. 1996. Harvest and storage losses associated with mid-size rectangular bales. *Applied Engineering in Agriculture,* **12:** 167–173.

Thorlacius, S.O. & Robertson, J.A. 1984. Effectiveness of anhydrous ammonia as a preservative for high-moisture hay. *Canadian Journal of Animal Science,* **64**: 867–880.

Undi, M. & Wittenberg, K.M. 1996. Effect of fungal biomass in mouldy alfalfa hay on preference by dairy calves with no previous exposure to mouldy feeds. *Journal of Dairy Science,* **79**: 1250–1254.

von Keyserlingk, M.A.G., Swift, M.L., Puchala, R. & Shelford, J.A. 1996. Degradability characteristics of dry matter and crude protein of forages in ruminants. *Animal Feed Science and Technology,* **57**: 291–311.

Wilkinson, S.M. 1981. Losses in the conservation and utilization of grass and forage. *Annals of Applied Biology,* **98**: 365–375.

Wittenberg, K.M. 1994. Nutritive value of high moisture alfalfa hay preserved with *Pediococcus pentosaceus*. *Canadian Journal of Animal Science,* **74**: 229–234.

Wittenberg, K.M. 1995. Efficacy of *Pediococcus pentosaceus* for alfalfa forage exposed to precipitation during field wilting. *Canadian Journal of Animal Science,* **75**: 303–308.

Wittenberg, K. M. 1997. Microbial and nutritive changes in forage during harvest and storage as hay. pp. 265–270, Session 14, *in:* Proceeding XVIII International Grassland Congress, Canada, June 1997.

Wittenberg, K.M., Moshtaghi-Nia, S.A., Mills, P.A. & Platford, R.G. 1989. Chitin analysis of hay as a means of determining fungal invasion during storage. *Animal Feed Science and Technology,* **27**: 101–110.

managed relatively intensively. Even although the level of supplementary feeding may be relatively high in dairy systems, grazed or conserved grass still accounts for 50–70 percent of the nutrient requirements of the cows, and 75–100 percent of requirements in most beef and sheep systems can be met from grazed or conserved grass.

Many of the world's arid and semi-arid rangelands are grazed by domestic livestock, and although the stocking density may be relatively low, animals provide livelihoods for pastoralists and are often central to their cultural heritage.

Compared with other sources of fodder for meat and milk production, grass, especially grazed grass, can provide a cheap source of feed (Table 7.2). The costs of grazed grass are generally about half of those of conserved hay or silage. In addition, the support-energy requirements for harvesting and storing conserved grass, especially as silage, are often considerable (Wilkins, 1990).

Table 7.2 Costs of different ruminant feeds

Feed	Cost (US$ tonne^{-1})
Grazed grass	36–56
Maize silage	48–60
Alfalfa haylage	70–100

Source: Hlukik and Smith, 1988.

While the above gives an indication as to the current state of meat and milk production from grasslands, the objective of this chapter is to look forward and to try to analyse future prospects. There are a number of recent trends and developments that might give some pointers as to future trends. Observers and commentators have identified three important drivers that are likely to have a bearing on the production of meat and milk from grass in the next 20 years. These are:

1. A general increase in global demand for milk and meat.
2. Concern about the environmental impact of agricultural practices.
3. An increasing concern by consumers about food quality, including food safety and animal welfare, especially in the developed world.

CHANGING GLOBAL DEMAND FOR ANIMAL PRODUCTS

There is a huge range in the amount of meat consumed by the population in different regions in the world. Meat consumption in the developed world is much higher than in developing countries. In the early 1990s, the per capita consumption of meat in the developed world was 78 kg, with higher amounts in the United States of America and lower amounts in Europe. In developing countries of Asia, 18 kg of meat was consumed per capita, while in Latin America consumption was 46 kg. In sub-Saharan Africa, the equivalent figure was only 9 kg (Delgado *et al.*, 1999).

Milk consumption per capita in Latin America is higher than in other developing countries: four times higher than in Africa and three times higher than in developing countries in Asia. However, even in Latin America, milk consumption per capita is about 40 percent less than in developed countries.

Per capita consumption in developed countries is likely to be relatively static over the next 20 years. The levels of consumption in developed countries are already relatively high and so the scope for further increases is limited. However, per capita consumption of meat and milk is forecast to increase considerably over the next 20 years in developing countries since demand for meat tends to increase more rapidly than disposable income (Ahmed and Gruhn, 1995). Thus, as countries become richer, meat consumption tends to increase quickly. The rate of increase in per capita income is much higher in some developing regions, e.g. China and Southeast Asia with rates of economic growth of 6–9 percent, whereas in developed countries economic growth rarely exceeds 3 percent and is often less. Urbanization is also thought to be a key factor in driving up demand for livestock products. The urban population in developing countries is growing at 3.5 percent per annum, compared with only 0.75 percent in developed countries, which are already highly urbanized (United Nations, 1995).

A further important factor is that changes in per capita consumption in the developing world are likely to have a much greater global impact than changes in the developed world. This is because, first, the developing countries have a much greater proportion of the world's population (about 77 percent in the 1990s), and, second, they have much higher rates of population growth (1.9 percent per annum, versus 0.4 percent in developed countries).

The International Food Policy Research Institute has developed a global food model (IMPACT; Delgado *et al*., 1999) and used it to predict the growth in demand for a range of foodstuffs in different regions of the world. Delgado *et al.* (1999) have produced estimates in the demand for different livestock products up to the year 2020. They predict considerable increases in demand for livestock products as a consequence of population growth, economic growth, higher income per capita and increased urbanization. Much of this increased demand is predicted to take place in China and Southeast Asia, where levels of economic growth are high. The IMPACT model predicts that in China, for example, meat consumption will reach 63 kg per capita, approaching the current consumption in developed countries. Total world meat production is predicted to increase by 63 percent by 2020 compared with the early 1990s. China alone is projected to account for almost 30 percent of global supply (including pig and poultry meat).

While much of the increased demand will be for white meat (pig meat and poultry), there will also be a considerable increase in the growth in demand for milk and beef (Table 7.3). Although the demand for beef will remain relatively constant up until 2020 in the developed world, there is predicted to be a doubling of beef production and a 2.5-fold increase in milk demand between 1993 and 2020. A number of assumptions are built into the IMPACT model, including the continued growth of the economies of developing countries, especially in Asia, and the continued supply of relatively cheap fossil fuel. Given the known levels of global fossil fuel reserves, this seems a reasonable assumption, at least over the twenty-year time period for the projections.

While there are a number of assumptions built into the IMPACT model and the absolute increases in consumption that are predicted might be questioned, there does appear to be

Table 7.3 Projected trends in production of selected livestock products

Region/Product	Projected annual growth in production (1993–2020)	Total production (million tonne)	
		1993	2020
Developed world			
Beef	0.6	35	38
Milk	0.4	348	371
Developing world			
Beef	2.6	22	44
Milk	3.2	164	401

considerable evidence to support the view that the global demand and production for meat and milk is likely to increase in the next 20 years. A major factor that will affect whether the predicted increases in demand can be satisfied will be the ability of the natural resources to support them, in the light of the increased demand that will be placed on those natural resources, upon which the production systems depend to cope with the environmental pressures that such an increased demand will place upon them.

ENVIRONMENTAL IMPACT OF GRAZING LIVESTOCK PRODUCTION

Types of grazing system

The form of the environmental impact of grazing livestock is determined to a large extent by the environmental conditions in which the grazing system is located, and especially the climatic conditions. From an environmental perspective, de Haan, Steinfeld and Blackburn (1997) classified grazing systems into a number of broad types, including:

- arid grazing systems;
- semi-arid grazing systems;
- subhumid tropical grazing systems;
- temperate grazing systems; and
- mixed farming systems.

Grazing systems are defined as livestock systems in which more than 90 percent of the dry matter fed to the animals comes from rangelands, pastures, annual forages and purchased fodder. In terms of total production, these grazing systems supply only 9 percent of global meat production. Livestock provide the sole source of income for 20 million pastoral families.

There are also mixed systems in which crop and livestock production are much more closely linked. In many cases, the contribution of grazed or conserved fodder from grassland is limited.

Arid and semi-arid grazing systems

Arid grazing systems have been defined as having an average growing season of less than 75 days, with rainfall limiting production. Semi-arid areas have a growing season of 75–180 days, with less variability in rainfall than in arid areas. Examples of these areas include parts of the Sahel and central Asia. These ecosystems are often important habitats for a range of plant and animal species and there is competition for resources between domestic animals and wildlife, although sometimes occupying different niches.

The higher rainfall in the semi-arid regions means that the grazing pressure on these resources can be considerable. The greater concentration of water points in semi-arid zones can also lead to high levels of local degradation. Historically, it was thought that many of these arid and semi-arid rangelands were "degraded" and that significant desertification had taken place. UNEP (1992), for example, suggested that about 20 percent of the world's rangelands had been degraded, and Gilmanov (1996) estimated that about one-third of the rangelands in the Former Soviet Union were degraded. More recently, there has been considerable debate as to both the definition and extent of degradation and desertification (e.g. Skarpe, 2000). Studies that have taken a long time frame (decades rather than years) suggest that many of the observed changes are climatic-driven rather than due to the effects of grazing livestock. The expansion of the Sahara desert in the 1970s and 1980s was probably a result of climatic changes rather than due to the impact of human activity.

The concept of degradation has also been questioned. The idea of degradation emerged from rangeland science in North America, based on the assumption that rangeland vegetation moved through a series of states to a climax in the absence of grazing, i.e. Clemensian succession. Degradation was then considered as the extent to which the rangeland differed from this climax state. This model still dominates much rangeland management policy, especially in the United States of America. In ecology, this model has given way to the concept that rangelands can exist in a range of stable states depending on weather conditions and management regimes. The "state and transition model" (Westoby, Walker and Noy-Meir, 1989) implies that there is no longer a simple state against which "degradation" can be measured. Rather the implication is that the desired state can only be defined in relation to a particular set of management objectives.

Behnke and Scoones (1993) argued that the condition of arid and semi-arid rangelands where there is considerable variation in rainfall is climatically driven rather than driven by grazing pressure, i.e. the "non-equilibrium model". However, recently Illius and O'Connor (1999) have questioned the view that rangeland systems can be conveniently classified as "equilibrium" or "non-equilibrium" and consider that there is a range of coupling mechanisms of plant and animal dynamics. Certainly Fernandez-Giminez and Allen-Diaz (1999) found that neither of these two models fitted data collected from Mongolia. Nevertheless, there are undoubted examples of degradation of arid rangelands in certain areas caused by grazing pressure, notably around settlements and water points where the vegetation can be completely absent as a consequence of pressure from grazing livestock.

Thus there needs to be a reassessment of the environmental impacts of livestock grazing on arid and semi-arid rangelands. In practical terms this is proving difficult, as currently the state and transitional model and the non-equilibrium models do not provide any conceptual basis for the development of monitoring techniques. Recently Stolte *et al.* (2004) have tested a field methodology for assessing temperate rangeland condition that is based on both a consideration of a desired state and a sound understanding of the interaction between the vegetation and the grazing herbivores. Whether such an approach is more generally applicable remains to be seen.

Sub-humid tropical grazing systems

In these zones, vegetation growth, stimulated by rainfall normally lasts for six to nine months, sometimes in two separate periods with one or two dry seasons. In the sub-humid tropics, the wetter climate leads to a range of diseases being endemic (e.g. river blindness, African sleeping sickness, tick-borne diseases, and tsetse fly, which spreads trypanosomiasis), which can limit the exploitation of the resources. With progress in tsetse and tick control, and with better medicinal treatment against animal diseases, many pastoralists from the semi-arid regions of Africa are entering wetter areas during the dry season to find grass regrowth, either after savannah burning or in the wet lowlands. Eradication or control programmes for tsetse fly in Zimbabwe and Cameroon, for example, have opened up new pasture areas to pastoralists. The main environmental issues in relation to grazing of these areas are the change in vegetation, especially bush encroachment, the dissemination of invasive plants, the long-term consequences of the bushfires on open pastures, and the consequences for wildlife.

Temperate grazing systems

Temperate grazing systems, based on natural permanent pastures, account for 13 percent of the world's pasture lands (de Haan, Steinfeld and Blackburn, 1997), mainly in China, northern United States of America and South America. Much of these temperate rangelands appear to be in a relatively healthy state (USDA, 1988), as domestic livestock have replaced the wild herbivores under which these ecosystems have evolved. However, there are areas of the world where the pressure from domestic livestock may be high, such as parts of the western United States of America, the Former Soviet Union, China and Mongolia. Zhong (1993) suggested that 40 percent of Inner Mongolia's rangelands were degraded.

In extensive grazing systems in Europe there is concern about the impact that grazing by domestic livestock has had on the biodiversity of what are often relatively fragile habitats. Thus, grazing by sheep and the replacement of natural vegetation with sown species has reduced the area of heather (*Calluna vulgaris*) in many parts of upland Britain (Mowle and Bell, 1995).

Mixed farming systems

Mixed farming systems occur in both the developed and developing world. They supply over 90 percent of the world's milk, 70 percent of the sheep and goat meat, and approximately 35 percent of the beef. They have the potential to be the most integrated of the livestock systems because of the opportunities for the efficient use of resources and in particular the integration of cropping and livestock systems to maintain soil fertility and biodiversity, better use of wastes and the opportunities for intensification of production without resorting to high levels of external inputs. However in developing countries with high population growth (e.g. the Sahel, Ethiopia, Nepal, Bangladesh) there is increased pressure on the pasture component of the system, with pasture land being converted to cropping, with a resultant increase in stocking rate on the remaining pasture. In the developed world, mixed systems usually entail pasture, grazed by ruminant livestock as part of a crop rotation, or the integration of the use of areas of natural pasture with cropping land. In the past few decades the levels of external inputs, especially fertilizers, has increased dramatically. Between the 1960s and 1990s, the amount of nitrogen fertilizer used by agriculture in the United Kingdom rose from about one-third of a million tonnes to over one-and-one-half million tonnes. While by 1986 some grassland was receiving over 300 kg N ha^{-1}, over half was receiving less than 100 kg ha^{-1} (Wilkins, 2000). However, nutrient surpluses have had a major impact on surface and groundwater, resulting in particular in high levels of nitrate in drinking water and eutrophication of surface waters. This has led to a series of nutrient control measures being applied to farms in some parts of the European Union and the United States of America. As mixed farming systems have become more intensive, with higher stocking densities and reliance on sown pasture with fewer plant species, biodiversity has probably decreased.

FOOD QUALITY, FOOD SAFETY AND ANIMAL WELFARE

In many developed countries, consumers are becoming increasingly aware of issues such as the quality and safety of food, as well as of the conditions under which animals are kept for food production. The impact that food safety can have on consumer confidence in products and on consumption was dramatically illustrated in the United Kingdom in 1996, when the British Government announced that there was a possibility that new variant Creutzfeldt Jacob Disease (CJD) in humans could be linked to Bovine Spongiform Encephalopathy (BSE). Beef consumption in the United Kingdom fell by 16 percent in 1996 compared with 1995 (Figure 7.1) and in the European Union by 10 percent, although by 1998 consumption levels in the United Kingdom had recovered.

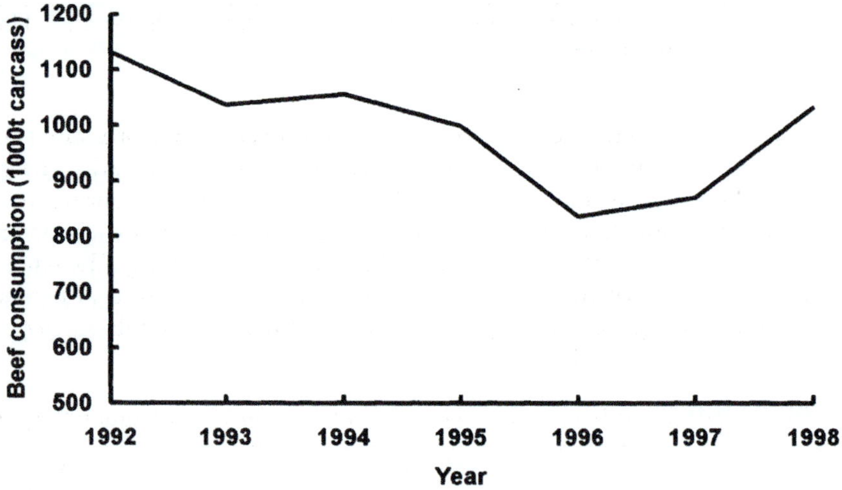

Figure 7.1 Beef consumption in the United Kingdom, 1992 to 1998.
Source: Eurostat, Luxembourg.

A recent survey of Irish consumers showed that 64 percent were concerned about beef. Concerns included freshness of food, BSE/CJD, antibiotic residues, hygiene standards and bacteria (Riordan, Cowan and McCarthy, 2002).

In developed countries, where consumers have high levels of disposable income, many are increasingly concerned about the conditions under which animals are kept for food production. Legislation often provides for a minimum standard of animal welfare that is considered by society as being acceptable.

Most concern has been expressed about intensive production methods for pigs and poultry, where animals are kept indoors at high stocking densities. At the same time, animal products that are produced from grass-based systems are seen as being more 'natural' by consumers, and to pose fewer welfare problems for the animals concerned.

Nutritive value of meat and milk from grass

The importance of animal products in the human diet has long been recognized. Much of the research on animal production in the early half of the twentieth century was stimulated by the need to increase the amount of animal protein in the diet of a mainly urban, industrialized population, with diseases such as rickets being prevalent in children in many of the industrialized towns in Britain, for example.

In the latter part of the twentieth century many human health problems, such as cardiovascular disease, have been linked to an overconsumption of fats, especially saturated fats, mainly derived from animal products such as meat and milk. Official government advice in many developed countries has been to reduce saturated fat intake, and considerable progress has been made through breeding and nutrition programmes in reducing the fat content in meat from cattle and sheep. In the 1960s, typical fat contents for beef and lamb as sold retail were 25 and 31 percent respectively, while by 2000 the equivalent figures were 5 and 8 percent.

The fatty acid composition of ruminant products has become the focus of considerable research attention in recent years, because of the link between saturated fat consumption and cardiovascular disease. The UK Department of Health, for example, has recommended a reduction in the intake of saturated fats and an increase in the intake of unsaturated fatty acids, especially the omega-3 polyunsaturated fatty acids (*n-3* PUFA). These are known to be beneficial, especially in protecting against heart disease. It is recommended that the ratio of polyunsaturated to saturated fatty acids (P:S) in the human diet should be about 0.4 and that there should be an increase of the relative intake of *n-3* compared with *n-6* PUFA (Department of Health, 1994).

The fatty acid composition of beef is partly determined by the animals' diet. Grass is rich in α-linolenic acid and recent research has shown that the fatty acid composition of beef from cattle which have been fed on a grass diet differs from that in cattle fed in concentrates (Table 7.4). In particular, the grass-fed animals had a higher concentration of α-linolenic acid (C18:3 n-3) and its n-3 long chain derivatives, while concentrate feeding resulted in higher amounts of linoleic acid (C18:2 n-6) and long-chain derivatives.

Table 7.4 Effect of diet on fatty acid composition (mg g^{-1} muscle) of longissimus dorsi muscle from cattle fed grass or concentrates

Fatty acid	Diet		P-value
	Grass	Concentrates	
C16 : 0	6.98	4.88	0.01
C18 : 0	4.56	2.79	0.001
C18 : 1 *n-9*	10.1	6.22	0.001
C18 : 2 *n-6*	0.65	1.62	0.001
C18 : 3 *n-3*	0.33	0.10	0.001
C20 : 4 *n-6*	0.21	0.44	0.001
C20 : 5 *n-3*	0.13	0.004	0.001
C22 : 6 *n-3*	0.02	0.01	0.001
Total	28.6	21.7	0.01

Source: Adapted from Enser *et al.*, 1998.

Meat and milk from ruminants also represents a major source of conjugated linoleic acids (CLA) in the human diet, which have a number of health promoting benefits such as anti-carcinogenic activity, anti-artherogenic activity, the ability to reduce the catabolic effect of immune stimulation, enhanced growth promotion and reduction in body fat (Banni and Martin, 1998). Recently, it has been shown that there is a positive relationship between the concentration of CLA in milk and the proportion of fresh forage in the diet of dairy cows (Ward *et al.*, 2003).

Recent research in Kenya has shown, for the first time using a controlled feeding intervention study, a link between meat and milk intake in children and a range of physical and behavioural measurements (Global Livestock CRSP, 2001). Children of 6 to 9 years of age were supplemented with meat or milk and a range of measures of cognitive function, physical measurements, physical activity and behaviour were monitored. Supplementation with meat promoted cognitive function, leadership behaviour and initiative, as well as physical activity and improved biochemical micronutrient status of iron, zinc, vitamin A, vitamin B12 and riboflavin, while supplementation with milk promoted linear growth and vitamin A, vitamin B12 and riboflavin status. This study highlights the importance of animal products in the diet of children, not only for physical development, but also in psychological development. These findings could have far-reaching consequences for development in developing countries, since learning capacity and the development of human potential may be being stifled by a lack of animal foods in the diet.

Thus in both the developed and developing world, the role of meat and milk in the human diet may receive more attention in the future. In the developed world there may be considerable health benefits from the inclusion in the human diet of meat and milk from cattle reared on grass. However, the extent to which this may significantly and practically improve human health remains to be determined, but may become an important factor in the promotion of meat from grass-based systems in the future. In developing countries there is clearly benefit from inclusion of animal products in situations where the diet is sub-optimal.

TECHNOLOGICAL DEVELOPMENTS

Genetic improvement of livestock

Considerable research and development effort has been devoted to genetic improvement in livestock in past decades. There have been some spectacular improvements in some production traits, but these, in the main, have been when animals have been kept in relatively intensive systems, such as in the pig and dairy cow sectors in developed countries. Pittroff, Cartwright and Kothmann (2002) have argued that in extensive systems, as exemplified by much of the world's grazing land, the opportunity for genetic improvement is much limited. Hammond (1947) stated that, in general, variability in quantitative characters is greatest under good nutritional conditions. Many grazing systems offer relatively low levels of nutrition, and animals may undergo annual cycles of weight gain and loss. In such circumstances, rates of reproduction are often limited by nutrition. Attempts to improve productive traits, such as milk yield or growth rate, may exacerbate the low levels of reproduction because of negative phenotypic correlations between these productive traits and reproductive traits, even although there may be no genetic correlation (Baptist and Carles, 1989). Thus, attempts to increase traits such as milk yield or growth rate in environments that offer relatively low levels of nutrition usually result in reduced reproductive success. Reproductive traits usually have low levels of heritability, and so attempts to improve them genetically are usually limited.

A further complicating factor is the variability in the environment in many grazed systems, even within a single location. Intra- and inter-annual variations in climate result in variability in the level of nutrition. The genetic ranking of animals that have been selected in one environment often changes in another, less favourable environment. Even at a single site, the effect of genotype × environment interactions can be important (Wright *et al.*, 1994).

There may be cases where genetic selection for disease resistance may be important, and could yield important benefits, but the interaction with other production traits will always need to be considered (Vagenas *et al.*, 2002).

It is not clear at this stage what role transgenics may play in the future. There is considerable public resistance to the introduction of transgenic crops, especially in Europe. Research on transgenic livestock has, to date, mainly been focused on the production of animals with the ability to express novel proteins in their milk for pharmaceutical use. Whether transgenic livestock will be acceptable in the food chain has yet to be tested, but given the limitations to genetic improvement in extensive grazing systems, discussed above, it is unlikely that transgenic technology in livestock will have an impact on the production of meat and milk from grassland in the next couple of decades.

Genetic improvement of plants

Advances in plant breeding have the potential to both increase productivity and improve the environmental impact of grassland-based systems. The development of varieties of grass with higher water soluble carbohydrate concentrations, for example, can lead to a better ratio of nitrogen and energy yielding compounds in the rumen. This can result in higher levels of production, increased efficiency of nitrogen use and reduced nitrogen excretion (Miller *et al.*, 2001). Transgenic technology could play an important part in the future, by allowing the introduction of genes into forage plants that confer traits that make them more suitable for harsh environments. The introduction of genes that enhance drought tolerance or cold tolerance (Thomas, 1997) could potentially have beneficial implications for livestock production from grassland.

Animal nutrition

One of the keys to the more efficient and sustainable production of meat and milk from grass-based systems is a better understanding of the nutritional resources available to grazing animals. While much is known about the way in which ruminants utilize nutrients for conversion to meat and milk, based on several decades of research (e.g. NRC, 1996; ARC, 1980), there are large gaps in our knowledge of the nutritional value of much of the world's grazing resources, especially tropical forages. Although conceptual models of diet selection have been developed (O'Reagain and Schwarz, 1995), in practical terms it is extremely difficult to predict both the quantity and the composition of the diet that grazing animals will select when faced with a complex array of plants. This lack of understanding and predictive ability represents a major limitation to the development of more efficient grazing systems. For temperate systems, based on relatively simple monocultures of ryegrass or simple mixtures of a few plant species, some progress has been made in the past two decades in understanding these complex plant-animal relationships, leading to the development of predictive models, but for much of the rest of the world's grasslands, such predictive models do not exist.

GRAZING MANAGEMENT

The increasing global demand for meat and milk, environmental concerns about the sustainability of intensive production systems and, in the western world at least, issues of food quality, safety and animal welfare are increasingly having an impact on the buying patterns of consumers. These factors, coupled with the fact that grass is a relatively cheap source of feed for ruminants and that grassland-based systems are seen as being environmentally sustainable and 'welfare friendly', suggest that there is considerable scope for the continuing production of meat and milk from grass-based systems. However, there is a real challenge to develop production systems that maximize the utilization of grass per unit of animal product, commensurate with sound environmental practice. This necessitates developing grazing management systems that maximize the use of grass and forage in the diet of animals.

The main aim of grazing management is to convert forage resources to animal products such as meat and milk, in the most efficient way possible. To achieve this, it is necessary to supply the animals with high quantities of high quality forage to ensure acceptable levels of animal performance. It is also important to ensure that the herbage is utilized effectively, with low levels of wastage. All this must be done within the context of sustainable farming systems, to ensure that the grazing resources are maintained and that the management practices to not have unacceptable deleterious impacts on the environment, in terms of soil, water, air and wildlife habitats.

At one level there are few management options available to managers for the management of grazing systems. The basic factors that can be altered by grazing management are (i) the species of livestock, (ii) the timing or seasonality of grazing, and (iii) the grazing pressure. The distribution of animals on grasslands is often quoted as an important factor, but as pointed out by Walker and Hodgkinson (1999), it can be affected by the three factors already mentioned. In practice, there are a multitude of ways in which these three principal factors can be manipulated, and grazing research has focused on how their manipulation affects both plants and animals.

In temperate regions, grassland systems are usually part of a mixed grassland-arable rotation or based on permanent pasture. The grass species, especially those used in sown swards, usually have relatively high nutrient contents, especially if they are grazed before the formation of seed heads (Frame, 1992). For example, if perennial ryegrass is defoliated regularly, the sward can be maintained in a vegetative state by continuous grazing, with a high dry matter digestibility. However, growth is often limited by low temperatures in winter and may often cease for six months or more in winter. In such situations the aim of grassland management is to maximize the utilization of grazed grass in summer, while ensuring sufficient grass can be conserved as hay or fodder to feed the herd or flock in winter.

The level of performance achieved by grazing livestock is primarily a consequence of their daily level of herbage intake. Efficient grazing systems should, therefore, aim to maximize the intake of grass by grazing cattle. In temperate swards, of relatively high digestibility, intake is not constrained by the digestibility of the diet and it is the three-dimensional structure of the sward canopy that determines the size of a grazing bite, and therefore, to a large extent, daily intake. The sward structural characteristics that determine bite size have been the subject of considerable research effort, and factors such as the bulk density of the grazed horizon, leaf-to-stem ratio, green leaf mass and the height of the pseudostem have all been implicated in determining bite size (Rook, 2000). However, many of these variables are highly correlated in a temperate sward, and they are often correlated with sward height. Sward height is therefore a simple measure that is related to daily herbage intake. When cattle grazed continuously on a perennial ryegrass dominated sward, maximum daily intake of herbage was achieved at a sward height of about 10–11 cm (Figure 7.2). At heights of below 10–11 cm, intake is progressively reduced (Wright, 1988), and this is reflected in the liveweight gain of the animals. Maximum liveweight gain of

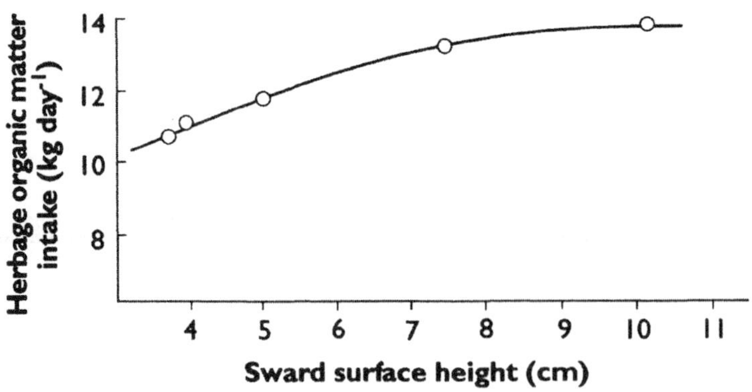

Figure 7.2 The herbage intake of beef cows continuously grazing swards of different heights.
Source: From Wright, 1988.

continuously grazing beef cows and growing/finishing cattle occur at sward heights of 8–10 cm. Above that height, liveweight gain decreases because of the reduction in the digestibility of the herbage, especially in late summer (Wright and Whyte, 1989; Swift *et al.*, 1989). For sheep, the maximum intake on perennial ryegrass dominated swards is around 4–6 cm (Figure 7.3).

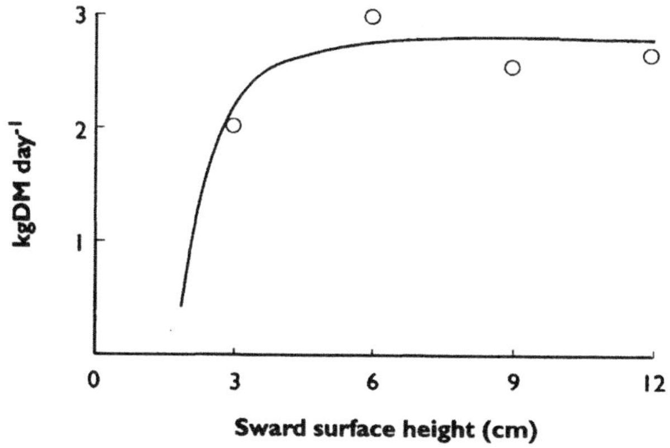

Figure 7.3 The herbage intake of ewes grazing swards of different heights.
Source: From Penning *et al.*, 1991.

This sward height information has been used to develop target sward height profiles for dairy cows (Mayne, Wright and Fisher, 2000), beef cattle (Wright, 1988) and sheep (Maxwell and Treacher, 1987) (Table 7.5). In spring, beef cattle should be turned out to pasture before sward height reaches 8 cm. If turnout is delayed until sward height is over 8 cm, it can be difficult to control sward height thereafter, because of high rates of herbage growth. In spring and early summer it is important to keep the sward surface height of continuously grazed swards to 7–8 cm because above that height there is increased development of flower heads in grass and this leads to a reduction in the digestibility and feeding value of the grass. Once the risk of seed head development has passed, then sward height can be allowed to increase to 10–12 cm in late summer. The same heights are recommended for mixed grass/white clover swards, although the clover component may benefit from a rest period followed by a conservation cut.

Table 7.5 Target sward heights (cm) for cattle and sheep grazing perennial ryegrass-dominated pasture

	Continuous stocking		Rotational grazing
	Early summer	Late summer	(post-grazing height)
Dairy cows	7–8	10–12	8–10
Growing/finishing cattle and beef cows	7–8	10–12	7–10
Sheep	4–6	6–8	—

Source: From Mayne *et al.*, 2000.

In New Zealand, mixed grass/white clover (*Trifolium repens*) swards have been central to the development of the pastoral agricultural industry. As concern grows about the impact of nutrient surpluses on surface water and groundwater, there is increased interest in the use of legumes in pasture in areas that have in recent decades relied on the application of high levels of nitrogen fertilizer Mixed grass/white clover swards are becoming more common, and offer considerable potential. In the United Kingdom, up to 280 kg nitrogen ha^{-1} $year^{-1}$ can be fixed by white clover in a mixed grass/white clover sward (Cowling, 1982), although a more usual figure might be 200 kg, and mixed grass/white clover swards with about 30 percent white clover can produce the same amount of herbage as all-grass swards receiving 200 kg nitrogen fertilizer ha^{-1} $year^{-1}$ (Morrison, 1981). Incorporation of white clover into a sward can improve the nutritive value of the pasture, with increased intake and nutrient supply, especially protein (Ulyatt *et al.*, 1980). Levels of animal performance are generally higher on grass/white clover swards than on all-grass swards (Davies and Hopkins, 1996) and Clark and Jans (1995) suggested that for each 10 percent increase in white clover in the sward, milk yield from dairy cows increases by 0.30–0.45 kg per day. Management systems based on mixed grass/white clover swards are being developed for environments where white clover is at the limit of its environmental range (Sibbald *et al.*, 2002) and mixed grass/legume systems will probably play a more important role in temperate grassland systems in the future. Several other legumes, such as red clover (*Trifolium pratense*) and birdsfoot trefoil (*Lotus corniculatus*) are currently being studied (Wilkins and Jones, 2000). Alfalfa (*Medicago sativa*) already plays a very important role as a source of conserved forage in many parts of the world.

The level of production of meat and milk from the arid grazing systems of the world is severely limited by climatic conditions. The levels of production rarely exceed that which can support subsistence levels, and the main grazing management tool is mobility of herd or flocks in response to the high spatial and temporal variability in rainfall. Herders move their animals, following water and pasture resources.

In semi-arid grazing systems, the level of biomass production is higher, although the stocking density is still relatively low. Traditionally, these areas have been grazed extensively by nomadic pastoralists, producing a range of meat, milk and fibre products, moving their animals seasonally. Where sedentarization of the population has occurred, this has often led ultimately to an increase in the stocking rate, sometimes stimulated by the ability to produce fodder for feeding during the dry season and an increase in the grazing pressure to the extent that the rangeland resources have become locally degraded.

The benefits of mobility in semi-arid grazing systems have been dramatically illustrated in Central Asia. In Kazakhstan, forced sedenterization of the nomads in the 1930s during collectivization in the Soviet Union led initially to a massive decline in animal numbers, due to starvation, as the animals were not moved to the traditional winter grazing areas. Gradually a form of centrally-organized flock mobility was re-introduced, with each collective having access to large areas of seasonally grazed pastures. At the break up of the Soviet Union sheep numbers collapsed in those countries where agriculture was privatized in the early 1990s, e.g. Kazakhstan (Behnke, 2003). Farmers did not have the resources to move their flocks to the traditional seasonal grazing areas and so most animals spent all year grazing around the villages, putting considerable pressure on the vegetation within 5–10 km of the village. Recently, larger farmers have started to become mobile and to move their flocks seasonally. These farmers have the resources, e.g. a truck and the ability to pay hired labour, to move their flocks of 300 or more. The benefits of doing so are twofold. The flocks that move seasonally incur much lower winter feed costs than those that stay around the village all year (US$ 0.24 compared with US$ 0.78 per sheep). The sheep in the mobile flocks also gain weight over winter (+3.7 kg) compared with a weight loss (–10.8 kg) in the non-mobile flocks (Kerven *et al.*, 2003). These differences demonstrate dramatically the benefits, from a production efficiency perspective, of flock mobility in these semi-arid systems.

Conservation of forage

Forage conservation will continue to play an important role in meat and milk production in some parts of the world to overcome seasonal shortages of feed supply. Forage can be conserved as hay, where weather conditions are such that the probability of rain occurring is relatively low. In higher rainfall areas, advances in silage-making technology in the past few decades have allowed a huge increase in the amount of silage made. For example, in the United Kingdom from 1970 to 1994, the amount of conserved fodder produced doubled from 8.2 to 16.3 million tonnes of dry matter with a sixfold increase in the amount of silage produced and a halving of the amount of hay made (Wilkinson, Wadephul and Hill, 1996). In Western Europe, about 70 percent of the fodder conserved is in the form of silage (Wilkins and Stark, 1992).

Recent advances in silage technology have included (i) the development of machinery, including mower design, conditioning and chopping equipment, and for producing big bale silage; (ii) the development of a range of chemical and biological additives to enhance and encourage fermentation; (iii) enhancement of silo design to reduce aerobic fermentation and effluent losses; and (iv) feeding technologies to reduce secondary fermentation.

While the costs per hectare of producing hay tend to be lower than for silage, the cost per unit of dry matter of fodder are similar because of higher yields of silage, related in part to lower field losses (Wilkins, 1990).

Future developments in silage making technology are likely to include the development of more effective inoculants, possibly by genetic engineering to produce lactic acid bacteria that can use a wider range of substrates, including cellulose. It may also be possible to breed varieties of grass that are particularly suited for ensiling, with, for example, a higher sugar content and lower protein degradation (Merry, Jones and Theodorou, 2000).

ORGANIC MEAT AND MILK PRODUCTION

In developed countries, where many consumers have high levels of disposable income, consumer concerns about food safety and quality and the environment have led to an increase in the production of organic or 'biological' farming. A simple definition of organic farming has eluded practitioners, commentators and scientists. As pointed out by Lampkin (1990), organic farming has often been defined by stating what it is not, but, as he points out, definitions such as 'farming without chemicals' misses out several key characteristics that are of fundamental importance. The most comprehensive definition of organic farming is provided by the guidelines to which producers have to adhere to qualify as 'organic farming'. Internationally, standards for organic farming are set by the International Federation of Organic Agriculture Movements. These standards are then interpreted by certifying authorities in different countries, taking account of local conditions and customs. Thus the precise standards may vary. For ruminant livestock production, the key standards are that no soluble mineral fertilizers can be used, animals may not be treated routinely with drugs, and there must be a minimum level of forage in the diet. The proportion of organic meat and milk sold in developed countries is increasing, partly due to an increase in the demand, but also in some cases partly due to financial incentives from governments to persuade farmers to convert production systems from conventional to organic because of the environmental benefits. The proportion of organic meat and milk production varies hugely in different countries, and although the market is increasing, it is not clear to what extent it will continue to do so.

The level of individual animal performance achieved in organic meat and milk systems is comparable to that in conventional systems, but the stocking rates are generally about 80 percent of those of conventional systems (Wright, Zervas and Louloudis, 2002). Thus because of the lower output per hectare, to be financially competitive with conventional systems the price for organic products must be higher than that for conventional products. For example, Figure 7.4 shows that for a beef production system in the United Kingdom, the price of organically produced beef must be at least 23% more than that of conventionally produced beef if the gross margins are to be comparable. While such premiums can be obtained, it is not clear if these can be sustained in the long term. While the price of organic milk in the United Kingdom is approximately 60 percent higher than conventional milk, the market may have reached saturation, as it is increasingly difficult for new organic producers to find a market.

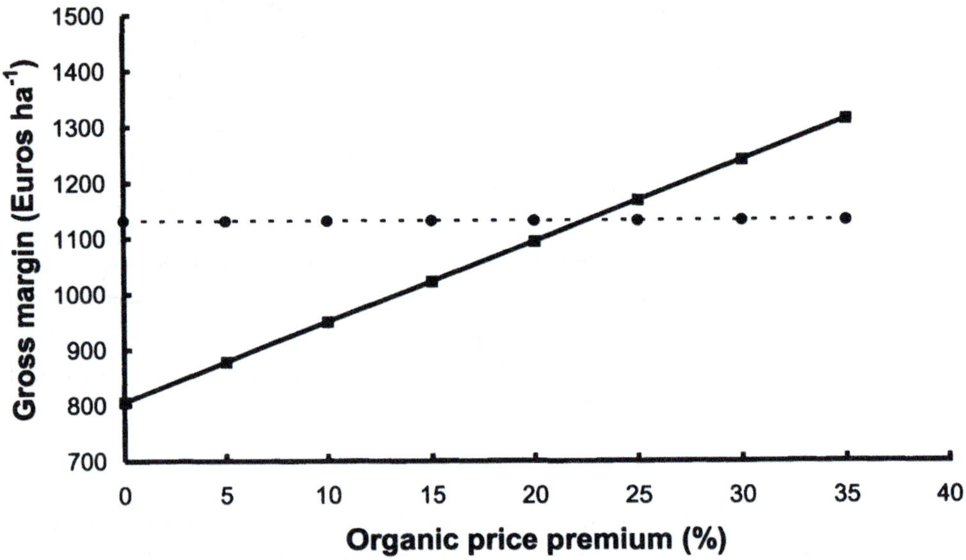

Figure 7.4 The effect of organic price premium on the gross margin from beef production.
Source: From Wright, Zervas and Louloudis, 2002.

Thus, while extensive production of meat and milk from extensive grassland systems are amongst the easiest to convert to organic production, because the levels of input have traditionally been relatively low, the long-term prospects for organic systems are not clear. As there is continued pressure to ensure that all livestock systems, and agriculture in general, develop in a way that has minimal environmental impact, the difference between organic and conventional systems may become less and so the growth in demand for organic products may slow.

CONCLUSIONS

There is considerable opportunity for producers of meat and milk to benefit from a number of future developments. In the developed world there is a real prospect that milk and meat, particularly from grassland-based systems, will increasingly be seen as an important component of a healthy human diet. Although this is not likely to result in significant increases in consumption, it may result in a stabilization of consumption. The projected increase in global demand potentially could result in opportunities for producers in developing countries. Whether they can take advantage of these opportunities will depend on a number of factors. These include the degree to which trade in meat and milk is liberalized and the extent to which markets are opened up. While there is a general move towards the reduction and removal of import tariffs, there may be a number of non-tariff barriers that prevent trade in agricultural products. These include sanitary standards, animal health and welfare requirements, and product quality and traceability (Holden, 2002). The increasing concerns about the environmental impact and animal welfare standards of intensive livestock production systems may provide increasing opportunities for the marketing of animal products from grass-based systems, which are seen by consumers as being more benign environmentally and offering a higher standard of animal welfare.

REFERENCES

ARC [Agricultural Research Council]. 1980. *The nutrient requirements of ruminant livestock.* Slough, UK: Commonwealth Agricultural Bureaux.

Ahmed, R. & Gruhn, P. 1995. Selected issues in the supply of and demand for red meat and poultry products in developing countries. *In:* R.T Wilson (ed). *Supply of livestock products to rapidly expanding urban populations.* Proceedings of the joint FAO/WAAO/KAAS symposium, Hoam Faculty Club, Seoul National University, Seoul, Korea, 16-20 May 1995. FAO, Rome.

Banni, S. & Martin, J.C. 1998. Conjugated linoleic acid and metabolites. pp. 261–302, *in:* J.L. Sebedio and W. Christie (eds). *Trans fatty acids in human nutrition.* Dundee, UK: The Oily Press.

Baptist, R. & Carles, A.B. 1989. How can litter size affect wool production with zero genetic and phenotypic correlations? pp. 124–130, *in: Proceedings of the Small Ruminant Collaborative Research Project Workshop.* Nairobi. VII,

Behnke, R.H. 2003. Reconfiguring property rights and land use in pastoral Kazakstan. *In:* C. Kerven and S. Akiner (eds). *Prospects for Pastoralism in Kazakhstan and Turkmenistan: From State Flocks to Private Farms.* Routledge Press (in press).

Behnke, R.H. & Scoones, I. 1993. Rethinking rangeland ecology: Implications for rangeland management in Africa. pp. 1–30, *in:* R.H. Behnke, I. Scoones and C. Kerven (eds). *Rangeland Ecology at Disequilibrium.* London: Overseas Development Institute.

Clark, D.A. & Jans, F. 1995. High forage use in sustainable dairy systems. pp. 497–526, *in:* M. Journet, E. Grenet, M.-H. Farce, M. Theriez and C. Demarquilly (eds). *Recent Developments in the Nutrition of Herbivores.* Proceedings of the VIth International Symposium on the Nutrition of Herbivores. Paris: INRA Editions.

Cowling, D.W. 1982. Biological nitrogen fixation and grassland production in the United Kingdom. *Philosophical Transactions of the Royal Society of London, Series B,* **296**: 397–404.

de Haan, C. Steinfeld, H. & Blackburn, H. 1997. *Livestock and the environment. Finding a balance.* European Commission Directorate General for Development.

Delgado, C., Rosegrant, M., Steinfeld, H., Ehui, S. & Courbois, C. 1999. *Livestock to 2020: the Next Food Revolution. IFPRI Food, Agriculture and the Environment Discussion Paper,* No. 28. International Food Policy Research Institute, Washington DC.

Davies, D.A. & Hopkins, A. 1996. Production benefits of legumes in grassland. pp. 234–246, *in:* D. Younie (ed). *Legumes in Sustainable Farming Systems. British Grassland Society Occasional Symposium,* No. 30.

Department of Health. 1994. Nutritional aspects of cardiovascular disease. *Report on health and social subjects,* No. 46. London: HMSO.

Enser, M., Hallet, K.G., Hewitt, B., Fursey, G.A.J., Wood, J.D. & Harrington, G. 1998. Fatty acid content and composition of UK beef and lamb muscle in relation to production system and implications for human nutrition. *Meat Science,* **49**: 323–341.

Fernandez-Giminez, M.E. & Allen-Diaz, B. 1999. Testing non-equilibrium models of rangeland vegetation dynamics in Mongolia. *Journal of Applied Ecology,* **36**: 871–885.

Frame, J. 1992. Feeding value of grass. pp. 146–160, *in: Improved Grassland Management.* Ipswich, UK: Farming Press.

Gilmanov, T.G. 1995. The state of rangeland resources in the newly independent states of the former USSR. pp. 10–13 (Vol. 2), *in:* N.E. West (ed). *Rangelands in a Sustainable Biosphere.* Proceedings of the Fifth International Rangeland Congress. Denver CO: Society for Range Management.

Global Livestock CRSP. 2001. Role of animal source foods to improve diet quality and growth and cognitive development in Kenyan school children. pp. 77–90, *in: Global Livestock CRSP Annual Report 2001*. Davis CA: University of California.

Hammond, J. 1947. Animal breeding in relation to nutrition and environmental conditions. *Biological Review of the Cambridge Philosophical Society*, **22**: 195–213.

Hlukik, J.G. & Smith, T.R. 1988. Economics of pasture systems. *Proceedings: Pasture in the Northeast Region of the United States*. NRAES-36. Cornell University, Ithaca, New York.

Hodgson, J. 1990. *Grazing Management: Science into Practice*. Harlow, UK: Longman.

Holden, S. 2002. How do trade regimes make livestock producers vulnerable? pp. 7–8, *in: Responding to the Increasing Global Demand for Animal Products*. Penicuik, UK: British Society of Animal Science.

Illius, A.W & O'Connor, T.G. 1999. When is grazing a major determinant of rangeland condition and productivity? pp. 474–476, *in:* D. Eldridge and D. Freudenberger (eds). *People of the Rangelands. Building the Future*. Proceedings of the VI International Rangeland Congress. Townsville, Australia, 19–23 July 1999.

Kerven, C., Alimaev, I., Behnke, R., Davidson, G., Franchois, L., Malmakov, N., Mathijis, E., Smailov, A. Temirbekov, S. & Wright, I. 2003. Retraction and expansion of flock mobility in Central Asia: Costs and Consequences. pp. 543–556, *in: Rangeland in the New Millennium*. Proceedings of VII International Rangeland Congress. Durban, South Africa, 28 July–1 August 2003.

Lampkin, N. 1990. *Organic Farming*. Ipswich, UK: Farming Press.

Mayne, C.S., Wright, I.A. & Fisher, G.E. 2000. Grassland management under grazing and animal response. pp. 247–291, *in:* A. Hopkins (ed). *Grass, its Production and Utilization*. 3rd ed. Oxford, UK: Blackwell Science.

Maxwell, T.J. & Treacher, T.T. 1987. Decision rules for grassland management. pp. 67–78, *in:* G.E. Pollot (ed). *Efficient sheep production from grassland. British Grassland Society Occasional Symposium*, No. 21.

Merry, R.J., Jones, R. & Theodorou, M.K. 2000. The conservation of grass. pp. 196–228, *in:* A. Hopkins (ed). *Grass, its Production and Utilization*. 3rd ed.. Oxford, UK: Blackwell Science.

Miller, L.A., Moorby, J.M., Davies, D.R., Humphreys, M.O., Scollan, N.D., Macrae, J.C. & Theodorou, M. K. 2001. *Grass and Forage Science*, **56**: 383–394.

Morrison, J. 1981. The potential of legumes for forage production. pp. 1.1–1.9, *in: Legumes and Fertilizers in Grassland Systems*. British Grassland Society, Winter Meeting 1981.

Mowle, A. & Bell, M. 1995. Rural policy factors and land use change. pp. 165–182, *in:* M.B. Usher and D.B.A Thompson (eds). *Ecological Change in the Uplands*. Special Publication No. 7 of the British Ecological Society. Oxford, UK: Blackwell Scientific Publications.

NRC (National Research Council). 1996. *Nutrient requirements of beef cattle*. 7th edition (revised). Washington DC: National Academy Press.

Nolan, A.J., Hope I. & Scott, R. 2002. Rapid assessment of grazing and trampling impacts on upland habitats. *in:* The Macaulay Institute Annual Report 2001. Macaulay Institute, Aberdeen, United Kingdom.

O'Reagain, P.J. & Schwartz, J. 1995. Dietary selection and foraging strategies of animals on rangeland. Coping with spatial and temporal variability. In: M. Journet, E. Grenet, M.-H. Farce, M. Theriez and C. Demarquilly (eds). *Recent Development in the Nutrition of Herbivores*. Proceedings of the VIth International Symposium on the Nutrition of Herbivores. Paris: INRA Editions.

Penning, P.D., Parsons, A.J., Orr, R.J. & Treacher, T.T. 1991. Intake and behaviour of responses by sheep to changes in sward characteristics under continuous grazing. *Grass and Forage Science*, **46**: 15–28.

Pittroff, W., Cartwright, T.C. & Kothmann, M.M. 2002. Perspectives for livestock on grazinglands. *Archivos Latinoamericanos de Producción Animal*, **10**: 133–143.

Riordan, N., Cowan, C. & McCarthy, M. 2002. Safety of Irish beef – concerns, awareness and knowledge of Irish consumers. *Journal of Food Safety*, **22**: 1–15.

Rook, A.J. 2000. Principles of foraging and grazing behaviour. pp. 229–246, *in:* A. Hopkins (ed). *Grass, its production and utilisation*. Oxford, UK: Blackwell Science.

Sibbald, A.R., Maxwell, T.J., Dalziel, A.J.I. & Agnew, R.D.M. 2002. The implications of controlling grazed sward height for the operation and productivity of upland sheep systems in the UK. 5. The effect of stocking rate and reduced levels of nitrogen fertilizer. *Grass and Forage Science*, **57**: 33–47.

Stolte, A.M., Nolan, A.J., Brewer, M.J., Duff, E.I, Elston, D.A. & Henderson, D.J. 2004. Rapid assessment of grazing and trampling impacts over large areas of rangeland; a sampling approach. *Biological Conservation* (in press).

Skarpe, C. 2000. Desertification, no-change or alternative states: Can we trust simple models of livestock impact in dry rangelands? *Applied Vegetation Science*, **3**: 261–268.

Swift, G., Lowman, B.G., Scott, N.A., Peebles, K., Nelson, D.R. & Hunter, E.A. 1989. Control of sward surface height and the growth of set-stocked finishing cattle. *Research and Development in Agriculture*, **6**: 91–97.

Thomas, H. 1997. Plants for all seasons. IGER Innovations. Institute of Grassland and Environmental Research, Aberystwyth, UK.

Ulyatt, M.J., Beever, D.E., Thomson, D.J., Evans, R.T. & Haines, M.J. 1980. Measurement of nutrient supply as pasture. *Proceedings of the Nutrition Society*, 39: 67A.

United Nations. 1995. *World Urbanization Prospects: the 1994 Revision*. United Nations, New York.

UNEP [United Nations Environment Programme]. 1992. *Status of Desertification and Implementation of the United Nations Plan of Action to Combat Desertification*. UNEP, Nairobi.

USDA [United States Department of Agriculture]. 1988. *Second RCA Appraisal: Soil, water and related resources on non-federal land in the US. Analysis of condition and trends*. Natural Resource Conservation Service, Washington DC.

Vagenas, D., Jackson, F., Russel, A.J.F., Merchant, M., Wright, I.A. & Bishop, S.C. 2002. Genetic control of resistance to gastro-intestinal parasites in crossbred cashmere goats: responses to selection, genetic parameters and relationships with production traits. *Animal Science*, **74**: 199–208.

Walker, J.W. & Hodgkinson, K.C. 1999. Grazing management: new technologies for old problems. pp. 424–430, *in:* D. Eldridge and D. Freudenberger (eds). *Proceedings of the VI International Rangeland Congress*. Townsville, Australia, 19–23 July 1999.

Ward, A.T., Wittenberg, K.M., Froebe, H.M., Przybylski, R. & Malcolmson, L. 2003. Fresh forage and solin supplementation on conjugated linoleic acid levels in plasma and milk. *Journal of Dairy Science*, **86**: 1742–1750.

Westoby, M., Walker, B. and Noy-Meir, I. 1989. Opportunistic management for rangelands not at equilibrium. *Journal of Range Management*, **42**: 131–138.

Wilkins, R.J. 1990. Grass utilisation for animal production. pp. 22–34 (Vol. 1), *in: Soil-Grassland-Animal Relationships*. Proceedings 13th General Meeting of the European Grassland Federation. Banska Bystica, June 1990.

Wilkins, R.J. 2000. Grassland in the twentieth century. IGER Innovations. Institute for Grassland and Environmental Research, Aberystwyth, United Kingdom.

Wilkins, R.J. & Stark, B.A. 1992. *Silage in Western Europe*. Second Edition. Canterbury, UK: Chalcombe Publications.

Wilkins, R.J. & Jones, R. 2000. Alternative home-grown protein sources for ruminants in the United Kingdom. *Animal Feed Science and Technology*, **85:** 23–32.

Wilkinson, J.M., Wadephul, F. & Hill, J. 1996. *Silage in Europe*. Lincoln, UK: Chalcombe Publications.

Wright, I.A. 1988. Suckler beef production. pp. 51–64, *in:* J. Frame (ed). *Efficient Beef Production from Grassland. British Grassland Society Occasional Symposium*, No. 22.

Wright, I.A., Jones, J.R., Maxwell, T.J., Russel, A.J.F. & Hunter, E.A. 1994. The effect of genotype × environment interactions on biological efficiency in beef cows. *Animal Production*, **58**: 197–207.

Wright, I.A. & Whyte, T.K. 1989. Effects of sward height on the performance of continuously stocked spring-calving beef cows and their calves. *Grass and Forage Science*, **44**: 259–266.

Wright, I.A., Zervas, G. & Louloudis, L. 2002. The development of sustainable farming systems and the challenges that face producers in the EU. pp. 27–37, *in:* I. Kyriazakis and G. Zervas (eds). *Organic Meat and Milk from Ruminants*. EAAP Publication, No. 106. Wageningen, The Netherlands: Wageningen Academic Publishers.

Zhong, M. 1993. Inner Mongolia grassland resources and its management. In: *Proceedings of International Symposium on Grassland Research*. Huhehot, China.

CHAPTER 8

New Zealand's Pastoral Industries: Efficient Use of Grassland Resources

John Hodgson[1], Keith Cameron[2], David Clark[3], Leo Condron[2], Tom Fraser[4], Mike Hedley[1], Colin Holmes[1], Peter Kemp[1], Richard Lucas[2], Derrick Moot[2], Steve Morris[1], Phillipa Nicholas[5], Nicola Shadbolt[1], Gavin Sheath[4], Ian Valentine[1], Garry Waghorn[4], Derek Woodfield[4]

ABSTRACT

The current status of New Zealand's pastoral industries is outlined in the context of climate and soil characteristics that ensure pasture growth at all seasons of the year in most regions. Traditionally, New Zealand's advantage for livestock products on world markets has been dependent on low feed costs resulting from the efficient use of grassland resources in year-round grazing systems. In recent years, changes in market opportunity have put increasing emphasis on improved animal performance and product specification. This has resulted in overall reduction in stocking rate in some livestock industries, with increasing emphasis on improvement in nutritive value in the choice of forage species and in the flexibility of management of pastures and supplementary forages. These changes have enhanced product value, at the risk of some decline in forage use efficiency and cost advantage. Recent developments in production technology are outlined, with emphasis on pasture production and nutritive value, and on animal proficiency and performance and on animal productive potential. Current land-use issues discussed relate to soil conservation, fertilizer use and water quality, and organic production systems.

[1]Massey University, Palmerston North, NZ
[2]Lincoln University, Canterbury, NZ
[3]Dexcel Co., Hamilton, NZ
[4]Agresearch Ltd. NZ
[5]Institute of Rural Studies, University of Wales, Aberystwyth, UK.

New Zealand's pastoral industries are committed to continuing improvement in production efficiency. Marketing strategy is expected to concentrate on food quality and unit value. Future market prospects will depend on the ability to maintain commercial advantage through efficiency of resource use, and on reputation for product quality, safety and reliability, reinforced by consumer perceptions of animal welfare, sustainable land use and environmental stability.

INTRODUCTION

The two main islands of New Zealand lie between 34°S and 47°S, between the Pacific Ocean and the Tasman Sea. The land area of the country is 268 675 km^2 with a total population of 4 million, giving a population density of 14.9 per km^2. The chain of mountains running roughly SW-NE through the length of the country is the product of continuing uplift from tectonic activity and associated volcanic activity. Much of the topography in the North Island is steep land. In both Islands there are extensive alluvial deposits associated with ice, water and wind movement, and in the North Island much of the land form is shrouded by layers of volcanic ash. As a consequence, most of the soils are relatively young and unstable. The mountains, rising to 3 764 m in the South Island and 2 797 m in the North Island, lie across the predominantly westerly wind flow. This results in a generally equable climate with well-distributed rainfall and temperatures adequate to maintain plant growth for most of the year. However, there are marked gradations in temperature with latitude and altitude, and a substantial decline in rainfall from west to east, reflecting the influence of the mountains on precipitation. The mean maximum temperature ranges from 15°C in the South to 20°C in the North, and variation in annual rainfall is from 9 000 mm on the western slopes of the southern mountains to 275 mm in the extreme rain shadow of their eastern flank.

The original climax vegetation of New Zealand was predominantly temperate forest and scrub, occupying up to 75 percent of the land area, with tussock grassland at high altitude and in drier areas. Extensive areas of forest had been burnt over by the original Polynesian peoples by the time of arrival of European settlers in the mid 1800s. Land clearance for farming continued until late in the twentieth century, by which time the indigenous forest cover had been reduced to 25 percent. Grassland now occupies 50 percent of the total land area, arable and horticultural land 2 percent, and planted exotic forest for timber production 5 percent.

New Zealand's suitability for pastoral production and its distance from export markets has resulted in a focus on livestock industries, with the major emphasis on efficient grazing systems. Year round grazing is the norm throughout most of the country. These advantages, together with reliance on legumes (mainly white clover (*Trifolium repens*)) to provide nitrogen, mean that the costs of livestock production are substantially lower than in most other developed countries, providing a strong base for competition in international food markets. New Zealand has relatively few natural resources apart from its climate and soils, so the country is heavily dependent on the pastoral industries for its economic welfare. Collectively, the crops, livestock, horticulture and forestry industries contribute 14 percent of GDP (17 percent if the service industries are taken into account). The pastoral industries are responsible for 45 percent of export earnings, (Table 8.1) with a very small proportion of the national production being absorbed by the home market.

Table 8.1 New Zealand's exports, year ending March 2002

	Value (millions)		% of total exports
	NZ$	US$	
Live animals	158	66	
Meat and meat products	4 514	1 891	
Dairy products	5 890	2 468	
Wool	940	394	
Other pastoral products[(1)]	2 557	1 069	
Total pastoral products	14 056	5 889	45
Horticultural products	2 110	884	
Other agricultural products[(2)]	456	191	
Total agricultural products	16 622	6 965	53
Forestry products	3 639	1 525	
Total primary product exports	20 261	8 489	65
Total New Zealand exports	31 112	13 036	100

Notes: (1) Includes deer velvet, caseinates, animal oils and fats, carpets, hides and skins. (2) Includes cereals, eggs, honey, flowers and plants.
Source: MAF, 2002.

Land use has always been dynamic in nature, in response to economic pressure and opportunity. Current medium-term changes involve rapid increase in the area devoted to dairying, and the removal for afforestation of marginal land from sheep production. A recent feature has been the development of deer units and forest blocks, both of limited scale, on many sheep and beef farms. Goat production is now a very small component of New Zealand farming. The land areas devoted to animal enterprises are shown in Table 8.2, and the national distribution of the main pastoral industries is illustrated in Figure 8.1. Changes in the number of animals in the main livestock categories in the last 15 years are shown in Table 8.3.

Table 8.2 Estimated land area of New Zealand in various pastoral enterprises

Pastoral enterprise	Hectares	%
Mixed sheep and beef	6 557 000	45.9
Predominantly sheep	3 044 000	21.3
Predominantly beef	1 953 000	13.7
Dairy	1 933 000	13.5
Deer farming	373 000	2.6
Others[(1)]	414 000	3.0
Total pasture	14 274 000	100

Notes: (1) Horses, goats, general grazing and dairy dry stock.
Source: Sanson and Pearson, 1997.

All government support for agriculture was phased out in the mid 1980s, and since that time the industry has had to compete for export sales on the international market (Sandrey and Reynolds, 1990). This has focused attention on the efficiency of resource use, with particular emphasis on maximizing the use of grazed grassland as the cheapest source of nutrients in ruminant production systems, and on the efficient use of labour and capital.

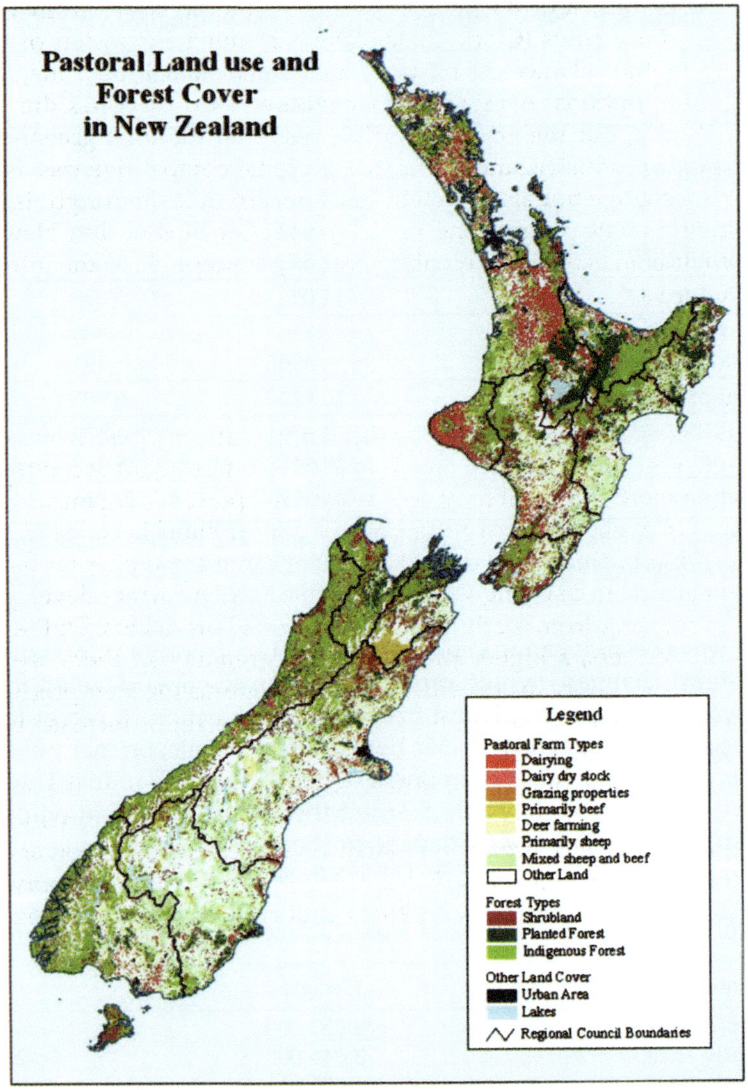

Figure 8.1 Pastoral land use and forest cover in New Zealand, 2002 (with acknowledgement to the Ministry of Agriculture and Forestry, Ministry for the Environment and Department of Conservation, and Terralink, New Zealand).

Table 8.3 Total livestock numbers in New Zealand 1985, 1995 and 2002 (thousand head)

	1985	1995	2002
Sheep	67 854	48 816	44 002
Dairy cattle	3 308	4 090	4 494
Beef cattle	4 613	5 183	4 787
Deer	320	1 179	1 800
Goats	427	256	175

Source: The Economic Service, 2003.

SOIL AND PASTURE RESOURCES

New Zealand grasslands can be conveniently divided on the basis of topography and elevation into three broad groups: high, hill, and flat to rolling country (Bryant and Sheath, 1987). While the area of each is similar (Table 8.4), they vary in the quantity of pasture produced, and the number and type of animals carried. The South Island high country is characterized by mountain terrain and low pasture production, especially during cold winter months. It supports mainly native tussock communities, and is used predominantly for sheep farming based on wool production. The intermediate hill country in both Islands is used for sheep, beef cattle and deer production. Flat to rolling country usually has good year-round pasture production, and supports almost all of New Zealand's dairy cattle, in addition to large numbers of sheep and beef cattle in intensive production systems. Many sheep and cattle are moved from hill farms to lowland farms for finishing.

Table 8.4 Area, pasture production and number of livestock carried on the three grassland regions

Region	Area (million hectares)	Pasture production (t DM ha^{-1})	Livestock (million)			su ha^{-1} (1)
			Sheep	Beef	Dairy	
High country	4.5	2.0	2.5	0.1	–	0.7
Hill country	5.0	7.0	20.3	1.9	0.2	7.5
Flat to rolling	4.5	11.0	26.7	2.9	3.4	14.0

Notes: (1) su = stock units. Since most farmers have both sheep and cattle, the stock unit (su) is the common unit used to express stocking rate, where 1 su = one breeding ewe and her lambs, equivalent to a forage intake of 550 kg DM $year^{-1}$. On this basis a breeding beef cow equates to 6 su, and a dairy cow to 8–9 su. From Coop (1965).
Source: Bryant and Sheath, 1987.

Within the area of improved pasture, flat to rolling land (slopes of 0°–15°) comprises approximately 55 percent, whereas 13 and 32 percent are hill pastures ranging in slope from 16°–20° and >21°, respectively. The soil orders on flat to rolling land are dominantly Fluvisols, Cambisols, Luvisols and Podzols in the South Island, and in the North Island Fluvisols, Luvisols, Cambisols, Acrisols, Gleysols and Podzols are mixed in the landscape with Andosols and Vitric Andosols (Bridges, 1997). These soil groups present only moderate limitations to pastoral farming. Soils orders of the hill country are mostly classed as steepland variants of Cambisols and Regosols. Although hill country with slopes of 21–26° can be cultivated by heavy machinery for pasture establishment and renovation, land with slopes greater than 21° is managed through the grazing pressure of sheep and cattle and aerial application of seeds, herbicides and fertilizers. The major limitation to pastoral farming on slopes greater than 16° is soil erosion.

With rainfall evenly distributed throughout all seasons in most areas, low winter and early spring (May-Sept.) evapotranspiration rates create seasonal drainage events ranging widely in volume depending upon location (Table 8.5). Large areas of Gleyic Cambisols, Gleyic Luvisols and Podzols have imperfect natural drainage caused by the formation of clay and iron pans, which prolongs the winter periods when soils are at or above field capacity and are vulnerable to stock treading damage. Areas of Distric and Eutric Cambisols and Fluvisols on the east coast of both Islands suffer from significant summer and autumn water deficits that limit pasture production, and these soils have significant irrigation potentials (Table 8.5). The total potential irrigable area is estimated at 2.56 million hectares of arable land (Sutton, 2002). Currently, 500 000 ha of agricultural and horticultural land is irrigated, either by flood irrigation or, increasingly, by more efficient spray and sprinkler systems. Examples of seasonal and annual variation in pasture production are illustrated in Figure 8.2.

Table 8.5 Simulated water volumes draining from the topsoil in the drainage season and the summer irrigation potential for selected sites in New Zealand

Location	Rainfall (mm)	Volume draining from topsoil (≈15 cm) (mm)	Irrigation potential (mm)	Number of days with drainage
North Island				
Whangarei (north)	1 510	773	218	77
Napier (east)	643	92	473	11
Ruakura (centre north)	1 221	382	85	49
New Plymouth (west)	1 707	923	127	97
Palmerston North (centre)	1 102	320	84	39
South Island				
Lincoln (east)	660	180	380	23
Alexandria (east)	535	22	205	4
Hokitika (west)	2 231	1 417	0	87
Gore (south centre)	845	221	77	38

Notes: The model of Moir *et al.* (2000) was used to generate simulated daily rainfall, evapotranspiration and soil water balances from long-term mean monthly climatological data (New Zealand Meteorological Service, 1973). The data is relevant to soils with silt loam texture and organic matter carbon contents of approximately 5 percent.

Irrigation potential is the irrigation volume required to meet evaporative demand and maintain the soil water deficit above –75 mm.

Aspect and altitude are particularly important in modifying temperature and soil moisture status in hilly areas. Sunny or north-facing aspects are warmer and drier than shady aspects, causing large differences in pasture composition and annual and seasonal DM production.

High producing pastures (Table 8.4) have a dominant grass component of perennial ryegrass (*Lolium perenne*) and a dominant legume component of white clover (*Trifolium repens*). However, pastures are not resown on a regular basis and so unsown species, both annual and perennial, often make an important contribution. C4 pasture species such as kikuyu (*Pennisetum clandestinum*) and paspalum (*Paspalum dilatatum*) contribute important amounts of feed during summer and autumn in northern regions. White clover has been the keystone to the development of New Zealand dairy pastures. Its ability to fix 100–350 kg N ha^{-1} $year^{-1}$ (Hoglund *et al.*, 1979) and provide a high quality feed during the warmer months has made it an ideal companion species for perennial ryegrass.

The major limitations to the establishment and maintenance of high producing pastures on flat to rolling land have been nutrient deficiencies, primarily nitrogen (N), phosphorus (P) and sulphur (S), with soil-specific, production-induced deficiencies of potassium (K), cobalt (Co), copper (Cu), molybdenum (Mo) and selenium (Se) (During, 1984; Floate, 1992). These deficiencies have been addressed through lime and fertilizer recommendations based on information gained from national field trials (e.g. Edmeades *et al.*, 1985; Sinclair *et al.*, 1997), which measured pasture growth responses to applied nutrients, and regular soil and herbage testing programmes. Decision-support software has been developed to estimate maintenance fertilizer requirements by mass-balance modelling of nutrient losses from the plant-available pool of nutrients in grazed pasture systems (Cornforth and Sinclair, 1982; Williams, Gregg and Hedley, 1990; Metherell *et al.*, 1997). Soil P and S status are adjusted to the needs of white clover. Single superphosphate (9% P, 12% S) remains the dominant phosphate fertilizer, accounting for 83 percent of the total tonnage. In some areas,

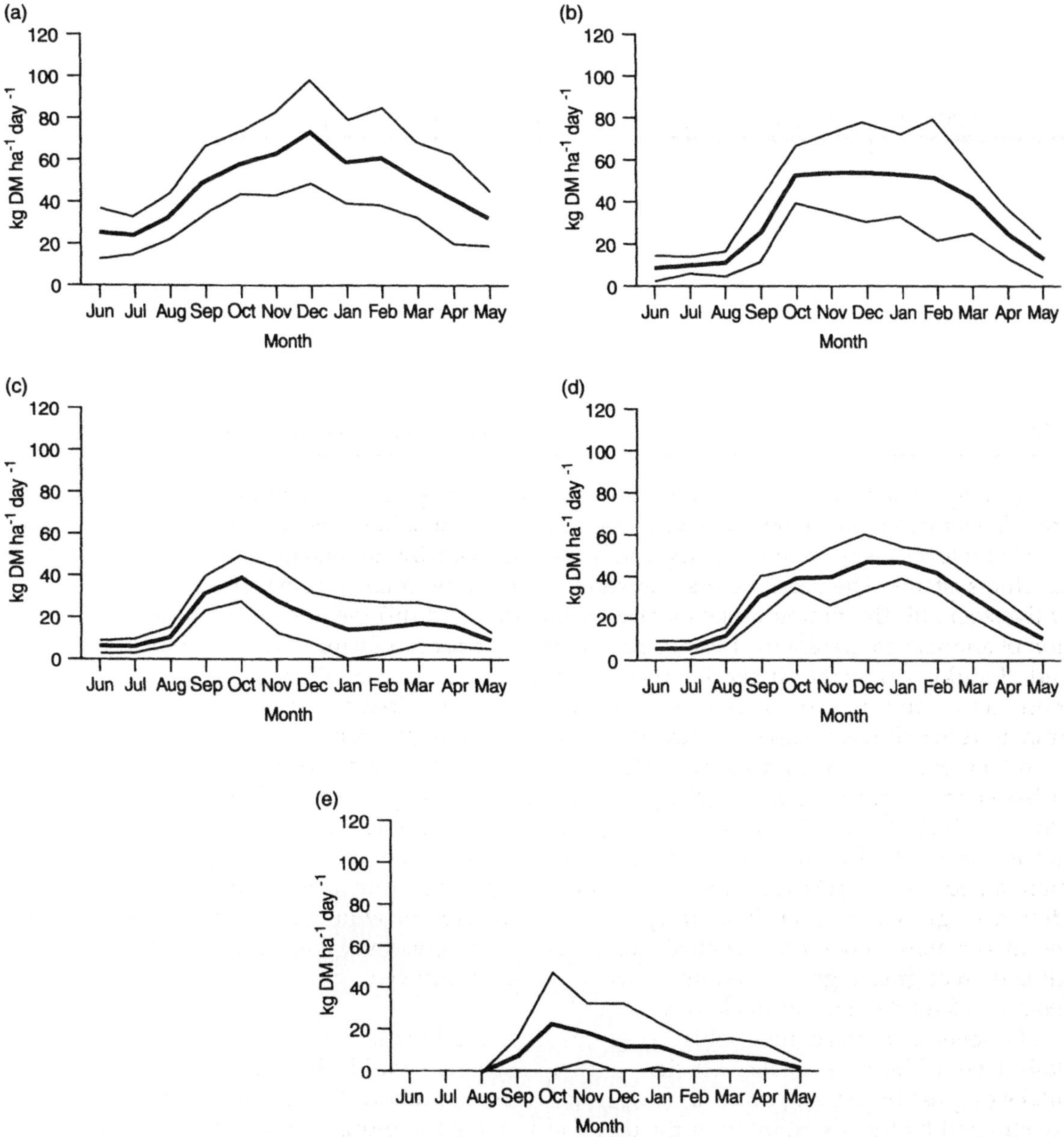

Figure 8.2 Seasonal variations in pasture production for contrasting locations in New Zealand: (a) Dargaville (North, North Island); (b) Winton (South, South Island); (c) Winchmore (East Coast, South Island) irrigated, and (d) unirrigated; (e) Poolburn (South Island high country). Bold lines show average values determined over 10–15 years; fine lines show ±one standard error (SE).
Source: McKenzie *et al.*, 1999.

increased magnesium deficiency has been induced as higher concentrations of Ca, K, SO_4 and Cl in lime and fertilizer have accelerated Mg loss from soils (McNaught *et al.*, 1973; Edmeades, Wheeler and Crouchley, 1985).

Hill pastures are an amalgam of contrasting slopes, aspects, soil parent materials and climate regimes. Current management, including aerial application of fertilizers, takes limited account of this great variation. Systems using computer-assisted soil mapping tools combined with an aerial topdressing plane fitted with a GPG/GIS controlled spreading unit have the potential to maximize the efficiency of pasture growth response to fertilizer and allow management plans to be developed to improve both the profitability and environmental sustainability of many hill country farms (Gillingham *et al.*, 1999).

The tussock grasslands occupy 5 million hectares in the South Island, comprising 3 million hectares of high country and 2 million hectares of hill country. Lying to the east of the Southern Alps, the grasslands are dominated by their physical environment and are the coldest pastoral lands in New Zealand. Even the lower-altitude inland basins have a 3-4 month dormant period in winter, with potential for frost in all seasons, and snow lies at high altitude for several months each winter. The soils are formed largely from mica schist or greywacke and the loess derived from these sandstone rocks (Eutric Cambisols).

Following European settlement in the 1850s, the grasslands were occupied as large sheep runs (O'Connor *et al.*, 1996). The predominant tall tussock (*Chionochloa* spp.) cover was quickly modified by burning and grazing by sheep and rabbits, and during the following 100 years of pastoralism the grasslands were depleted of cover, native species and soil nutrients. The tall tussock only survived at high altitude, and elsewhere has been replaced by short tussock (*Poa laevis, Festuca novae-zelandiae*), mat plants and bare ground. At the same time, introduced grasses and herbs have spread through the grassland, mainly species tolerant of low fertility. Concentration on the use of improved pastures and forage crops on lower altitude flats and terraces has resulted in a doubling of livestock numbers in the last 50 years.

DAIRY PRODUCTION

Dairying systems in New Zealand are largely based on the efficient use of grazed pasture, which contributes about 90 percent of animal feed requirements. Milk production is based on a seasonal calving cycle in which the majority of cows calve in a limited period in late winter-early spring. This cycle means that peak feed demand nine to ten weeks after calving coincides with peak pasture growth in mid spring, with some pasture surplus at this time and deficits in summer and winter (Figure 8.3). Reliance on grazed pasture means that production per cow is limited, judged by international standards, but production per hectare of grassland is high, reflecting intensive grazing management and high stocking rates (Table 8.6). The technical efficiency of the industry is indicated by the steady increase in stocking rate, production per cow and per hectare over many years; changes over the last ten years are shown in Table 8.6.

New Zealand has a history of dairy farm system research using farmlets and conducted over at least one whole lactation. This research has developed feed budget, pasture assessment and cow condition scoring techniques that allow a herd manager to keep both pasture cover (the average pasture present on-farm, expressed as kg DM ha^{-1}) and cow condition within bounds optimal for both present and future production. The calculation of the correct stocking rate coupled with long-term and short-term pasture and herd management decisions allows for the profitable production of milk from the lowest-cost feed source (see Holmes *et al.*, 2002). The result of this efficient use of grazed grassland is that production costs (NZ$ 4.00 per kg milk solids (US$ 1.68) at current prices) are only half of these in Europe and USA, where grazed pastures make up only half or less of the diet.

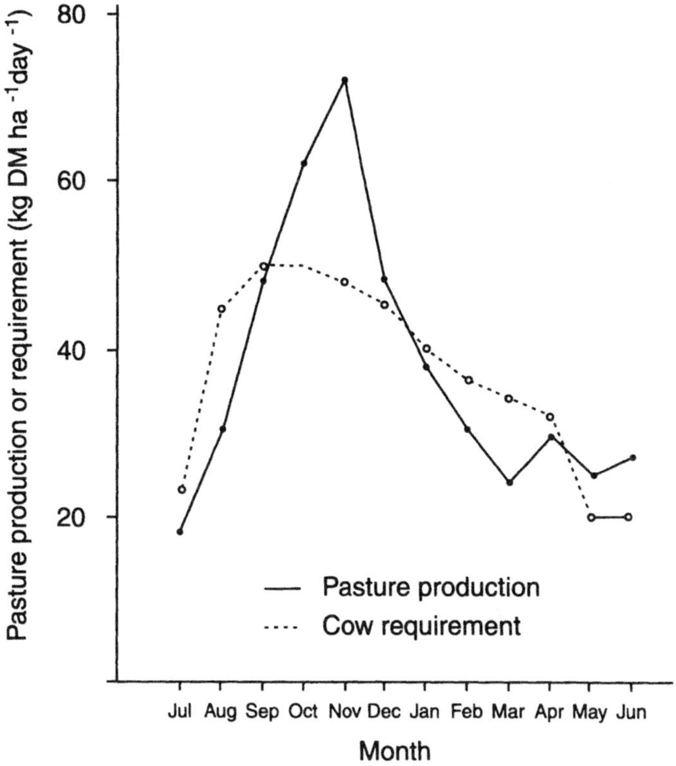

Figure 8.3 Seasonal curves of pasture production and feed demand for dairy cows stocked at 3.3 cows ha^{-1} in the central North Island.
Source: Holmes *et al.*, 2002.

Table 8.6 Changes in herd size, stocking rate, production per cow and per hectare (kg milk solids) in the New Zealand dairy industry, 1992/3 to 2001/2

		1992/3	2001/2
Herd size	(cows)	180	271
Stocking rate	(cows ha^{-1}))	2.5	2.67
Milk solids	(kg)		
per cow		259	307
per ha		653	824

Source: LIC, 2003

The New Zealand dairy industry has increased in size and undergone important structural changes in the past decade. From 1991/92 to 2001/02, dairy cow numbers increased from 2.44 million to 3.69 million, land area for dairy farming increased from 1.0 million hectares to 1.4 million hectares, and milk processed increased from 7 454 million litres to 12 998 million litres (LIC, 2003). A feature of this expansion has been the conversion of land to dairying from dryland sheep and cropping in the South Island provinces of Canterbury and Otago, and intensive lamb production in Southland.

This rapid expansion of milk production in New Zealand might suggest an industry that is very confident of the future, and indeed the industry has accepted a goal of 4 percent annual increase in productivity. The confidence is justified by the long-term growth projections of world demand for dairy produce. However, New Zealand's current 45 percent of the world trade in dairy produce is absolutely dependent on maintaining a lower cost of milk production than international competitors. Recent developments in New Zealand have put this competitiveness at risk. Almost all farmers now purchase extra feed for their farm. This includes off-farm grazing for replacement stock, buying maize silage, grain, by-products, pasture silage and hay and a range of New Zealand or overseas feedstuffs. Most of these options lead to increases in the average feed cost on farm. Where pasture utilization is at optimal levels, this strategy offers ways of increasing profitability per hectare, but in many cases pasture use is not optimal and substitution of low-cost pasture by expensive imported feeds means that pasture is wasted and profits decline.

The increased use of bought-in supplementary feed and nitrogen fertilizer has occurred because of a recognition that unirrigated perennial ryegrass-white clover pastures are unable to grow much more than 15 tonne DM ha^{-1} $year^{-1}$ (Hodgson, 1989). Maize as silage now plays an important role in many North Island dairy farms, with increasing interest in using cereal silages in the South Island. Average maize silage yields in the north are 21 tonne DM ha^{-1} $year^{-1}$ and additional feed can be grown with an annual ryegrass or cereal crop in the winter. However, maize silage cannot form the major portion of a cow's ration because the protein content of 80 g kg^{-1} DM is only half of that required for moderate milk yields.

From 1980 to 2000 the percentage of Holstein Friesian cows in the New Zealand herd with North American Holstein genes has increased from 2 to 38 percent. Harris and Kolver (2001) concluded that North American derived Holsteins had higher milk volume and milk protein yield but lower survival and poorer fertility than New Zealand Holsteins. Failure of the former to get in calf and maintain a 365-day calving interval appeared to be a key reason for the reduced survival of these cows. Many farmers have increased their use of supplements and nitrogen fertilizer in an effort to feed the North American Holstein cows better, but there is a continuing debate about the merits of this approach in an environment of low milk price and high feed costs.

SHEEP, BEEF CATTLE AND DEER PRODUCTION

There are few specialized sheep, deer or cattle farms. Having sheep and cattle or deer on the same farm increases management flexibility because it increases the ability to preferentially feed some stock while maintaining high levels of grazing pressure with other classes. The role of the breeding cow continues to be important in the sustainability of hill country farms where the contour requires that the prevention of pasture deterioration and weed ingress is primarily a function of stock pressure and grazing management.

Pasture provides over 95 percent of the diet of sheep, deer and beef cattle, but sheep and beef cattle farms have marked variations in annual and seasonal pasture production (Figure 8.4) and sward composition, requiring careful adjustment in stock and pasture management. In the wet North Island hill country the challenge is to winter sufficient stock to then be able to utilise the high levels of pasture production over the spring and summer. This means that a high proportion of the stock farmed are breeding ewes and cows which can be fed at maintenance in the winter, while lactation demand and the development of grazing activity in lambs and calves will increase spring and summer pasture demand. Increasing the proportion of finishing stock on the farm will lead to greater pasture demand in the winter and decreased demand in the summer. The key to success is to adjust stock numbers in the autumn to achieve pasture cover and animal liveweight targets at the start of winter. This then enables feed to be rationed throughout the winter to transfer pasture into the more productive period after lambing and calving. This rationing will be best achieved under a rotational grazing regime. The aim is then to fully feed ewes and cows during lactation. For ewes and lambs target sward conditions and animal intakes at lambing and during lactation are best obtained under continuous stocking.

In the drier areas of the East Coast the requirement is to achieve maximum production through the spring and early summer and then to significantly reduce demand over the summer when drought reduces pasture production (Figure 8.4). This suits lamb production providing lambing can take place early enough in the spring to enable lambs to be marketed before the onset of the summer dry. Cattle breeding is less suitable in these areas due to lactation demand over the summer, but cattle finishing systems can be successfully operated provided supplementary feed is available over the winter and finishing stock are sold early in the summer.

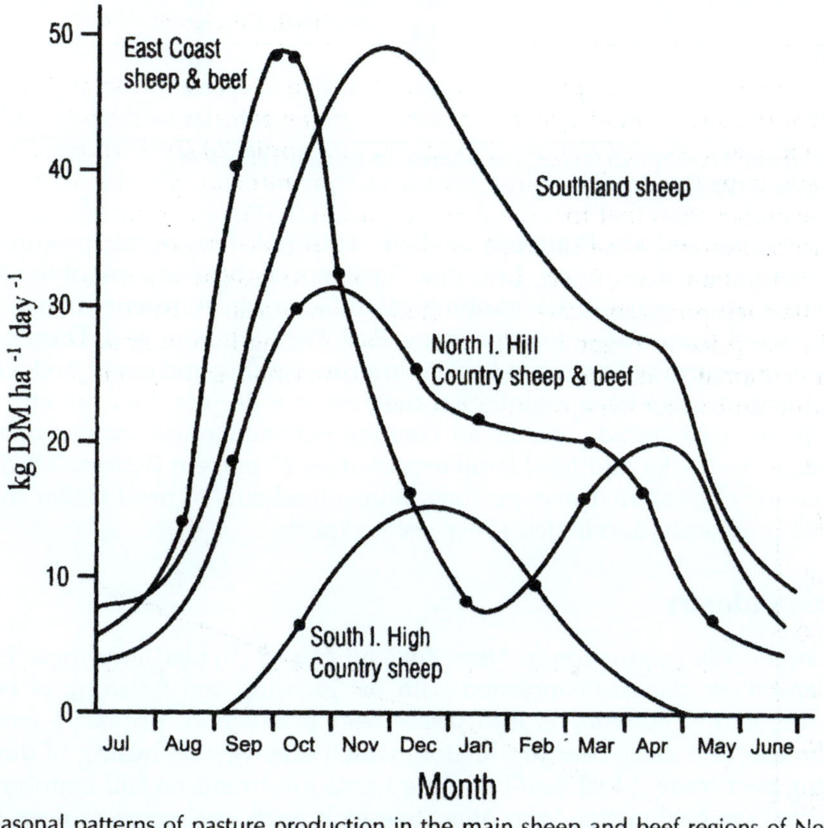

Figure 8.4 Seasonal patterns of pasture production in the main sheep and beef regions of New Zealand. Source: Matthews *et al.*, 1999.

The sheep industry

Sheep farming in New Zealand has in the past been extensive in nature, with sheep being farmed on the high or hill country at stocking rates around 8–12 sheep per hectare, mostly with no supplementary feed and no housing. Some key performance statistics for sheep and cattle farms are given in Table 8.7. Annual lambing percentages (lambs surviving to weaning per 100 ewes mated) have increased significantly over the last 15 years, from 98 percent in 1987 to 125 percent in 2002, and range from 88 percent on high country to 138 percent on South Island intensive finishing country.

Table 8.7 Average statistics for three types of sheep and cattle farms (2000–2001)

		SI high country	NI hill country	NI finishing
Area per farm	(ha)	9 246	421	261
Sheep	('000)	9.4	2.5	1.6
Stocking rate	(su ha^{-1})	1.1	9.8	12.2
Percentage su as cattle		18	41	47
production	(kg ha^{-1})			
– meat (sheep & beef)		11	187	289
– wool		4.4	30	36.4
Lambing (%)		88	110	116
Calving (%)		84	80	86
Labour units		2.7	1.6	1.4

Notes: SI = South Island; NI = North Island. su = stock units (see Table 8.4).
Source: Adapted from MWIES, 2002.

Most of the sheep farmed are Romneys or their crossbred derivatives, producing meat and wool suitable for carpet manufacturing. Less than 5 percent of the sheep are of the Merino breed, which produce fine wools suitable for clothing manufacturing. A recent change has been the increased use of breed composites, incorporating East Friesian, Finn and Texel genes into the existing Romney, Coopworth or Perendale flock to improve fertility and meat production potential.

The meat and wool sectors have maintained their returns despite the dramatic reductions in stock numbers. In the past decade, the mean lambing percentage has increased by 18 percent, mean carcass weight by 2.9 kg and total lamb exported by 23 percent (Fraser, 2000). At the same time the meat industry has shifted to exporting value-added cuts of meat rather than the frozen carcasses that had previously dominated sheep meat exports.

The beef cattle industry

The systems of beef cattle production in New Zealand can be divided into those involving beef breeding cows and those that are concerned with the growing and finishing of beef cattle. An alternative division of the industry is into prime beef production, primarily from steers and heifers of beef breeds, and dairy beef production, which involves the raising of dairy-bred bulls for the processing beef trade. Most beef-breeding herds are found on hill country farms in the North Island, usually in conjunction with other livestock, such as sheep and deer. Growing and finishing farms for beef production, in contrast, are mainly on the lowlands, where the cattle can be finished on high quality pastures.

The ratio of sheep to cattle on a stock unit basis (see Table 8.4) varies from 40:60 in Northland to 80:20 in Southland. Under extensive conditions, cattle can be considered as complementary to sheep but, when subdivision and stocking rate increase on an individual farm, it is true to say that the role of cattle changes from complementary to competitive. Breeding and finishing systems are often combined. Most beef cattle in New Zealand are on farms where the highest proportion of income is derived from some other enterprise, usually sheep.

Bull beef farms are more likely to be specialized. Typically, spring-born Friesians are purchased as weaned calves in October and November at four months of age and 100 kg liveweight. The bulls are sold for slaughter over the period December to April, when 14 to 18 months of age and 500 to 580 kg liveweight (260 to 300 kg carcass weight). To achieve this weight, an average liveweight gain of over 1.10 kg head^{-1} day^{-1} is required for the entire period. Best economic returns are obtained at stocking rates of between 2.5 and 3.5 bulls per hectare; higher rates fail to achieve adequate animal performance (McRae, 1988; Cassells and Mathews, 1988). Friesian bulls are now being farmed on hill country rather than in specialized bull farms, and Hereford-Friesian heifer calves from the dairy herd are being used as a source of high producing beef breeding cows for once-bred heifer systems, and for heifer beef finishing (Morris *et al.*, 1992).

Beef production is highly seasonal, with peak slaughter production from January to May, and is sensitive to variation in pasture production. This means that processing plants are underutilized for long periods, and export flows are seasonal.

The deer industry

Wild deer have been captured from mountain and forest areas during the last 30 years to establish an increasingly valuable farmed-deer industry based on two main products: meat (venison) from one-to-two-year old animals, which is largely sold into markets in Europe; and antlers (velvet) from mature animals sold into the speciality markets for health products in Asia. Farmed venison has established a high reputation as a red meat with a low fat (around 9 percent) content. Targets for venison production are 90–100 kg live weight (50–60 kg carcass) at 12–15 months of age. Average yields of velvet vary from 1 kg in two-year-olds to 2.5 kg in five-to-eight-year-old stags, and exceptional stags can produce up to 4.5 kg.

Virtually all farmed deer (mainly red deer, with increasing numbers of North American elk, or wapiti, used as crossing sires) are run in separate units on existing sheep and cattle farms. Deer have adapted well to enclosure and confinement, and to grass-clover pastures, intensive stocking and rotational grazing. A complication to farming deer is that hinds do not calve until late November or early December, after the spring flush of pasture growth. Therefore their maximum feed requirement is in the summer when both pasture quality and growth may be declining. Relative to their sheep and cattle counterparts, stags being finished for meat production have a high feed requirement during winter. The seasonal feed demand of the deer herd is therefore considerably out of phase with pasture growth.

PASTURES

Grazing management

Successful grazing management depends on knowledge of feed requirements, pasture availability and the integration of different classes of animals into the system. Fitting the system together has required a clear understanding of pasture growth patterns and animal requirements, and strategies to overcome periods of feed surpluses and deficiencies. Conscious rationing of pastures to animals means that pasture can be accumulated and used at a later date. This strategy is most often used in the period leading into and during winter, to enable a build up of pasture mass for grazing at lambing and calving in spring. Late spring feed surpluses can be conserved as hay or silage and fed out during the winter, or during the summer, as is the case in summer dry regions. These principles are illustrated in the sections of this chapter dealing with dairy, sheep, beef and deer production respectively, and more detailed examples are given in Sheath and Clark (1996).

New Zealand pastoral systems have traditionally been based on rotational grazing of perennial ryegrass and white clover pastures at high stocking rates to ensure efficient pasture utilization, and tight control of calving and lambing dates to match seasonal variation in pasture supply and demand (Matthews, Hodgson and White, 1999). These principles remain in force, but in recent years there has been increasing concern about the impacts of such management strategies on pasture production and stability and animal performance, and associated environmental risks. As a consequence, there has been a progressive tendency to reduce stocking rates and put more emphasis on animal performance targets to meet market specifications (Matthews, Hodgson and White, 1999). This change has been accompanied by greater emphasis on the control of pasture condition as a determinant of herbage production and nutritive value, and of animal performance. Though the emphasis is still predominantly on maximizing the use of grazed pasture, increased attention is paid to the importance of using forage-based supplements and alternative pasture species to balance variations in pasture quantity and quality, and the use of objective pasture assessment techniques to provide a solid basis for management decisions.

Pasture production and persistence

Historically, pastures in New Zealand have been long-term in nature, based on perennial ryegrass and white clover. However, as production systems have become more specialized, there has been an increasing use of different forage species to accommodate different managements and environments. This change has been associated with the shift to the higher pasture quality needed to achieve good animal performance to meet market specifications.

In dairying and livestock finishing systems there has been greater use of short-term (1–3 years) pastures, with the emphasis on high quality and out-of-season production. Short-term ryegrasses (Italian (*Lolium multiflorum*) and hybrid (*Lolium hybridum*)), red clover (*Trifolium pratense*) and herbs like chicory (*Cichorium intybus*) and plantain (*Plantago lanceolata*) are increasingly being used in these systems. Use of both phosphate and nitrogen fertilizer may be increased to accommodate these changes. Livestock breeding enterprises, in contrast, place the emphasis on long-term pastures, with a reliance on endophyte-safe perennial ryegrass, tall fescue (*Festuca arundinacea*) mixed with white clover and, in the drier areas, subterranean clover (*Trifolium subterraneum*). In the South Island hill areas with limiting soil nutrient status, alternative legumes are Caucasian clover (*Trifolium ambiguum*) and *Lotus pedunculatus*. Lucerne (*Medicago sativa*) is also showing a resurgence as a crop for conservation or sheep grazing in the drier eastern regions, especially in the South Island, with particular value for dairy cattle as a balance for the low crude protein content of maize silage. Forage brassicas are also grown extensively as supplements for both sheep and cattle in summer and winter.

Persistence of perennial pastures has become a source of concern. Recent evidence (Smith *et al.*, 1998; Daly *et al.*, 1999; Knowles, Frazer and Daly, 2003) shows that some species in long-term pastures on sheep and beef farms throughout New Zealand are starting to decline within four years of establishment. In summer-dry areas, white clover may disappear from pastures after the first drought, and its reappearance is dependent on buried seed. Also, perennial ryegrass cultivars may survive only if they are not grazed during the drought period. Taking into account the wide range of environmental conditions in the country, it is perhaps surprising that the reliance on perennial ryegrass and white clover has not resulted in greater problems, particularly in view of the practice of evaluating new pasture cultivars in high fertility, easy environments under ideal management conditions.

Pasture quality and ruminant nutrition

The introduction of better quality species and cultivars, together with management practices that lead to improving pasture quality, are now the main drivers for increasing animal performance. High pasture quality is achieved by maintaining emphasis on maximizing legume content, maximizing grass green leaf and minimizing senescent and dead material. It has long been accepted that legume content has a major influence on feed quality, but it is only more recently that the importance of the effect of senescent material on herbage quality has been recognized.

Progress has been achieved in selection of grasses to provide good nutrition for ruminants. New cultivars are able to overcome limitations to animal performance caused by deleterious fungal endophytes and oestrogenic compounds (Waghorn, Adams and Woodfield, 2002), and mixtures of grasses, legumes, herbs and silages offer opportunities for creating forage-based mixed rations (Burke, Waghorn and Chaves, 2002) able to meet the requirements of animals of high genetic potential. Selective breeding has created ryegrasses with late maturation, which should improve intake during spring grazing. Current research aims to increase concentrations of rapidly digestible constituents – such as lipids and soluble carbohydrates – in plant dry matter, using both conventional breeding and genetic engineering approaches. The problem of slowly degradable plant fibre has not been subject to serious research, although any increase in proportions of lipids and soluble carbohydrates may reduce the amount of fibre in forages.

The condensed tannins present in *Lotus* spp. and sulla (*Hedysarum coronarium*), and in the flowers of white clover and some red clovers, have become the focus of an extensive research effort to exploit their benefits for ruminant production and welfare. Benefits include prevention of bloat, reduced impact of intestinal worms, increased milk production and milk protein concentration in sheep and cows, and increases in liveweight gain in lambs and lambing percentage (Woodward *et al.*, 1999). More recently, condensed tannins have been shown to reduce the production of enteric methane (Waghorn, Tavendale and Woodfield, 2002). All these effects are influenced by both the concentration and the chemical structure of the tannins in the diet.

The endophytic fungi (*Neotyphodium lolii*) growing in association with perennial ryegrass have had substantial effects on the productivity of grazing systems through the production of alkaloids like peramine, which protects the plant against attack by Argentine stem weevil (*Listronatus bonariensis*) and other insects, and lolitrem and ergovaline, which adversely affect animal health by stimulating tremorgenic and heat stress responses, respectively. Isolation of these active agents, combined with the identification and selection of strains of endophyte that synthesize peramine but not lolitrem or ergovaline (Woodfield and Matthew, 1999), represent major advances in pasture persistence and animal health and production.

Development of forages with improved nutritive value and fewer deleterious compounds (Waghorn, Adams and Woodfield, 2002), together with appropriate nutritional management, has allowed New Zealand farmers to improve animal productivity and profitability. An increasing amount of research is now being directed toward product quality, especially understanding flavours and flavour precursors associated with pasture feeding.

Plant improvement

Intensification in New Zealand farming systems is placing increasing pressure on the performance of traditional grass-legume pastures (Clark, Matthew and Crush, 2001). Genetic improvements in herbage yield, nitrogen fixation and animal performance have been well documented in the main forage legumes (Woodfield, 1999), as have improvements in herbage yield, rust resistance and heading date of the main forage grasses (Woodfield, 1999; Westwood and Norriss, 1999; Easton *et al.*, 2002). These improvements are due to selection for improved yield potential and to reduction in the effects of yield limiting factors, such as pest and diseases.

Plant improvement effort has focused on ryegrasses and white clover. A continuum of ryegrass cultivars – from persistent perennial ryegrasses through to annual Italian ryegrasses – have been produced, with heading date and ploidy (diploid versus tetraploid) level providing the main points of differentiation (Charlton and Stewart, 2000). Cultivars of a range of other grasses (tall fescue, cocksfoot (*Dactylis glomerata*), timothy (*Phleum pratense*), phalaris (*Phalaris aquatica*) and brome (*Bromus stamineus*)) have also been developed. During the past decade, the use of tall fescue in permanent pastures and Italian ryegrass, hybrid ryegrass, chicory and forage brassicas in specialist finishing pastures has increased dramatically. White clover cultivars suited to the full range of environmental conditions and farming systems have been released. White clover and a range of other legumes (red clover, Caucasian clover, birdsfoot trefoil (*Lotus corniculatus*), greater lotus (*Lotus pedunculatus*) and several annual legumes) are incorporated into pastures to improve nutritive value and intake rates. More recently, cultivars of herbs such as chicory and plantain have been commercialized for similar reasons.

Forage quantity, quality and seasonal distribution remain the core targets in New Zealand breeding programmes. Plant factors that are considered important determinants of pasture quantity and quality include establishment vigour, regrowth following grazing, seasonal growth, drought tolerance and soluble carbohydrate content. Persistence is also a prominent target, with selection for a range of morphological traits and various biotic and abiotic stress tolerances. Breeding objectives have increasingly diversified to include improved protein-energy balance, increased by-pass protein levels, leaf properties affecting forage intake, and manipulation of compounds that affect animal health and welfare, fertility, and product flavour and texture.

Genetic resistance to pests and diseases is an important part of integrated pest management strategies. The ability of races of disease organisms such as rusts to evolve and overcome existing resistance mechanisms requires constant attention. Significant improvement has been made for resistance to crown and stem rusts in perennial ryegrass and cocksfoot (Easton *et al.*, 2002). Sources of resistance to three nematode species in white clover have been identified and incorporated into current breeding programmes (Barrett *et al.*, 2001). The recent arrival of clover root weevil (*Sitona lepidus*) (Barratt, Barker and Addison, 1996) has seriously compromised white clover performance in northern New Zealand. No sources of resistance have been identified to date, but sources of tolerance have been identified in both red clover and white clover (Eerens *et al.*, 2001).

ISSUES

Soil conservation

Only 8.4 million hectares out of the total 16 million hectares of farmland in New Zealand are regarded as being suitable for sustaining the soil resource without soil conservation measures

(Wilkinson, 1999). Pastoral farming is the main land use on the remaining 7.6 million hectares that are rated as unsustainable without soil erosion controls like tree planting or land retirement (MOE, 1997). The current trend is for the area used for hill pastoral farming to decrease as land that is too steep, badly eroded or poorly accessible is either left to return to indigenous shrub or forest, or is planted to exotic forest, usually *Pinus radiata*. Also, there has been an increase in the numbers of pastoral farmers planting woodlots, protecting indigenous forest remnants by fencing them from livestock, and using bush and tussock areas for recreational activities (e.g. Wallace, 2002) such as walking, wildlife conservation, bird watching, hunting and fishing. Legislation for the conservation of soil resource and the protection of water quality (see next section) is embodied in the Resource Management Act (1991), which has wide-ranging implications for resource conservation and environmental management.

The integration of trees into pastoral farming in New Zealand is vital for the development of sustainable pastoral systems, as well as for meeting societal demands for sustainable land use, shelter for animal welfare, clean waterways, protection of aesthetic landscapes, and flora and fauna conservation. On erosion-prone hills, the most effective soil conservation method that still enables pastoral farming to occur is to plant widely spaced trees (20–150 trees ha^{-1}), with poplars (*Populus* spp.) and willows (*Salix* spp.) being the most widely used (Wall *et al.*, 1997). Secondary advantages are the moderate nutritive value of the leaves and fine stems of these species, and the timber potential of poplar (McGregor *et al.*, 1999). Silvipastoral systems based on widely spaced *Pinus radiata* trees have been found to be ineffective in providing either persistent pastures or good quality timber (Hawke, 1997). Future use of *Pinus radiata*, the predominant species in commercial forestry, is likely to be restricted to high density plantations for erosion protection of retired pastoral land and as managed woodlots on pastoral farms.

Eucalyptus, *Alnus* and *Acacia* species have also been used in silvipastoral systems, but their more specific environmental requirements compared with poplars and willows, and the need to establish them from seedlings, has restricted their use. There has been research on tree fodder blocks based on browse shrubs (e.g. *Chamaecytisus palmensis*, *Atriplex* sp., *Dorycnium rectum*) for supplementary feed during summer, but with only modest uptake by farmers over the last 30 years (Lambert, Jung and Costall, 1989).

The integration of the use of poplars and willows into hill pastoral systems is now being encouraged by Regional Councils but, despite their effectiveness in controlling soil erosion, analysis suggests that the economic benefit is marginal (Parminter, Dodd and Mackay, 2001). An issue for the future management of pastoral farming on hills will be determining how the cost of soil erosion and the management of water catchments is balanced between pastoral farmers and the rest of society.

With New Zealand's all-grass wintering systems, stock are often grazed intensively on pasture throughout winter and early spring, when soil moisture levels are near or above field capacity. This reduces soil strength and increases vulnerability to treading damage. The on-site consequence of treading is a loss of soil structure and a reduction in pasture production (Singleton, Boyes and Addison, 1999). The off-site impact is an increase in nutrient and sediment runoff, which affects water quality in drainage streams (Nguyen *et al.*, 1998; Lambert *et al.*, 1985). Grazing systems that incorporate areas where animals can be kept off pasture are becoming more common, and decision-support models are being developed (Finlayson *et al.*, 2002).

About 80 percent of an average high country property is government land held under long-term pastoral lease. This includes almost all the higher altitude unimproved land where vegetation depletion and soil erosion may occur under extensive pastoralism. At the same time, pastoral leases pose a restriction on pasture improvement and other land uses, such as forestry, on flats and lowlands. From 1998, a change in the law permits run holders to have their pastoral lease reviewed, with the opportunity to acquire the freehold of the lower hills and flats, but to retire from grazing the higher altitude steeplands, including areas of historic, scientific, ecological and cultural significance. These are retained in Crown ownership and control.

Water quality

The quality of New Zealand rivers and lakes is very high by international standards, and they generally contain lower concentrations of nutrients and other dissolved material than rivers in many other countries (MOE, 1997). This is because most rivers are short and steep, which gives them less distance to accumulate pollutants, and catchments are relatively small. Pollution from point sources (e.g. dairy farms, piggeries and factories) has declined over the past 20–30 years as treatment systems have been improved and land treatment has been encouraged in preference to stream discharges. However, pollution due to non-point sources, such as pastoral farming, remains a concern. The use of fertilizer nitrogen has steadily increased on the more intensive farms, and regional councils (e.g. Environment Waikato, 1996) and the New Zealand Fertiliser Manufacturers Association (1998), concerned over a possible repeat of the European groundwater nitrate problem, have moved to introduce recommended rates of nitrogen application that do not exceed 200 kg N ha^{-1}.

Nitrate concentrations in water are generally less than 4 mg NO_3-N $litre^{-1}$ (Hoare and Rowe, 1992), but there is a wide range. The concentration of NO_3^- in some shallow aquifers has been found to exceed the limit for drinking water (11.3 mg NO_3-N $litre^{-1}$) set by the Ministry of Health (Hoare, 1986; Hanson, 2002), but deep aquifers generally have very low NO_3^- concentrations (Bowden, 1986). Grazed pasture and cropland are considered to be major sources of NO_3^- contamination. In some localized areas, point sources (e.g. dairy factory waste, piggery waste) and market gardens have been identified as also contributing to the NO_3^- load.

Dissolved reactive P (<1.00 μm) levels range from 0.003 to 0.01 mg P $litre^{-1}$ (median 0.005 mg P), well below the established 'suitability threshold' of 0.015-0.030 mg P $litre^{-1}$. However 40 percent of 177 New Zealand lakes examined contained high to very high levels of total N (> 0.30 mg $litre^{-1}$) and total P (> 0.02 mg $litre^{-1}$) and were considered to be either eutrophic or hypereutrophic (MOE, 1997). Pastoral farming is the dominant land use in the catchments of nutrient enriched lakes, which are found mainly in the North Island.

Exclusion of stock from waterways is being advocated by Regional Councils and the New Zealand Dairy Industry (Currie and Loganathan, 2002) as a method of minimizing nutrient transfer to waterways. Fenced and planted riparian strips are also advocated for the same reasons. However their long-term efficacy in nutrient capture from run-off waters is questionable (Cooper, Smith and Smith, 1995). Preliminary findings indicate that it is impossible to have infertile waters draining fertile soils and, in most soils, available phosphate concentrations that exceed agronomic optima will certainly contribute P-enriched runoff and drainage water to surface waters.

Loss of surplus nutrients from intensively managed pastoral systems to surface waters and groundwater is considered to be a primary factor that can be mitigated to some extent by careful nutrient budgeting. Decision-support software (e.g. Overseer™, Ledgard, Edgecombe and Roberts, 1999) has been produced to allow nutrient inputs through fertilizer, supplementary feeds and management of grazing animals to be considered in calculating nutrient balances at the paddock and farm scale.

Climate change

Climate change threatens an industry that is reliant on moderate temperatures and evenly distributed rainfall to grow pasture. There are a number of possible outcomes. These include the effect of increased CO_2 concentration on pasture yields, the effect of higher temperatures on the balance of C3 and C4 species in pasture and hence pasture nutritive value, and the effect on the frequency and intensity of El Niño events (Basher, 2000).

Considerable research effort is currently being directed at quantifying the carbon budgets (e.g. Tate *et al.*, 2000) and greenhouse gas emissions from pasture soils under different grazing systems (de Klein *et al.*, 2001). Preliminary results indicate that short-term intensification of pasture production (improvement through increased soil fertility) appears to lead to greater carbon dioxide emissions (Tate *et al.*, 2000). Greater emissions of the greenhouse gas, nitrous oxide, have also been measured from higher fertility dairy pastures than from lower fertility sheep pastures (Saggar *et al.*, 2002).

New Zealand has recently ratified the Kyoto Protocol, thereby agreeing, by 2012, to reduce greenhouse gas emissions to 1990 levels. New Zealand is unusual in that methane emissions are greater than CO_2 emissions, and 85 percent of these emissions are from ruminants, with 21 percent coming from dairy cows (Clark, 2002). The continued expansion of the dairy industry, together with a substantial reliance on pasture, means that it will be a major challenge to find cost-effective mitigation options. A move to concentrate-based diets has the potential to reduce methane per unit of milk, but the effect of such a move on the total greenhouse gas emissions associated with the dairy industry, and on its international competitiveness, is far from clear.

Allied with climate change but also existing as a concern in its own right is that of biosecurity. Despite a major government programme on all aspects of biosecurity, New Zealand pastoral industries, and the national economy, remain extremely susceptible to the introduction of animal or plant pests or diseases.

ORGANIC PASTORAL FARMING

Organic exports from New Zealand to date have been dominated by fresh fruit sales, but production of organic meat, milk and fibre has been small. In 2001/02, fresh fruit accounted for 77 percent of organic exports, while meat and wool only accounted for 7 percent. Organically managed pastures make up less than 1 percent of the approximately 13.5 million hectares of grassland in New Zealand. The proportion of the 13 600 dairy farmers and the 4 500 deer farms that are organic is only 0.4 percent. Expansion in dairying will be dependent upon current developments in contracts for collecting and processing from a critical mass of organic producers. Of the 15 740 commercial sheep and beef cattle farmers, a higher proportion are organic and there are successful organic marketing groups and supply chains providing good premiums for their products. While the total contribution of meat and wool is small, there is currently a rapid increase in supply from this sector (Organics New Zealand, 2002).

Conventional New Zealand pastoral farming systems are some of the most suitable in the world for ease of conversion to organic production. There are favourable climatic and soil conditions for good pasture growth, there is already a reliance on clover for N-fixation and maximizing milk production from forage, and being self-sufficient in feed production is already part of the psyche of most conventional farmers. Moving away from a reliance on soluble fertilizers, herbicides and pesticides and developing a better understanding of soil biology and soil-plant-animal interactions are the main challenges for those converting to organic production.

Very little published research is available on organic pastoral farming systems in New Zealand, and those studies available tend to concentrate on soil rather than plant aspects (Reganold *et al.*, 1993; Reganold, 1995; Niezen *et al.*, 1996; Richardson, 1997; Condron *et al.*, 2000). A case study carried out on dairy pastures in New Zealand (Macgregor, 2002) showed yields of 21.5 tonne DM ha^{-1} $year^{-1}$ on a conventional farm and 16 tonne DM ha^{-1} $year^{-1}$ on a paired farm in conversion to organic production, with stocking rates of 2.2 and 2.3 cows ha^{-1} for the conventional farms and 1.8 and 1.5 cows ha^{-1} for the in-conversion farms. A review of organic dairy pasture systems carried out in 2002 (Nicholas & Bauer-Eden, 2002) identified a number of research gaps for New Zealand. These included the effects of organic management on soil physical characteristics, soil biology, pasture composition, pasture productivity and animal health; how soils and pastures interact; the value of herbs in pastures; and how animal effluent should be best used. In another New Zealand study (Mackay *et al.*, 2002) a group of pastoral farmers identified the need for a better understanding of the soil under organic management (fertility, nutrients, biological and physical management), methods for control of internal parasites and lungworm, and methods for the control of woody weeds.

PROSPECTS

Profit margins have been maintained in the dairy and livestock sectors by the use of more intensive farming methods, increased application of fertilizers, new and better pasture types and improved management techniques, introduction of new breeds and cross-breds, and increasingly producing to specification. However, productivity needs to further increase if New Zealand is to continue to compete in the international market place. The dairy industry is committed to continuing increase in productivity, and recent developments in the sheep industry (i.e. reduced numbers but increased production by improved lambing percentages and higher carcass weights and increased processing of carcases), provide examples for future productivity gains.

In the sheep and beef industries more development is likely to occur in the future toward specialized breeding and finishing units. Increased networking will occur between breeders and finishers, and farmers will increasingly align themselves with particular supply chains and processors. Breeders will continue to operate in the predominantly steeper country, while finishers will operate in easier country, producing a large amount of forage year round by using short-term pastures such as Italian and hybrid ryegrasses, winter and summer brassicas, chicory, red clover and lucerne.

In the dairy industry, there are important future opportunities in the use of pastures for forage production and environmental protection. In forage production there is an urgent need for low cost feeds of higher nutrition value. A possibility involves the redesign of existing dairy systems based on perennial ryegrass and white clover to include tannin-containing legumes (25 percent of area) and cereal grain (25 percent of area) (Clark, 2002). This option would allow lower stocking rates and a higher feeding value diet, with significant savings in total farm costs.

A second approach involves biotechnology to modify existing perennial ryegrass by producing high levels of non-structural carbohydrate or increased lipid content (Roberts, Hancock and Woodfield, 2002), or reducing specific components of lignin (Clark and Wilson, 1993). Higher intake potential would allow stocking rates to be reduced, with a greater proportion of consumed feed going into milk production rather than maintenance. A less fibrous feed would also reduce methane loss as a proportion of gross energy intake, giving the double benefit of increasing milk yield at the same time as reducing methane losses per unit of feed intake.

An extremely valuable modification of white clover would be the incorporation of the tannin biosynthetic pathway (Morris and Robbins, 1997). A tannin-containing white clover would provide protein protected from excessive rumen degradation. This would increase the production of high value milk protein and reduce the loss of nitrogen in urine, and thereby reduce environmental damage from nitrate leaching and nitrous oxide emissions.

Diet directly influences flavour and texture properties of milk and meat, as well as greenhouse gas emissions from ruminants. Compounds contributing to the grassy flavour of pasture-fed animal products are being identified for use in selection programs to improve consumer acceptability. Other compounds, such as lipids, are being manipulated to add value to animal products from pastures. Plant attributes, like condensed tannins that reduce methane and nitrous oxide, are emerging as breeding targets, because approximately 55 percent of total New Zealand methane and nitrous oxide emissions arise from land-based industries.

Increasingly, molecular biology is playing a role in New Zealand breeding programmes. Transgenic white clovers with resistance to white clover mosaic virus and with resistance to porina have been successfully developed but have not yet been commercialized (Voisey *et al.*, 2001). Linkage maps with a high density of DNA markers have been developed for white clover and perennial ryegrass (Barrett *et al.*, 2001). This has opened the way for cost-effective marker-assisted selection, introgression of traits from wild relatives, and identification of heterotic groups for hybrid cultivar development. Substantial genomics research efforts are also underway in ryegrass and white clover, which have the potential to link economically important phenotypic characteristics with the genes responsible for them. They will also increase our understanding of the physiological and metabolic processes underlying important traits. Breeding programmes with elite germplasm will be strongly positioned to capture scientific advances that arise from the current biotechnology and genomics era. This integration of genomics and conventional improvement has the potential to meet the ongoing demands of intensification in New Zealand farming systems.

All of these developments will be dependent upon the maintenance of innovation and technical competence in the pastoral industries. However, there are already serious concerns about the loss of government support for applied research in the pastoral sector (Caradus, 2002), and about the decline in the numbers of young people involved in post-secondary education and training programmes appropriate for employment in the agricultural industry (Morriss *et al* 2001).

New Zealand will continue to promote and market its livestock products on the basis of quality, and to position itself at the top end of the market. The industry will think more in terms of quality not quantity, and value not volume. The clean, green and natural production images, seen by the two million or so tourists that visit New Zealand each year, will almost certainly be an advantage for food exports as they compete in the high-priced "niche" markets of the world. However, future market access will be heavily dependent upon the sensitivity of sophisticated overseas consumer groups to aspects of food safety (including genetic manipulation), environmental conservation and animal welfare.

CONCLUSIONS

This chapter has emphasized the technical efficiency of the pastoral industries in New Zealand, and in particular their ability to maximize the effective use of grazed grassland as a feed resource for ruminant animals. In the long term this must be a priority in the contribution of the world's grasslands to human food needs and environmental stability.

Also emphasized has been the dynamic nature of these industries, and in particular the rapid changes taking place in the flexible use of forages, and in integrated land use systems, to maximize business opportunity. There is a risk that these changes may result in some loss of competitive advantage in feed costs per unit of production, which, together with its reputation for product quality, safety and reliability, determines New Zealand's success in export markets. These pressures seem certain to be reinforced by concerns about food safety, the sustainable use of land and water resources, and about environmental stability at both national and global scales. These factors will determine the future of the pastoral industries in New Zealand, and their ability to continue to compete successfully in open markets with other countries with similar climatic advantages.

REFERENCES

Barratt, B.I.P., Barker, G.M., & Addison, P.J. 1996. *Sitona lepidus* (Coleoptera: Curcurlionidae) a potential clover pest to New Zealand. *New Zealand Entomologist,* **19**: 23–30.

Barrett, B., Griffiths, A., Mercer, D., Ellison, E., Faville, M., Easton, H.S. & Woodfield, D.R. 2001. Marker-assisted selection to accelerate forage improvement. *Proceedings of the New Zealand Grassland Association,* **63**: 241–245.

Basher, R. 2000. The impacts of climate change on New Zealand. pp. 121–142, *in:* A. Gillespie and W.C.G. Burns (eds). *Climate change in the South Pacific: Impacts and Responses in Australia, New Zealand and Small Island States.* Boston, MA: Kluwer Academic Publishers.

Bowden, M.J. 1986. *The Christchurch Artesian Aquifers.* North Canterbury Catchment Board, Christchurch, New Zealand. 159p.

Bridges, E.M. 1997. *World Soils.* 3rd edition. Cambridge, UK: Cambridge University Press. 176p.

Bryant, A.M. & Sheath, G.W. 1987. The Importance of grazing management to animal production in New Zealand. pp. 13–17, *in: Proceedings of the 4th AAAP Animal Science Congress.* Hamilton, New Zealand, February 1987.

Burke, J.L., Waghorn, G.C. & Chaves, A.V. 2002. Improving animal performance using forage-based diets. *Proceedings of the New Zealand Society of Animal Production,* **62**: 267–272.

Caradus, J. 2002. Generating wealth from our primary resources. pp. 1–11, *in:* J.S. Rowarth and G.P. Sutherland (eds). *Preserving New Zealand's Wealth-Generating Capacity.* Proceedings of the Primary Resources Forum, June 2002. *Institute of Agricultural Science and the Royal Society of New Zealand Miscellaneous Series,* No. 62.

Cassells, J.B. & Matthews P.N.P. 1988. Bull beef production. *Proceedings of the New Zealand Grassland Association,* **49**: 47–51.

Charlton, D. & Stewart, A. 2000. Pasture and forage plants for New Zealand. *Grassland Research and Practice Series,* No. 8. 74p.

Clark, D.A. 2002. Forage supply systems for pasture based dairying. pp. 30–37, *in: Facing the Challenges of the Supply Chain in 2010.* Proceedings of a Conference at Dookie College, Shepparton, Australia, April 2002. University of Melbourne, Institute of Land and Food Resources.

Clark, D.A. & Wilson, J.R. 1993. Implications of nutritive value improvements for plant performance and grassland management. pp. 543–549, *in: Proceedings of the XVII International Grassland Conference.* New Zealand and Queensland. New Zealand Grassland Association, Tropical Grassland Society of Australia, New Zealand Society of Animal Production, Australian Society of Animal Production – Queensland Branch, New Zealand Institute of Agricultural Science. February 1993.

Clark, D.A., Matthew, C. & Crush, J.R. 2001. More feed for New Zealand dairy systems. *Proceedings of the New Zealand Grassland Association,* **63**: 283–288.

Condron, L. M., Cameron, K.C., Di, H.J., Clough, T.J., Forbes, E.A., McLaren, R.G. & Silva, R.G. 2000. A comparison of soil and environmental quality under organic and conventional farming systems in New Zealand. *New Zealand Journal of Agricultural Research,* **43**: 443–466.

Coop, I.E. 1965. A review of the ewe equivalent system. *New Zealand Agricultural Science,* **1**(3): 13–88.

Cooper, A.B., Smith, C.M. & Smith, M.J. 1995. Effects of riparian set-aside on soil characteristics in an agricultural landscape: implications for nutrient transport and retention. *Agriculture, Ecosystems & Environment,* **55**: 61–67.

Cornforth, I.S. & Sinclair, A.G. 1982. Model for calculating maintenance phosphate requirements for grazed pastures. *New Zealand Journal of Experimental Agriculture,* **10**: 53–61.

Currie, L.D. & Loganathan, P. 2002. *Dairy Farm Soil Management*. Massey University, Palmerston North.

Daly, M.J., Fraser, T.J., Perkins, A. & Moffat, C.M. 1999. Farmer perceptions of reasons for perennial ryegrass persistence and the relationship of these with management practice, species composition, and soil fertility. *Proceedings of the New Zealand Grassland Association,* **61**: 9–15.

de Klein, C.A.M., Sherlock, R.R., Cameron, K.C. & van der Weerden, T.J. 2001. Nitrous oxide emissions from agricultural soils in New Zealand – a review of current knowledge and directions for future research. *Journal of the Royal Society of New Zealand,* **31**: 543–574.

During, C. 1984. *Fertilisers and Soils in New Zealand Farming*. Wellington: NZ Government Printer.

Easton, H.S., Amyes, J.M., Cameron, N.E., Green, R.B., Kerr, G.A., Norriss, M.G. & Stewart, A.V. 2002. Pasture plant breeding in New Zealand: where to from here? *Proceedings of the New Zealand Grassland Association,* **64**: 173–179.

Edmeades, D.C., Wheeler, D.M. & Crouchley, G. 1985. Effects of liming on soil magnesium on some soils in New Zealand. *Communications in Soil Science and Plant Analysis* **16**: 727-739.

Edmeades, D.C., Pringle, R.M., Mansell, G.P., Shannon, P.W., Ritchie, J. & Stewart, K.M. 1985. Effects of lime on pasture production on soils in the North Island of New Zealand. 5. Description of a lime recommendation scheme. *New Zealand Journal of Experimental Agriculture,* **13**: 47–58.

Eerens, J.P.J., Cooper, B.M., Willoughby, B.E. & Woodfield, D.R. 2001. Searching for clover root weevil (*Sitona lepidus*) resistance/tolerance – A progress report. *Proceedings of the New Zealand Grassland Association,* **63**: 177–181.

Environment Waikato. 1996. Guidelines for farm dairy effluent and sludge application to pasture. Hamilton East, Waikato, New Zealand.

Finlayson, J.D., Betteridge, K., Mackay, A., Thorrold, B., Singleton, P. & Costall, D.A. 2002. A simulation model of the effects of cattle treading on pasture production on North Island, New Zealand, hill land. *New Zealand Journal of Agricultural Research,* **45**: 255–272.

Floate, M.J.S. 1992. Guide to Tussock Grassland Farming. Invermay, NZ: AgResearch. 127p.

Fraser, T. 2000. Presidential address. *Proceedings of the New Zealand Grassland Association,* **62**: 1–2.

Gillingham, A.G., Maber, J., Morton, J. & Tuohy, M. 1999. Precise aerial fertiliser application on hill country. *Proceedings of the New Zealand Grassland Association,* **61**: 221–226.

Hanson, C.R. 2002. Nitrate concentrations in Canterbury groundwater – a review of existing data. Environment Canterbury, Christchurch, New Zealand. Report No. R02/17.

Harris, B.L. & Kolver,E.S. 2001. A review of holsteinization in intensive pastoral dairy farming in New Zealand. *Journal of Dairy Science,* **84** Electronic supplement E56-E61.

Hawke, M.F. 1997. The changing focus of pastoral agroforestry in temperate zones. Vol. 1: 121–128, *in: Proceedings of XVIII International Grassland Congress*. Winnipeg, and Saskatoon, Canada, June 1997.

Hoare, R.A. 1986. Groundwater nitrate in the Hamilton Basin. Waikato Valley Authority, Tech. Rep. 1986/16. 18p.

Hoare, R.A. & Rowe, L.K. 1992. Water quality in NZ. p. 211, *in:* M.P. Mosley (ed). *Waters of New Zealand*. Christchurch, New Zealand: New Zealand Hydrological Society and Caxton Press.

Hodgson, J. 1989. Increases in milk production per cow and per hectare: pasture production. *Dairyfarming Annual,* **41**: 76–82.

Hoglund, J.H., Crush, J.R., Brock, J.L., Ball, P.R. & Carran, R.A. 1979. Nitrogen fixation in pasture. XII. General discussion. *New Zealand Journal of Experimental Agriculture,* **7**: 45–51.

Holmes, C.W., Brookes, I.M., Garrick, D.J., Mackenzie, D.D.S., Parkinson, T.J. & Wilson, G.F. 2002. *Milk Production from Pasture – Principles and Practices*. Palmerston North, New Zealand: Massey University. 601p.

Knowles, I.M., Fraser, T.J. & Daly, M.J. 2003. White clover: plant loss in drought and subsequent recovery following rain. pp. 37–42, *in:* D.J. Moot (ed). *Legumes for Dryland Pastures*. Proceedings of a New Zealand Grassland Association (Inc.) Symposium, Lincoln University, November 2003. Grassland Research and Practice Series, No. 11.

Lambert, M.G., Devantier, B.P., Nes, P. & Penny, P.E. 1985. Losses of nitrogen, phosphorus, and sediment in runoff from hill country under different fertiliser and grazing management regimes. *New Zealand Journal of Agricultural Research,* **28**: 371–379.

Lambert, M.G., Jung, G.A. & Costall, C.A. 1989. Forage shrubs in North Island hill country. 1. Forage production. *New Zealand Journal of Agricultural Research,* **32**: 477–483.

Ledgard, S.F., Edgecombe, G.A. & Roberts, A.H.C. 1999. Application of the nutrient budgeting model OVERSEER™ to assess management options and regional council Consent requirements on a Hawke's Bay dairy farm. *Proceedings of the New Zealand Grassland Association,* **61**: 227–231.

Livestock Improvement Corporation. 2003. Dairy Statistics 2001-2002.

Macgregor N. 2002. Case-study of New Zealand dairy farms in transition. pp. 114–123, *in:* F. Van Steensel *et al.* (eds). *A Review of New Zealand and International Organic Land Management Research – Relevant to Soil, Dairy Pasture and Orchard Management in New Zealand*. New Zealand: Bio Dynamic Farming and Gardening Association.

Mackay, A., Doak, M., Rhodes, A.P., Burborough, T., Jackman, J., Redward, D., Williams, P., Pearse, T., Pollard, J. & Nicholas, P.K. 2002. Understanding the costs and risks of conversion to organic production systems. Ministry of Agriculture and Forestry. *MAF Technical Paper* No. 2002/1.

Matthews, P.N.P., Hodgson, J. & White, J.G.H. 1999. Livestock farming systems in New Zealand. pp. 133–152, *in:* J. White & J. Hodgson (eds). *New Zealand Pasture and Crop Science*. Auckland, NZ: Oxford University Press.

McGregor, E., Mackay, A.D., Dodd, M.B. & Kemp, P.D. 1999. Silvopastoralism using tended poplars on New Zealand hill country: The opportunities. *Proceedings of the New Zealand Grassland Association,* **61**: 85–89.

McKenzie, B.A., Kemp, P.D., Moot, D.J., Matthew, C. & Lucas, R.J. 1999. Environmental effects on plant growth and development. pp. 29–44, *in:* J. White & J. Hodgson (eds). *New Zealand Pasture and Crop Science*. Auckland, NZ: Oxford University Press.

McNaught, K.J., Dorofaeff, F.D., Ludecke, T.E. & Cottier, K. 1973. Effect of potassium fertilizer, soil magnesium status, and soil type on uptake of magnesium by pasture plants from magnesium fertilisers. *New Zealand Journal of Experimental Agriculture,* **1**: 329–347.

McRae A.F. 1988. Bull beef production. The Tuapaka experience. *Proceedings of the New Zealand Grassland Association,* **49**: 41–45.

Metherell, A.K., Thorrold, B.S., Woodward, S.J.R., McCall, D.G., Marshall, P.R., Morton, J.D. & Johns, K.L. 1997. A decision-support model for fertiliser recommendations for grazed pasture. *Proceedings of the New Zealand Grassland Association,* **59**: 137–140.

Ministry for the Environment. 1997. *The State of New Zealand's Environment*. Wellington, New Zealand: The Ministry for the Environment.

Ministry of Agriculture and Forestry. 2002. Situation and Outlook for New Zealand Agriculture and Forestry.

Moir, J.L., Scotter, D.R., Hedley, M.J. &, Mackay, A.D. 2000. A climate driven, soil fertility dependent, pasture production model. *New Zealand Journal of Agricultural Research,* **43**: 491–500.

Morris, P. & Robbins, M.P. 1997. Manipulating condensed tannins in forage legumes. pp. 147–173, *in:* B.D. McKersie and D.C.W. Brown (eds). *Biotechnology and the improvement of forage legumes*. Wallingford, UK: CABI.

Morris, S.T., Parker, W.J., Purchas, R.W. & McCutcheon, S.N. 1992. Dairy crossbreeding alternatives to improve New Zealand beef production. *Proceedings of the New Zealand Grasslands Association,* **54**: 18–22.

Morriss, S., Tipples, R., Townshend, W., Mackay, B. & Eastwood, C. 2001. People make the difference. pp. 79–89, *in:* J.S. Rowarth and G.P. Sutherland (eds). *The Role of the Primary Resources in New Zealand's Economic Revival*. Proceedings of the Primary Resources Forum. June 2001. *Institute of Agricultural Sciences and the Royal Society of New Zealand Miscellaneous Series*. No. 61.

MWIES [Meat and Wool Innovation Limited Economic Service]. 2002. The New Zealand Sheep and Beef Farm Survey 2000-2001. Publication No. G2235. Meat and Wool Innovation Limited Economic Service, Wellington, New Zealand.

New Zealand Fertiliser Manufacturers Association. 1998. *Code of Practice for Fertiliser Use*. New Zealand Fertiliser Manufacturers Research Association, Auckland, New Zealand.

New Zealand Meteorological Service. 1973. Summaries of climatological observations to 1970. New Zealand Meteorological Service Miscellaneous Publication, No. 143. Government Printer, Wellington, New Zealand.

Nguyen, M.L., Sheath, G.W., Smith, C.M. & Cooper, A.B. 1998. Impact of cattle treading on hill land. 2. Soil physical properties and contaminant run-off. *New Zealand Journal of Agricultural Research,* **41**: 279–290.

Nicholas, P. & Bauer-Eden, H. 2002. Organic dairy pasture systems – A review. pp. 50–65, *in:* F. Van Steensel *et al.* (eds). *A Review of New Zealand and International Organic Land Management Research – Relevant to Soil, Dairy Pasture and Orchard Management in New Zealand*. New Zealand: Bio Dynamic Farming and Gardening Association.

Niezen, J.H., Charleston, A.G., Hodgson, J., Mackay, A.D. & Leathwick, D.M. 1996. Controlling internal parasites in grazing ruminants without recourse to anthelmintics: approaches, experiences and prospects. *International Journal for Parasitology,* **26**: 983–992.

O'Connor, K.F., Scott, D., Williams, W.M. & Nordmeyer, A.H. 1996. Issues and options in high country farming. *Proceedings of the New Zealand Grassland Association,* **58**: 133–159.

Organics New Zealand. 2002. Organic Production and Marketing Information Package. Organic Products Exporters of New Zealand Inc (OPENZ), Christchurch, New Zealand.

Parminter, I., Dodd, M.B. & Mackay, A.D. 2001. Economic analysis of poplar planting on steep hill country. *Proceedings of the New Zealand Grassland Association,* **63**: 127–130.

Reganold, J.P. 1995. Soil quality and profitability of biodynamic and conventional farming systems: a review. *American Journal of Alternative Agriculture,* **10**: 36–45.

Reganold, J.P., Palmer, A.S., Lockhart, J.C. & Macgregor, A.N. 1993. Soil quality and financial performance of biodynamic and conventional farms in New Zealand. *Science,* **260**: 344–349.

Resource Management Act. 1991. Published under the authority of the New Zealand Government, Wellington, New Zealand.

Richardson, A.G. 1997. The potential for commercial organic sheep farming in the South Island. Kellogs Leadership Group. See http://www.nzmeat.co.nz

Roberts, N.J., Hancock, K.R. & Woodfield, D.R. 2002. Genetic modification to create novel high quality forages. *Proceedings of the New Zealand Society of Animal Production,* **62:** 278–281.

Saggar, S., Andrew, R., Tate, K., Rodda, N., Hedley, C. & Townsend, J. 2002. Measurements and modelling of nitrous oxide emissions from dairy pastures. pp. 201–215, *in:* L.D. Currie and P. Loganathan (eds). *Dairy Farm Soil Management.* Occasional report No.15. Fertilizer & Lime Research Centre, Massey University, Palmerston North.

Sandrey, R. & Reynolds, R. (eds). 1990. *Farming without Subsides: New Zealand Recent Experience.* New Zealand Ministry of Agriculture and Fisheries, Wellington, New Zealand.

Sanson, R. & Pearson, A. 1997. Agribase – a national spatial farm database. *Proceedings of 8th International Symposium on Veterinary Epidemiology and Economics,* Epidémiologie et Santé Animale 31-32, 12.16.1-12.16.3.

Sheath, G.W. & Clark, D.A. 1996. Management of grazing systems: temperate pastures. pp. 301–323, *in:* J. Hodgson and A.W. Illius (eds). *The Ecology and Management of Grazing Systems.* Wallingford, UK: CABI.

Sinclair, A.G., Johnstone, P.D., Smith, L.C., Roberts, A.H.C., O'Connor, M.B. & Morton, J.D. 1997. Relationship between pasture dry matter yield and soil olsen P from a series of long-term field trials. *New Zealand Journal of Agricultural Research,* **40:** 559–567.

Singleton, P.L., Boyes, M. & Addison, B. 1999. Long-term changes in soil physical properties under dairying in the Waikato and Northland, New Zealand. pp. 299–309, *in: Proceedings of the Workshop Best Soil Management Practices for Production.* 10–11 February 1999. Fertilizer and Lime Research Centre, Massey University, Palmerston North.

Smith, D.R., Fraser, T.J., Moffat, C.A. & Daly, M.J. 1998. Effects of soil fertility on the persistence of dryland cocksfoot and tall fescue pastures. *Proceedings of the New Zealand Grassland Association,* **60:** 265–270.

Sutton, J. 2002. Dairy Farmers of New Zealand Meeting. Wellington, 18 June 2002.

Tate, K.R., Scott, N.A., Parshotam, A., Brown, L., Wilde, R.H., Giltrap, D.J., Trustrum, N.A., Gomez, B. & Ross, D.J. 2000. A multi-scale analysis of a terrestrial carbon budget: is New Zealand a source or sink of carbon? *Agriculture, Ecosystems & Environment,* **82:** 229–246.

The Economic Service. 2003. Annual Review of the New Zealand Sheep and Beef Industry, 2001-2002.

Voisey, C.R., Dudas, B., Biggs, R., Burgess, E.P.J., Wigley, P.J., McGregor, P.G., Lough, T., Beck, D.L., Forster, R.L.S. & White, D.W.R. 2001. Transgenic pest and disease resistant white clover plants. pp. 239–250, *in:* G. Spangenberg (ed). *Molecular breeding of forage crops.* Dordrecht, The Netherlands: Kluwer Academic Publishers.

Waghorn, G.C., Adams, N.R. & Woodfield, D.R. 2002. Deleterious substances in grazed pastures. pp. 333–356, *in:* M. Freer and H. Dove (eds). *Sheep Nutrition.* CSIRO Publishing. Collingwood.

Waghorn, G.C., Tavendale, M.H. & Woodfield, D.R. 2002. Methanogenesis from forages fed to sheep. *Proceedings of the New Zealand Grasslands Association,* **64:** 167–171.

Wall, A.J., Mackay, A.D., Kemp, P.D., Gillingham, A.G., Edwards, W.R.N. 1997. The impact of widely spaced soil conservation trees on hill pastoral systems. *Proceedings of the New Zealand Grassland Association,* **59:** 171–177.

Wallace, D.G. 2002. Ecotourism – The Maungatautari example. *Proceedings of the New Zealand Grassland Association,* **64:** 11–16.

Westwood, C.T. & Norriss, M.G. 1999. Liveweight changes in lambs grazing six perennial ryegrass cultivars. *Proceedings of the New Zealand Grassland Association,* **61:** 31–35.

Wilkinson, A.G. 1999. Poplars and willows for soil erosion control in New Zealand. *Biomass and Bioenergy,* **16:** 263–274.

Williams, P.H., Gregg, P.E.H. & Hedley, M.J. 1990. Mass balance modelling of potassium losses from grazed dairy pasture. *New Zealand Journal of Agricultural Research,* **33:** 661–668.

Woodfield, D.R. 1999. Genetic improvements in New Zealand forage cultivars. *Proceedings of the New Zealand Grassland Association,* **61:** 3–7.

Woodfield, D.R. & Matthew, C. 1999. Ryegrass endophyte: an essential New Zealand symbiosis. New Zealand Grassland Association, Palmerston North. *Grassland Research and Practice Series,* No. 7. 168p.

Woodward, S.L., Auldist, M.J., Laboyrie, P.J. & Jansen, E.B.L. 1999. Effect of *Lotus corniculatus* and condensed tannins on milk yield and milk composition of dairy cows. *Proceedings of the New Zealand Society of Animal Production,* **59:** 152–155.

CHAPTER 9

Organic Grassland: Principles, Practices and Potential

David Younie[1] and Ton Baars[2]

ABSTRACT

This chapter explains the general principles and legislative background to organic farming. The practical implementation of these principles is then reviewed in detail in relation to grassland management. In particular, the organic approach to the major management challenges faced by grassland farmers is examined; nutrient supply to the crop, control of weeds; and maintenance of animal health. These are areas that, in a conventional situation, would be dealt with primarily through inputs of fertilizers or agrochemicals, but which, in the organic situation, must be addressed using an ecological approach, working with biological and ecological processes.

For nitrogen (N) nutrition, the importance of biological N-fixation by herbage legumes and enhancing N-mineralization is emphasized. For phosphorus and potassium supply, the priority should be an efficient nutrient cycling system, in order to minimize losses from the farm. Control of perennial weeds such as dock and thistle is a difficult challenge. The farmer needs to base the control strategy on altering the management factors that cause the problem. Specific weed control practices are suggested. A particular management focus for organic grassland farmers is the maintenance of good health in grazing animals, especially the control of internal parasites such as intestinal worms in young animals. A range of management options is suggested.

The potential of a range of herbage species for sowing in organic grassland is described and the impact of organic grassland on biodiversity and the environment (nutrient losses and energy consumption) is reviewed. It is recognized that productive, legume-based sown pastures are unlikely to be species rich, and N losses to the environment will depend on the overall supply of N in the system. Potential livestock output per hectare, based on research on organic temperate grassland, is reviewed, and the major issues and challenges for the future discussed.

[1]SAC, Craibstone Estate, Bucksburn, Aberdeen AB21 9YA, United Kingdom.
[2]Louis Bolk Institute, Hoofdstraat 24, 3972 LA Driebergen, The Netherlands.

INTRODUCTION

Public interest in the methods employed in food production increased considerably during the final decade of the twentieth century, fuelled primarily by concerns about food safety and quality, animal welfare and environmental impact. Discussions have been centred principally in the developed world, where agricultural systems tend to be most intensive. As a result of these concerns, the demand for organic food has increased dramatically, and support schemes to encourage farmers to convert to organic were introduced in the European Union in 1993.

Further evidence of this trend is provided by the adoption of EU Regulation 2092/91 in 1991, which effectively defined organic production in the EU. Similarly, the FAO Codex Alimentarius Guideline was published in 2000 (and revised in 2001), fulfilling a similar purpose and facilitating trade globally as the EU Regulation does in Europe (CEC, 1991; FAO, 2001).

By virtue of such instruments, organic farming ('ecological' or 'biological' agriculture in continental Europe) is a clearly defined system of production that has food quality, human health, environmental, animal welfare and socio-economic aims – aims that derive more from a consumer perspective than from a producer perspective. The result is that organic food has a very strong brand image in the eyes of consumers and thus can command higher prices for retailers and farmers than for conventionally produced food.

Grassland, and in particular forage legumes, plays a major role in almost all successful organic systems, because nitrogen (N) is the most important nutrient for most crops, and organic farming principles and standards place strong emphasis on building soil fertility with minimal use of non-renewable resources.

PRINCIPLES OF ORGANIC FARMING

The detailed husbandry standards of organic farming are primarily based on the principles of enhancement and exploitation of the natural biological cycles in soil (e.g. N_2-fixation, nutrient cycling), in crops (e.g. manipulation of competitive ability of crops and populations of natural predators of crop pests) and in livestock (e.g. rumen digestion in ruminants, development of natural immunity in young animals, interruption of host-pathogen relationships). In addition, there is strong emphasis on optimizing animal welfare, avoidance of pollution and improvement of wildlife habitats on the farm.

The aim is to work **with** natural processes, rather than seek to dominate them, as is often the case in intensive systems, and to minimize the use of non-renewable natural resources, such as the fossil fuel used for manufacture of fertilizers and pesticides (IFOAM, 1998).

EU Regulation 2092/91 (CEC, 1991) sets out the minimum standards for organic crop production in Europe and defines how certification procedures must operate. The important elements in this in relation to grassland are as follows:

- seeds must have been grown organically (for at least one generation). Because of the shortage of organically grown herbage seed, EU Regulation 2092/91 currently (2003) contains a derogation that permits the use of seed from conventional production if organically grown seed of that species is not available;
- manurial supply to crops must be based on a balanced rotation including legumes, and effective nutrient cycling systems within the farm;

- only fertilizers from an approved list are permitted, and then only in certain circumstances. Generally, these approved fertilizers are either manures or mineral fertilizers such as lime and rock phosphate, that are slowly soluble in the soil. Soluble mineral fertilizers are, for the most part, prohibited. Use of conventional manure is restricted to situations where its use can be justified to the certification body (e.g. where a soil nutrient deficiency exists);
- livestock must have access to pasture during the growing season;
- ruminant livestock must be fed a diet that is at least 60 percent green forage, on a daily dry matter basis (i.e. maximum 40 percent concentrates).
- until August 2005, a derogation permits the use of up to 10 percent of non-organic feedstuffs in the annual dry matter intake of ruminants (25 percent on a daily basis). No feedstuffs are permitted that originate from genetically engineered crops; and
- livestock health policy must be based on preventative management strategies – aimed at minimizing disease challenge and maximizing the animal's ability to withstand the challenge; no routine treatment of healthy animals with drugs, except in the case of a known farm problem. However, chemotherapy of individual sick animals **is** permitted, although withdrawal periods are extended. Maximum of two courses of treatment permitted per individual per annum.

It can be seen from the above that the central feature of organic farming is an ecological approach to the management of the system. Soil and manure management, and grassland and livestock husbandry, are modified in order to maximize biological N_2 fixation and soil N mineralization, to minimize nutrient losses from the farm, and to minimize disease and parasite problems in livestock. Ultimately the objective is to achieve moderate levels of output from healthy and stress-free animals living in harmony with their environment, with little or no veterinary drug input, little or no fertilizer input, and little or no pesticide input.

In contrast, conventional systems have at their disposal a wide range of fertilizers, crop protection chemicals and veterinary drugs, which enables the farmer to overcome to a large extent the natural ecological constraints of soil, crop and livestock ecosystems. The objective in conventional systems has tended to be to maximize output per unit area, whereas the organic farmer has deliberately chosen to restrict output per unit area in order to preserve fossil energy, minimize chemical input, and enhance biodiversity and animal welfare.

BOTANICAL COMPOSITION AND BIODIVERSITY IN ORGANIC PASTURES

Nitrogen is the single most important nutrient required for herbage growth. In organic systems, nitrogen is also supplied from mineralized soil organic N and from application of organic manures, but the ultimate source of N input to the system is atmospherically-derived N_2 fixed by legumes.

In organic swards managed for production, white clover (*Trifolium repens*) is by far the most appropriate and widely used forage legume in temperate maritime climates, because of its adaptability to a range of management and soil fertility conditions. It is persistent, can thrive at a soil pH of 5.8 (mineral) or 5.5 (peat) as well as in more alkaline soils, and can be used for management regimes ranging from continuous sheep grazing (for which small-leaved varieties are most suitable) to lax defoliation, including cutting (for which larger leaved varieties are most suitable). Red clover (*T. pratense*) can also grow in moderately acid soils, like white clover, and is highly productive, but is not persistent and should only be sown in mixtures intended for short-term leys (up to three years), primarily for cutting. It is probably the best species for use as a one- or two-year green manure. Alfalfa (*Medicago sativa*) and sainfoin (*Onobrychis sativa*) also have considerable potential as crops for conservation, but are suitable only for soils with a high pH and, in the case of alfalfa, good drainage. In very acid soils, of pH 5 and less, bird's foot trefoil (*Lotus corniculatus*) and alsike clover (*T. hybridum*) are alternatives to white clover and red clover respectively.

Although perennial ryegrass (*Lolium perenne*) has been criticised as being more appropriate for intensive high-N systems than for organic or low-input systems, it is undoubtedly the most suitable species for ley farming in temperate maritime conditions, given its ease of establishment, yield potential, persistence and quality characteristics. Tetraploid varieties are known to promote a higher clover content in the sward than diploid varieties, because of their more open growth habit, and should constitute at least 75 percent of the perennial ryegrass component in organic seed mixtures. The main weakness with perennial ryegrass is its lack of winterhardiness in situations with severe winters, and so in Scandinavia, continental Europe and Canada, species such as timothy (*Phleum pratense*), cocksfoot (*Dactylis glomerata*) or meadow fescue (*Festuca pratensis*) have more potential.

As indicated elsewhere in this chapter, maximizing production per hectare is not always the primary goal in organic farming, and this is true also in terms of choice of species for organic pastures. Apart from productive legumes and grass species, there may be merit, particularly in terms of animal nutrition and health, in offering a wider range of herbage species to livestock. In discussions on botanical composition in grassland, three different sward types can be identified: (i) short-term leys; (ii) permanent swards with a production goal; and (iii) permanent swards managed specifically for their nature conservation value (floral and faunal diversity objectives). Evidence for a division of permanent grassland into the two categories listed above can be seen in the survey of 91 organic grassland sites in Schleswig-Holstein by Worner and Taube (1995), who found that the index of stand value (for agriculture) was closely related to the presence of perennial ryegrass and white clover. The percentage of legume in the sward increased with intensity of utilization, and the content of *T. repens* tended to be negatively correlated with species diversity. Similarly, Younie and Armstrong (1996) observed relatively little species diversity in an intensively managed and highly productive organic sward over a period of nine years from sowing. The swards were sown initially with a mixture of *L. perenne*, *P. pratense* and *T. repens*. After nine years, over 95 percent of ground cover in this sward comprised *L. perenne* and *T. repens*.

Sown pastures, therefore, are usually poor in species diversity, but have a major advantage in terms of yield and quality. Tables 9.1 and 9.2 compare three pasture types in the Netherlands (natural (N), old (O) and sown grass/clover (G/C) pastures).

Table 9.1 Production parameters in three different types of organic pasture in Friesland province[(1)]

Sward type	Mean no. of cuts	Output per hectare				Content	
		Total herbage t DM ha^{-1}	Clover t DM ha^{-1}	Energy GJ ha^{-1}	N kg ha^{-1}	Energy MJ kg^{-1} DM	N g kg^{-1} DM
N	2	6.0	0.04	35.81	106	5.96	18
O	3.9	8.3	0.75	50.89	224	5.82	27
G/C	5	10.3	3.88	66.22	337	6.44	33

Key: N = natural sward; O = old sward; G/C = grass–clover sward.
Note: (1) Data are means of measurements on three different swards of each sward type, measured over two years.
Source: Ennik *et al.*, 1982.

Table 9.2 Number of cuts per year and species number in three different types of organic pasture[(1)]

Sward type	Mean no. of cuts	Number of herbage species in each sward				
		Total	Grasses	Legumes	Herbs	Sedges
N	2	42	19	2	18	3
O	3.9	37	16	3	17	1
G/C	5	13	6	1	6	0

Key: N = natural sward; O = old sward; G/C = grass–clover sward.
Note: (1) Data are means of measurements on three different swards of each sward type, measured over two years.
Source: Ennik *et al.*, 1982.

Sown pastures had a higher clover yield and a higher N yield than old swards. Total yield and the protein content of natural pastures were lower still. This herbage could not be used as the main fodder for high yielding dairy cows, mainly because of the low protein content. The potential number of cuts per year was greatest with grass–clover swards, but the number of species in the old organic pastures was higher (Table 9.2). The input of N must be limited to between 50 and 150 kg N ha^{-1} yr^{-1} in order to maintain the species diversity of this old *Lolio-Cynosuretum* community. Below 50 kg N ha^{-1} yr^{-1}, and especially under wet conditions and a cutting regime, the natural grasslands were even more botanically diverse. Phosphate status was also correlated with species diversity (negatively) and primary production (positively). Oomes, Korevaar and Altena (1980) showed that a low level of soil potassium was also an important prerequisite for high species diversity.

Although it is clear that there is a correlation between sward composition and sward productivity in terms of livestock output per hectare, farmers often decide to keep their old pastures. The cost of reseeding is high and there is a risk of damaging soil structure in heavy clay soils and peat areas, but farmers also have personal interests, for example in the maintenance of meadow birds in their grasslands. In addition, the maintenance of a high health status in grazing animals and the enhancement of nutrient cycling are regarded by some farmers as important priorities and equal to maximizing output per hectare. Species-rich permanent pastures have an important role to play in achieving these objectives. As long-term biotopes, they are well adapted to the local environment and can offer the advantages associated with a well-developed genotype–environment interaction (compared with the short-term ley, which inevitably employs foreign genotypes). Species-rich permanent pastures constitute a bank of genetic resources as well as a resource for wildlife (insects, butterflies, birds). De Boer (1990) concluded that the potential for meadow birds is highest in extensive grasslands (receiving less than 50 kg N ha^{-1} yr^{-1}), in combination with long rest periods (longer than 60 days).

A number of herb species are deep rooting and have a better mineral status than "improved" sown species (Jones, 1990; Swift *et al.*, 1993; Fisher and Baker, 1996), and therefore potentially contribute to the health resource of farm animals and to an enhanced nutrient cycle within the farm. They also tend to be more drought resistant than grasses and their tap root systems may also lead to a more open soil structure.

Wilman and Derrick (1994) reported higher calcium and magnesium status in lambs fed a range of herbs, including ribwort (*Plantago lanceolata*) and dandelion (*Taraxacum officinale*), compared with perennial ryegrass. Similarly, again in comparison with perennial ryegrass, Younie *et al.* (2001) demonstrated significant improvements in calcium, sodium, magnesium, copper and vitamin

B_{12} (cobalt) status of lambs from feeding a 100 percent chicory diet (*Cichorium intybus*). Several trace elements (selenium, zinc and copper) are important in relation to the immune reaction (Van der Brug, 1996). Baars and Opdam (1998) found a relationship between trace element supplementation and somatic cell counts in organic dairy heifers fed a grass–red clover diet. In addition, some herbs, particularly those with a high content of condensed tannins (CT), have been shown to exhibit anti-worm properties when fed to lambs (see below). A number of authors have reviewed the potential of a range of forage herbs, including Elliot (1914), Foster (1988) and Frame, Fisher and Tiley (1994). The herbs with the greatest agronomic, nutritive and medicinal potential are chicory, ribwort, birdsfoot trefoil and, to a lesser extent, burnet and yarrow.

Despite the potential advantages of herbs, there has been little effort to integrate them into intensively managed sown pastures (Foster, 1988). This is changing to some extent, at least in New Zealand and Australia, where breeding effort has led to the commercial release of varieties of chicory and ribwort (Moloney and Milne, 1993; Rumball, 1986). Perhaps the main reason for the low level of use of forage herbs is their generally low contribution to sward biomass, largely resulting from low seed rates, poor establishment and persistence. This may be due, at least in part, to their inability to persist under intensive management regimes that involve frequent utilization by both cutting and grazing – regimes to which vigorously tillered grass species such as perennial ryegrass are ideally suited. In contrast, as indicated above, herbs can persist satisfactorily in natural or low-input extensively managed swards, grazed at stocking densities low enough to allow herbage plants to flower and set seed, and with companion species less competitive than perennial ryegrass.

A major challenge for the future, therefore, is to develop management techniques that can ensure that herbs can make a predictable contribution in a mixed, intensively managed sward. Research in New Zealand has indicated that chicory should be managed according to a rotational grazing regime, and that, compared with 8-week regrowth intervals, use of 4-week intervals increases the ratio of leaves to stems, thus enhancing the palatability and digestibility of the material (Clark, Anderson and Berquist, 1990; Li, Kemp and Hodgson, 1997). New thinking is also required in relation to the most appropriate companion species and mixtures for forage herbs. Umrani (1998) has shown that perennial ryegrass, with its vigorous, densely tillered growth habit, is a major competitor. It significantly reduced root length, root weight and shoot weight of individual herb plants. The upright and less well tillered growth habit of timothy is less competitive and, coupled with the N-fixing ability of white clover, makes for a more suitable mixture. Such timothy–herb–clover combinations will be best sown in strips or as separate swards, with the main bulk of the field sown to a ryegrass-based mixture, given that perennial ryegrass, with all its other advantages, will probably remain the mainstay of productive organic grassland, at least in temperate maritime climates.

Thus, while production and biodiversity tend to be incompatible, opportunities for segregation or integration of different sward types can be considered. This can be approached at landscape level, farm level or field level. Within a farm or within a landscape, it is possible to maintain more extensive permanent grasslands (with or without animal manure) alongside more intensively used sown grasslands. In many situations, of course, there is no dispute about which is the more appropriate choice. On the one hand, land of inherently low nutritional status (e.g. peats or pure sands) may not have the potential for high levels of production. On the other hand, short-term grass–clover leys that are part of a crop rotation will have little opportunity to develop into species-rich pastures, although, as outlined above, it may also be possible to develop productive sown pastures that incorporate a different range of herbage species from that used at present, for specific purposes, such as de-worming paddocks. Similarly, at field level, the two approaches can be married by the management of grassland strips or field margins and boundaries for nature conservation, improved mineral nutrition or anti-worming areas.

CONTROL OF PERENNIAL WEEDS IN ORGANIC GRASSLAND

While a diverse range of herbage species is desirable in some contexts, pernicious weeds such as dock (*Rumex obtusifolius*), ragwort (*Senecio jacobea*) or creeping thistle (*Cirsium arvense*) can rarely be seen as desirable sward components. Control of these perennial weeds presents a major agronomic challenge in long-term organic grassland, since herbicide use is prohibited. As in conventional grassland, the best method of weed control is to maintain a dense, vigorous sward that offers little opportunity for weed seeds to become established. In practice this is often not possible, particularly where cutting for hay or silage is necessary, since this opens up the sward and allows an opportunity for weeds to flower and set seed.

Where the site is ploughable, adoption of a regular programme of grass renewal (i.e. a ley system, using 3–5-year leys) will minimize the risk of the problem developing in the first place. This ley policy also has other major advantages in the organic farming context, such as the annual provision of swards uncontaminated with stomach worm larvae, and the opportunity to exploit fertility built up in the soil by growing arable cash crops or annual forage crops for finishing lambs. However, there is an expense involved in reseeding and there will be situations where site conditions prevent regular ploughing and reseeding.

Ragwort is effectively controlled in long-term grassland by sheep grazing in winter and early spring, but this is not effective for dock and creeping thistle. With these weeds, other approaches are necessary. A dock infestation becomes worse much more rapidly under a cutting than under a grazing management regime (Courtney, 1985), and so continuous cutting for hay or silage should be avoided in permanent grassland, if possible. Regular topping of docks in grazed swards, and soil aeration to improve grass growth and cause physical damage to dock roots, may also limit the increase in dock density (Hopkins *et al.*, 1997). At low levels of dock infestation, hand digging should be undertaken in order to prevent dock numbers from increasing. When dock-infested grassland is destroyed prior to establishment of the following crop, rotovating the turf to a depth of 8–10 cm in early to mid-summer, followed by tine cultivation and desiccation of the surface material for 8–10 weeks before ploughing, has been successful in farm practice (M. Measures, pers. comm.). In addition to killing the crowns of mature dock plants, this practice essentially creates a stale seedbed, encouraging germination of dock seeds that are subsequently killed by the further cultivation. Cropping with a whole-crop silage mixture of cereals before reseeding with grass provides a further opportunity to reduce the dock seedbank through the allelopathic and competitive characteristics of the cereal crop. Ideally, this should be combined with harvesting of the mature dock tap roots during seedbed preparation for the cereal crop, e.g. by ridging and use of a potato stone separator or potato elevator harvester to bring the roots to the surface. Creeping thistle can be controlled by the competition created by the dense canopy of a specialist red clover mixture managed for early-cut silage. The creeping thistle plants will be depleted within one or two seasons under this regime. Topping at an early growth stage (15–20 cm) is reported as being more effective than topping later, immediately prior to flowering (M. Dock Gustaffson, pers. comm.). Inevitably, a weed control strategy involving a combination of these approaches is likely to be required for long-term control of these weeds.

PESTS AND DISEASES IN ORGANIC GRASSLAND

Organic farmers in maritime NW Europe rarely express concern about pests and diseases in grassland, although clearly stem eelworm (*Ditylenchus dipsaci*) (in red clover) and Sclerotinia clover rot (*Sclerotinia trifoliorum*) could potentially be major problems. Slug damage can be a problem where clover varieties are used that contain low levels of cyanogenic glucosides, and in oversowing situations (Baars, 2002). There is little that can be done to control these problems within the constraints of the organic standards, except by selection of resistant varieties and by rotation, i.e. alternating between legume species where appropriate rather than basing the rotation entirely on one species. Pests and diseases are greater threats with alfalfa than with red and white clovers. Verticillium wilt (*Verticillium alba-atrum*) and alfalfa weevil (*Hypera* spp.) are probably the most important, but Frame, Charlton and Laidlaw (1998) list 22 diseases and 20 pests of alfalfa. Varietal resistance and rotation are again the main approaches in controlling these problems in organic farming.

MAINTAINING LIVESTOCK HEALTH AT GRASS

Stomach worm infection in lambs is the most common health problem in organic lambs (Halliday *et al.*, 1991; Roderick, Short and Hovi, 1996). Grazing management has a major role to play in minimizing worm infection (Younie *et al.*, 2003). It should be the main component of worm control strategy, and management decisions must be designed primarily to minimize parasite infection. Essentially, this involves restricting the stocking rate of susceptible animals, particularly young stock, and minimizing the contamination of pasture with parasite larvae and the parasite challenge to these animals, for example, by:

- having an enterprise mix on the farm that includes both cattle and sheep (since the parasites that affect one species do not, in general, affect the other);
- having a mix of adult stock (which are largely immune to parasite attack) as well as young stock on the farm;
- alternating cattle, sheep, or forage conservation, or a combination, from year to year on any one field (which will minimize the contamination of pasture with worm larvae over winter);
- avoiding high stocking rates per se; or
- regular reseeding, which, where possible, will provide uncontaminated pasture in the first year.

Parasite challenge to susceptible animals is likely to be least on lowland, mixed farms with a mix of cattle and sheep enterprises, where opportunities exist to utilize crop stubbles, crop residues and annual forage crops as well as grassland; where young, larvae-free reseeds are available on an annual basis; where all grass fields can be mown for conservation and included in alternating clean grazing systems; and where cattle can provide an effective dilution of the parasite challenge to lambs. Conversely, parasite and disease problems are likely to be greatest on the sheep-only upland farm, where opportunities to take forage conservation cuts are limited and a clean grazing system is difficult to establish.

Even where it is possible to establish an alternating clean grazing system, it is not always effective in controlling parasitic gastroenteritis in lambs, primarily because of the pasture contamination caused by the ewes themselves at lambing (the peri-parturient egg rise) (Halliday, Gray and Younie, 1994). There is a need to consider other approaches to reducing the problem, e.g. breeding for resistance to parasites in the sheep flock, exploiting the differential parasite burdens of different pasture species (Keatinge *et al.*, 2002; Niezen *et al.*, 1998a, b; Thamsborg, Roepstorff and Larsen, 1999). For example, a number of herbage species, primarily high-condensed-tannin species such as *Lotus pedunculatus* and *L. corniculatus*, but also chicory, have been shown to reduce the effects of worm infection in young animals (Barry and McNabb, 1999; Molan *et al.*, 2000). The suggested mechanisms for this include (i) a direct toxic effect of high-condensed-tannin on the parasite; (ii) an improvement in protein or mineral status of the animal, or both; and (iii) an adverse affect of these herbage species on the survival of the parasite in the sward. Further parasitological and agronomic research is necessary to determine how to utilize and integrate these species into the grassland system to best effect (e.g. as separate anti-worming paddocks for use at critical times, or as an integral component of the main seed mixture).

Grassland and ruminant livestock management on organic farms should be planned to make the maximum use of grazing. Apart from considerations of the cost of conserving grass (and the concomitant nutrient removal from the soil, see below) and the cost of organic concentrates, health problems tend to be fewer in animals kept outdoors, particularly where calving and lambing are undertaken outdoors (provided weather and soil conditions permit). Stress, which can contribute to metabolic disorders (e.g. hypomagnesaemia) or breakdown in natural immunity, should be minimized at all times in the grazing situation, for example by appropriate construction of gateways and feeding stances, so as to minimize poaching (and hence potential foot problems), by provision of access to shelter (e.g. by hedging along field boundaries), and by careful grouping, feeding and handling of animals.

Bloat is often mentioned as a potential risk in legume-based grassland systems. While the risk exists (in the grazing situation, but not with silage, and with cattle, but much less so with sheep), it is a relatively infrequent occurrence and can be predicted (and therefore avoided) to a large extent. The risk is greatest in the second half of the growing season (in NW maritime Europe), if animals are offered relatively rank herbage with a high clover content and low dry matter content (e.g. silage aftermaths or around dung pats, on wet or dewy mornings). Hungry animals should never be introduced to such herbage in these conditions. Animals should be introduced to aftermaths soon after the crop has been removed, when the regrowth is very short, so that the clover grows up to the animals rather than vice versa. In high-risk situations, a straw buffer feed reduces the risk. Dairy cows should be fed a buffer feed of forage after milking before returning to pasture. However, poloxalene bloat preventative, fed with a carrier feed such as barley or brewer's grains, is the most effective insurance. The risk of bloat is also substantially reduced in legumes with a high tannin content, e.g. sainfoin, and so considerable effort is now under way to breed anti-bloating clovers with a high tannin content, although varieties produced by transgenic procedures (genetic engineering) would not be permitted in organic farming.

ORGANIC FARMING SYSTEMS AND NITROGEN FIXATION

As indicated earlier, nitrogen is the single most important nutrient required for herbage growth. In organic systems, N is supplied from mineralized soil organic N and from applications of organic manures, but the ultimate source of N input to the system is atmospherically-derived N fixed by herbage legumes. The N resource in organic farming is, therefore, a renewable resource in comparison with fossil-fuel derived artificial N fertilizers. Whitehead (1995), quoting seven series of trials in the United Kingdom, reported amounts of fixed N ranging from 0 to 445 kg N ha^{-1} yr^{-1}, with an overall average of 152 kg N ha^{-1} yr^{-1}. Other workers have reported N fixation input to grassland as a fertilizer-N equivalent value, ranging between 100 and 250 kg N ha^{-1} yr^{-1} for temperate European conditions. The lower end of this range is associated with soils of high potential for N mineralization, continuous grazing management and smaller-leaved white clovers, while the upper end of the range is associated with soils of low potential for N mineralization, cutting or rotational grazing management, large-leaved white clovers and red clover or alfalfa (Reid, 1970; Morrison, 1981; Frame, 1990; Davies and Hopkins, 1996). Halberg, Kristensen and Kristensen (1995) estimated atmospherically derived N inputs on six organic mixed farms in Denmark. Their estimates, which include a figure of 21 kg N ha^{-1} from precipitation, range from 39 to 56 kg N ha^{-1} yr^{-1} in permanent grass and from 120 to 408 kg N ha^{-1} yr^{-1} in grass-clover leys (the latter figure of 408 kg N ha^{-1} being derived from alfalfa).

The level of N fixation is notoriously difficult to measure accurately (Wood, 1996) but is accepted to be primarily dependent on the clover content of the sward, particularly in the early years of a ley (Frame and Newbould, 1986; Van der Meer and Baan Hofman, 1989). However, this is not a linear relationship. In addition to contributing to herbage growth, a proportion of the atmospherically fixed N is stored in roots and stubble, or is immobilized in soil organic matter, i.e. it contributes to the longer-term build up of soil organic matter and soil N status. Simultaneously, a proportion of the N in the soil organic pool is mineralized and becomes available for uptake by pasture plants. As the supply of N from mineralized soil organic matter increases (e.g. in the later years of a ley, or in swards with a very high clover content), the relative contribution from N fixation to the overall supply of N to the sward declines (Younie, 1992; Davies and Hopkins, 1996). Baars (2002) has reported that, in the Netherlands, between 40 and 65 kg N is fixed per tonne of white clover DM, but as clover content in the sward increases above 50–60 percent in DM, N fixation decreases to 7 kg N per tonne of white clover DM because of the increased amount of N in the soil and the lack of companion grass to act as a sink for this soil N.

Halberg, Kristensen and Kristensen (1995) estimated that, on Danish organic farms, the amount of surplus N available annually for immobilization in the soil and for loss to the environment was between 18 and 231 kg N ha^{-1} on grass-clover leys and between 3 and 59 kg N ha^{-1} on permanent grass. The equivalent figure calculated by Watson and Younie (1995a), for an organic beef farmlet, was 103 kg N ha^{-1}, compared with 216 kg N ha^{-1} for a conventional high-N-fertilizer system. On a whole-farm basis, the average N surplus for 14 commercial organic farms calculated by Halberg, Kristensen and Kristensen (1995) was 124 kg N ha^{-1} yr^{-1}, while for the comparable sixteen conventional farms the N surplus was 240 kg N ha^{-1} yr^{-1}.

Thus, herbage legumes, normally in the form of grass-clover leys, play a vitally important role in organic systems through the capture of the atmospheric N resource, enabling its subsequent exploitation by ruminant livestock and arable crops. In a self-contained organic farming system, the proportion of grass-clover ley or herbage legume crop in the rotation will determine the total N input, which will in turn determine arable crop yields (Younie *et al.*, 1995). The ideal proportion of fertility-building crop (i.e. legume) in the rotation has never been definitively determined. A minimum of 50 percent is often suggested, but good crop yields have been obtained in good soils, at least over the short term, with a stockless rotation containing only one year in five of red clover green manure, plus a grain legume (Cormack, 1997). Nitrogen fixation rates tend to be higher in the early years of a grass–clover ley (e.g. Kristensen, Høgh-Jensen and Kristensen, 1995), which would suggest that short-term (two- to three-year) leys are a more efficient way, compared with longer-term leys, of utilizing the atmospheric N resource within the overall context of fertility building and exploitation. Johnson *et al.* (1994) have also suggested that the optimum length of ley in terms of its residual value to following crops is three years.

Soil structure

A well aerated soil is an essential aim in organic farming because of the beneficial effect this has on soil N mineralization and the reduced risk of N loss through denitrification. The positive influence of grassland on soil structure is well established (Tisdall and Oades, 1980; Eder and Harrod, 1996) and it has been suggested that legumes in particular enhance the soil aggregation process (Angers and Carter, 1996). Reganold *et al.* (1993), in a paired-farm comparison between biodynamic and conventional farms in New Zealand, found that pasture soils on biodynamic farms had significantly lower bulk density and lower penetration resistance than pastures on conventional farms, indicating better soil structure and less compaction. These authors suggested no reasons for these differences, but a higher clover content in the swards may have contributed.

Although a grass-clover ley will undoubtedly improve soil structure relative to the arable cropping phase, soil compaction does occur in grassland situations, including organic farming, and indeed is of particular importance in the organic context, given the heavy reliance on mineralized N in organic farming. Heavy silage-making equipment, high stocking densities, particularly when soil conditions are wet (e.g. around feeding troughs in winter), will all cause compaction and reduce soil N availability, both to the grass–clover ley itself and possibly to the following arable crop. The effect of compaction on N offtake in unfertilized grass (reflecting N mineralization), caused by a typical sequence of silage harvesting operations, is shown in Table 9.3.

Table 9.3 Effect of compaction on annual N off-take (kg ha^{-1}) from three silage cuts (zero N fertilizer input)

Soil condition	1988	1989	1990	Average
Non-compacted	41.3	34.7	39.5	38.5
Typically compacted	25.8	27.1	25.7	26.2

Source: Douglas and Crawford, 1993.

Outdoor pig production can also cause serious soil compaction, which can have an adverse residual effect on the subsequent arable crop (Edwards and Watson, 1997). As an indication of development of anaerobic zones in the soil, Petersen and Eriksen (1999) also measured a high level of denitrification from grazing areas with outdoor sows. The strategy for moving huts and siting feeding areas must aim to avoid compaction (Larsen and Kongsted, 1999). In the further development of organic pig farms (as opposed to conventional farms), it is very important that grazing systems for the pigs are developed so that the pigs actually utilise the grass sward as a source of nutrients. This may at the same time lead to outdoor pig production systems with much smaller negative effects on soil compaction and nutrient losses. There is need for further research, within the context of organic farming, to quantify these effects and to examine practical methods of ameliorating soil compaction in organic grassland.

Soil biological activity

A high degree of soil biological activity is associated with good soil structure and soil N mineralization. Manipulation and enhancement of this biological resource is a major aim of organic farming. As a result of the permanent green cover and vigorous root growth afforded by grassland, by the absence of soil disturbance, and by the consequent increase in soil organic matter, soil biological activity is greater under grassland than under arable cropping. This is one of the primary reasons for the inclusion of grass–clover leys in organic rotations. Earthworm populations increase during the ley phase and decline under cropping as a result of increased frequency of cultivation, as shown in organic crop rotation experiments in NE Scotland (G. Armstrong, unpublished information).

Likewise, from the same set of experiments, Watson *et al.* (1996) showed that soil microbial biomass and activity were greater under grass–clover than under arable crops. These results are not unique and have been mirrored elsewhere (e.g. Follett and Schimel, 1989; Weil, Lowell and Shade, 1993).

Because of the higher investment costs and lower profitability of livestock systems relative to cropping systems, stockless organic rotations are an attractive proposition economically. However, an unanswered question remains about the level of nitrogen supply (and therefore crop yield) that such systems can sustain with a reduced proportion of herbage legume and grass–clover ley in the rotation. Part of this question relates to the maintenance of soil biological activity in a system with high cultivation and cropping frequency. It would seem logical that the nearer to a continuous cropping system we approach, and the shorter the ley or green manure phase, the lower the level of soil biological activity, the lower the soil organic matter content and the lower the potential for soil N mineralization. Indeed, the long-term Rodale Farming Systems Trial in the USA has shown this to be the case: soil biological activity was greater in the livestock-based organic system than in the stockless system (Wander and Traina, 1996). It is important, therefore, that research on stockless rotations should consider soil biology responses.

Higher soil biological activity (e.g. earthworm populations) has sometimes been reported in organic grassland (Figure 9.1) compared with conventional grassland (Reganold and Palmer, 1995). The addition of more organic matter in the form of animal manures will improve the food supply for earthworms and conversely the use of biocides in conventional agriculture may reduce earthworm populations, but it is conceivable that the application of mineral fertilizer in conventional agriculture, by virtue of increasing the food supply, will enhance earthworm populations (Younie and Armstrong, 1996). However, Baars (2002) has shown that, compared with slurry and farmyard

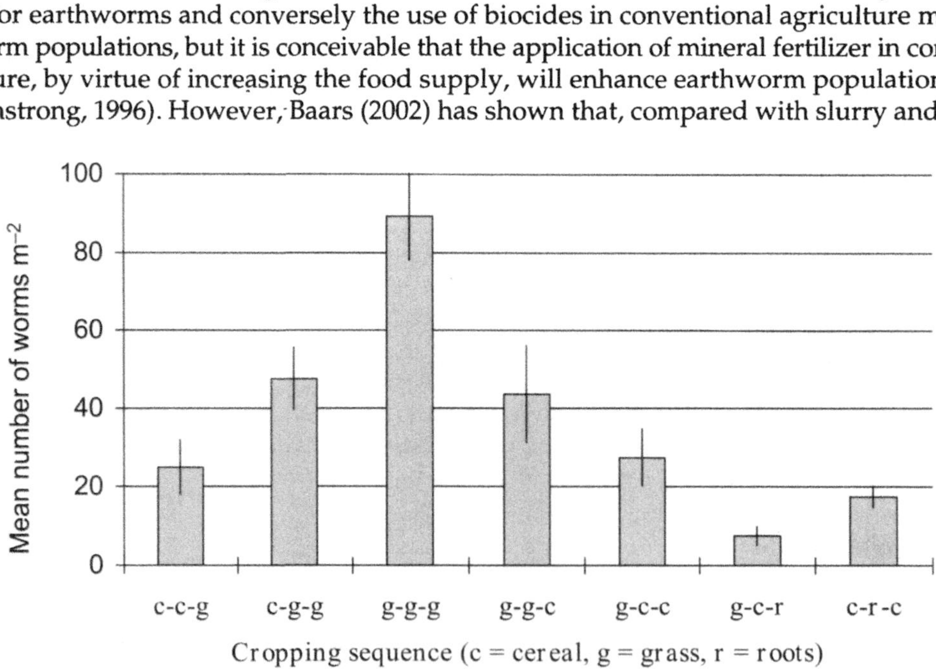

Figure 9.1 Effect of cropping sequence on population of *Lumbricus terrestris*.

manure (FYM), superphosphate and muriate of potash do reduce earthworm populations, possibly as a result of lower soil pH. In these studies, application of FYM resulted in the highest density of earthworms, particularly those worm species that move vertically into deeper soil layers.

MAINTAINING SOIL NUTRIENT STATUS

One of the major aims of organic farming is to minimize the use of non-renewable resources. It follows, therefore, that soil nutrient status must be maintained as far as possible through management means and by efficient nutrient cycling within the farm. In grassland, the main priority concerns the removal of nutrients from fields cut for hay or silage. With a herbage yield of 10 tonne DM ha^{-1} yr^{-1}, nutrient offtake in the herbage will be in the region of 80 kg P_2O_5 ha^{-1} yr^{-1} and 260 kg K_2O ha^{-1} yr^{-1} (Fowler, Watson and Wilman, 1993; Younie *et al.*, 1998). This can obviously have a major effect on soil K and P content and consequently on herbage and crop yield.

In a systems comparison in Germany, soil P content decreased by 39 percent over five years in the organic system (Weber *et al.*, 1993). Similarly, in two out of three organic and conventional paired-farm comparisons in New Zealand, soil P content in pasture soils was significantly lower on the organic farms, reflecting negative P budgets on these farms (Nguyen, Haynes and Goh, 1995). Soil P content in the top 5 cm in a sandy soil in the Netherlands, managed organically, fell over five years by 25 percent under cutting management and by 18 percent under grazing+cutting management, even where 50 kg P_2O_5 ha^{-1} yr^{-1} was applied as FYM (Younie and Baars, 1997).

Figure 9.2 shows the effect of first-cut silage yield on exchangeable soil K content nine months later, in grass–white clover swards in an organic crop rotations trial on a sandy loam soil (10 percent clay) in NE Scotland. The relationship is significant at the $P<0.001$ level.

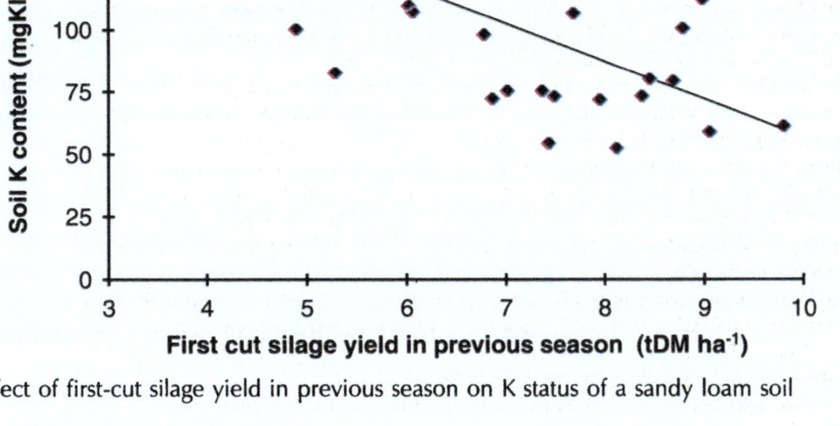

Figure 9.2 Effect of first-cut silage yield in previous season on K status of a sandy loam soil

Similarly, on a sandy soil in the Netherlands, Baars (2002) reported reductions of 62 percent and 39 percent in soil K content over three years in, respectively, cut and cut+grazed organic grass–clover leys yielding 9 to 11 t DM ha^{-1}. In contrast, on similarly managed swards on clay soils, no decline in soil K content was detected.

Clovers are poor competitors for soil P and K compared with grass. The reductions in K content in the experiments of Baars (2002) described above caused a significant loss of clover from the swards. This was also shown in an experimental situation on a sandy soil by Van der Meer and Baan Hofman (1999). Fothergill, Davies and Morgan (1995) have also shown the significant effect of P and K on white clover content in upland pasture. Low clover content resulting from restricted K input has also been observed in swards in a long-term trial on sandy loam soil in NE Scotland (Younie, unpublished information). This adverse effect of low soil P and K content on clover content will restrict the input of atmospherically fixed N in the system and have a knock-on effect on herbage yield. Newton and Stopes (1995) reported that soil P status was the best predictor of herbage production in a survey of organic farms.

It is essential, therefore, that organic grass–clover swards intended for conservation be given high priority in the distribution of manures, particularly on sandy soils. In a field trial in which a range of manure and silage effluent treatments were applied to an organic grass–clover sward on loamy sand soil (5 percent clay) for two cuts of silage, it was estimated that, in order to maintain a zero change in soil K content, an annual application of 233 kg K ha^{-1} (282 kg K_2O ha^{-1}) was necessary (Younie *et al.*, 1998). The same trial indicated that where two cuts were being harvested, a single spring application of 20 t FYM ha^{-1} was less effective in maintaining soil K content than two applications of 10 t ha^{-1}, one for each cut. Baars (2002) showed that K yield and K concentration per kilogram of herbage DM were affected by the total K application (Table 9.4). Phosphorus yield and P concentrations, however, were much less affected by P inputs.

Table 9.4 Relationship between mineral yield (Y) or mineral concentration (C) and mineral input (I) (for each equation, n = 9).

	Yield (Y) (kg ha^{-1})	Concentration (C) (g in herbage DM)
Potassium (K)	Y = 152 + 0.49 I (R^2 = 0.88)	C = 17.2 + 0.03 I (R^2 = 0.83)
Phosphorus (P)	Y = 38 + 0.09 I (R^2 = 0.38)	C = 3.8 + 0.00 I (R^2 = 0.02)

Source: Baars, 2002

At the same time, careful consideration needs to be given to K application strategy in order to minimize the risk of luxury uptake. Luxury uptake of K can result from large spring applications of manure, and therefore manure and effluent applications should be spread evenly throughout the season, perhaps even with an emphasis towards late-season application, although this increases the risk of leaching loss of N and K.

Because of the labile nature of K in manure and to a lesser extent in soils, the risk of loss during the nutrient cycling process itself is substantial, in contrast to the situation with P. Berner (1986) estimated losses of 50–60 percent of total K from FYM heaps on commercial organic farms. Nolte and Werner (1994) estimated that 78 percent of whole-farm K losses were lost during the internal nutrient cycling process, but only 25 percent of P losses, and suggested that the K balance could be improved substantially by improving manure handling and storage procedures, such as by covering manure heaps and avoiding application of manures in autumn and winter.

Apart from the specific issue of manure management, practical issues of management and land type often limit the efficiency with which nutrients can be recycled within the farm. Ideally, in order to prevent excessive nutrient offtake from individual fields, management of any one field should alternate annually between cutting and grazing, rather than continuous cutting. This alternating management also has benefits in terms of weed control (see earlier). However, in practice this is difficult to achieve on many farms, e.g. dairy farms, where the cow grazing pastures must be located within walking distance of the milking parlour, or on upland farms where a restricted choice of flat fields for cutting results in some fields being cut every year. On some farms, all stock are fed and wintered outside, e.g. on rough ground. While this may be healthy for the livestock, it means that nutrients are being transferred from silage cutting fields to the wintering land. On farms with arable crops as well as grassland and livestock, inevitably there will be transfer of nutrients, of K in particular, from grassland to arable cropping land, given that FYM is one of the major sources of nutrient for arable and vegetable crops in organic systems. However, in a mixed ley–arable system the rotation of crops means that these nutrients will ultimately be uniformly distributed around the farm.

Nevertheless, even if 100 percent efficiency in nutrient cycling were possible, there would still be a considerable loss of P and K in crop and livestock products sold off the farm and in leaching losses in the field (on biodynamic farms, Nolte and Werner (1994) estimated 14 percent of K losses and 73 percent of P losses were in products sold off the farm). The inference is, therefore, that soil P and K in grassland will be depleted over time, or that supplies of P and K must be brought into the system. In the Netherlands, where neighbouring specialist organic arable farms and organic dairy farms have developed partnerships for transfer of feed and manure, there is a view that soil P and K content on grass–clover swards on dairy farms should be maintained by mineral fertilizer (see below) and that the manure is better targeted towards arable land (Baars, 2002).

The rate and extent of depletion of soil P and K, and therefore the need for inputs, will depend on the level of crop yield and nutrient offtake and the level of soil nutrient reserves. Some clay soils have large reserves of potassium ions and have the capacity to release K into the exchangeable fraction, and hence support good crop growth over a long period of time. Sandy or coarse-textured soils with a low clay content, in contrast, have a low reserve of non-exchangeable K and will be unable to sustain high production levels without K inputs.

Rock phosphate is permitted as a nutrient input in organic farming and is a relatively effective source of P, but unfortunately there are few effective sources of K input that are acceptable in organic systems. Simpson and Stopes (1991) assessed a range of potassium fertilizers for organic farming and found that rock potash was a relatively ineffective source of potash, whereas lime kiln dust gave useful responses. Potassium sulphate, potassium salts (kainit, sylvinite) and sugar beet by-product (Cummulus K) are effective, but under EU Regulation 2092/91, before these products can be used the producer must prove to the certification body that there is a need. A further potential source of potash is in straw bought in for bedding, but, assuming a stocking rate of one Livestock Unit per forage hectare, this will supply only about 13 kg K ha^{-1} if one tonne of barley straw is purchased annually per Livestock Unit. Further research is necessary to assess the effectiveness of approved external sources of these nutrients, potassium in particular.

THE IMPACT OF ORGANIC GRASSLAND ON THE ENVIRONMENT

Care for the environment, including wildlife habitats, and minimal use of non-renewable resources are enshrined as principles of organic farming. One of the strategic objectives at farm level is to minimize leaching of nutrients into surface and ground waters. For the organic farmer in particular, this is important from a production, as well as from an environmental, point of view. The extent to which organic farming can be regarded as a sustainable system can also be assessed by the potential for reduced use of fossil fuel resources and for enhanced biodiversity. Grassland plays an important role in these elements of organic philosophy. Aspects of botanical composition and biodiversity in organic grassland were discussed earlier in this chapter, and so the discussion below focuses on nutrient loss to the environment and on energy use.

Nutrient emissions to the aquatic environment

It is now accepted that the risk of nitrate-N leaching from grassland is influenced more by the total quantity of N circulating in the system, rather than by the source of the N. Swards with similar stock carrying capacity will result in similar environmental losses of N, whether based on N derived from N-fixation by legumes or supplied as fertilizer N (Titchen and Philipps, 1996). A number of comparative studies have shown that the N surplus, and hence the potential for nitrate-N leaching, is less from clover-based systems, including organic grassland, than from intensively fertilized grassland (Halberg, Kristensen and Kristensen, 1995; Watson and Younie, 1995a; Tyson *et al*., 1996). However, Schils (2002) showed an increased loss of nitrogen as white clover content in the sward increased, especially where the sward was grazed after September and in combination with high proportions of *Poa annua* and *Taraxacum officinale* (dandelion) in the sward.

Maximizing green cover of soil over winter is encouraged in organic farming, not only to ensure uptake of surplus mineral N from the soil and so minimize leaching losses, but also to minimize the risk of soil erosion. The inclusion of a ley in the crop rotation obviously goes some way towards achieving this objective, but other forage species are often also sown specifically as winter covers during the arable cropping phase of the rotation. Species such as mustard (*Sinapis alba*) and *Phacelia* are effective scavengers of soil N and are commonly used as cover crops, but graminaceous species such as grazing rye (*Secale cereale*) and Italian ryegrass (*Lolium multiflorum*) can also be used. At farm level, however, the farmer must balance the perceived benefits in terms of the amount of N saved, or prevented from leaking to the environment, against the establishment costs, including seed costs. The extent of any benefit will depend not only on the species used as a cover crop, but also on how well it establishes, and on how much mineral N is present in the soil initially. In this latter regard, the winter following the first arable crop in the rotation is normally when soil N is at its highest level in the cropping phase of the rotation and therefore when winter cover crops offer the greatest potential benefit. A further consideration in relation to establishment of winter covers is the length of the growing season available for establishment, following the harvest of the main crop. Effective establishment of winter covers can be achieved from direct sowing following early harvested crops such as early potatoes or vegetables, or in locations with a long growing season; however, in areas with a short growing season, or following a late-harvested cereal, for example, direct sowing of winter cover crops can often result in poor establishment (Watson and Younie, 1995b). In these situations, undersowing the winter cover with the previous cereal crop will give better establishment.

In the context of organic mixed ley-arable systems, the management of the transition phase from ley to arable cropping, the most critical point for N leaching in organic rotations, is of crucial importance. It is clear that autumn ploughing (usually for establishment of a winter cereal) leads to more leaching of N than spring ploughing. Cobb *et al*. (1997) measured leaching losses of between 119 and 132 kg N ha^{-1} in organic winter wheat in England following the ploughing of leys. A switch to the production of spring crops would reduce losses significantly (Philipps and Stopes, 1995), but winter wheat has considerable economic advantages over spring cereals, and autumn ploughing is essential for satisfactory seedbed preparation for heavy clay soils. In any case, this transition phase from ley to arable cropping represents only 15–20 percent of the rotation and the lower losses from the other phases of the organic rotation reduce the average losses over the whole rotation to 10–20 kg N ha^{-1} per annum (Philipps and Stopes, 1995).

Date of ploughing of the ley (or indeed of a grass winter cover crop) relative to the sowing date of the following crop is important not only in relation to minimizing leaching loss, but also in controlling N immobilization by the incorporated ley vegetation. If a ley (or any other vegetation type with a high carbon:nitrogen ratio) is ploughed too close to the sowing date of the crop, N can be immobilized and this can significantly reduce crop N uptake and yield. Younie and Watson (1995) measured a 1.3 t ha^{-1} reduction in grain yield of spring barley following a March ploughing of an organic grass–clover ley, compared with a January ploughing date. A delay of at least one month is necessary between the ploughing of a ley or a grass winter cover crop and the sowing of the subsequent arable crop.

Nutrient emissions to the atmosphere

Ammonia volatilization from livestock systems makes a major contribution to acid rainfall. For example, in the Netherlands, concern about ammonia volatilization has led to the introduction of compulsory covering of slurry storage facilities. Ruminant animals play a major role in organic systems and so the potential leakage of N into the atmosphere as ammonia from associated livestock housing, manure storage and land spreading cannot be dismissed as insignificant. Halberg, Kristensen and Kristensen (1995) estimated ammonia losses from organic dairy farms in Denmark to vary over the same range (22 to 52 kg N per Livestock Unit per annum) as comparable conventional dairy farms, although stocking rates on organic farms tend to be lower than on conventional farms.

In addition to ammonia volatilization, grassland and ruminant livestock systems represent a major source of the greenhouse gases methane (CH_4) and nitrous oxide (N_2O), and carbon dioxide (CO_2) is released in the manufacture of inputs such as fertilizers and the burning of fossil fuels on farms. Bakken *et al.* (1994) compared the emissions of greenhouse gases from high intensity and low intensity (organic) dairy farms in Denmark (Table 9.5).

They calculated that, on a per-hectare basis, the organic system produced less CO_2 equivalents; but because of the lower output per hectare of the organic system, the amount of CO_2 equivalent

Table 9.5 Effect of grassland farming intensity on global warming gas emissions (expressed as t CO_2 equivalent ha^{-1} per annum)

Gas type	High intensity farm	Low intensity farm (organic)
N_2O	1.8	1.0
CH_4	5.0	4.1
Energy	2.3	1.4
Total	9.1	6.5

Source: From Bakken *et al.*, 1994.

per unit of milk produced was almost identical for the two types of farm. However, the energy consumption on the organic farm used by Bakken *et al.* (1994) for their study may be unusually high because thermal weed control was used in the fodder beet crop. N_2O emissions from grassland on the organic farm were only one third that from the conventional farm, as a result of lower fluxes from the soil (due to lower N levels) and the absence of emissions from the industrial production of nitrate. The largest contributor of greenhouse gases was CH_4, originating primarily from rumen digestion (contributing approximately 60 percent of CO_2 equivalents), and the organic farm produced only 17 percent less per hectare than the conventional farm, simply mirroring the lower stocking rate.

Thus, organic grassland systems do appear, on a per hectare basis, to produce less pollution of the aquatic and atmospheric environments, although to some extent this is simply a reflection of lower stocking rates. Since the main sources of pollution from organic livestock farming are derived from integral components within ruminant livestock systems (e.g. volatilization of ammonia, methane from rumen digestion), rather than from excessive nutrient or energy inputs, the main approach towards reducing emission of nutrients and greenhouse gases must be the further improvement of internal nutrient cycling (e.g. slurry injection in early spring rather than surface spreading in early winter).

Energy use in organic grassland

Energy consumption is discussed in the previous section in the context of CO_2 release and the contribution of organic grassland to global warming. It is also pertinent to discuss the influence of organic grassland systems in terms of consumption of non-renewable fossil fuel resources. Refsgaard *et al.* (1998), in a modelling exercise based on case studies, compared the energy utilization of the crop enterprises on fourteen organic and seventeen conventional dairy farms in Denmark. Using as a basis for comparison the energy productivity of a given crop (quantity of yield per unit of energy required), they found that grass–clover leys and alfalfa on organic farms were approximately three times more productive than those on conventional farms (Table 9.6).

Table 9.6 Model of energy productivity in grass–clover swards in organic and conventional dairy farms; non-irrigated sand, Denmark

		Conventional	Organic
51% of grass–clover harvested for silage:			
Sum, direct energy	(MJ ha^{-1})	2 745	2 454
Sum, indirect energy	(MJ ha^{-1})	10 922	1 378
Total energy requirement	(MJ ha^{-1})	13 667	3 832
Yield	(SFU ha^{-1})	6 000	5 200
Energy productivity	(SFU MJ^{-1})	0.44	1.36
100% harvested for silage:			
Energy productivity	(SFU MJ^{-1})	0.37	0.84
100% grazed:			
Energy productivity	(SFU MJ^{-1})	0.57	7.03

Key: SFU = Scandinavian Feed Unit (= feeding value of 1 kg barley)
Source: Refsgaard *et al.*, 1998.

Silage harvesting had the biggest demand for direct energy (i.e. diesel fuel) in both organic and conventional systems (51–53 percent), but organic farms consumed more direct energy for manure handling, particularly for dry manure (17 percent and 6 percent for organic and conventional systems, respectively). However, the main difference in energy consumption between the two farm types was in indirect energy use, in which conventional grassland exceeded organic grassland by a factor of eight, primarily because of the energy required in the manufacture of fertilizer N. Grass+clover was more energy productive than fodder beet or whole-crop silage, and increasing the forage proportion grazed rather than ensiled improved the productivity still further.

Organic grassland appears to be relatively energy efficient, principally as a result of the non-use of N fertilizer. Scope for further improvements in energy utilization appears to lie primarily in modifications to the silage making process and manure handling and application.

Adoption of high dry matter silage systems (to minimize the amount of water that has to be moved) and big bales (to minimize the amount of chopping involved) may offer opportunities for reducing the energy costs associated with silage making. There is a need to model the energy costs of the manure handling process; composting will increase costs but subsequent spreading may be easier and incur lower energy costs.

LIVESTOCK SYSTEMS FOR ORGANIC GRASSLAND

Given satisfactory levels of soil P, K and pH status (and soil moisture supply), the level of herbage production and livestock output from organic grassland is largely dependent on its legume content. In this regard it equates closely with conventional low-external input grass+clover swards. A considerable body of research literature now exists on legume-based grassland systems and most of this is relevant to organic farming (e.g. Frame, Charlton and Laidlaw, 1998).

Published data on herbage production measured under organic farming conditions is presented in Table 9.7. These generally reflect the conditions under which the swards were monitored, namely site class, legume species, cutting date or regrowth interval. The data of Halberg and Kristensen (1997) included a comparison with conventional farms. Corrected for differences in climate, the "organic" yield of grass+clover swards was 12–14 percent lower. In the comparison of Younie and Wightman (1992), the organic system yielded 16–18 percent less than the comparable conventional system, which received 270 kg N ha^{-1} yr^{-1}.

Factors relating to livestock nutrient supply (from grassland and supplementary feeds) have a major influence on the choice of livestock system on organic farms. Organic standards specify that ruminant animals should obtain most of their nutrient supply from forage. Calving in spring or early summer rather than in autumn is an ideal system since the peak period for nutrient requirement in the animal matches the grass growing season. With sheep, early lambing in January is not an appropriate system, given the level of concentrate feeding involved. In contrast, lambing later than normal, in late April or May, to coincide with the later spring growth of clover-based organic swards, is well suited to organic farming and reduces the input of concentrates and the overall cost of production. However, in practice the marketing implications (i.e. when the animal is likely to be ready for slaughter) must also be taken into account together with these technical factors when selecting a date for calving and lambing.

Milk yield is very important for farm profitability in organic dairying. It is often difficult to satisfy expected nutritional requirements in winter diets, particularly the need of high yielding dairy cows for absorbable protein. In this respect, it is interesting that Mogensen and Kristensen (1999)

Table 9.7 Production per hectare from organic grassland

Sward type	Herbage production (t DM ha^{-1}) Annual (grazing)	Annual (mainly cut)	1st cut silage	Stocking rate LU(5) per forage ha	Reference
G/WC(1)	10.1	10.1		1.6	[1]
Varied	8.2/7.0(6)			1.6	[2]
Unspecified				1.6	[3]
G/WC	7.2		4.2	1.7	[4] Case (i)
G/WC	7.1	8.8		1.6	[4] Case (ii)
G/RC(2)		14			[4] Case (iii)
Unspecified				1.7	[5]
G/WC		13.0(4)			[6]
Unspecified				1.8	[7]
G/WC	6.4			1.9	[8] Case (i)
G/RC, L(3)		10.9			[8] Case (ii)
G/WC	8.7	8.9	5.9	2.1	[9]
G/WC		8.0	5.5		[10]
G/WC		7.6	5.4	1.2	[11]
G/WC		9.3–10.7	2.7–3.4		[12]
G/RC		12.6	4.5		[12]
G/WC		7.7–15.4			[13]

Notes: (1) G/WC = Grass-white clover. (2) G/RC = Grass-red clover. (3) L = Lucerne (Alfalfa). (4) Cut to ground level. (5) LU = Livestock Unit, where 1 LU = One Friesian or Holstein cow producing 6 000 litre milk per annum. (6) Calculated as Scandinavian Feed Unit (SFU) × 1.35.

Sources: [1] Fowler, Watson and Wilman, 1993. [2] Halberg and Kristensen, 1997. [3] Houghton and Poole. 1990. [4] Jones, Bowling and Haggar, 1996, reporting three situations. [5] Lampkin and Measures, 1999. [6] Newton and Stopes, 1995. [7] Redman, 1991. [8] Van der Meer and Baan Hofman, 1989, reporting two situations. [9] Younie and Wightman, 1992. [10] Younie, 1996. [11] Younie, 1997. [12] Baars, 2002. [13] Schils, 2002.

found no effect of protein supplementation in organic dairy herds fed a diet with a very high proportion of grass+clover silage. This underlines the importance of highly digestible, high quality grass+clover silage for winter feeding. However, in summer, pastures with a high white clover content can give an excessively high protein content in the diet, resulting in an energy:protein imbalance. Milk yield and nitrogen efficiency are improved if supplementary fodder crops with high energy content, such as maize silage or cereal whole-crop silage, are fed during this period, which usually occurs after June in NW Europe. Measurement of the amount of urea in milk is an effective means of determining the level of protein excess in the diet (Van Eekeren, 2000).

Profitability of livestock enterprises is a function of physical output multiplied by unit sale price, minus input costs. Physical output per head from organic beef and sheep enterprises compares well with equivalent conventional grass-based enterprises (Younie and Mackie, 1996; Younie, 1997), with beef animals reaching slaughter condition in 17 to 21 months. In contrast, the average milk yield per cow from organic dairy herds, at between 5 000 and 6 000 kg cow^{-1} yr^{-1}, is normally lower than from conventional units (Houghton and Poole, 1990; Redman, 1991; Lampkin, 1994; Weller, Cooper and Padel, 1996; Krutzinna, Boehncke and Herrmann, 1996), reflecting the generally lower levels of concentrate input in organic systems, although milk yields of 7000 kg cow^{-1} yr^{-1} are achievable. The high cost of purchased organic concentrates can result in relatively high feed costs per head, particularly in dairy and finishing beef enterprises, and although this is offset by reduced forage costs, total variable costs per head are generally only slightly lower than in conventional enterprises.

Fluctuations in unit sale price of both organic and conventional products make detailed financial comparisons difficult and valid only in the short term. Where organic premiums can be achieved, gross margin per head for organic livestock enterprises is normally substantially higher than the conventional equivalent. On a per hectare basis, organic systems that are clover-based will normally support a stocking rate and livestock output equivalent to a conventional system receiving up to about 180–200 kg N ha^{-1} yr^{-1} and should therefore achieve similar or better gross margins compared with conventional enterprises, even without the benefit of organic premium prices. With the added bonus of higher prices, organic systems have the potential to be substantially more profitable on a per hectare basis than conventional enterprises (e.g. Keatinge, 1997; Younie, 1997). At the same time, compared with intensively fertilized conventional enterprises receiving more than 200 kg N ha^{-1} yr^{-1}, organic premiums are probably essential if organic systems are to match equivalent conventional systems on gross margin per hectare.

For example, under Danish conditions, financial analysis showed a considerably higher profitability from organic milk production in the period 1993–1997 compared with conventional production. During this period, the organic milk price at the farmgate was 30–40 percent higher than the conventional milk price (Danish Central Advisory Service, 1998). It is estimated that an organic premium of 15–20 percent is necessary for organic milk production to maintain a level of profitability equivalent to conventional milk production.

A FUTURE PERSPECTIVE ON ORGANIC GRASSLAND

In fulfilling the principles of organic farming, the two major strategic objectives of farmers are to maximize the efficiency of internal nutrient cycling, thus minimizing the need for nutrient inputs, and to minimize problems by modifying the management system. These objectives are much easier to achieve in the sphere of grassland-based livestock systems than in the arable farming sphere: a wide range of herbage legumes is available to ensure an adequate N supply; there is built-in nutrient cycling by animals eating the herbage; and management opportunities exist for minimizing weed and pest problems.

However, this is not to say that further challenges do not exist for organic grassland farmers. These challenges exist not only at the farm system level, but also derive from external pressures. The organic movement in general is faced with pressure to dilute its standards because of global competition as organic becomes more accepted as a mainstream system. Pressure on the standards also comes from the increasing reliance on genetic manipulation techniques in breeding programmes and on the targeting of the organic sector by agricultural ancillary industries with new input products.

There are other pressures that require discussion within the organic movement, such as the issue of farm specialization and nutrient inputs. The ideal organic farm is a mixed grass+livestock+arable farm in which nutrients are cycled efficiently via herbage, crop by-products and livestock. The need for imported nutrients in such a system depends primarily on the proportion of arable crop in the rotation (and hence the proportion of nutrients that leave the farm). However, as in conventional agriculture, there is a trend towards specialization in organic farming, and there is a need to consider how best to integrate livestock farms with arable farms. Given the importance of soil organic matter, there may be a case for transfer of manure from livestock farms to neighbouring partner arable farms, with P and K removal from the livestock farms compensated for by the importation of permitted mineral P and K fertilizers. A discussion needs to be initiated within the organic movement to discuss the acceptability of approaches such as this. Linked with this is the issue of use of human waste in organic farming, a practice that is currently prohibited but that in theory would close the nutrient cycle.

At the farm system level, organic livestock should be in harmony with their environment. This includes relationships with other stock, with the nutritional and physical environments to which they are exposed, and the interaction they have with humans. In this context, there is a need for further development and adoption of grassland management techniques for parasite control in livestock. A better understanding is required of the complex links between parasite epidemiology, weather conditions, botanical composition of the sward, sward structure, stocking rate and stock handling. Another area where grassland management has a role in animal health is in mineral nutrition, particularly the balance of minerals in the herbage. In conventional farming, mineral imbalances are sometimes caused through fertilizer applications. There should be a lower risk in organic farming, where external nutrient inputs are minimized, but nevertheless greater knowledge and understanding is required. This may involve greater acceptance and use of unimproved natural pastures or, in sown swards, the use of seed mixtures with a wider range of species. Such swards and mixtures would also offer an opportunity for self-medication by animals. This is a concept that is claimed to apply to animals in the wild, but it may also apply in domesticated animals if the opportunity existed.

While the agronomy of the major legumes is now sufficiently well understood to provide reliable herbage production guidelines for most farmers, there is still a need for more reliable forage conservation techniques in marginal or extreme areas (e.g. regions at high altitude or high latitude; regions with high or low rainfall), for example by the introduction of new varieties or species and better soil management, in particular through improved soil aeration. Control of perennial weeds, particularly docks, is still one of the major problem areas in organic grassland and simple and reliable control techniques have still to be developed. Given the complexity of vegetation dynamics in grass swards, it is likely that simple solutions are not possible and that a weed control strategy involving a number of approaches is necessary, but nevertheless we still need to find the most efficient control techniques possible.

REFERENCES

Angers, D.A. & Carter, M.R. 1996. Aggregation and organic matter storage in cool, humid agricultural soils. pp. 193–211, *in:* M.R. Carter and B.A. Stewart (eds). *Structure and Organic Matter Storage in Agricultural Soils*. Boca Raton, Florida: CRC Press, Inc.

Baars, T. 2002. Reconciling scientific approaches for organic farming research. Part II: Effects of manure types and white clover cultivars on the productivity of grass-clover mixtures grown on a humid sandy soil. PhD thesis, Wageningen University, The Netherlands. 346p.

Baars, T. & Opdam, A. 1998. Trace elements and somatic cell counts in heifers. *In:* Th. Wensing (ed). *Production Diseases in Farm Animals*. Wageningen, The Netherlands: Wageningen Press. 297p.

Bakken, L., Refsgaard, K., Christensen, S. & Vatn, A. 1994. Energy use and emission of greenhouse gases from grassland agricultural systems. pp. 361–376, *in:* L. 't Mannetje and J. Frame (eds). *Grassland and Society*. Proceedings 15th General Meeting, European Grassland Federation. Wageningen, June 1994.

Barry, T.N. & McNabb, W.C. 1999. The implications of condensed tannins on the nutritive value of temperate forages fed to ruminants. *British Journal of Nutrition*, **81**(4): 263–272.

Berner, A. 1986. Einfluß des Sickersaftes von Kompostmieten auf die Umwelt [The effect of compost effluents on the environment]. Forschungsinstitut für Biologischen Landbau, Oberwil, Switzerland.

de Boer, P.B. 1990. Aangepaste landbouw: ecologische en landbouwkundige effecten op weidebedrijven [Adapted agriculture: ecological and agronomic effects on grassland farms]. *COAL-Publikatie*, No. 52. 232 pp.

CEC [Commission of the European Communities]. 1991. Council Regulation No 2092/91 on organic production. Official Journal of the European Communities, L198, 1–15.

Clark, D.A., Anderson, C.B. & Berquist, T. 1990. Growth rates of 'Grasslands Puna' chicory (*Cichorium intybus* L.) at various cutting intervals and heights and rates of nitrogen. *New Zealand Journal of Agricultural Research*, **33**(2): 213–217.

Cobb, R., Feber, R., Hopkins, A. & Stockdale, E. 1997. The Organic Farming Study. *Elm Farm Research Centre Bulletin*, No. 33(December 1997): 8–10.

Cormack, W. 1997. Testing the sustainability of a stockless arable rotation on a fertile soil in eastern England. pp. 127–135, *in:* J. Isart and J.J. Llerena (eds). *Resource Use in Organic Farming.* Proceedings of Third Workshop of the European Network for Scientific Coordination in Organic Farming (ENOF), University of Ancona, Italy, June 1997.

Courtney, A.D. 1985. Impact and control of docks in grassland. pp. 120–127, *in:* J.S. Brockman (ed). *Weeds, Pests and Diseases of Grassland and Herbage Legumes. British Grassland Society Occasional Symposium,* No. 18, / *British Crop Protection Council Monograph,* No. 29.

Danish Central Advisory Service. 1998. [Production-economy-ecology] (in Danish). 15p.

Davies, D.A. & Hopkins, A. 1996. Production benefits of legumes in grassland. pp. 234–246, *in:* D. Younie (ed). *Legumes in Sustainable Farming Systems. British Grassland Society Occasional Symposium,* No. 30.

Douglas, J.T. & Crawford, C.E. 1993. The response of a ryegrass sward to wheel traffic and applied nitrogen. *Grass and Forage Science,* **48:** 91–100.

Eder, G. & Harrod, T.R. 1996. Influence of grassland on soil erosion. aggregate stability and water quality. pp. 683–693, *in:* G. Parente, J. Frame and S. Orsi (eds). *Grassland and Land Use Systems.* Proceedings 16th General Meeting of European Grassland Federation, Grado, Italy, September 1996. Published as Grassland Science in Europe, Vol. 1.

Edwards, S. & Watson, C.A. 1997. An approach to investigating the environmental impact of an outdoor pig production system. pp. 335–340, *in:* J.T. Sorensen (ed). *Livestock Production Systems – More than food production.* EAAP Publication, No. 89. Wageningen, The Netherlands: Wageningen Pers.

Elliot, R.H. 1914. *The Clifton Park System of Farming.* 4th ed. London: Faber and Faber.

Ennik, G.C., Baan Hofman, T., Wieling, H. & Altena, H.J. 1982. Graslandproductie zonder kunstmeststikstof op het bedrijf van de familie Cuperus te Boksum (Fr) [Grassland production without nitrogen fertilizer on the farm of the Cuperus family in Boksum (Friesland province)]. CABO-verslag 42, Wageningen. 32p.

FAO/WHO Codex Alimentarius Commission. 2001. Guidelines for the Production, Processing, Labelling and Marketing of Organically Produced Foods. (GL 32 – 1999, Rev. 1 – 2001). See: http://www.fao.org/organicag/doc/glorganicfinal.pdf.

Fisher, G. & Baker, L. 1996. The chemical composition of forb species in grassland. pp. 429–431, *in:* G. Parente, J. Frame & S. Orsi (eds). *Grassland and Land Use Systems.* 16th meeting of European Grassland Federation. Grado, Italy, September 1996.

Follett, R.F. & Schimel, D.S. 1989. Effect of tillage practices on microbial biomass dynamics. *Soil Science Society of America Journal,* **53:** 1091–1096.

Foster, L. 1988. Herbs in pastures. Development and research in Britain, 1850–1984. *Biological Agriculture and Horticulture,* **5:** 97–133.

Fothergill, M., Davies, D.A. & Morgan, C.T. 1995. The effect of extensification of upland pasture on white clover. pp. 171–172, *in:* G.E. Pollott (ed). *Grassland into the 21st Century. British Grassland Society Occasional Symposium,* No. 29.

Fowler, S.M., Watson, C.A. & Wilman, D. 1993. N, P and K on organic farms: herbage and cereal production, purchases and sales. *Journal of Agricultural Science* (Cambridge), **120:** 353–360.

Frame, J. 1990. The role of red clover in United Kingdom pastures. *Outlook on Agriculture,* **19:** 24–55.

Frame, J. & Newbould, P. 1986. Agronomy of white clover. *Advances in Agronomy,* **40:** 1–88.

Frame, J., Charlton, J.F.L. & Laidlaw, A.S. 1998. *Temperate Forage Legumes.* Wallingford, UK: CAB International. 336 pp.

Frame, J., Fisher, G.E.J. & Tiley, G.E.D. 1994. Wildflowers in grassland systems. pp. 104–114, *in:* R.J. Haggar and S. Peel (eds). *Grassland Management and Nature Conservation. British Grassland Society Occasional Symposium,* No. 28.

Halberg, N., Kristensen, E.S. & Kristensen, I.S. 1995. Nitrogen turnover on organic and conventional mixed farms. *Journal of Agricultural and Environmental Ethics,* **8**(1): 30–51.

Halberg, N. & Kristensen, I.S. 1997. Expected crop yield loss when converting to organic dairy farming in Denmark. *Biological Agriculture and Horticulture,* **14**: 25–41.

Halliday, G., Ramsay, D.A., Scanlan, S. & Younie, D. 1991. *A survey of organic livestock health and treatment.* Kintail Land Research Foundation, Glasgow, in association with Scottish Agricultural College. 45 pp.

Halliday, G.J., Gray, D. & Younie, D. 1994. Some animal health problems on a small organic farm in Scotland. *Actes du Congrés de l'Élevage en Agriculture Biologique,* Brioude, France, October, 1994. Institut Technique de l'Agriculture Biologique, Rue de Bercy, Paris.

Hopkins, A., Jones, E.L, Bowling, P.J., & Johnson, R.H. 1997. Cultural methods of dock control in permanent pasture. pp. 39–40, *in:* Fifth Research Conference, British Grassland Society, Newton Abbot, September 1997.

Houghton, M. & Poole, A.H. 1990. *Organic milk production*. Genus Information Unit Report, No. 70. 31 pp.

IFOAM [International Federation of Organic Agriculture Movements]. 1998. *Basic Standards of Organic Agriculture*. Tholey-Theley, Germany: IFOAM.

Johnson, A.E., McEwen, J., Lane, P.W., Hewitt, M.V., Poulton, P.R. & Yeoman, D.P. 1994. Effects of one- to six-year-old ryegrass-clover leys on soil nitrogen and on the subsequent yields and fertilizer nitrogen requirements of the arable sequence winter wheat, potatoes, winter wheat, winter beans (*Vicia faba*) grown on sandy loam soil. *Journal of Agricultural Science*, (Cambridge), **122:** 73–89.

Jones, L. 1990. Forage chicory at a site in south-east England. pp. 67–69, *in:* G. Pollott (ed). *Milk and Meat from Forage Crops. British Grassland Society Occasional Symposium*, No. 24.

Jones, E.L., Bowling, P.J. & Haggar, R.J. 1996. Forage yields and quality during conversion. pp. 57–65, *in:* R.J. Haggar and S. Padel (eds). *Conversion to organic milk production*. Institute of Grassland and Environmental Research Technical Review, No. 4.

Keatinge, R. 1997. Organic hill farming: Does it pay? *New Farmer and Grower*, No. 55 (Autumn 1997): 24–25.

Keatinge, R., Jackson, F., Kyriazakis, I. & Cork, S.A. 2002. Developing parasite control strategies in organic systems. pp. 341–345, *in: Proceedings of the UK Organic Research 2002 Conference: Research in Context*. Aberystwyth, Wales, March 2002

Kristensen, E.S., Høgh-Jensen, H. & Kristensen, I.S. 1995. A simple model for estimation of atmospherically-derived nitrogen in grass-clover systems. *Biological Agriculture and Horticulture*, **12**: 263–276.

Krutzinna, C., Boehncke, E. & Herrmann, H.-J. 1996. Organic milk production in Germany. *Biological Agriculture and Horticulture*, **13:** 351–358.

Lampkin, N.H. 1994. Economics of organic farming in Britain. pp. 71–89, *in:* N.H. Lampkin and S. Padel (eds). *The economics of organic farming: An international perspective*. Wallingford, UK: CAB International.

Lampkin, N. & Measures, M. 1999. *Organic Farm Management Handbook*. 3rd ed. University of Wales, Aberystwyth, and Elm Farm Research Centre. 163p.

Larsen, V.A., Kongsted, A.G., 1999. Management of outdoor sow herds – basis for improvements. pp. 273-278, *in:* D. Gagnaux, R. Racord, A. Gibon, J.R. Poffet and A.R. Sibbald, A.R. (eds). *Livestock farming systems, Integrating animals science advances into the search for sustainability*. Proceedings 5th International Livestock Farming Systems Symposium, Switzerland, August 1999. EAAP Publication, No. 97.

Li, G.D., Kemp, P.D. & Hodgson, J. 1997. Regrowth, morphology and persistence of 'Grasslands Puna' chicory (*Cichorium intybus* L.) in response to grazing frequency and intensity. *Grass and Forage Science*, **52**(1). 33–41.

Mogensen, L. & Kristensen, T. 1999. Self-supply of organic feed on private dairy farms in Denmark. Proceedings NJF-seminar 303. *Ecological Animal Husbandry in the Nordic Countries*, Horsens, Denmark, September 1999.

Molan, A.L., Duncan, A., Barry, T.N. & McNabb, W.C. 2000. Effects of condensed tannins and sesquiterpene lactones extracted from chicory on the viability of deer lungworm larvae. *Proceedings of the New Zealand Society of Animal Production*, **60**: 26–29.

Moloney, S.C. & Milne, G.D. 1993. Establishment and management of 'Grasslands Puna' chicory used as a specialist, high quality forage herb. *Proceedings of the New Zealand Grassland Association*, **55:** 113–118.

Morrison, J. 1981. The potential of legumes for forage production. pp. 1.1–1.9, *in: Legumes and Fertilisers in Grassland Systems*. British Grassland Society Winter Meeting, 1981.

Newton, J. & Stopes, C. 1995. Grassland productivity on organic farms, 1992–1994. *Elm Farm Research Centre Bulletin*, No. 18: 2–6.

Nguyen, M.L., Haynes, R.J. & Goh, K.M. 1995. Nutrient budgets and status in three pairs of conventional and alternative mixed cropping farms in Canterbury, New Zealand. *Agriculture, Ecosystems and Environment*, **52:** 149–162.

Niezen, J.H., Waghorn, T.S. & Charleston, W.A.G. 1998a. Establishment and fecundity of *Ostertagia circumcincta* and *Trichostrongylus colubriformis* in lambs fed lotus (*Lotus pedunculatus*) or perennial ryegrass (*Lolium perenne*). *Veterinary Parasitology*, **78:** 13–21.

Niezen, J.H., Robertson, H.A., Waghorn, G.C. & Charleston, W.A.G. 1998b. Production, faecal egg counts and worm burdens of ewe lambs which grazed six contrasting forages. *Veterinary Parasitology*, **80:** 15–27.

Nolte, C. & Werner, W. 1994. Investigations on the nutrient cycle and its components on a biodynamically-managed farm. *Biological Agriculture and Horticulture*, **10**: 235–254.

Oomes, M.J.M., Korevaar, H. & Altena, H.J. 1980. Production and botanical composition of extensively used grasslands. CABO-reports No. 30, Wageningen. 28p.

Petersen, S.O. & Eriksen, J. 1999. Denitrification losses from outdoor piglet production: Temporal and spatial variability. pp 23-26, *Proceedings 10th Nitrogen Workshop*. Copenhagen, August 1999.

Philipps, L. & Stopes, C. 1995. The impact of rational practice on nitrate leaching losses in organic farming systems in the United Kingdom. pp. 123–134, *in: Nitrogen Leaching in Ecological Agriculture*. Bicester, UK: AB Academic Publishers.

Redman, M. 1991. Organic dairy costings: Update. *New Farmer and Grower*, **33:** 20–21.

Refsgaard, K., Halberg, N. & Kristensen, E.S. 1998. Energy utilisation in crop and dairy production in organic and conventional livestock production systems. *Agricultural Systems*, **57**(4): 599–630.

Reganold, J.P. & Palmer, A.S. 1995. Significance of gravimetric versus volumetric measurements of soil quality under biodynamic, conventional, and continuous grass management. *Journal of Soil and Water Conservation*, **50**(3): 298–305.

Reganold, J.P, Palmer, A.S, Lockhart, J.C. & Macgergor, A.N. 1993. Soil quality and financial performance of biodynamic and conventional farms in New Zealand. *Science*, Washington, **260:** 344–349.

Reid, D. 1970. The effects of a wide range of nitrogen application rates on the yields from a perennial ryegrass sward with and without white clover. *Journal of Agricultural Science* (Cambridge), **74:** 227–240.

Roderick, S., Short, N. & Hovi, M. 1996. Organic livestock production: Animal health and welfare research priorities. Veterinary Epidemiology and Economics Research Unit, Department of Agriculture, University of Reading.

Rumball, W. 1986. 'Grasslands Puna' chicory (*Cichorium intybus* L.). *New Zealand Journal of Experimental Agriculture*, **14:** 165–171.

Schils, R.L.M. 2002. White clover utilisation on dairy farms in the Netherlands. PhD thesis, Wageningen University, The Netherlands. 172p.

Simpson, D. & Stopes, C. 1991. The requirement and availability of potassium fertilisers in organic agriculture. *Ecology and Farming*, **3:** 14–16.

Swift, G., Davies, D.H.K., Tiley, G.E.D. & Younie, D. 1990. The nutritive value of broad-leaved weeds and forage herbs in grassland. *Scottish Agricultural College, Technical Note*, No. 223.

Thamsborg, S.M., Roepstorff, A. & Larsen, M. 1999. Integrated and biological control of parasites in organic and conventional production systems. *Veterinary Parasitology*, **84:** 169–186.

Titchen, N.M. & Philipps, L. 1996. Environmental effects of legume-based grassland systems. pp. 257–261, *in:* D. Younie (ed). *Legumes in Sustainable Farming Systems. British Grassland Society Occasional Symposium*, No. 30.

Tisdall, J.M. & Oades, J.M. 1980. The effect of crop rotation on aggregation in a red-brown earth. *Australian Journal of Soil Resources*, 18: 423–433.

Tyson, K., Stone, A.C., Scholefield, D. & Jarvis, S.C. 1996. A comparison of animal output and nitrogen leaching losses from grassland receiving 200 kg N/ha/annum and grass/clover receiving no fertiliser nitrogen. pp. 273–274, *in:* D. Younie (ed). *Legumes in Sustainable Farming Systems. British Grassland Society Occasional Symposium*, No. 30.

Umrani, A.P. 1998. Sustainable approaches for rangeland management and livestock production in Pakistan. PhD Thesis, University of Aberdeen, Aberdeen, Scotland.

Van der Brug, Y. 1996. *The effect of animal nutrition on clinical and subclinical mastitis in dairy cows.* Faculty of Veterinary Medicine, Utrecht University, The Netherlands. 80 pp.

Van der Meer. H.G. & Baan Hofman, T. 1989. Contribution of legumes to yield and nitrogen economy of leys on a biodynamic farm. pp. 25–36, *in:* P. Plancquaert and R.J. Haggar (eds). *Legumes in Farming Systems.* Developments in Plant and Soil Science, **37**.

Van der Meer, H.G. & Baan Hofman, T. 1999. Effects of low-emission slurry application techniques and periods of application on the yield and nitrogen economy of a mixed sward of perennial ryegrass and white clover. pp. 195–216, *in:* H.G. Van der Meer (ed). *Reducing inputs and losses of nitrogen and energy on dairy farms.* Final Report, Project AIR 3 CT92-0332.

Van Eekeren, N. 2000. Balancing summer rations of dairy cows by means of urea content in bulk tank milk. pp. 555–557, *in:* K. Soegaard, C. Ohlsson, J. Sehested, N.J. Hutchings and T. Kristensen (eds). *Grassland Farming – Balancing environmental and economic demands.* 18th General Meeting of the European Grassland Federation, Aalborg, Denmark. May 2000. Published as Grassland Science in Europe, Vol. 5.

Wander, M.M. & Traina, S.J. 1996. Organic matter fractions from organically and conventionally managed soils: 1. Carbon and nitrogen distribution. *Soil Science Society of America Journal*, **60:** 1081–1087.

Watson, C.A. & Younie, D. 1995a. Nitrogen balances in organically and conventionally managed beef production systems. pp. 197–199, *in:* G.E. Pollott (ed). *Grassland into the 21st Century. British Grassland Society Occasional Symposium*, No. 29.

Watson, C.A. & Younie, D. 1995b. The contribution of winter cover crops to the sustainability of agricultural systems. pp. 194–196, *in:* G.E. Pollott (ed). *Grassland into the 21st Century. British Grassland Society Occasional Symposium*, No. 29.

Watson, C.A., Ritz, K., Younie, D. & Franklin, M. 1996. Nitrogen and soil biomass dynamics in ley/arable crop rotations. pp. 43–50, *in:* J.H. Clarke, D.H.K. Davies, P.M.R. Dampney, R.J. Froud-Williams, P.J. Griffith, A. Lane, L. Sim and D.B. Stevens (eds). *Rotations and Cropping Systems.* Published as Aspects of Applied Biology, **47**.

Weber, S., Pabst, K., Schulte-Corne, H., Westphal, R. & Gravert, H.O. 1993. Funfjahrige Untersuchung zur Umstellung auf okologische Milcherzeugung. 1. Produktionstechnik [Five years studies on the conversion to ecological milk production. 1: Production techniques]. *Zuchtungskunde*, **65** (5): 325–337.

Weil, R.R., Lowell, K.A. & Shade, H.M. 1993. Effects of intensity of agronomic practices on a soil ecosystem. *American Journal of Alternative Agriculture*, **8:** 5–14.

Weller, R.F., Cooper, A. & Padel, S. 1996. Animal production during conversion. pp. 84–96, *in:* R. Haggar and S. Padel (eds). *Conversion to organic milk production.* Institute of Grassland and Environmental Research Technical Review, No. 4.

Whitehead, D.C. 1995. *Grassland Nitrogen.* Wallingford, UK: CAB International. 397 pp.

Wilman, D. & Derrick, R.W. 1994. Concentration and availability to sheep of N, P, K, Ca, Mg and Na in chickweed, dandelion, dock, ribwort and spurrey, compared with perennial ryegrass. *Journal of Agricultural Science* (Cambridge), **122:** 217–223.

Wood, M. 1996. Nitrogen fixation: How much and at what cost? pp. 26–35, *in:* D. Younie (ed). *Legumes in Sustainable Farming Systems. British Grassland Society Occasional Symposium,* No. 30.

Worner, M. & Taube, F. 1995. Artenzusammensetzung des Dauergrunlandes im okologischen Landbau – eine Erhebung auf norddeutschen Praxisflachen. [Species composition of permanent grassland in ecological agriculture – a survey on North German farms]. pp. 623–626, *in:* Kongressband 1992 Gottingen. Vortrage zum Generalthema des 104. VDLUFA-Kongresses vom 14–19.9.92 in Gottingen: *Okologische Aspekte extensiver Landbewirtschaftung.* 1992.

Younie, D. 1992. Potential output from forage legumes in organic systems. pp. 116–129, *in:* A. Peeters and V. van Bol (eds). *Potential and limits of organic farming.* Proceedings of an EC Workshop, Louvain-la-Neuve, Belgium, September 1992.

Younie, D. 1996. A comparison of crop rotations in organic farming: agronomic performance. pp. 379–382, *in:* J.H. Clarke, D.H.K. Davies, P.M.R. Dampney, R.J. Froud-Williams, P.J. Griffith, A. Lane, L. Sim and D.B. Stevens (eds). *Rotations and Cropping Systems.* Published as Aspects of Applied Biology, 47.

Younie, D. 1997. Organic Beef and Sheep Systems. Paper presented at *Grass is Greener.* British Grassland Society Winter Meeting 1997. Great Malvern, November 1997.

Younie, D. & Armstrong, G. 1996. Botanical and invertebrate diversity in organic and intensively fertilised grassland. pp. 35–44, *in:* J. Isart and J. Llerena (eds). *Biodiversity and Land Use: the Role of Organic Farming.* Proceedings of first Workshop of the European Network for Scientific Coordination in Organic Farming (ENOF). Bonn, December 1995.

Younie, D. & Baars, T. 1997. Resource use in organic grassland: the central bank and the art gallery of organic farming. pp. 43–60, *in:* J. Isart and J. Llerena (eds). *Resource Use in Organic Farming.* Proceedings of Third Workshop of European Network for Scientific Coordination in Organic Farming (ENOF). University of Ancona, Italy, June 1997.

Younie, D. & Hermansen, J. 2000. The role of grassland in organic farming. pp. 493–510, *in:* K. Soegaard, C. Ohlsson, J. Sehested, N.J. Hutchings and T. Kristensen (eds). *Grassland Farming – Balancing environmental and economic demands.* 18th General Meeting of the European Grassland Federation. Aalborg, Denmark. May, 2000. Published as Grassland Science in Europe, Volume 5.

Younie, D. & Mackie, C.K. 1996. Factors affecting profitability of organic, low-input and high-input beef systems. pp. 879–882, *in:* G. Parente, J. Frame and S. Orsi (eds). *Grassland and Land Use Systems.* 16th meeting of European Grassland Federation. Grado, Italy, September 1996.

Younie, D. & Watson, C.A. 1995. Crop growth and nitrogen recovery following ploughing of grass/clover swards. pp. 206–208, *in:* G. Pollott (ed.). *Grassland into the 21st Century: Challenges and Opportunities. British Grassland Society Occasional Symposium,* No. 30.

Younie, D. & Wightman, P.S. 1992. Herbage production over eight years from clover-based and intensively fertilised swards under grazing and silage management. pp. 99-100. Proc. Third Research Meeting, British Grassland Society. September 1992.

Younie, D., Thamsborg, S.M., Ambrosini, F. & Roderick, S. 2003. Grassland management and parasite control. pp. 309-328 *In*: M. Vaarst, S. Roderick, V. Lund and W. Lockeretz (eds). *Animal Health and Welfare in Organic Agriculture.* Wallingford, UK: CABI.

Younie, D., Umrani, A.P., Gray, D. & Coutts, M. 1997. Influence of chicory or ryegrass diets on trace element status of lambs. Paper presented at Royal Agricultural Society of England Conference: *Organic Farming – Science into Practice.* Royal Agricultural Society of England, Stoneleigh, November 1997.

Younie, D., Umrani, A.P., Gray, D. & Coutts, M. 2001. Effect of chicory or perennial ryegrass diets on mineral status of lambs. pp. 278–280, *in:* J. Isselstein, G. Spatz and M. Hoffman (eds). *Organic Grassland Farming,* Proceedings Occasional Symposium of European Grassland Federation, July 2001, Witzenhausen, Germany. Published as Grassland Science in Europe, Vol. 6.

Younie, D., Watson, C.A., Ramsay, D.A. & Mackie, C.K. 1995. Systems interactions and the role of grassland in sustainability. pp. 314–316, *in:* G.E. Pollott (ed). *Grassland into the 21st Century. British Grassland Society Occasional Symposium,* No. 29.

Younie, D., Wilson, J.F., Reid, C. & Wightman, P.S. 1998. Maintenance of soil potassium in organic farming silage systems. pp. 289–292, *in:* G. Nagy and K. Peto (eds). *Ecology and Grassland.* 17th General Meeting of the European Grassland Federation, Debrecen, Hungary, May, 1998. Published as Grassland Science in Europe, Vol. 3.

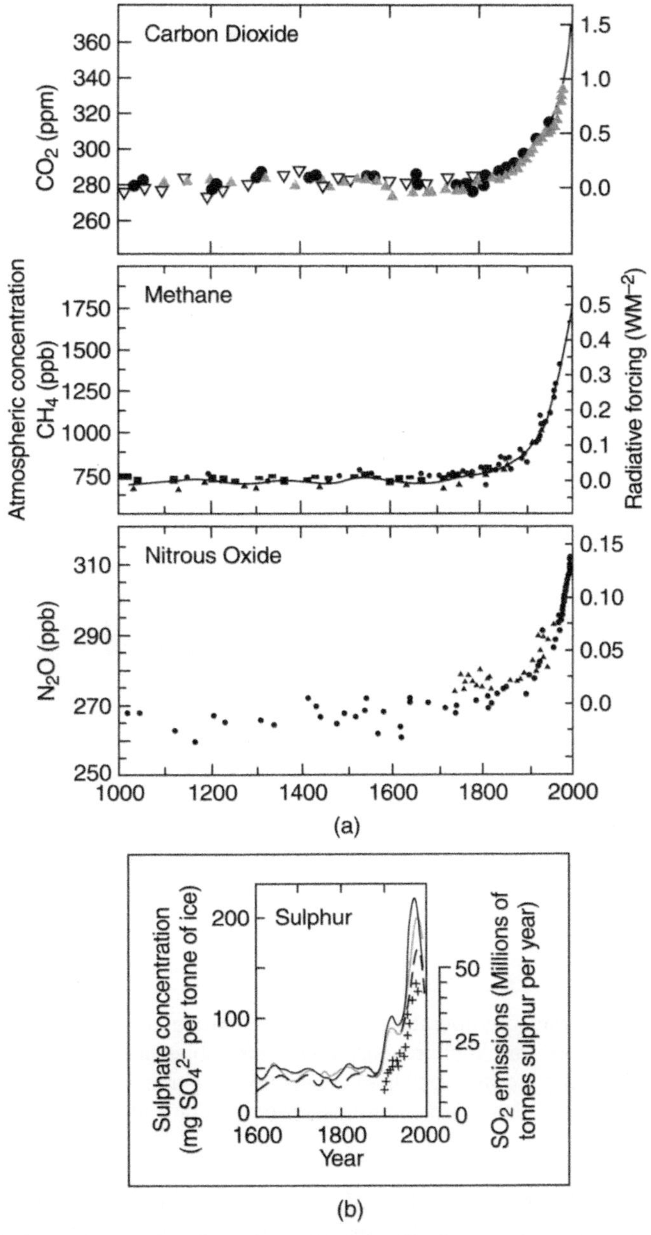

Figure 10.1 Records of changes in atmospheric gases. (a) Atmospheric concentrations of CO_2, CH_4 and N_2O over the past 1 000 years. Ice core data for several sites in Antarctica and Greenland (shown by different symbols) are supplemented with data from atmospheric samples taken over the past few decades (shown by line for CO_2 or incorporated into the curve for CH_4). The estimated radiative forcing is indicated on right-hand scale. (b) Sulphate concentration in several Greenland ice cores and total SO_2 emissions from sources in the USA and Europe (crosses).

Source: Figure reprinted with permission from the Intergovernmental Panel on Climate Change (Albritton *et al.*, 2001; Figure 8).

effects on radiative transfer. Despite the cooling due to aerosols, the net effect of gas emissions is one of warming.

Greenhouse gas emissions resulting from human activities occur against a background of variable solar radiation cycles, natural emissions of greenhouse gases (volcanic activity), and other natural changes in climate forcing unrelated to greenhouse gases, like the El Niño-Southern Oscillation (ENSO) phenomenon (ocean-atmospheric interactions in the Pacific tropics, which affect world climates). Evaluating the influence of anthropogenic gas emissions on past and projected future climates must account for these other natural anomalies. A modelling exercise by Mitchell *et al.* (2001) included both anthropogenic and natural forcings to evaluate the performance of a coupled ocean-atmosphere climate model; the model performed well when compared with measured temperature anomalies from the late 1800s until 2000 (Figure 10.2). The results suggest we may have a good degree of confidence in our ability to predict future scenarios climate change induced by greenhouse gases.

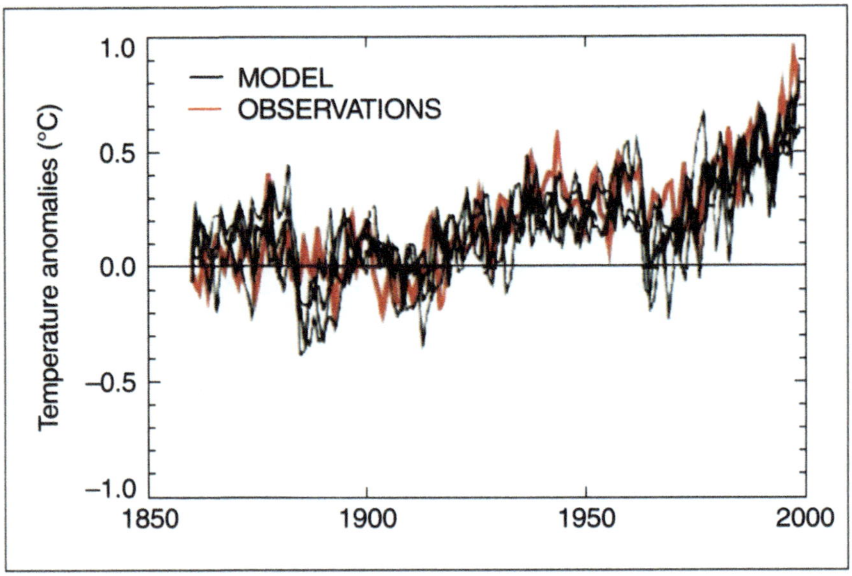

Figure 10.2 Global mean surface temperature anomalies relative to the 1880 and 1920 mean from the instrumental record (red line) compared with ensembles of four simulations with a coupled ocean-atmosphere climate model (black line). Simulations include both anthropogenic and natural forcings. Source: Figure reprinted with permission from the Intergovernmental Panel on Climate Change (Albritton *et al.*, 2001; Figure 15c).

The construction of future scenarios of climate change relies on projected patterns in demographic, economic and technical forces that affect greenhouse gas and sulphur emissions. Since it is impossible to develop one or even several scenarios that capture the many possibilities for Earth's continued human development, the Intergovernmental Panel on Climate Change utilizes up to 40 different scenarios as evaluation tools to predict a range of possible future climates. In their latest report, Albritton *et al.* (2001) present results from numerous models, which were run for 35 different scenarios, and concluded that globally averaged surface temperatures would increase by between 1.4° and 5.8°C from 1990 to 2100, with a wider range and warmer temperatures for terrestrial surfaces, especially in winter and in the Northern Hemisphere, where warming is predicted to be maximal. A similar exercise was done for precipitation patterns, and the results suggest increased summer and winter precipitation at higher latitudes. Increased winter precipitation is also predicted over northern mid-latitudes, tropical Africa and Antarctica, and in the summer for southern and eastern Asia. Decreases in winter precipitation are predicted for Australia, Central America and southern Africa.

A summary of some of the major predictions and confidences in them are given in Table 10.1. The models predict warmer global temperatures, more hot days over most land areas, increased heat index, and more intense precipitation events. Since more intense storms will lead to increased surface runoff and erosion (Campbell, Stafford Smith and McKeon, 1997), increased rainfall in some areas may not translate into more mesic conditions. In fact, with warming added to the picture, there will probably be increased continental drying and associated risk of drought (Table 10.1).

Table 10.1 Estimates of confidence in observed and projected changes in extreme weather and climate events. The table depicts an assessment of confidence in observed changes in extremes of weather and climate during the latter half of the twentieth century (left column) and in projected changes during the twenty-first century (right column). This assessment relies on observational and modelling studies, as well as physical plausibility of future projections across all commonly used scenarios and is based on expert judgment. Table reprinted with permission from the Intergovernmental Panel on Climate Change.

Confidence in observed changes (latter half of the 20th century)	Changes in phenomenon	Confidence in projected changes (during the 21st century)
Likely	Higher maximum temperatures and more hot days over nearly all land areas	Very likely
Very likely	Higher minimum temperatures, fewer cold days and frost days over nearly all land areas	Very likely
Very likely	Reduced diurnal temperature range over most land areas	
Likely, over many areas	Increase of heat index over land areas(1)	Very likely, over most areas
Likely, over many Northern Hemisphere mid- to high-latitude land areas	More intense precipitation events(2)	Very likely, over many areas
Likely, in a few areas	Increased summer continental drying and associated risk of drought	Likely, over most mid-latitude continental interiors (lack of consistent projections in other areas)
Not observed in the few analyses available	Increase in tropical cyclone peak wind intensities(3)	Likely, over some areas
Insufficient data for assessment	Increase in tropical cyclone mean and peak precipitation intensities(3)	Likely, over some areas

Notes: (1) Heat index: A combination of temperature and humidity that measures effects on human comfort.
(2) For other areas there are either insufficient data or conflicting analyses.
(3) Past and future changes in tropical cyclone location and frequency are uncertain.
Source: Albritton *et al.*, 2001; Table 4.

PRIMARY PRODUCTION RESPONSES

Carbon dioxide

Carbon dioxide (CO_2) is a substrate for photosynthesis. Increases in its concentration above present atmospheric levels result in higher photosynthesis rates in plants and plant communities, and to a lesser extent, lead to greater aboveground primary production (Drake, Gonzàlez-Meler and Long, 1997; Jackson *et al.*, 1994; Lawlor and Mitchell, 1991; Mielnick *et al.*, 2001; Morgan *et al.*, 2001, 2004; Owensby *et al.*, 1996). However, the degree of plant response to CO_2 can vary between species or environmental conditions (Porter, 1993; Wand *et al.*, 1999), so information on species or functional group responses is important in predicting responses of plant communities and ecosystems to rising atmospheric CO_2 concentrations.

Differences in photosynthetic pathway may be a useful indicator of a plant's potential to respond to CO_2. Plants with the C_3 photosynthetic pathway are the most abundant and widely distributed class of plants, and have a photosynthetic metabolism that is unsaturated at present atmospheric CO_2 concentrations. In contrast, photosynthesis of the second most abundant class of plants, those with the C_4 pathway, is saturated near present atmospheric CO_2 concentrations of near 370 µlitre litre^{-1} air. Because of these differing photosynthetic sensitivities to CO_2, early predictions were that increasing CO_2 would tend to give C_3 plants a photosynthetic advantage over C_4 plants in future CO_2-enriched atmospheres (Bazzaz, 1990; Pearcy and Ehleringer, 1984). However, subsequent experimentation revealed that, while the relative photosynthetic and growth responses are on average less in C_4 than C_3 species, the differences are often not as great as expected based on their underlying biochemistry (Bowes, 1993; Porter, 1993; Wand *et al.*, 1999).

There are a couple of explanations of why elevated CO_2 does not induce large differences in photosynthetic and growth responses between C_3 and C_4 plants. First and foremost, raising CO_2 induces stomatal closure in most herbaceous species, regardless of photosynthetic class (Field, Jackson and Mooney, 1995; Kimball and Idso, 1983; Wand *et al.*, 1999). This reduces transpiration, resulting in higher plant and soil water potentials, and results in enhanced plant biomass yields through improved water use efficiency (Drake, Gonzàlez-Meler and Long, 1997; Knapp, Hamerlynck and Owensby, 1993; Morgan *et al.*, 2001; Niklaus, Spinnler and Körner, 1998; Owensby *et al.*, 1997; Sindhøj *et al.*, 2000; Volk, Niklaus and Körner, 2000). Such water relations responses may be most important in native grassland and rangelands, where water scarcity is an important limiting factor in production. Second, photosynthetic acclimation, a downward adjustment of photosynthetic capacity in C_3 plants exposed long-term to elevated CO_2, can reduce or even eliminate the photosynthetic enhancement achieved under elevated CO_2 (Lee *et al.*, 2001; Sage, Sharkey and Seeman, 1989; Sage, 1994). Acclimation lessens CO_2 response differences between C_3 and C_4 species.

Some plants may respond more to CO_2 due to different strategies in balancing C assimilation and its utilization for growth and maintenance. Porter (1993) noted that herbaceous C_3 crop plants were more responsive than wild C_3 species to CO_2 enrichment, and also that CO_2-induced growth responses were greater in fast-growing compared with slow-growing wild species. He suggested that plants with stronger C sinks may transport reduced C compounds more effectively away from leaves and reduce metabolic feedbacks that lower photosynthetic activity, and, ultimately, reduce growth response to CO_2.

Legumes are one of the most responsive functional groups to CO_2, in part because of their ability to fix atmospheric N (Hebeisen *et al.*, 1997; Jablonski, Wang and Curtis, 2002; Lüscher *et al.*,

2000; Porter, 1993; Reich *et al.*, 2001; Wand *et al.*, 1999). Insufficient N appears to be one of the major causes behind photosynthetic acclimation and reduced growth responses of plants to elevated CO_2 (Lee *et al.*, 2001; Morgan *et al.*, 2001; Sage, Sharkey and Seeman, 1989), so legumes may have an advantage over non-legumes by virtue of their ability to fix atmospheric N into organic compounds. Not only does the N fixation benefit the plant by providing a strong sink for carbohydrates, but it also provides the N needed for biosynthesis of proteins, nucleic acids and other vital plant compounds. However, legumes may not be as responsive to CO_2 enrichment in soils with low available phosphorus (Stöcklin, Schweizer and Körner, 1998; Warwick, Taylor and Blum, 1998), or when soil N is in abundant supply (Geeske *et al.*, 2001).

As a group, forbs tend to be more responsive than grasses (Leadley *et al.*, 1999; Morgan *et al.*, 2001; Owensby *et al.*, 1999; Porter, 1993; Potvin and Vasseur, 1997; Reich *et al.*, 2001; Teyssonneyre *et al.*, 2002). However, no satisfactory explanation has yet been proposed to explain apparent high CO_2 sensitivity in this diverse group of plants.

Growth of plants at elevated CO_2 may also affect their respiration. This has important consequences for the atmosphere as well as for plants, since as much as 50 percent of C fixed in C_3 photosynthesis may be respired back into the atmosphere (Farrar, 1985). There is some evidence that plant respiration may be reduced in CO_2-enriched environments (Bunce, 1990; Drake, Gonzàlez-Meler and Long, 1997; Rogers *et al.*, 1997). However, since CO_2 often increases the root : shoot ratio in plants, and root systems have higher specific respiration rates than shoots, whole plant respiration may tend to increase as CO_2 rises (Rogers *et al.*, 1997). Unfortunately, much plant respiration research has not been conducted in a manner to evaluate the particular mechanism behind the plant responses (Rogers *et al.*, 1997), or to evaluate yet whether interspecific differences may be an important aspect of variable species responses to CO_2.

Temperature and CO_2

Temperature is a significant factor in setting the boundaries on growing seasons and plant communities, as well as determining growth rates, plant developmental responses and the availability and utilization of resources like water, radiation and nutrients. Warmer climates will lengthen growing seasons and will accelerate plant development; growth rate will increase up to a point, and then decline as super-optimal temperatures are reached. A meta-analysis by Rustad *et al.* (2001) suggests that global warming may enhance primary production in this century by an average of 19 percent in tundra, grassland and forested biomes, and the response is likely to be greater in the colder ecosystems, particularly in the tundra.

There is considerable evidence that CO_2 and temperature responses are not additive, so it is useful to understand warming in the context of rising CO_2. For instance, the stimulatory effect of CO_2 enrichment on plant growth is often enhanced at warmer temperatures (Cure, 1985; Greer *et al.*, 2000). For C_3 plants this stimulation may be related to CO_2-induced changes in photorespiration (Long, 1991) and activity of RuBisco, the enzyme responsible for fixation of CO_2 into organic compounds (Bowes, 1993); changes in both tend to increase photosynthesis more at high than low temperatures (Greer *et al.*, 2000; Sage, Sharkey and Seeman, 1989). For many plants, warmer temperatures enhance carbohydrate sink demand (Arp, 1991; Farrar and Williams, 1991), which may stimulate the CO_2 growth response (Greer *et al.*, 2000; Newton *et al.*, 1994). For these reasons, the effect of elevated CO_2 on primary production in temperate pastures is likely to be highly seasonal (Newton *et al.*, 1994; Nösberger, Blum and Fuhrer, 2000).

Many of the present predictions for combined CO_2 and temperature responses of rangelands, mostly based on simulation models, suggest increasing CO_2 and warming will lead to greater primary production of most grasslands (Baker *et al*., 1993; Coughenour and Chen, 1997; Neilson *et al*., 1998; Parton *et al*., 1995; Riedo, Gyalistras and Fuhrer, 2000, 2001), although the degree of response varies considerably depending on the particular ecosystem and other environment conditions. Warmer temperatures may be especially important at enhancing relative CO_2 production responses at high altitudes and at high- and mid-latitude regions where low temperatures currently limit CO_2 production responses (Körner, 1996; Riedo, Gyalistras and Fuhrer, 2001; Rounsevell, Brignall and Siddons, 1996).

Precipitation, temperature and CO_2

Ninety percent of the variability in rangeland primary production is accounted for by variation in annual precipitation (Campbell, Stafford Smith and McKeon, 1997). Production is therefore likely to follow any changes in precipitation patterns to the extent they impact upon seasonal soil water dynamics. In more mesic and productive grasslands, the consequences of altered rainfall will probably be less, but will still be important since seasonal productivity is commonly limited by distribution and amounts of precipitation (Knapp, Briggs and Koelliker, 2001).

While increased precipitation in many parts of the world may lead to more production, these changes must be considered in the context of global warming, which will tend to offset the precipitation responses by desiccation (Polley *et al*., 2000; Rounsevell, Brignall and Siddons, 1996). Changes in the seasonal distribution and intensity of rainfall will affect seasonal soil water dynamics and the efficiency of plant use, and may have a larger impact on rangelands than changes in precipitation amounts (Giorgi *et al*., 1998). More intense storms are likely to lead to increased surface runoff and erosion and concentration of water in smaller portions of the landscape (Campbell, Stafford Smith and McKeon, 1997). The negative impact of increased global temperatures through desiccation may be especially important in present arid and semi-arid regions of the world, especially in Central America, Australia, the Near East, Southeast Asia and southern Africa, where precipitation may be little affected or may decline (Albritton *et al*., 2001). However, as mentioned previously, rising atmospheric CO_2 will lead to increased water use efficiency and improved water relations. Thus, the overall hydraulic and plant productivity response to changes in precipitation will depend on complicated interactions with temperature and CO_2.

PLANT REPRODUCTIVE AND RECRUITMENT RESPONSES

In addition to growth responses, reproductive and plant recruitment response to CO_2 enrichment and associated climate change can have important impacts on productivity, species composition and, ultimately, biodiversity (Edwards, Clark and Newton, 2001; Grünzweig and Körner, 2001).

There are a number of ways in which CO_2 may effect recruitment of new individuals. In terms of direct plant responses, elevated CO_2 is often, but not always (Farnsworth and Bazzaz, 1995; Smith *et al*., 2000), associated with increases in seed number (Edwards, Clark and Newton, 2001; Grünzweig and Körner 2001; Jablonski, Wang and Curtis, 2002; Lawlor and Mitchell, 1991; Smith *et al*., 2000), size or weight (Edwards, Clark and Newton, 2001; Jablonski, Wang and Curtis, 2002; Lawlor and Mitchell, 1991; Newton, 1991; Steinger, Gall and Schmid, 2000). These responses, however, vary with both species and environmental conditions (Jablonski, Wang and Curtis, 2002; Grünzweig and Körner, 2001); often responses in one reproductive trait (seed number or size) offset or cancel out another. Further, CO_2-induced changes in seed number, size and dispersal become important only to the extent that seedling recruitment is limited by seed number, which may often be the case for arid and semi-arid grasslands, but less so for mesic grasslands (Turnbull, Crawley and Rees, 2000). Seed size may not necessarily be a good indicator of reproductive success, especially in CO_2-enriched atmospheres where larger seeds produced may sometimes be depleted in nutrients (Jablonski, Wang and Curtis, 2002; Steinger, Gall and Schmid, 2000). Other factors, such as competitive plant interactions (Bazzaz and McConnaughay, 1992) and seed predation (Arnone *et al.*, 1995), can prevent the establishment of simple relationships between seed number, seed size and recruitment. Nevertheless, CO_2-induced changes and species differences in seed dispersal and seed size can be important in affecting recruitment of new individuals (Edwards, Clark and Newton, 2001). Ultimately, the outcome will depend on interactions between reproduction and the microsite conditions that determine germination and establishment of the seedling.

Proper soil temperature and moisture for germination and establishment of seedlings is an important factor in the establishment of grassland species (Defosse, Bertiller and Robberecht, 1997; Minnick and Coffin, 1999; Wester, 1995). Thus, any changes in temperature or precipitation patterns that affect soil water dynamics have the potential to affect germination and recruitment processes. Higher CO_2 also has the potential to enhance recruitment indirectly through its improvement of soil water content resulting from reduced plant transpiration.

BELOW-GROUND RESPONSES

In native grasslands, the majority of plant production occurs below ground, and root biomass pools often exceed above-ground pools by a factor of two to five (Arnone *et al.*, 2000; Stanton, 1988). In more productive grassland systems, growth responses to CO_2 are often greater and more consistent in below-ground than above-ground plant organs (Newton, 1991; Rogers, Runion and Krupa, 1994; Rogers *et al.*, 1997). Further, the long-term cycling of nutrients, their availability for plant growth, and the dependence of soil processes on dynamic root responses may control, in large part, the long-term response of ecosystems to CO_2 enrichment and climate change (Arnone *et al.*, 2000; Curtis *et al.*, 1994; Norby and Jackson, 2000; Thornley and Cannell, 2000). An understanding of these complex but important interactions is required in considering ecosystem responses and possible management options for grazing lands.

Root responses

Increasing global air temperatures will lead to higher soil temperatures (Schlesinger and Andrews, 2000). As a result, root production and mortality should increase, assuming soil water and nutrients remain at sufficient levels (Norby and Jackson, 2000). In natural systems, conditions of sufficient water and nutrients are often not met, making the prediction of root dynamic responses to increased temperature difficult. The seasonality of root activity will also be affected by climate change (Reeder, Franks and Milchunas, 2001), with root activity commencing earlier in the growing season, and continuing later as growing seasons extend.

In rangelands where plant production is limited mostly by water, any change in soil water dynamics, whether resulting directly from altered precipitation patterns or indirectly from warmer temperatures or increased CO_2, will alter root activity. In arid and semi-arid rangelands, increased water will enhance root activity. In more mesic grasslands, increased precipitation may have both positive and negative consequences for root activities. Under dry conditions, increased nutrient mineralization and root activity will be important features of belowground system responses to increased water (Mosier, 1998; Rogers *et al.*, 1999) leading to enhanced plant growth. Under wetter conditions, negative consequences for roots may occur with water additions, such as leaching of mobile nutrients, loss of N through de-nitrification, or soil erosion (Rounsevell, Evans and Bullock, 1999).

Although elevated CO_2 can significantly enhance production of belowground organs in productive grasslands (Newton, 1991; Nösberger, Blum and Fuhrer, 2000; Rogers, Runion and Krupa, 1994), the response of less productive native grasslands is less certain. About half of native grassland studies exhibit little or no response in root biomass to growth at elevated CO_2, while the other half exhibit substantial increases in root biomass under elevated CO_2 (Arnone *et al.*, 2000). These variable responses will probably result from the differing soil and environmental conditions of the experiments, and the consequences of those conditions on plant growth and on root-shoot partitioning (Larigauderie, Hilbert and Oechel, 1988).

Nutrient cycling and soil feedbacks

Soil feedbacks may ultimately determine ecosystem responses to CO_2 and climate change. Perhaps the strongest evidence for this has been the almost universal decline in shoot N concentration that occurs in CO_2-enrichment studies, especially those in which CO_2-induced production changes are evident (Drake, Gonzàlez-Meler and Long, 1997). While part of this response may be due to the enhanced resource-use efficiency that plants experience under elevated CO_2, there is good evidence that soils cannot release N quickly enough to keep pace with CO_2-induced production increases (Zak *et al.*, 2000). Most research has focused on the N cycle, but nutrient immobilization and release processes also involve other essential nutrients.

Briefly, CO_2-enhanced production will lead to increased carbon substrates entering the soil, which will fuel microbial growth and initially immobilize soil nutrients (Figure 10.3). However, the additional microbial biomass will eventually cycle and release more nutrients back into the soil through mineralization (Morgan, 2002). Thus, the balance between these and other competing processes can result in increased, decreased or no change in plant available nutrients (Zak *et al.*, 2000). In the long term, soil nutrient cycles may adjust to changes in organic inputs, and nutrient availability will increase to meet new demands in CO_2-enriched atmospheres (Thornley and Cannell, 2000); the time this takes to occur will differ among ecosystems due to differences in species composition and soil type (Reeder, Franks and Milchunas, 2001), as well as to differences in climate.

The degree to which the N, or any other nutrient, cycle is altered due to CO_2-enhanced plant production changes may account for much of the variability among studies and ecosystems in their responses to CO_2. For instance, the increase in root-shoot ratios in many CO_2-enrichment studies has been interpreted as an indirect response to CO_2-induced plant N deficits, which causes plants to shift more substrates below ground to acquire more N (Larigauderie, Hilbert and Oechel, 1988; Newton, 1991; Rogers, Runion and Krupa, 1994; Rogers *et al.*, 1997, 1999; Stulen and den Hertog, 1993). The acclimation of plants that often occurs under long-term CO_2 enrichment, whereby photosynthetic and growth responses become less responsive over time, appears to be

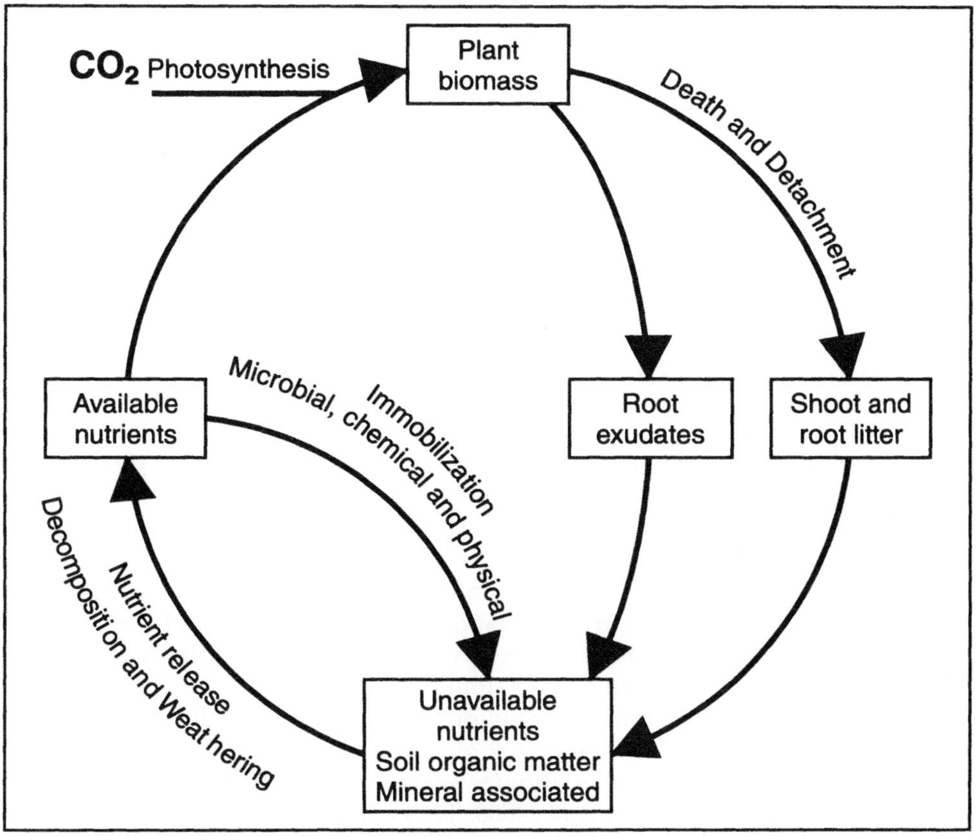

Figure 10.3 Nutrient Cycling Feedbacks. While CO_2 enrichment may lead to increased photosynthesis and enhanced plant growth, the long-term response will depend on nutrient cycling feedbacks. Litter from decaying plants and root exudates enters a large soil nutrient pool that is unavailable to plants until they are broken down and released by microbial activity. Soil microbes may also fix available nutrients into new microbial biomass, thereby temporarily immobilizing them. The balance between these and other nutrient release and immobilization processes determines available nutrients and ultimate plant response.

Source: Figure reprinted with permission from *Science* (Morgan, 2002).

related to the build up of carbohydrates that occurs more readily under limited soil N (Arp, 1991; Sage, 1994).

Warmer temperatures may lead to increased organic matter decomposition rates in mesic grasslands, but the same may not hold for dry rangelands. Globally, warmer regions are known to exhibit higher rates of soil respiration and organic matter decomposition compared with colder regions (Epstein, Burke and Lauenroth, 2002; Rustad *et al.*, 2001). However, decomposition is typically modelled by an interaction of temperature and soil water (Esser, 1992; Parton *et al.*, 1987, 1995; Rastetter *et al.*, 1991). As water becomes limiting, decomposition and respiration are predicated less by temperature and more by soil water content (Epstein, Burke and Lauenroth, 2002), with lower soil water resulting in reduced decomposition rates. To the extent it affects soil water dynamics, rising atmospheric CO_2 will also affect decomposition processes (Mosier, 1998; Rogers *et al.*, 1999).

PLANT COMMUNITY CHANGES

While responses of plant community productivity to CO_2 and climate change are predicted to be important in many ecosystems, long-term changes in plant community composition may be even more important with regard to ecological impacts and effects on ecosystem goods and services (Körner, Bazzaz and Field, 1996; Leadley and Körner, 1996; Polley *et al.*, 2000; VEMAP Members, 1995).

Climatic changes

The distribution of plant types is affected strongly by spatial and temporal patterns of precipitation and temperature (Epstein *et al.*, 1997; Paruelo and Lauenroth, 1996). Water is the most important determinant of geographical plant distribution (Stephenson, 1990) and productivity (Sala *et al.*, 1988; Webb *et al.*, 1978). Any changes in climate that affect soil water dynamics will have an equally important effect on plant communities. Although precipitation is predicted to increase over much of the globe, the concentration of more rainfall into fewer rainfall events will lead to more uneven spatial distribution of water across the landscape, and more drought in some regions (Campbell, Stafford Smith and McKeon, 1997). In rangelands, where plant community responses are commonly event-driven, changes in timing and amount of precipitation will have especially important consequences for plant community responses (Polley *et al.*, 2000). Patterns of precipitation that significantly alter the vertical distribution of water will correspondingly alter the proportion of shallow-rooted versus deep-rooted species (Ehleringer *et al.*, 1991).

Temperature is also important in the distribution and functioning of plant communities (Epstein *et al.*, 1997; Field and Forbe, 1990; Pyankov *et al.*, 2000; Terri and Stowe, 1976). The increased presence of C_4 grasses toward tropical latitudes is one of the simplest and yet most successful models of plant distribution (Pyankov *et al.*, 2000: Terri and Stowe, 1976). Modelling exercises that consider both temperature and precipitation aspects of climate change predict an increase in relative abundance of C_4 grasses and a decrease in C_3 grasses in North and South American grasslands and shrublands (Coffin and Lauenroth, 1996; Epstein *et al.*, 1997; Epstein, Burke and Lauenroth, 2002; VEMAP Members, 1995). In Europe, latitudinal vegetation gradients in calcareous grasslands exhibit good correlations between species composition and summer temperature (Duckworth, Bunce and Malloch, 2000), suggesting that present-day transitions in grassland communities from north to south may be good indicators of future plant community changes to warming.

Predictions of plant community changes for shrublands are less certain than for grasslands. There is some evidence that warming or precipitation increases, or a combination, may lead to an increase in subtropical shrublands, but a decrease in temperate shrublands (Epstein, Burke and Lauenroth, 2002; VEMAP, 1995). However, these predictions are uncertain given our incomplete understanding of shrubland ecology. The encroachment of shrubs into former semi-arid and arid grasslands has been attributed to livestock grazing, fire suppression, increased atmospheric CO_2, and climate change (Archer, Schimel and Holland, 1995; Bond and Midgley, 2000; Moore *et al.*, 2001; Polley *et al.*, 1997; Roques, O'Connor and Watkinson, 2001). Unfortunately, sorting out the relative importance of these competing forces is difficult at best, and predictions based on them may be unreliable.

The long-term responses of grazing land plant communities to warming and precipitation changes will be determined to a large extent by soil nutrient cycling and water dynamic feedbacks,

interseeding legumes into rangelands as a means to facilitate this process, as well as a means to sequester C in the soil, thereby abating greenhouse gas emissions (Mortenson, Schuman and Ingram, 2003). However, research needs to determine whether such increases in legumes will have effects on the exchange of other trace gases like N_2O or methane, which contribute to greenhouse gas warming. Selection of appropriate legume species and the development of improved germplasm and management practices may be required to maximize the benefits of such legume usage for rangelands.

Climate change will no doubt lead to altered animal usage. Camels, sheep and goats are better adapted to dry regions, and will probably become more important in regions that experience increased drought (Squires and Sidahmed, 1997). Goats may become economically more practical in subtropical rangelands experiencing more climatic variability (Souza Neto *et al.*, 1998). A high degree of genetic diversity already exists within livestock species, so warming in temperate regions may be accommodated by selecting livestock from more tropical climates (Squires and Sidahmed, 1997). Attention has been given in the past to adapting the genetics of livestock to targeted environments, like the tropics (Bonsma, 1949; Rötter and Van De Geijn, 1999). However, there is little information on how these adapted animals would perform with the further warming predicted for this century, and virtually no information on how such selected or altered livestock will perform in future rangelands that will undoubtedly undergo significant plant compositional changes (Squires and Sidahmed, 1997).

Implications for rangeland livestock management systems

Two types of grazing operations prevail on rangelands: pastoralism and ranching (Squires and Sidahmed, 1997). Pastoralism is a practice dating to ancient times, and traditionally centred on herding domestic livestock, but may include some cropping where feasible. Pastoralists typically maintain herds as insurance against extreme drought or hardship. It is a system that in modern times has tended to exist on marginal lands, and is currently in crisis due to a complex interaction of ecological, socio-economic and population pressure problems (Squires and Sidahmed, 1997). Commercial ranching, a more recently developed enterprise, tends to be more economically motivated (Squires and Sidahmed, 1997). Private ownership of land is common, operational scale is large, and improvements in the way of fencing, improved pastures and access to water and other resources are common.

The impacts of climate change are likely to differ between pastoralists and commercial ranchers. For instance, Smith and Lazo (2001) concluded that climate change will have less impact on livestock yields than on forage yields, in part because there currently exists excess capacity for livestock, or that some expansion of rangelands is possible. However, this may not apply to some pastoral groups that are already constrained by human population growth and decreasing grazing lands (Squires and Sidahmed, 1997). Further, pastoral grazing often involves complex social arrangements in the seasonal movement of animals across the landscape (Parish and Funnell, 1999). Although pastoralists are characterized by an ability to adapt to extreme climatic events (Mortimore and Adams, 2001), their complex socio-economic structures and dwindling land resources in some regions may make them more susceptible than ranchers to changes brought about by climate change (Squires and Sidahmed, 1997; Turner, 1993). Climate changes that affect production and plant community composition may challenge social arrangements that have developed over hundreds of years, and further complicate modern-day attempts to protect the landscape while preserving important cultural and societal arrangements as human populations expand.

Other uses for rangelands

In addition to livestock grazing, rangelands provide many other goods and services, including biodiversity, tourism, fuelwood (Gitay *et al.*, 2001; Watkinson and Ormerod, 2001), and more recently, C stores (Meeting *et al.*, 2001; Moore *et al.*, 2001; Schuman, Herrick and Janzen, 2001). Undoubtedly, climate change, government policy and economics will alter the balance among these uses, and may shift management from one enterprise to another. Commercial ranching is already moving away from intensive livestock grazing and towards urbanization, game ranching and transfer of homelands back to indigenous peoples (Polley *et al.*, 2000). Some rangelands will be taken out of production and converted to cropping, as human populations increase and crop production on marginal lands becomes economically more feasible. The contribution livestock make to climate change through methane production is an important problem that will no doubt be considered in future research and management options (Gitay *et al.*, 2001; McCrabb and Hunter, 1999). However, less intensive livestock practices may be encouraged due to the lower net contribution of rangeland to methane production, and thus to greenhouse gas warming compared with more intensive livestock operations (Subak, 1999).

Productive grasslands

Production in many temperate humid grasslands is likely to remain unchanged (Rounsevell, Brignall and Siddons, 1996), or may increase due to increased CO_2 and warmer temperatures, which will extend growing seasons, and in some cases increase primary production. Because water limitations are generally less severe in productive grasslands, the negative impact of warmer temperatures on desiccation will be less of a problem than in semi-arid and arid rangelands. However, temperature increases in excess of 4°C are likely to have negative consequences because of increased drought stress and reduced growing season lengths (Baethgen, 1997; Rounsevell, Brignall and Siddons, 1996). Subtropical and tropical grasslands may be more vulnerable to climate change than cooler, temperate grasslands. A major risk in subtropics and tropics would be a decline in forage quality (Polley *et al.*, 2000), increased heat stress in livestock, and – in dry zones or regions with periodic droughts–increased incidences of drought or more severe droughts (Baethgen, 1997). However, predicted increased precipitation in some regions may counter the negative response to temperature. As for rangelands, management options may have much to do with altered species composition, changed stocking rates and systems, and, in some situations, different grazing animals or breeds. The options in productive grasslands will tend to be greater since manipulations are economically more practicable than in rangelands.

In semi-natural grasslands with high species diversity, species composition is expected to gradually change due to rising CO_2 (Teyssonneyre *et al.*, 2002) and changes in climate (Duckworth, Bunce and Malloch, 2000). Defoliation can affect the balance of functional group and species composition, so management practices developed to deal with rising CO_2 and climate change will have to balance animal productivity responses with management that maintains or enhances biodiversity (Teyssonneyre *et al.*, 2002). As with rangelands, an understanding of rising CO_2 and climate change impacts on grassland invasions will be required to develop viable management practices to combat those invasions (Buckland *et al.*, 2001; Davis, Grime and Thompson, 2000; Watkinson and Ormerod, 2001).

CONTRASTING RANGELANDS AND PRODUCTIVE GRASSLANDS

While many of the same issues brought on by rising atmospheric CO_2 and climate change are likely to occur in rangeland and productive grasslands, differences between these two types of systems are likely to affect how they are manifested and addressed. For instance, livestock grazing practices for improved pastures rely on the introduction of adapted species, as well as on ecological considerations of maintaining the optimal balance of desired species that is important for both rangelands and productive grasslands. Sowing legumes into productive grasslands is already an accepted practice for improving N nutrition of pastures. Thus, the increased development and use of legumes to combat N limitations in a future CO_2-enriched world should be a relatively easy technology to adapt (Gitay *et al*., 2001; Schenk, Jäger and Weigel, 1997). The same may not be true for rangelands, where establishment and subsequent control of introduced species is more problematic. In general, much of the species changes in improved pastures will be due to conscious management decisions to sow or nurture particular adapted species. In rangelands, species issues will involve management of the ecological changes and states resulting from combined climate change and past management to ensure that future desired plant communities result.

CONCLUSIONS

Our present understanding of atmospheric CO_2 and climate change and their impact on grazing lands is modest. We know that atmospheric trace gas concentrations are rising and that significant changes in the Earth's climate are already under way, and we are beginning to understand how these global changes will manifest in different world regions. They will have important consequences for grazing lands, leading ultimately to management modifications, and in some instances, changes in enterprise. In addition to the ecological effects, significant socio-economic consequences will surely follow. But our ability to predict and prepare for possible future scenarios of ecological and management responses is poor.

Our challenge as we look to the future of world grazing lands is to acknowledge that these complex ecosystem responses to CO_2, climate change and other aspects of global change – like increased N deposition or higher levels of ozone – are under way, and to conduct research to understand what has already occurred, as well as develop appropriate technologies and decision-support systems to deal with future changes. The success of global change research programmes will depend in large part on our ability to involve ranchers, pastoralists and other land managers in the design of new research projects to develop proactive strategies and technologies to prepare for the future (Campbell *et al*., 2000). Where past research has focused on the responses of ecosystems to changes in one or two environmental factors, future research must consider the multi-factorial nature of global changes. More work is needed to understand the complex below-ground processes that will ultimately govern long-term responses. Animal responses and their effects on grazing systems need more attention if we are to realistically simulate combined global change and management responses. And we must advance our capability to respond by identifying low-risk, realistic strategies that allow adaptation, but make sense in the face of uncertainty concerning the degree and rate of global changes and ecosystem responses. Grazing land agriculture has always been characterized by change and adaptability – despite our increasing knowledge about the Earth's environment, there is little reason to indicate that it will be any different in the future.

ACKNOWLEDGEMENTS

The author would like to thank Dana Blumenthal, Dan LeCain, Jean Reeder, Robin Kelly and Justin Derner, who made many useful contributions that helped greatly in developing and clarifying this manuscript.

REFERENCES

Ackerly, D.D. & Bazzaz, F.A. 1995. Plant growth and reproduction along CO_2 gradients: non-linear responses and implications for community change. *Global Change Biology*, **1**: 199–207.

Adams, R.M., Hurd, B.H., Lenhart, S. & Leary, N. 1998. Effects of global climate change on agriculture: an interpretive review. *Climate Research*, **11**: 19–30.

Albritton, D.L., Meira Filho, L.G., Cubasch, U., Dai, X., Ding, Y., Griggs, D.J., Hewitson, B., Houghton, J.T., Isaksen, I., Karl, T., McFarland, M., Meleshko, V.P., Mitchell, J.F.B., Noguer, M., Nyenzi, B.S., Oppenheimer, M., Penner, J.E., Pollonais, S., Stocker T. & Trenbath. K.E. 2001. Technical Summary. *In:* J.T. Houghton, Y. Ding, D.J. Griggs, M. Noguer, P.J. van der Linden, X. Dai, K. Maskell and C.A. Johnson (eds). *Climate Change 2001: The Scientific Basis.* Contribution of Working Group I to the Third Assessment Report of the Intergovernmental Panel on Climate Change. Cambridge, UK, and New York, NY: Cambridge University Press. 881p.

Archer, S., Schimel, D.S. & Holland, E.A. 1995. Mechanisms of shrubland expansion: land use, climate or CO_2? *Climate Change*, **29**: 91–99.

Arnone III, J.A., Zaller, J.G., Spehn, E.M., Niklaus, P.A., Wells, C.E. & Körner, C. 2000. Dynamics of root systems in native grasslands: effects of elevated atmospheric CO_2. *New Phytologist*, **147**: 73–86.

Arnone, J.A., Zaller, J.G., Ziegler, C., Zandt, H. & Körner, C. 1995. Leaf quality and insect herbivory in model tropical plant communities after long-term exposure to elevated atmospheric CO_2. *Oecologia*, **104**: 72–78.

Arp, W.J. 1991. Effects of source-sink relations on photosynthetic acclimation to elevated CO_2. *Plant, Cell & Environment*, **14**: 869–875.

Baethgen, W.E. 1997. Vulnerability of the agricultural sector of Latin America to climate change. *Climate Research*, **9**: 1–7.

Baker, B.B., Hanson, J.D., Bourdon, R.M. & Eckert, J.B. 1993. The potential effects of climate change on ecosystem processes and cattle production on US rangelands. *Climatic Change*, **25**: 97–117.

Bazzaz, F.A. 1990. The response of natural ecosystems to the rising global CO_2 levels. *Annual Review of Ecological Systematics*, **21**: 167–196.

Bazzaz, F.A. & McConnaughay, K.D.M. 1992. Plant-plant interactions in elevated CO_2 environments. *Australian Journal of Botany*, **40**: 547–563.

Bianca, W. 1965. Reviews of the progress of dairy science. Section A: Physiology. Cattle in a hot environment. *Journal of Dairy Research*, **32**: 291–345.

Bond, W.J. & Midgley, G.F. 2000. A proposed CO_2-controlled mechanism of woody plant invasion in grasslands and savannas. *Global Change Biology*, **6**: 865–869.

Bonsma, J.C. 1949. Breeding cattle for increased adaptability to tropical and subtropical environments. *Journal of Agricultural Sciences*, **39**: 204–210.

Bowes, G. 1993. Facing the inevitable: plants and increasing atmospheric CO_2. *Annual Review of Plant Physiology and Plant Molecular Biology*, **44**: 309–332.

Buckland, S.M., Thompson, K., Hodgson, J.G. & Grime, J.P. 2001. Grasslands invasions: effects of manipulations of climate and management. *Journal of Applied Ecology*, **38**: 301–309.

Bunce, J.A. 1990. Short- and long-term inhibition of respiratory carbon dioxide efflux by elevated carbon dioxide. *Annals of Botany*, **65**: 637–642.

Campbell, B.D., Stafford Smith, D.M. & GCTE Pastures and Rangelands Network members. 2000. A synthesis of recent global change research on pasture and rangeland production: reduced uncertainties and their management implications. *Agriculture, Ecosystems and Environment*, **82**: 39–55.

Campbell, B.D., Stafford Smith, D.M. & McKeon, G.M. 1997. Elevated CO_2 and water supply interactions in grasslands: a pastures and rangelands management perspective. *Global Change Biology*, **3**: 177–187.

Coffin, D.P. & Lauenroth, W.K. 1996. Transient responses of North-American grasslands to changes in climatic. *Climatic Change*, **34**: 269–278.

Coughenour, M.B. & Chen, D.-X. 1997. Assessment of grassland ecosystem responses to atmospheric change using linked plant-soil process models. *Ecological Applications,* **7**: 802–827.

Cure, J.D. 1985. Carbon dioxide doubling responses: a crop survey. pp. 99–116, *in:* B.R. Strain and J.D. Cure (eds). *Direct Effects of Increasing Carbon dioxide on Vegetation.* U.S. Department of Energy, Washington, D.C.

Curtis, P.S., O'Neill, E.G., Terri, J.A., Zak, D.R. & Pregitzer, K.S. 1994. Belowground response to rising atmospheric CO_2: implications for plants, soil biota and ecosystem processes. *Plant and Soil,* **165**: 1–6.

Davis, M.A., Grime, J.P. & Thompson, K. 2000. Fluctuating resources in plant communities: a general theory of invisibility. *Journal of Ecology,* **88**: 528–534.

Defosse, G.E., Bertiller, M.B. & Robberecht, R. 1997. Effects of topography, soil moisture, wind and grazing on *Festuca* seedlings in a Patagonian grasslands. *Journal of Vegetation Science,* **8**: 677–684.

Drake, B.G., Gonzàlez-Meler, M.A. & Long, S.P. 1997. More efficient plants: a consequence of rising atmospheric CO_2? *Annual Review of Plant Physiology and Plant Molecular Biology,* **48**: 609–639.

Duckworth, J.C., Bunce, R.G.H. & Malloch, A.J.C. 2000. Vegetation gradients in Atlantic Europe: the use of existing phytosociological data in preliminary investigations on the potential effects of climate change on British vegetation. *Global Ecology and Biogeography,* **9**: 187–199.

Edwards, G.R., Clark, H. & Newton, P.C.D. 2001. The effects of elevated CO_2 on seed production and seedling recruitment in a sheep-grazed pasture. *Oecologia,* **127**: 383–394.

Ehleringer, J.R., Phillips, S.L., Schuster, W.S.F. & Sandquist, D.R. 1991. Differential utilization of summer rains by desert plants. *Oecologia,* **88**: 430–434.

Epstein, H.E., Burke, I.C. & Lauenroth, W.K. 2002. Regional patterns of decomposition and primary production rates in the U.S. Great Plains. *Ecology,* **83**: 320–327.

Epstein, H.E., Lauenroth, W.K., Burke, I.C. & Coffin, D.P. 1997. Productivity patterns of C_3 and C_4 functional types in the U.S. Great Plains. *Ecology,* **78**: 722–731.

Esser, G. 1992. Implications of climate change for production and decomposition in grasslands and coniferous forests. *Ecological Applications,* **2**: 47–54.

Farnsworth, E.J. & Bazzaz, F.A. 1995. Inter- and intra-generic differences in growth, reproduction, and fitness of nine herbaceous annual species grown in elevated CO_2 environment. *Oecologia,* **104**: 454–466.

Farrar, J.F. 1985. The respiratory source of CO_2. *Plant, Cell and Environment,* **8**: 427–438.

Farrar, J.F. & Williams, M.L. 1991. The effects of increased atmospheric carbon dioxide and temperature on carbon partitioning, source-sink relations and respiration. *Plant, Cell & Environment,* **14**: 819–830.

Field, T.R.O. & Forbe, M.B. 1990. Effects of climate warming on the distribution of C_4 grasses in New Zealand. *Proceedings of the New Zealand Grassland Association,* **51**: 47–50.

Field, C.B., Jackson, R.B. & Mooney, H.A. 1995. Stomatal responses to increased CO_2: implications from the plant to the global scale. *Plant, Cell and Environment,* **18**: 1214–1225.

Geeske, J., Chapin III, F.S., Chiariello, N.R., Thayer, S.S. & Field, C.B. 2001. Species-specific responses of plant communities to altered carbon nutrient availability. *Global Change Biology,* **7**: 435–450.

Giorgi, R., Meehl, G.A., Kattenberg, A., Grassl, H., Mitchell, J.F.B., Stouffer, R.J., Tokioka, T., Weaver, A.J. & Wigley, T.M.L. 1998. Simulation of regional climate change with global coupled climate models and regional modelling techniques. pp. 427–437, *in:* R.T. Watson, M.C. Zinyowerra, R.H. Moss and D.J. Dokken (eds). *The Regional Impacts of Climate Change: an Assessment of Vulnerability.* New York, NY: Cambridge University Press.

Gitay, H., Brown, S. Easterling, W., Jallow, B., Antle, J., Apps, M., Beamish, R., Chapin, T., Cramer, W., Frangi, J., Laine, J., Erda, L., Magnuson, J., Noble, I., Price, J., Prowse, T., Root, T., Schulze, E., Sirotenko, O., Sohngen, B. & Soussana, J. 2001. Ecosystems and their goods and services, *In:* J.J. McCarthy, O.F. Canziani, N.A. Leary, D.J. Dokken and K.S. White (eds). *Climate Change 2001: Impacts, Adaptation and Vulnerability.* Contribution of Working Group II to the Third Assessment Report of the Intergovernmental Panel on Climate Change Cambridge, UK, and New York, NY: Cambridge University Press. 1032p.

Greer, D.H., Laing, W.A., Campbell, B.D. & Halligan E.A. 2000. The effect of perturbations in temperature and photon flux density on the growth and photosynthetic responses of five pasture species to elevated CO_2. *Australian Journal of Plant Physiology,* **27**: 301–310.

Grünzweig, J.M. & Körner, C. 2001. Growth, water and nitrogen relations in grassland model ecosystems of the semi-arid Negev of Israel exposed to elevated CO_2. *Oecologia,* **128**: 251–262.

Hanson, J.D., Baker, B.B. & Bourdon, R.M. 1993. Comparison of the effects of different climate change scenarios on rangeland livestock production. *Agricultural Systems,* **41**: 487–502.

He, J.-S., Bazzaz, F.A. & Schmid, B. 2002. Interactive effects of diversity, nutrients and elevated CO_2 on experimental plant communities. *Oikos,* **97**: 337–348.

Hebeisen, T., Lüscher, A., Zanetti, S., Fischer, B.U., Hartwig, U.A., Frehner, M., Hendrey, G.R., Blum, G. & Nösberger, J. 1997. Growth response of *Trifolium repens* L. and *Lolium perenne* L. as monocultures and bi-species mixture to free air CO_2 enrichment and management. *Global Change Biology*, **3**: 149–160.

Jablonski, L.M., Wang, X. & Curtis, P.S. 2002. Plant reproduction under elevated CO_2 conditions: a meta-analysis of reports on 79 crop and wild species. *New Phytologist*, **156**: 9–26.

Jackson, R.B., Sala, O.E., Field, C.B. & Mooney, H.A. 1994. CO_2 alters water use, carbon gain, and yield for the dominant species in a natural grassland. *Oecologia*, **98**: 257–262.

Joel, G., Chapin III, F.S., Chiariello, N.R., Thayer, S.S. & Field, C.B. 2001. Species-specific responses of plant communities to altered carbon and nutrient availability. *Global Change Biology*, **7**: 435–450.

Kimball, B.A. & Idso, S.B. 1983. Increasing atmospheric CO_2: effects on crop yield, water use and climate. *Agricultural Water Management*, **7**: 55–72.

Klinedinst, P.L., Wilhite, D.A., Hahn, G.L. & Hubbard, K.G. 1993. The potential effects of climate change on summer season dairy cattle milk production and reproduction. *Climatic Change*, **23**: 21–36.

Knapp, A.K., Briggs, J.M. & Koelliker, J.K. 2001. Frequency and extent of water limitation to primary production in a mesic temperate grassland. *Ecosystems*, **4**: 19–28.

Knapp, A.K., Hamerlynck, E.P. & Owensby, C.E. 1993. Photosynthetic and water relations responses to elevated CO_2 in the C_4 grass *Andropogon gerardii*. *International Journal of Plant Science*, **154**: 459–466.

Körner, C. 1996. The response of complex multispecies systems to elevated CO_2. pp. 20–42, *in:* B.H. Walker & W.L. Steffen (eds). *Global Change and Terrestrial Ecosystems*. Cambridge, UK: Cambridge University Press.

Körner, C. 2002. Grasslands in a CO_2-enriched world. *Grassland Science in Europe*, **7**: 611–624.

Körner, C., Bazzaz, F.A. & Field, C.B. 1996. The significance of biological variation, organism interactions and life histories in CO_2 research. pp. 415–429, *in:* H.A. Mooney and G.W. Koch (eds). *Community, Population and Evolutionary Responses to Elevated Carbon Dioxide Concentration*. San Diego, CA: Academic Press.

Larigauderie, A., Hilbert, D.W. & Oechel, W.C. 1988. Effect of CO_2 enrichment and nitrogen availability on resource acquisition and resource allocation in a grass, *Bromus mollis*. *Oecologia*, **77**: 544–549.

Lawlor, D.W. & Mitchell, R.A.C. 1991. The effects of increasing CO_2 on crop photosynthesis and productivity: a review of field studies. *Plant, Cell & Environment*, **14**: 807–818.

Leadley, P.W. & Körner, C. 1996. Effects of elevated CO_2 on plant species dominance in a highly diverse calcareous grassland, pp. 159–175, *in:* C. Körner and F. Bazzaz (eds). *Carbon Dioxide, Populations and Communities*. San Diego, CA: Academic Press.

Leadley, P.W., Niklaus, P.A., Stocker, R. & Körner, C. 1999. A field study of the effects of elevated CO_2 on plant biomass and community structure in a calcareous grassland. *Oecologia*, **118**: 39–49.

Lee, T.D., Tjoelker, N.G., Ellsworth, D.S. & Reich, P.B. 2001. Leaf gas exchange responses of 13 prairie grassland species to elevated CO_2 and increased nitrogen supply. *New Phytologist*, **150**: 405–418.

Lilley, J.M., Bolger, T.P., Peoples, M.B. & Gifford, R.M. 2001. Nutritive value and the nitrogen dynamics of *Trifolium subterraneum* and *Phalaris aquatica* under warmer, high CO_2 conditions. *New Phytologist*, **150**: 385–395.

Long, S.P. 1991. Modification of the response of photosynthetic productivity to rising temperature by atmospheric CO_2 concentrations: has its importance been underestimated? *Plant, Cell & Environment*, **14**: 729–740.

Lüscher, A., Hartwig, U., Suter, D. & Nösberger, J. 2000. Direct evidence that symbiotic N_2 fixation in fertile grassland is an important trait for a strong response of plants to elevated atmospheric CO_2. *Global Change Biology*, **6**: 655–662.

Lüscher, A., Hendrey, G.R. & Nösberger, J. 1998. Long-term responsiveness to free air CO_2 enrichment of functional types, species and genotypes of permanent grassland. *Oecologia*, **113**: 37–45.

McCrabb, G.J. & Hunter, R.A. 1999. Prediction of methane emissions from beef cattle in tropical production systems. *Australian Journal of Agricultural Research*, **50**: 1335–1339.

Meeting, F.B., Smith, J.L., Amthor, J.S. & Izaurralde, R.C. 2001. Science needs and new technology for increasing soil carbon sequestration. *Climatic Change*, **51**: 11–34.

Mielnick, P.C., Dugas, W.A., Johnson, H.B., Polley, H.W., & Sanabria, J. 2001. Net grassland carbon flux over a subambient to superambient CO_2 gradient. *Global Change Biology*, **7**: 747–754.

Minnick, T.J. & Coffin, D.P. 1999. Geographic patterns of simulated establishment of two *Bouteloua* species: implications for distributions of dominants and ecotones. *Journal of Vegetation Science*, **10**: 343–356.

Mitchell, J.F.B., Karoly D.J., Hegerl G.C., Zwiers F.W., Allen M.R. & Marengo J., 2001. Detection of Climate Change and Attribution of Causes. *In:* J.T. Houghton, Y. Ding, D.J. Griggs, M. Noguer, P.J. van der Linden, X. Dai, K. Maskell and C.A. Johnson (eds). *Climate Change 2001: The Scientific Basis*. Contribution of Working Group I to the Third Assessment Report of the Intergovernmental Panel on Climate Change. Cambridge, UK, and New York, NY: Cambridge University Press. 881p.

Moore, J.L., Howden, S.M., McKeon, G.M., Carter, J.O. & Scanlan, J.C. 2001. The dynamics of grazed woodlands in southwest Queensland, Australia, and their effect on greenhouse gas emissions. *Environmental International*, **27**: 147–153.

Morgan, J.A. 2002. Looking beneath the surface. *Science*, **298**: 1903–1904.

Morgan, J.A., LeCain, D.R., Mosier, A.R. & Milchunas, D.G. 2001. Elevated CO_2 enhances water relations and productivity and affects gas exchange in C_3 and C_4 grasses of the Colorado shortgrass steppe. *Global Change Biology*, **7**: 451–466.

Morgan, J.A., Mosier, A.R., Milchunas, D.G., LeCain, D.R., Nelson, J.A. & Parton, W.J. 2004. CO_2 enhances productivity, alters species composition, and reduces digestibility of shortgrass steppe vegetation. *Ecological Applications*, in press.

Mortenson, M.C., Schuman, G.E. & Ingram, L.J. 2003. Carbon sequestration in rangelands interseeded with yellow-flowering alfalfa (*Medicago sativa* subsp. *falcata*). *Environmental Management*, in press.

Mortimore, M.J. & Adams, W.M. 2001. Farmer adaptation, change and 'crisis' in the Sahel. *Global Environmental Change*, **11**: 49–57.

Mosier, A.R. 1998. Soil processes and global change. *Biological Fertility and Soils*, **27**: 221–229.

Neilson, R.P., Prentice, I.C., Smith, B., Kittel, T. & Viner, D. 1998. Simulated changes in vegetation distribution under global warming. pp. 439–456, *in:* R.T. Watson, M.C. Zinyoweere and R.H. Moss (eds). *The Regional Impacts of Climate Change: an Assessment of Vulnerability*. New York, NY: Cambridge University Press.

Newton, P.C.D. 1991. Direct effects of increasing carbon dioxide on pasture plants and communities. *New Zealand Journal of Agricultural Research*, **34**: 1–24.

Newton, P.C.D., Clark, H., Bell, C.C., Glasgow, E.M. & Campbell, B.D. 1994. Effects of elevated CO_2 and simulated seasonal changes in temperature on the species composition and growth rates of pasture turves. *Annals of Botany*, **73**: 53–59.

Niklaus, P.A., Leadley, P.W., Schmid, B. & Körner, C. 2001. A long-term field study on biodiversity × elevated CO_2 interactions in grassland. *Ecological Monographs*, **71**: 341–356.

Niklaus, P.A., Spinnler, D. & Körner, C. 1998. Soil moisture dynamics of calcareous grassland under elevated CO_2. *Oecologia*, **117**: 201–208.

Nösberger, J., Blum, H. & Fuhrer, J. 2000. Crop ecosystem responses to climatic change: Productive grasslands. pp. 271–291, *in:* K.R. Reddy and H. F. Hodges (eds). *Climate Change and Global Crop Productivity*. Wallingford, UK, and New York, NY: CABI Publishing.

Norby, R.J. & Jackson, R.B. 2000. Root dynamics and global change: seeking an ecosystem perspective. *New Phytologist*, **147**: 3–12.

Ojima, D.S. & Lackett, J.M. and the Central Great Plains Steering Committee and Assessment Team. 2002. Preparing for a changing climate: The potential consequences of climate variability and change – Central Great Plains. Report for the US Global Change Research Program, Colorado State University, 103p.

Owensby, C.E., Cochran, R.C. & Auen, L.M. 1996. Effects of elevated carbon dioxide on forage quality for ruminants. pp. 363–371, *in:* C. Körner and F. Bazzaz (eds). *Carbon Dioxide, Populations and Communities*. Physiological Ecology Series. Academic Press, Inc.

Owensby, C.E., Ham, J.M., Knapp, A.K., Rice, C.W., Coyne, P.I. & Auen, L.M. 1996. Ecosystem-level responses of tallgrass prairie to elevated CO_2. pp. 147–162, *in:* G.W. Koch and H.A. Mooney (eds). *Terrestrial Ecosystem Response to Elevated CO_2*. London, UK: Academic Press.

Owensby, C.E., Ham, J.M., Knapp, A.K., Bremer, D. & Auen, L.M. 1997. Water vapour fluxes and their impact under elevated CO_2 in a C_4-tallgrass prairie. *Global Change Biology*, **3**: 189–195.

Owensby, C.E., Ham, J.M., Knapp, A.K. & Auen, L.M. 1999. Biomass production and species composition change in a tallgrass prairie ecosystem after long-term exposure to elevated atmospheric CO_2. *Global Change Biology*, **5**: 497–506.

Parish, R. & Funnell, D.C. 1999. Climate change in mountain regions: some possible consequences in the Moroccan High Atlas. *Global Environmental Change*, **9**: 45–58.

Parsons, D.J., Armstrong, A.C., Turnpenny, J.R., Matthews, A.M., Cooper, K. & Clark, J.A. 2001. Integrated models of livestock systems for climate change studies. *Global Change Biology*, **7**: 93–112.

Parton, W.J., Schimel, D.S., Cole, C.V. & Ojima, D.S. 1987. Analysis of factors controlling soil organic matter levels in Great Plains grasslands. *Soil Science Society of America Journal*, **51**: 1173–1179.

Parton, W.J., Scurlock, J.M.O., Ojima, D.S., Schimel, D.S., Hall, D.O. & Scopegram Group Members. 1995. Impact of climate change on grassland production and soil carbon worldwide. *Global Change Biology*, **1**: 13–22.

Paruelo, J.M. & Lauenroth, W.K. 1996. Relative abundance of plant functional types in grasslands and shrublands of North America. *Ecological Applications*, **6**: 1212–1224.

Pearcy, R.W. & Ehleringer, J. 1984. Comparative ecophysiology of C_3 and C_4 plants. *Plant, Cell & Environment*, **7**: 1–13.

Polley, H.W., Mayeux, H.S., Johnson, H.B. & Tischler, C.R. 1997. Viewpoint: atmospheric CO_2, soil water and shrub/grass ratios on rangelands. *Journal of Range Management*, **50**: 278–284.

Polley, H.W., Morgan, J.A., Campbell, B.D. & Stafford Smith, M. 2000. Crop Ecosystem Responses to Climatic Change: Rangelands. pp. 293–314, *in:* K.R. Reddy and H.F. Hodges (eds). *Climate Change and Global Crop Productivity.* Wallingford, UK, and New York, NY: CABI Publishing.

Porter, H. 1993. Interspecific variation in the growth response of plants to an elevated ambient CO_2 concentration. *Vegetatio,* **104/105**: 77–97.

Potvin, C. & Vasseur, L. 1997. Long-term CO_2 enrichment of a pasture community: Species richness, dominance, and succession. *Ecology,* **78**: 666–677.

Pyankov, V.I., Gunin, P.D., Tsoog, S. & Black, C.C. 2000. C_4 plants in the vegetation of Mongolia: their natural occurrence and geographical distribution in relation to climate. *Oecologia,* **123**: 15–31.

Rastetter, E.B., Ryan, M.G., Shaver, G.R., Melillo, J.M., Nadelhoffer, K.J., Hobbie, J.E. & Aber, J.D. 1991. A general biogeochemical model describing the responses of the C and N cycles in terrestrial ecosystems to changes in CO_2, climate, and N deposition. *Tree Physiology,* **9**: 101–126.

Reeder, J.D., Franks, C.D. & Milchunas, D.G. 2001. Root biomass and microbial processes. pp. 139–166, *in:* R.F. Follett, J.M. Kimble and R. Lal (eds). *The Potential of U.S. Grazing Lands to Sequester Carbon and Mitigate the Greenhouse Effect.* New York, NY: Lewis Publishers.

Reich, P.B., Tilman, D., Craine, J., Ellsworth, D., Tjoelker, M.G., Knops, J., Wedin, D., Naeem, S., Bahauddin, D., Goth, J., Bengtson, W. & Lee, T.D. 2001. Do species and functional groups differ in acquisition and use of C, N and water under varying atmospheric CO_2 and N availability regimes? A field test with 16 grassland species. *New Phytologist,* **150**: 435–448.

Riedo, M., Gyalistras, D. & Fuhrer, J. 2000. Net primary production and carbon stocks in differently managed grasslands: simulation of site-specific sensitivity to an increase in atmospheric CO_2 and to climate change. *Ecological Modelling,* **134**: 207–227.

Riedo, M., Gyalistras, D. & Fuhrer, J. 2001. Pasture responses to elevated temperature and doubled CO_2 concentration: assessing the spatial pattern across an alpine landscape. *Climate Research,* **17**: 19–31.

Rogers, H.H., Runion, G.B. & Krupa, S.V. 1994. Plant responses to atmospheric CO_2 enrichment with emphasis on roots and the rhizosphere. *Environmental Pollution,* **83**: 155–189.

Rogers, H.H., Runion, G.B., Krupa, S.V. & Prior, S.A. 1997. Plant responses to atmospheric carbon dioxide enrichment: Implications in root-soil-microbe interactions. *In: Advances in Carbon Dioxide Effects Research.* ASA Special Publication, No. 61. ASA, CSSA, SSSA, Madison, WI, USA. 34p.

Rogers, H.H., Runion, G.B., Prior, S.A. & Tolbert, H.A. 1999. Response of plants to elevated atmospheric CO_2: Root growth, mineral nutrition, and soil carbon, pp. 215–244, *in:* Y. Luo and H.A. Mooney (eds). *Carbon Dioxide and Environmental Stress.* New York, NY: Academic Press.

Roques, K.G., O'Connor, T.G. & Watkinson, A.R. 2001. Dynamics of shrub encroachment in an African savanna: relative influences of fire, herbivory, rainfall and density dependence. *Journal of Applied Ecology,* **38**: 268–280.

Rötter, R. & Van De Geijn, S.C. 1999. Climate change effects on plant growth, crop yield and livestock. *Climatic Change,* **43**: 651–681.

Rounsevell, M.D.A., Brignall, A.P. & Siddons, P.A. 1996. Potential climate change effects on the distribution of agricultural grassland in England and Wales. *Soil Use and Management,* **12**: 44–51.

Rounsevell, M.D.A., Evans, S.P. & Bullock, P. 1999. Climate change and agricultural soils: impacts and adaptation. *Climatic Change,* **43**: 683–709.

Rustad, L.E., Campbell, J.L., Marion, G.M., Norby, R.J., Mitchell, M.J., Hartley, A.E., Cornelissen, J.H.C., Gurevitch, J. & GCTE-NEWS. 2001. A meta-analysis of the response of soil respiration, net nitrogen mineralization, and aboveground plant growth to experimental ecosystem warming. *Oecologia,* **126**: 543–562.

Sage, R.F. 1994. Acclimation of photosynthesis to increasing atmospheric CO_2: The gas exchange perspective. *Photosynthesis Research,* **39**: 351–368.

Sage, R.F., Sharkey, T.D. & Seeman, J.R. 1989. Acclimation of photosynthesis to elevated CO_2 in five C_3 species. *Plant Physiology,* **89**: 590–596.

Sala, O.E., Parton, W.J., Joyce, L.A. & Lauenroth, W.K. 1988. Primary production of the central grasslands region of the United States. *Ecology,* **69**: 40–45.

Schenk, U., Jäger, H.-J. & Weigel, H.-J. 1997. The response of perennial ryegrass/white clover mini-swards to elevated atmospheric CO_2 concentrations: effects on yield and fodder quality. *Grass and Forage Science,* **52**: 232–241.

Schlesinger, W.H. & Andrews, J.A. 2000. Soil respiration and the global carbon cycle. *Biogeochemistry,* **48**: 7–20.

Schlesinger, W.H., Reynolds, J.F., Cunningham, G.L., Huenneke, L.F., Jarrell, W.M., Virginia, R.A. & Whitford, W.G. 1990. Biological feedbacks in global desertification. *Science,* **247**: 1043–1048.

Scholes, R.J. & Archer, S.R. 1997. Tree-grass interactions in savannas. *Annual Review of Ecology and Systematics,* **28**: 517–544.

Schuman, G.E., Herrick, J.E. & Janzen, H.H. 2001. The dynamics of soil carbon in rangelands. pp. 267–290, *in:* R.F. Follett, J.M. Kimble and R. Lal (eds). *The Potential of U.S. Grazing Lands to Sequester Carbon and Mitigate the Greenhouse Effect.* Boca Raton, FL: Lewis Publishers.

Sindhøj, E., Hansson, A.-C., Andrén, O., Kätterer, T., Maarissink, M. & Pettersson, R. 2000. Root dynamics in a semi-natural grassland in relation to atmospheric carbon dioxide enrichment, soil water and shoot biomass. *Plant and Soil,* **223**: 253–263.

Sionit, N. & Patterson, D.T. 1984. Responses of C_4 grasses to atmospheric CO_2 enrichment. I. Effect of irradiance. *Oecologia,* **65**: 30–34.

Smith, J.B. & Lazo, J.K. 2001. A summary of climate change impact assessments from the U.S. country studies program. *Climatic Change,* **50**: 1–29.

Smith, S.D., Huxman, T.E., Zitzer, S.F., Charlet, T.N., Housman, D.C., Coleman, J.S., Fenstermaker, L.K., Seemann, J.R. & Nowak, R.S. 2000. Elevated CO_2 increases productivity and invasive species success in an arid ecosystem. *Nature,* **408**: 79–82.

Souza Neto, J. de, Conner, J.R., Stuth, J.W., Hamilton, W.T. & Richardson, J.W. 1998. Management strategies and climate impact change on rangelands. *Pesq. agropec. bras., Brasília,* **33**: 1533–1541.

Squires, V.R. & Sidahmed, A. 1997. Livestock management in dryland pastoral systems: Prospects and problems. *Annals of Arid Zone,* **36**: 79–96.

Stanton, N.L. 1988. The underground in grasslands. *Annual Review of Ecology and Systematics,* **19**: 573–589.

Steinger, T., Gall, R. & Schmid, B. 2000. Maternal and direct effects of elevated CO_2 on seed provisioning, germination and seedling growth in *Bromus erectus*. *Oecologia,* **123**: 475–480.

Stem, E., Mertz, G.A., Stryker, J.D. & Huppi, M. 1988. Changing animal disease patterns introduced by the greenhouse effect: Report of a preliminary study to the Environmental Protection Agency, North Grafton, MA, USA. Tufts University School of Veterinary Medicine. 76p.

Stephenson, N.L. 1990. Climatic control of vegetation distribution: the role of the water balance. *American Naturalist,* **135**: 649–679.

Stöcklin, J., Schweizer, K. & Körner, C. 1998. Effects of elevated CO_2 and phosphorus addition on productivity and community composition of intact monoliths from calcareous grassland. *Oecologia,* **116**: 50–56.

Stulen, I. & den Hertog, J. 1993. Root growth and functioning under atmospheric CO_2 enrichment. *Vegetatio,* **104/105**: 99–115.

Subak, S. 1999. Global environmental costs of beef production. *Ecological Economics,* **30**: 79–91.

Terri, J.A. & Stowe, L.G. 1976. Climatic patterns and the distribution of C_4 grasses in North America. *Oecologia,* **23**: 1–12.

Teyssonneyre, F., Picon-Cochard, C., Falcimagne, R. & Soussana, J.-F. 2002. Effects of elevated CO_2 and cutting frequency on plant community structure in a temperate grassland. *Global Change Biology,* **8**: 1034–1046.

Thompson, G.E. 1973. Review of the progress of Dairy Science – Climatic physiology of cattle. *Journal of Dairy Research,* **40**: 441–473.

Thornley, J.H.M. & Cannell, M.G.R. 2000. Dynamics of mineral N availability in grassland ecosystems under increased [CO_2]; hypotheses evaluated using the Hurley Pasture Model. *Plant and Soil,* **224**: 153–170.

Turnbull, L.A., Crawley, M.J. & Rees, M. 2000. Are plant populations seed-limited? A review of seed sowing experiments. *Oikos,* **88**: 225–238.

Turner, M. 1993. Overstocking the range: a critical analysis of the environmental science of Sahelian Pastoralism. *Economic Geography,* **69**: 402–421.

VEMAP Members. 1995. Vegetation/ecosystem modelling and analysis project: comparing biogeography and biogeochemistry models in continental-scale study of terrestrial ecosystem responses to climate change and CO_2 doubling. *Global Biogeochemical Cycling,* **9**: 407–437.

Vitousek, P.M., Aber, J.D., Howarth, R.W., Likens, G.E., Matson, P.A., Schindler, D.W., Schlesinger, W.H. & Tilman, D.G. 1997. Human alteration of the global nitrogen cycle: sources and consequences. *Ecological Applications,* **7**: 737–750.

Volk, M., Niklaus, P.A. & Körner, C. 2000. Soil moisture effects determine CO_2 responses of grassland species. *Oecologia,* **125**: 380–388.

Wand, S.J.E., Midgley, G.F., Jones, M.H. & Curtis, P.S. 1999. Responses of wild C_4 and C_3 grass (Poaceae) species to elevated atmospheric CO_2 concentrations: a meta-analytic test of current theories and perceptions. *Global Change Biology,* **5**: 723–741.

Warwick, K.R., Taylor, G. & Blum, H. 1998. Biomass and compositional changes occur in chalk grassland turves exposed to elevated CO_2 to two season in FACE. *Global Change Biology,* **4**: 375–385.

Watkinson, A.R. & Ormerod, S.J. 2001. Grasslands, grazing and biodiversity: editors' introduction. *Journal of Applied Ecology,* **38**: 233–237.

Webb, W.S., Szarek, S.T., Lauenroth, W.K. & Kinerson, R.S. 1978. Primary production and abiotic controls in forests, grassland and desert. *Ecology*, **64**: 134–151.

Wester, D.B. 1995. Seed physiology and germination. pp. 168–238, *in:* D.J. Bedunah and R.E. Sosebee (eds). *Wildland plants: physiological ecology and developmental morphology*. Denver, CO: Society for Range Management.

Whitford, W.G., Martinez-Turanzas, G. & Martinez-Meza, E. 1995. Persistence of desertified ecosystems: explanations and implications. *Environmental Monitoring and Assessment*, **37**: 319–332.

Zak, D.R., Pregitzer, K.S., King, J.S. & Holmes, W.E. 2000. Elevated atmospheric CO_2, fine roots and the response of soil microorganisms: a review and hypothesis. *New Phytologist*, **147**: 201–222.

CHAPTER 11

Biodiversity in Grasslands. Current Changes and Future Scenarios

Martín R. Aguiar[1]

ABSTRACT

Loss of biodiversity is a pressing problem for the biosphere. Current estimations for one of the attributes of biodiversity – species richness – indicate that extinction rates are higher than in the recent past and are still increasing. The environmental drivers of species extinction rates are also changing other structural and functional attributes of biodiversity. The main causes of biodiversity loss are global; land use and land cover change has been proposed as having the biggest impact. The grassland biome, which includes a wide range of ecosystem types, from humid prairies to arid shrub–grass steppes, has been subjected to particularly intense pressure for the production of food and fibres, so current extinction rates of grassland species are expected to remain high or even increase. The main aim of this chapter is to consider current processes and future scenarios for grassland biodiversity. As one considers the future, it is useful to address three questions regarding biodiversity loss. First, what are the effects of the main drivers of biodiversity loss, and what are the main trends in the biodiversity of grasslands? Second, how will biodiversity continue to change after extinctions take place? Third, how can we manage grassland biodiversity in the transition to the new scenario? Currently, there is a lot more information available to answer the first question than there is information to address the other two. Nevertheless, it is necessary to address the second and third questions in order to develop proactive strategies for managing the transition to new scenarios.

[1]IFEVA and Cátendra de Ecología, Facultad de Agronomia, Universidad de Buenos Aires. Av. San Martin 4453, Buenos Aires (1417). Argentina

INTRODUCTION

On the verge of the sixth major extinction event in the geological history of the Earth (Myers and Knoll, 2001), this chapter reviews both current and probable near-future trends in grassland biodiversity. Concern about biodiversity derives from two certainties. First, that changes in major drivers of ecosystem structure and functioning are global and affect all biomes. Second, that these changes will affect humanity through the erosion of the life-support systems of the Earth. In other words, there will be a reduction in ecosystem capability to provide natural resources and to process wastes (i.e. ecosystem services). Grasslands have been and still are central to the production of food and fibres for human use (Redman, 1999).

The main land uses in grasslands—agronomy and animal husbandry—involve four main activities:

- extraction of resources (e.g. mineral nutrients);
- changes in energy and material transfers (e.g. individual plant growth and nutrient mineralization);
- changes in species composition (e.g. addition of crops and weeds) and substances (i.e. agrochemicals, principally pesticides and fertilizers); and
- changes in the disturbance regime (e.g. cultivation, grazing).

These activities introduce major changes in the structure and functioning of ecosystems (see, for example, Aguiar *et al.*, 1996). However, superimposed on land use changes, grasslands are experiencing changes in other drivers, such as climate, atmospheric composition, and non-planned species exchanges, and these changes combine to threaten ecosystem integrity on a global scale (Vitousek *et al.*, 1997a).

Biodiversity has been identified as one of the best descriptors of ecosystem condition. Biodiversity is the sum of total biotic variation, from gene to landscape, but researchers have primarily focused on species diversity (Huston, 1994). This has resulted in the use of biodiversity and species richness as synonyms and a consequent need to relate species richness with ecosystem functioning (Chapin *et al.*, 2000). Over the past decades, ecologists have studied the determinants of species diversity (e.g. Hutchinson, 1959; MacArthur and Wilson, 1967; Tilman and Pacala, 1993; Huston, 1994; Hubbell, 2001) and the connection with ecosystem functioning (Chapin *et al.*, 2000). The issues are far from completely resolved, since many processes at different temporal and spatial scales contribute to or reduce species diversity and change ecosystem functioning (Huston, 1994; Lawton, 1995; Stohlgren *et al.*, 1999). Nevertheless, our current understanding confirms that human appropriation of Earth's ecosystems promotes the extinction of species at higher rates (Lawton, 1995; Rosenzweig, 1995) than in the past, in addition to extensive changes in ecosystem functioning (Vitousek *et al.*, 1997a).

Management for biodiversity conservation requires the integration of social and ecological systems into regional systems (cf. Chapin and Whiteman, 1998). Management of regional systems requires understanding of the main interactions between social systems and ecosystems. Scientists working in biophysical and social science, land managers, stakeholders and decision-makers should reach a common understanding in order to address the major causes of biodiversity loss, and to plan for the future use of natural resources. An integrated view of the social-ecological system is one of the main components of the new generation of projects studying biocomplexity. Biocomplexity theory deals with the study of complex systems and proposes complementing the study of parts of systems with the study of systems as a whole.

This chapter summarizes the main scenarios foreseen for the coming decades concerning the changes in biodiversity in grasslands. Written for grassland managers, the chapter starts by introducing a framework for discussing biodiversity. This framework allows the reader to escape from the usual trap of confounding biodiversity with species richness. Subsequently it addresses the question: "What are the scenarios of grassland biodiversity over the next few decades?" To answer this, the main drivers of changes in biodiversity and their impact on biodiversity are analysed. The chapter then addresses the question: "What will happen after extinction occurs?" In particular, "What types of species will dominate landscapes?" and "What will be the main pattern of evolution for the remaining species?" In the final section, the emerging concept of biocomplexity is discussed, and the main challenges that grassland ecologists will face during the next decades are proposed.

BIODIVERSITY AS A HIERARCHY

Biodiversity is, in some way, a fuzzy term, one that includes the sum of all biotic variation at three levels: intraspecific genetic variation; species diversity; and ecosystem diversity (Purvis and Hector, 2000). Being such a multidimensional system attribute, biodiversity cannot be reduced to a single number, such as species richness, so there is the need for a series of indicators that allows us to identify and measure the different attributes of biodiversity. Noss (1990) proposed a hierarchical framework to study the attributes of biodiversity. He proposed three complementary sets of attributes of biodiversity: compositional; structural; and functional (Table 11.1). The compositional attributes refer to the identity and variety of the elements included in the target system. They range from intraspecific genetic diversity (i.e. alleles), to landscape types (i.e. proportion of habitats). The structural attributes refer to the physical organization of the elements. They range from genetic structure (i.e. effective population size) to landscape structure (i.e. patchiness and connectivity among patches). The functional attributes include the ecological and evolutionary processes that organize ecosystems. They include gene flow, matter and energy exchanges, and disturbance regimes at the landscape level. In summary, biodiversity is a hierarchical concept that can be evaluated based on three complementary sets of attributes. Each set of attributes includes a variety of indices that can be measured.

Table 11.1 Biodiversity set of attributes according to Noss (1990). The three sets of attributes can be studied at four levels, from landscapes to genes. Levels are nested in a hierarchy and attributes are related by their effects on each other.

Biodiversity sets of attributes		
Compositional	Structural	Functional
Landscape types	Landscape patterns	Landscape processes (disturbance regime, land use trends)
Communities	Habitat structure	Interspecific interactions, ecosystem processes
Species, populations	Population structure (size distribution, sex ratio, spatial)	Demographic processes
Genes	Genetic structure	Genetic processes

Why is the Noss framework relevant to biodiversity assessment and monitoring in grasslands? There are two issues worth considering. First, the framework helps us to move away from using just a single attribute, such as composition, and a single indicator: diversity or richness. Considering the structural and functional attributes of biodiversity helps us to include essential aspects for understanding and monitoring ecosystem functioning. For example, at the population level, an important indicator of the population structure is spatial organization of genotypes. Spatial organization also affects other indicators, such as interspecific interactions (e.g. competition or facilitation) among plant populations. Another benefit of this framework is that it promotes the use of hierarchy theory. In other words, it promotes consideration of the levels above and below a target level of inquiry, from temporal and spatial perspectives. For example, fully understanding the effect of grazing on an individual species depends not only on knowledge of the interactions of that species with other species in the plant community (i.e. a functional attribute above the species level), but also awareness of the variability of ecotype diversity (i.e. a compositional attribute in the hierarchy below the species level).

GLOBAL CHANGE AND BIODIVERSITY SCENARIOS IN GRASSLANDS

Grasslands, in common with other major biomes, are experiencing the effects of major global changes. There are various possible scenarios of change in biodiversity for the next century, according to the report by Sala *et al.* (2000). The report discusses the sensitivity of biomes to the different global changes. They screened five drivers of change: land use; climate; nitrogen deposition; biotic exchanges; and atmospheric CO_2. A projected scenario of biodiversity results from multiplying the relative importance of the change in the drivers, by the impact on biodiversity of those changes.

Grasslands show intermediate effects from most of the drivers, except for land use. Changing land use will have a major impact on biodiversity, since it reduces habitat availability. Climate changes are expected to have less impact on grasslands – which are located mostly at temperate and intermediate latitudes – compared with higher latitude biomes, such as tundra (Walter, 1984). Nitrogen deposition will certainly have a major impact in N-limited systems. In general, grasslands are not the most N-limited biome, so the impact of this driver was assumed to be intermediate. Biotic introductions are also intermediate, because while mild habitat conditions facilitate establishment of invaders, existing levels of species diversity constrain resource availability. Finally, atmospheric changes in CO_2 will have a relatively high impact on grasslands because of the frequent mixed composition in terms of C_3 and C_4 species. In general, however, it is considered that these drivers have only an intermediate impact on biodiversity.

Since none of these drivers acts alone, the report proposes that it is possible to consider three types of interaction: additive; response to only the driver with the maximum effect; and multiplicative. After considering these interactions, grasslands were ranked in the upper half of the biomes studied, in terms of altered biodiversity. This means that grasslands are ecosystems where some of the largest changes in biodiversity are expected to take place. Other systems, such as Mediterranean scrub or savannahs, usually linked to grasslands under the category of "rangelands", ranked close to grasslands when the additive or the multiplicative model of interaction among drivers was assumed. The following use Noss' framework, presented in the previous section, to discuss the consequences of changes in the drivers that Sala *et al.* (2000) identified as those most important for grassland biodiversity, namely land use, climate, nitrogen deposition, biotic exchanges and atmospheric CO_2.

LAND USE EFFECTS ON BIODIVERSITY

Grassland "land use" includes all human activities involving grasslands. These activities range from grazing and management of domestic animals, to land conversion to crops, forest or urbanization. This section discusses the effect on biodiversity of grazing and land conversion to crops and forests. Since grazing and cropping in grasslands have been extensively discussed elsewhere, the objective is to focus on Noss' framework in order to organize and review the information that already has been presented. The objective is therefore twofold. First, to review the ways in which land use affects grassland biodiversity, and, second, to use the hierarchical and multiperspective model proposed by Noss.

Grazing by domestic herbivores

Domestic herbivores are selective grazers and promote changes in plant community composition (i.e. the composition attributes of biodiversity). This was soon noted by rangeland ecologists. Dykterhuis (1949) proposed a theory of rangeland management based on this knowledge. According to this theory, by changing stocking rates it was possible to manage plant community composition. This statement proved, however, not to be valid for all conditions and ecosystems. New knowledge enriched the model of rangeland management. For example, it is widely accepted that transition from one community composition to another may not be linear. In other words, changes in plant community composition may not be gradual, but triggered as disturbance reaches a certain threshold. The currently accepted model of grazing effects on plant dynamics indicates that a given ecosystem includes several states (i.e. different species compositions). The transition from one state to another is not solely the result of stocking rates. For example, variability between years in climate and in disturbance regime also play major roles in determining the occurrence of change (Westoby, Walker and Noy-Meir, 1989).

In addition to community composition, other attributes of biodiversity change as a result of grazing. In arid and semi-arid ecosystems, vegetation is frequently organized in a dual-phase mosaic composed of patches with high vegetation cover (or plant density) dispersed in a matrix of low plant cover (Figure 11.1) (Aguiar and Sala, 1999). This horizontal organization of plant canopy was reported in an early study (Noy-Meir, 1981), but its origin is not fully comprehended. A growing number of studies enrich our knowledge of the processes that maintain this particular community pattern. More importantly, ecologists are using the concept of a dual-phase mosaic to create and build communities with a particular structure (Boeken and Shachak. 1994) (Figure 11.1, left). This structural dimension in plant community biodiversity is gaining recognition as an issue needing consideration.

Many ecosystem and population processes are influenced by the spatial organization of the vegetation mosaic (Aguiar and Sala, 1999). For example, in banded vegetation in Mexico and Australia, water availability and hydrological balance are controlled by the vegetation pattern (Figure 11.1, centre). The matrix functions as a source area that transfers water to the vegetation patches, allowing greater primary production, or a particular flora (Cornet, Dehlhoume and Montaña, 1988; Montaña, López and Mauchamp, 1990; Gutiérrez *et al.* 1993; Ludwig and Tongway 1995). The effects of grazing on spatial structure are not clear. Since grazing generates a pattern in vegetation (Sala, 1988), the effect of grazing depends on the vegetation pattern being analysed (banded or spotted mosaic) (Aguiar and Sala, 1999). In some cases, such as in the banded vegetation in Australia, it has been suggested that overgrazing disrupts the mosaic (Ludwig and Tongway,

Figure 11.1 Structural biodiversity in vegetation mosaics at the landscape and community levels.
On the left is a schematic example of a vegetation mosaic with a regular pattern in agro-ecosystems.
The centre image is a banded pattern, such as those described in semi-arid landscapes with a gradual slope, found in Australia, Africa or North America. Arrows indicate the direction of water flow in the inter-band areas. Water infiltrates in the bands.
The mosaic on the right is a patch pattern as described in semi-arid ecosystems of South and North America.

1995). A recent review of the effects of grazing on spatial structure of plant communities found that currently there is no strong evidence that grazing disrupts vegetation patterns (Adler, Raff and Lauenroth, 2001). In summary, a spatially-explicit view of community and ecosystems allows new perspectives for processes (i.e. dispersal, population dynamics, and plant–plant and plant–animal interactions) and functions (i.e. soil organic matter dynamics, and water and nutrient circulation). Whereas structural diversity has been recognized at the landscape level, structural diversity has been ignored at the community level, even though long proposed (Watt, 1947).

From the compositional perspective, grazing has a major impact not only on species composition (as already discussed), but also at the population level. Population genetic diversity and structure of preferred species can change as grazing drives plant density down (Schemske *et al.*, 1994). Rangeland theory assumes that, with proper management, species reduced by overgrazing can recover if the physical environment has not changed drastically. In other words, there are no demographic constraints to recovery for the few remaining plants. From the genetic point of view, however, the situation may be different. As the size of a population decreases, genetic variability is reduced. Genetic drift, the loss of genetic diversity that occurs in small size populations, contributes to the process (Schemske *et al.*, 1994). Grazing can exert strong selection pressure, promoting traits that increase tolerance or avoidance of herbivory. In small populations, genetic drift is a common process, and erodes genetic variability in a random fashion, without increase in fitness. Low genetic diversity levels may therefore constrain population sustainability by reducing the capacity of the population to cope with environmental variability. Essentially, adequate management can promote an increase in population numbers, but not necessarily an increase in genetic variability, and therefore the limited genetic diversity of the recovering population can restrain population viability.

Most reviews of biodiversity mention genetic diversity as one of the most pressing problems associated with human-induced global changes. However, for grasslands, there is not enough information on this factor. More information is needed in order to understand the relationship between grazing and genetic variability. More information is needed on the structural and functional attributes of genetic diversity. In other words, it is necessary not only to know how much variability exists in a population, but also how it is spatially distributed (i.e. structural attributes of genetic diversity). Also, it is necessary to address the genetic processes, such as migration of genes. As plant density decreases, distances between individuals increase. Two

pattern of invasion and change of fire regime seems to be common. According to them, biological invasions have caused more species extinction than climate change or atmospheric composition change. They stress the fact that biological invasions change not only the community diversity (compositional attributes of biodiversity) but also the functioning of ecosystems (functional attributes of biodiversity). Other ecosystems processes such as energy fluxes and material cycles are also affected by invasions.

AFTER EXTINCTION

Previous sections analysed scenarios for different attributes of biodiversity. They also presented examples in which changes in the drivers promoted changes in the various attributes of biodiversity. This section abandons the use of the three sets of attributes of biodiversity proposed by Noss and focuses on species diversity alone. It moves on one step and assumes that diversity reduction has effectively occurred. According to current estimations, the extinction rate is higher than in the recent past, and it is increasing. What will occur after extinction? Will extinction alter the evolutionary processes by which diversity is generated? Will any particular types of species dominate future ecosystems? Will biodiversity eventually recover? What will be the recovery rates? These are relevant questions, if we hope to develop proactive management strategies for the "new world." Sponsored by the National Academy of Science of the United States of America, a colloquium discussed these issues (*The Future of Evolution* [contributed papers can be downloaded free from the *Proceedings of the National Academy of Science* Internet site]). Experts in different fields evaluated the information recovered from, for example, the geological record. They also evaluated current knowledge on species requirements and species strategies in relation to environmental templates in new ecosystems. The direction of evolution is not predictable, but knowing what happened in the past may indicate how evolution might proceed (Myers and Knoll, 2001).

Various patterns of evolution might occur after extinctions, since major disruptions are taking place in very diverse biomes, such as tropical forests, grasslands, wetlands, coral reefs and estuaries. Several ideas summarize the current view of the future (Myers and Knoll, 2001). As extinction proceeds, an outburst of speciation may occur because of a large number of vacant niches and the fragmentation of populations. A few ecological types may possibly account for most of the speciation because of the predominance of human-dominated ecosystems throughout the biosphere. For example, it is possible that opportunistic and generalist species may proliferate. Disruption in particular ecosystems, such as tropical forests, may affect evolutionary patterns in other ecosystems. It has been proposed that tropical forests are the engines for new species, and their disappearance may hamper future evolution in neighbouring grasslands. As we evaluate future evolution, it is necessary to consider not only environmental changes but also biotic changes. For example, extinction will affect the "biodisparity" within the biota. In other words, extinction can preferentially affect species with particular morphology and physiology. In this sense, future patterns of evolution will develop from an incomplete biota. Because of anthropogenic effects it is probable that higher taxa and guilds are candidates for extinction, and for this reason, re-diversification might be limited.

Recovery rates (i.e. the velocity of rebuilding species diversity) after extinction are difficult to estimate from the geological record. These records are also not completely appropriate for this endeavour. Perturbations that caused extinction events were probably intense, but short-lived. In contrast, we are currently facing global changes that may alter the biosphere and the major drivers of extinction over longer time periods. Jablonski (2001) indicates that recovery would be slow in terms of a human time scale. However, many aspects of recovery, such as the spatial context (e.g. source–sink dynamics) and the reorganization of ecosystems after extinction, increase the complexity and decrease the predictability of the recovery process. More geographical analyses of the fossil record may shed light on the issue. Erwin (2001) found no clear relationship between the magnitude of extinction and the rate of recovery. Theoretical modelling confirmed that empty niches should refill with new species rapidly, but empirical studies showed complex dynamics. Erwin submits that, since extinction may disrupt ecosystem structure and function, it is necessary to accept that recovery will involve the concurrent redevelopment of species diversity and ecosystem processes. In this sense, some studies reveal a time lag in recovery, as well as the reappearance of groups that disappeared again subsequently in the recovery process.

Other evidence indicates that rates of diversification after extinction may not be slow. Cowling and Pressey (2001) present the case of the Cape Floristic Region in South Africa, where species diversification occurred rapidly after climate change in the late Pliocene. During that time, the climate became Mediterranean, and fires became frequent. In the Cape Floristic Region, diversification did not occur randomly among ecological groups. Cowling and Pressey (2001) indicate that, for example, species from the woody groups are "low [stature], fire killed (i.e. non-sprouting) shrubs with poorly dispersed seeds, small and short-lived seed banks, and insect-pollinated flowers." According to them, all these characteristics lead to rapid generation turnover, which promotes diversification. A short-lived seed bank favours non-overlapping generations and promotes the expression of novel characteristics in each generation. Fire induces population fragmentation and local population extinction, and short dispersal distances induce isolation and diversification of populations in different habitats.

Tilman and Lehman (2001) discuss the particular case of plant species extinction and the subsequent diversification in the grasslands biome. They use a set of well-established theoretical ideas to address two issues. First, what will be the expected changes in the factors that constrain plant population growth, and what will be the consequences of the predicted changes? Second, assuming a certain type of new environment, what will be the probable types or guilds of plant species evolving from the remaining guilds? Plants respond to major constraints by adjusting allocation of resources to roots, stems, leaves and seeds (recruitment of new individuals). Most grasslands or woody–grass ecosystems are strongly constrained by soil resources (e.g. nitrogen, water). The plant types adapted to these constraints allocate an important amount of growth to roots. The consequence is that these species guilds are poor competitors for light, or poor recruiters or dispersers, or a combination. As discussed above, two of the major environmental changes are an increase in nitrogen availability and environmental fragmentation. In grasslands, most species are not suited to successfully compete in this new environment. They either lack the ability to compete for light or lack dispersal capacity to overcome human-created barriers or fragmentation. Few species will succeed in this environment. These species are usually classified as weedy because of their ability to respond rapidly to resource availability, their poor competitive ability, and their good dispersal ability (Tilman and Lehman, 2001).

To address the second issue, Tilman and Lehman (2001) use a simple model that considers a trade-off between competitive ability and dispersal ability. They assume that after extinction a single type of a plant species remains. This type, as already discussed, is a good disperser but a poor competitor. Their model projects that, after enough generations, two types of species will dominate: one will be a very poor competitor but a good disperser, whereas the other will be a good competitor, but a poor disperser. After a very long time (0.5 million years) the model predicts the evolution of more than 20 species types, scattered along a competition-dispersal gradient.

The model does not consider other factors that can alter these results, such as periodic invasions of species that can displace the species evolving *in situ*. As was reviewed above, biotic exchange would be a predominant process in the future. There is evidence that, in the past, invasive species altered the evolution pathway of native species (Mooney and Cleland, 2001). Alteration of evolution derives from many possible processes, including competitive exclusion, niche displacement, hybridization, introgression, predation and extinction itself. It is also necessary to say that invaders contribute to biodiversity and also evolve in response to the physical and biotic environment.

It is clear that estimating post-extinction patterns and rates of evolution of species diversity (compositional attribute of biodiversity) is a hard task and our predictive capabilities are still poor. Trying to include in the picture the other attributes of biodiversity (i.e. structural and functional) is a remaining task. A complementary strategy would be to monitor changes in the three attribute sets as a way to gain an understanding of the entire process. As several participants in the colloquium agreed, a proactive response to the extinction process is necessary (Cowling and Pressey, 2001; Novacek and Cleland, 2001; Woodruff, 2001; Ehrlich, 2001; Bazzaz, 2001). Monitoring and adaptive management are complementary tasks. Adaptive management is not an easy task due to the many issues at stake (social, economic, cultural and technical), and also due to our currently poor forecasting power (Clark *et al.*, 2001) and because of the science-policy gap (Bradshaw and Borchers, 2000). These topics will be discussed in the next section.

MANAGEMENT OF BIOCOMPLEXITY—THE NEXT FRONTIER

Human pressure on the biosphere will increase because of both population growth and increasing per capita consumption. This pressure will not only be intense but also widespread in grasslands, as crop and animal husbandry occurs mostly in this biome. As discussed above, any project oriented to managing the biodiversity transition must include, among other drivers of global change, human-driven land-use change as a major component of the scenario (Meyer and Turner, 1992). Since the social system affects and is affected by ecosystem structure and functioning, there is increasing interest in studying the interactions between social and natural systems (Redman, 1999; Chapin and Whiteman, 1998; Bradshaw and Bekoff, 2001).

Social systems and ecological systems have interacted at least since the beginning of human existence. In this sense, the study of ancient human impact on ecosystems could be used as the prologue to the design of sound management strategies. Redman (1999) integrates archaeological and historical records in a model of the interactions between natural and social systems. He identifies three different phases in human systems: expansion, intensification and abandonment. In this conceptual model, human society is initially constrained by factors from the social system and the natural systems. Humans promote changes in land use and land cover, acting on landscape organization and biological and biogeochemical processes. As a consequence, human society is released from the constraints and initiates a phase of human expansion. Paralleling changes in the natural system, the human system also becomes more complex as new social groups develop (e.g. administrators, religious organizations and military). More resources are needed to support the expansion and the new functions in the society. There is also the need to accumulate reserves to meet society needs in bad years. All these changes result in intensification of natural resource exploitation. Pressures on ecosystems move them to a structure and function beyond their resilience, leading to the next phase: collapse and abandonment. One important feature of this model is that the social group that makes the decisions about natural resource management is not the same group that is involved in working the land, i.e. decision-makers have no direct contact with natural resources.

Based on the present review, grasslands, as socio-ecological systems, should be observed from three different perspectives: biophysical, productive-technical, and socio-economic (Figure 11.2).

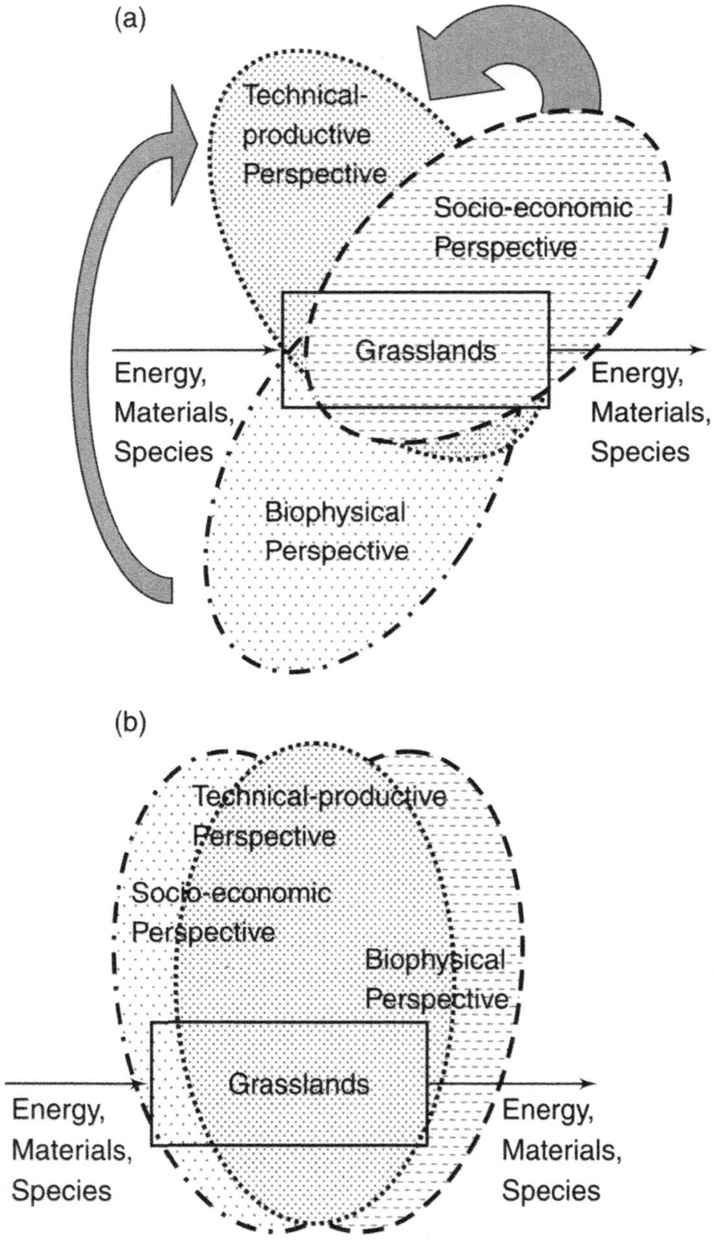

Figure 11.2 Schematic representation of the three perspectives from which grasslands can be analysed.
(a) Currently the biophysical perspective has little influence on the technical-productive and socio-economic perspectives, being the least important determinant of the way that grasslands are utilized.
(b) Represents a complementary evaluation of grassland management.

The biophysical perspective is the one with which most ecologists are familiar. Biodiversity is studied to understand its relationship with the physical and biotic environment. The productive-technical perspective refers to all the management, technology and tools developed to modify grassland structure and function. It includes, for example, the selection of species genotypes, control of natural flora or fauna, and modification of soil conditions through tillage, fertilization or irrigation. The socio-economic perspective includes the social, political, economic, cultural and religious or spiritual aspects of grasslands. These three perspectives should be complementary, although not always true in practice. The biophysical perspective, in theory, may influence the productive-technical perspective, since the biological and physical components constrain what can be done and achieved. In practice, productive-technical decisions are mostly associated with the socio-economic perspective. For example, commercial strategy is a rather large determinant in management decisions. On many occasions, it goes against what the biophysical perspective indicates (sometimes even against what the technical-productive perspective indicates). For example, the development of a new herbicide is an effective weed control, but in time (i.e. several generations of the weed species or several cultivation cycles) the weed community changes its composition, or the weed population develops resistance, thus reducing the herbicide's value for crop production. Delaying the appearance of herbicide resistance includes the cost associated with incomplete or imperfect weed control, but it results in a long-term benefit. Incomplete weed control permits survival of susceptible genotypes and delays the dominance of resistance in the weed population. Another example, at a global scale, is the failure to fully implement the Kyoto protocol on CO_2 emissions. Here, policy decisions may hamper efforts to adjust technical-productive procedures that reflect the biophysical perspective (Durand, 2002).

Conserving and managing biodiversity both need the integration of all three perspectives in a complementary way (Figure 11.2b). This is not an easy task and requires a new framework, rooted in the understanding that a collaboration of social and biophysical sciences is necessary (Bradshaw and Bekoff, 2001). That framework also needs to incorporate the idea that socio-ecological systems are complex systems (Anonymous, 1999). A new area of interest is the study of biocomplexity. According to the US National Science Foundation,

> "biocomplexity refers to phenomena that arise as a result of dynamic interactions that occur within biological systems, including humans, and between these systems and the physical environment. From individual cells to ecosystems, these systems exhibit properties that depend not only on the individual actions of their components, but also on the interactions among these components and between these components and the environment."

The complexity of these systems derives not necessarily from them having many components. On the contrary, the main reason is that their behaviour cannot be predicted from the knowledge of their parts and feedback among the system components (Zimmer, 1999). Complexity derives from different system properties, such as variability in their components, context-dependent processes, non-linear processes, time-delay responses, and chaotic behaviour. Both social and ecological systems have these characteristics. For example, communities and populations are composed of different species and genotypes, respectively. Usually, biological responses have short time lags, if we consider morphological, physiological or behavioural responses. However, time lags become long if we consider numerical responses, such as birth and mortality rates. Also, many population and ecosystem processes are non-linear.

The study of socio-ecological systems requires a new approach that includes not only the study of the parts but also the study of the whole. It also requires a major collaborative effort of social and biophysical sciences. Complexity theory can bridge the gap between sciences, since both social and biophysical systems are complex (Anonymous, 1999; Bradshaw and Bekoff, 2001). It also requires that biophysical sciences are considered as part of society and therefore objects of study themselves.

Another issue that hampers the integration of the three perspectives of grasslands systems is that the power of prediction of biophysical models decreases as spatial and temporal scales increase from the local paddock to the global, and from days to decades, respectively. Uncertainty in forecasts is at the root of the science-policy gap. Uncertainty and complexity are part of the normal lives of scientists, whereas "society and policy-makers seek certainty and deterministic solutions" (Bradshaw and Borchers, 2000). In this scenario of uncertainty, it is difficult to apply a standard decision-making process in social-ecological systems (Walker *et al.*, 2002). An alternative approach is a resilience analysis and management that seeks to define a desirable configuration of systems that reinforces system resilience, which is the capacity to cope with future shocks. Resilience results from three characteristics: the amount of change that the system can undergo without altering its configuration; the capability of self-organization; and the capacity to learn and adapt. In this framework, system stakeholders are key players in identifying both threats to the system and management practices that maintain and promote resilience. Their experiences are used to define thresholds in system attributes (i.e. desirable structure and functions, such as debt–income ratio). For example, perception of the desertification process by stakeholders may be very vague, but income reduction is readily perceived. Connecting with some key features of an ecosystem, e.g. woodiness or presence of weed seeds in wool, may be easier than comprehending any indicator of biodiversity, such as the ones presented here. Participation of stakeholders in the analysis fosters a change in perception of uncertainty from ignorance to knowledge. This is a key change in order to reduce the science-policy gap.

CONCLUSIONS

Future scenarios of biodiversity indicate that grasslands, as with other biomes, will suffer from biodiversity loss or erosion. The main causes of biodiversity losses are global changes; however, land use (a regional-scale phenomenon) might have the biggest impact. As we move into this new scenario with diminished biodiversity, it is useful to address three aspects of the biodiversity erosion. First, "What are the main trends in the biodiversity of grasslands, and what are the effects of the main drivers of biodiversity loss?" Second, "What are the changes expected in the biodiversity of grasslands after extinctions have occurred." Third, "How can we manage the biodiversity of grasslands in the transition to the new scenario?"

Biodiversity includes many levels of organization, and many temporal and spatial scales. It is a useful and necessary concept to address the main characteristics of ecosystems. In time, this concept has become vague, as scientists commonly use "biodiversity" as a synonym for species richness. Noss (1990) proposed a framework that allows us to monitor and manage biodiversity. This framework for biodiversity has three sets of attributes: compositional, structural and functional. Adopting this three-attribute-set framework allows us to address the many changes that are occurring at different spatio-temporal scales and in biological hierarchies. It also ensures that we include the spatial structure of ecological systems in our analyses. Our current understanding points to the fact that spatial structure is a key determinant of many processes, across populations, communities and ecosystems. Many examples show that land cover and land use changes are eroding all sets of attributes, not just the compositional.

It is necessary to begin to understand the future patterns of evolution and change of ecological systems as many species disappear, landscapes becomes fragmented, communities are invaded, and nutrient cycles are modified. A simulation exercise, based on the compositional attributes alone, allows a first approximating of the evolutionary paths. In the case of plants, it seems that weedy species will predominate. However, in time it is possible that new species will evolve from this group.

Since grasslands will continue to be heavily used ecosystems, it is necessary to implement new, sound management practices to address biodiversity. Humankind is a component of eco-social systems, and therefore management needs to be based on a complementary use of biophysical, technical and socio-economic perspectives. In order to address a complementary view of grasslands, more interactions between social and ecological scientists are necessary. Complex system theory, and the biocomplexity framework in particular, provides a template for studying these interactions. Accepting that uncertainty is a key attribute of both social and ecological systems is the next frontier in studies of biocomplexity and biodiversity.

ACKNOWLEDGEMENTS

This review was funded by the Food and Agriculture Organization of the United Nations (FAO); Agencia Nacional de Promoción Científica y Tecnológica of Argentina (PICT99 01-6641); University of Buenos Aires; and Fundación Antorchas. Martín Oesterheld and Howard Epstein improved this work with insightful comments.

REFERENCES

Adler, P., Raff, D. & Lauenroth, W.K. 2001. The effect of grazing on the spatial heterogeneity of vegetation. *Oecologia*, **128**: 465–479.

Aerts, R. & Berendse, F. 1988. The effect of increased nutrient availability on vegetation dynamics in wet heathlands. *Vegetatio*, **76**: 63–69.

Aguiar, M.R., Paruelo, J.M., Sala, O.E. & Lauenroth, W.K. 1996. Ecosystem responses to changes in plant functional types: An example from the Patagonian steppe. *Journal of Vegetation Science*, b: 381–390.

Aguiar, M.R. & Sala, O.E. 1999. Patch structure, dynamics and implications for the functioning of arid ecosystems. *Trends in Ecology and Evolution*, **14**: 273–277.

Anonymous. 1999. Complex systems. *Science*, **284**: 79–109.

Bazzaz, F. 2001. Plant biology in the future. *Proceedings of the National Academy of Science (USA)*, **98**: 5441–5405.

Bernardos, J.N., Viglizzo, E:F., Jouvet, V., Lértora, F.A., Pordomino, S.J. & Cid, F.D. 2001. The use of EPIC model to study the agroecological change during 93 years of farming transformation in the Argentine pampas. *Agricultural Systems*, **69**: 215–234.

Boeken, B. & Shachak, M. 1994. Desert plant communities in human made patches – implications for management. *Ecological Applications*, **4**: 702–716.

Bradshaw, G.M. & Borchers, J. 2000. Narrowing the science-policy gap: Uncertainty as information. *Conservation Ecology*, **4**. (see http://www.cosecol.org/vol4/iss1/art7).

Bradshaw, G. M. & Bekoff, M. 2001. Ecology and social responsibility: the re-embodiment of science. *Trends in Ecology and Evolution*, **16**: 460–465.

Briggs, J.M., Hoch, G. & Johnson, L. 2002. Assessing the rate, mechanisms, and consequences of the conversion of Tallgrass Prairie to *Juniperus virginiana* forest. *Ecosystems*, **5**: 578–586.

Burke, I.C., Elliott, E. & Cole, C.V. 1995. Influence of microclimate, landscape position, and management on soil organic matter in agroecosystems. *Ecological Applications*, **5**: 124–131.

Chaneton, E., Perelman, S., Omacini, M. & León, R. 2002. Grazing, environmental heterogeneity, and alien plant invasions in temperate Pampa grasslands. *Biological Invasions*, **4**: 24–24.

Chapin, F.S. & Whiteman, G. 1998. Sustainable development of the boreal forest: interaction of ecological, social, and business feedbacks. *Conservation Ecology [online]*, **2**(2): 12. (see http://www.consecol.org/vol2/iss2/art12).

Chapin III, F.S., Zaveleta, E.S., Eviner, V.T., Naylor, R.L., Vitousek, P.M., Lavorel, S., Reynolds, H.L., Hooper, D.U., Sala, O.E., Hobbie, S.E., Mack, M.C. & Díaz, S. 2000. Consequences of changing biotic diversity. *Nature*, **405**: 234–242.

Clark, J.S., Carpenter, S.R., Barber, M., Collins, S., Dobson, A., Foley, J., Lodge, D., Pascual, M., Pielke, R., Pizer, W., Pringle, C., Reid, W.V., Rose, K.A., Sala, O.E., Schlesinger, W., Wall, D.H. & Wear, D. 2001. Ecological forecasts: an emerging imperative. *Science*, **293**: 657–660.

Cornet, A., Dehlhoume, J. & Montaña, C. 1988. Dynamics of striped vegetation patterns and water balance in the Chihuahuan Desert. pp. 221–231, *in:* H. During, M. Werger and J. Willens (eds). *Diversity and pattern in plant communities*. The Hague: SPB Academic Publishing.

Cowling, R. & Pressey, R. 2001. Rapid diversification: Planning for an evolutionary future. *Proceedings of the National Academy of Science (USA)*, **98**: 5452–5457.

D'Antonio, C. and Vitousek, P. 1992. Biological invasions by exotic grasses, the grass/fire cycle, and global change. *Annual Review of Ecology and Systematics*, **23**: 63–87.

Durand, F. 2002. Ceguera ante una amenaza mortal. [Blindness against a mortal threat]. *Le Monde diplomatique*, **42**: 28–29.

Dykterhuis, E.J. 1949. Condition and management of range land based on quantitative ecology. *Journal of Range Management*, **2**: 104–115.

Ehrlich, P. 2001. Intervening in evolution: Ethics and actions. *Proceedings of the National Academy of Science (USA)*, **98**: 5477–5480.

Erwin, D. 2001. Lessons from the past: Biotic recoveries from mass extinctions. *Proceedings of the National Academy of Science (USA)*, **98**: 5399–5402.

Ghersa, C.M. & León, R.J.C. 1999. Successional changes in agro-ecosystems of the rolling pampa. pp. 487–502, *in:* L.R. Walker (ed). *Ecosystems of disturbed ground*. (Ecosystems of the World, vol. 16). Amsterdam, The Netherlands: Elsevier.

Ghersa, C.M., de la Fuente, E., Suárez, S. & León, R.J.C. 2002. Woody species invasion in the Rolling Pampa grasslands, Argentina. *Agriculture, Ecosystems and Environment*, **88**: 271–278.

Gutiérrez, J., Meserve, P., Contreras, L., Vasquez, H. & Jaksic, F. 1993. Spatial distribution of soil nutrients and ephemeral plants underneath and outside the canopy of *Porleria chilensis* shrubs (Zygophyllaceae) in arid coastal Chile. *Oecologia*, **95**: 347–352.

Hall, A.J., Rebella, C.M., Ghersa, C.M. & Culot, P.H. 1992. Crop systems of the pampas. pp. 413–449, *in:* C.J. Pearson (ed). *Field crop ecosystems*. (Ecosystems of the World, vol. 18.) Amsterdam, The Netherlands: Elsevier.

Huston, M. 1994. *Biological diversity: The coexistence of species on changing landscapes*. Cambridge UK: Cambridge University Press. 681p.

Hubbell, S.P. 2001. The unified neutral theory of biodiversity and biogeography. Princeton, NJ: Princeton University Press. 375p.

Hutchinson, G. 1959. Homage to Santa Rosalia; or, why are there so many kinds of animals. *American Naturalist*, **93**: 145–159.

Jablonski, D. 2001. Lessons from the past: Evolutionary impacts of mass extinction. *Proceedings of the National Academy of Science (USA)*, **98**: 5393–5398.

Jackson, R.B., Sala, O.E., Field, C.S. & Mooney, H.A. 1994. CO_2 alters water use, carbon gain, and yield for the dominant species in a natural grassland. *Oecologia*, **98**: 257–262.

Jobbagy, E. & Jackson, R. 2003. Patterns and mechanism in soil acidification in the conversion of grasslands to forest. *Biogeochemistry*, **54**: 205–229

Lauenroth, W.K. & Coffin, D.P. 1992. Belowground processes and the recovery of semiarid grasslands from disturbance. *In:* M.K. Wali (ed). *Ecosystem Rehabilitation*, Vol. 2. The Hague: SPB Academic Publishing.

Lawton, J.H. 1995. Population dynamic principles. pp. 147–163, *in:* J.H. Lawton and R.M. May (eds). *Extinction rates*. Oxford, UK: Oxford University Press.

Ludwig, J. & Tongway, D. 1995. Spatial organization of landscape and its function in semiarid woodland, Australia. *Landscape Ecology*, **10**: 51–63.

MacArthur, R. & Wilson, E. 1967. *The theory of island biogeography*. Princeton, NJ: Princeton University Press.

Mack, R. 1981. Invasion of *Bromus tectorum* L. into western North America: an ecological chronicle. *Agroecosystems*, **7**: 145–165.

Mazia, C.N., Chaneton, E.J., Ghersa, C.M. & León, R.J.C. 2001. Limits to tree species invasion in pampean grassland and forest plant communities. *Oecologia*, **128**: 594–602.

Melgoza, G., Nowak, R.S., & Tausch, R.J. 1990). Soil water exploitation after fire: Competition between *Bromus tectorum* (cheatgrass) and two native species. *Oecologia*, **83**: 7–13.

Meyer, W. & Turner, B.L. 1992. Human population growth and global land-use/cover change. *Annual Review of Ecology and Systematics*, **23**: 39–61.

Milchunas, D.G. & Lauenroth, W.K. 1993. Quantitative effects of grazing on vegetation and soils over a global range of environments. *Ecological Monographs*, **63**: 327–366.

Montaña, C, López, J. & Mauchamp, S. 1990. The response of two woody species to the condition created by a shifting ecotone in an arid ecosystem. *Journal of Ecology*, **78**: 789–798.

Mooney, H. & Hobbs, R. (eds). 2000. *Invasive species in a changing world*. Washington, DC: Island Press.

Mooney, H. & Cleland, E. 2001. The evolutionary impact invasive species. *Proceedings of the National Academy of Science (USA)*, **98**: 5446–5451.

Myers, N. & Knoll, A. 2001. The biotic crisis and the future of evolution. *Proceedings of the National Academy of Science (USA)*, **98**: 5389–5392.

Noss, R.F. 1990. Indicators for monitoring biodiversity: a hierarchical approach. *Conservation Biology*, **4**: 355–364.

Novacek, M. & Cleland, E. 2001. The current biodiversity extinction event: Scenarios for mitigation and recovery. *Proceedings of the National Academy of Science (USA)*, **98**: 5366–5470.

Noy-Meir, I. 1981. Spatial effects in modeling of arid ecosystems. pp. 411–432, *in:* D. Goodall and R. Perry (eds). *Arid ecosystems: Structure, functioning, and management*. Vol. 2. Cambridge, UK: Cambridge University Press.

Ong, C.K., Corlet, J.E., Singh, R.P. & Black, C.R. 1991. Above- and-below ground interactions in agroforestry systems. *Forest Ecology and Management*, **45**: 45–56.

Paruelo, J.M. 2002. Temperate grasslands. *In:* J. Canadel and H. Mooney (eds). *The earth system: Biological and ecological dimensions of global environmental change*. (Encyclopaedia of Global Environmental Change, Vol. 2). New York, NY: John Wiley.

Potvin, C. & Vasseur, L. 1997. Long-term CO_2 enrichment of a pasture community: species richness, dominance, and succession. *Ecology*, **78**: 666–677.

Purvis, A. & Hector, A. 2000. Getting the measure of biodiversity. *Nature*, **405**: 212–219.

Redman, C. 1999. *Human impact on ancient environments*. Tucson AZ: University of Arizona Press. 239p.

Rosenzweig, M. 1995. *Species diversity in space and time*. Cambridge, UK: Cambridge University Press.

Rusch, G. & Oesterheld, M. 1997. Relationship between productivity, and species and functional group diversity in grazed and non-grazed pampas grasslands. *Oikos*, **78**: 519–526.

SAGPYA [Secretaría de Agricultura, Ganadería, Pesca y Alimentación]. 1999. Ochos años de implementación de plantacionaes forestales (RPPF) [Eight years of implementation of the forestry promotion programme]. *SAGPYA Forestal*, **13**: 11–13.

Sala, O.E. 1988. The effect of herbivory on vegetation structure. pp. 317–330, *in:* M.J.A. Werger, P.J.M. van der Aart, H.J. During and J.T.A. Verhoeven (eds). *Plant form and vegetation structure*. The Hague: SPB Academic Publishing.

Sala, O.E., Chapin III, F.S., Armesto, J.J., Berlow, E., Bloomfield, J., Dirzo, R., Huber-Sanwald, E., Huenneke, L.F., Jackson, R., Kinzig, A., Leemans, R., Lodge, D., Mooney, H.A., Oesterheld, M., Poff, N.L., Sykes M.T., Walker, B.H., Walker, M. & Wall, D.H. 2000. Global biodiversity scenarios for the year 2100. *Science*, **287**: 1770–1776.

Schemske, D., Husband, B., Ruckelshaus, M., Goodwille, C., Parker, I. & Bishop, J. 1994. Evaluating approaches to the conservation of rare and endangered plants. *Ecology*, **75**: 584–606.

Senft, R.L., Coughenour, M.B., Bailey, D., Rittenhouse, L., Sala, O. & Swift, D. 1987. Large herbivore foraging and ecological hierarchies. *BioScience*, **37**: 789–799.

Silvertown, J. 1980. The dynamics of a grassland ecosystem: botanical equilibrium in the Park Grass Experiment. *Journal of Applied Ecology*, **17**: 491–504.

Stohlgren, T.J., Binkley, D., Chong, G., Kalkhan, M.A., Schell, L.D., Bull, K.A., Otsuki, Y., Newman, G., Bashkin, M. & Son, Y. .1999. Exotic plant species invade hot spots of native plant diversity. *Ecological Monographs*, **69**: 25–46.

Tilman, D. 1987. Secondary succession and pattern of plant dominance along experimental nitrogen gradients. *Ecological Monographs*, **57**: 1898–214.

Tilman, D. & Pacala, S. 1993. The maintenance of species richness in plant communities. pp. 13–25, *in:* R. Ricklefs and D. Schluter (eds). Species diversity in ecological communities. Chicago IL: University of Chicago Press.

Tilman, D. & Lehman, C. 2001. Human-caused environmental change: impacts on plant diversity and evolution. *Proceedings of the National Academy of Sciences (USA)*, **98**: 5433–5440.

Turner, M., Gardner, R. & O'Neill, R. 2001. *Landscape ecology in theory and practice*. New York NY: Springer Verlag.

Viglizzo, E.F., Lértora, F.A., Pordomino, S.J., Bernardos, J.N., Roberto, Z.E. & Del Valle, H. 2001. Ecological lessons and applications from one century of low external-input farming in the pampas of Argentina. *Agriculture, Ecosystems and Environment*, **83**: 65–81.

Vitousek, P., Mooney, H., Lubchenko, J. & Melillo, J.M. 1997a. Human domination of Earth's ecosystems. *Science*, **277**: 494–499.

Vitousek, P.M., Aber, J.D., Howarth, R.W., Likens, G.E., Matson, P.A., Schindler, D.W., Schlesinger, W.H. & Tilman, D. 1997b. Human alteration of the global nitrogen cycle: sources and consequences. *Ecological Applications*, 7: 737–750.

Walker, B., Carpenter, S., Anderis, J., Abril, N., Cumming, G., Janssen, M., Lebel, L., Norberg, J., Oeterson, G. & Pritchard, R. 2002. Resilience management in social-ecological systems: a working hypothesis for participatory approach. *Conservation Ecology*, 6: 14. [online] (See: http://www.concecol.org/vol6/iss1/art14)

Walter, H. 1984. *Vegetation of the Earth*. 3rd edition. Berlin: Springer Verlag.

Wand, S., Midgley, G. & Jones, M. 1999. Responses of wild C4 and C3 grass (Poaceae) species to elevated atmospheric CO_2 concentration: a Meta-analytic test of current theories and perspective. *Global Change Biology*, **5**: 723–741.

Watt, A. 1947. Pattern and process in plant community. *Journal of Ecology*, **35**: 1–22.

West, N.E. 1979. Basic synecological relationships of sagebrush-dominated lands in the Great Basin and the Colorado Plateau. pp. 33–41, *in* Anon. *The Sagebrush Ecosystem: A Symposium*. Utah State University, College of Natural Resources, Logan, Utah.

Westoby, M., Walker, B. & Noy-Meir, I. 1989. Opportunistic management for rangelands not at equilibrium. *Journal of Range Management*, **42**: 266–274.

Woodruff, D. 2001. Declines of biomes and biotas and the future of evolution. *Proceedings of the National Academy of Science (USA)*, **98**: 5477–5480.

Zimmer, C. 1999. Life after chaos. *Science*, **284**: 83–86.

CHAPTER 12

Biotechnological Approaches to Temperate Forage Improvement

C.J. Pollock[1], M.O. Humphreys, P. Morris, M.T Abberton and M.W Humphreys

ABSTRACT

This review considers the current status and future opportunities for biotechnology in the improvement of forages. It concentrates on temperate forages, although potential applications extend across the full range of livestock systems. The review summarizes progress in genome analysis, mapping, marker-assisted selection, introgression and transformation in both grasses and legumes. In general, there has been substantial "proof of principle" for all these approaches, and novel varieties are in development. The application of these approaches has coincided with increased knowledge of the factors affecting forage quality, and the expectation is that the focus of forage improvement will shift in concert with the implementation of biotechnology.

INTRODUCTION

The scientific improvement of forage species began later than that of the major arable crops, and still presents a unique set of challenges. The complexities of ruminant digestion mean that it is difficult to define, in a form that breeders can utilize, heritable traits that affect quality, palatability or intake characteristics. Additionally, as well as being grazed directly, forages can be cut and fed fresh, cut and dried, or cut and ensiled, and all these treatments affect the chemical composition and nutritional value of the product.

The challenges do not end there. Many forages are perennial, and improvement strategies need to balance increases in useable production with maintenance of persistency. Generally, the nutritional quality of vegetative tissues is higher than that of the flowering organs, so there is a

[1]All authors are with the Institute of Grassland and Environmental Research (IGER), Plas Gogerddan, Aberystwyth SY23 3EB, Wales, UK.

further need for breeders to balance reproductive performance in terms of seed production against the leafiness of the grazed or conserved sward. Finally, many forage species have breeding systems that themselves cause problems, such as apomixy or obligate outbreeding.

Despite these challenges, breeders have made significant advances in terms both of annual production and quality (Table 12.1). These improvements have been achieved largely through the selection of elite material out of populations, selective cross-breeding and incorporation of novel traits from exotic accessions. There has also been some breeding of intraspecific and intrageneric hybrids. Biotechnology has the potential to have significant impacts on forage breeding, but, because of the challenges outlined above, that potential is only just beginning to be exploited. This chapter outlines the approaches currently being introduced into our breeding programme and considers their potential and any limitations. Our targets are almost exclusively temperate forage grasses and legumes, concentrating on perennial ryegrass (*Lolium perenne*) and white clover (*Trifolium repens*). The majority of the examples cited will, therefore, be drawn from these species, but the approaches are generic and equivalent programmes exist in other countries concentrating on other forages, both temperate and tropical.

Table 12.1 Breeding advances in forage crops

Species	Average genetic gain			
	Annual yield (dry Matter)(1)	Digestibility(2)	Intake(2)	Crude protein(2)
Lolium perenne	3.8	1.0	14.7	6.5
Dactylis glomerata	3.3	2.4		
Trifolium repens	4.9			
Medicago sativa	4.1	3.3		2.7

Notes: (1) percent decade^{-1} (after Humphreys, 1999). (2) percent generation cycle^{-1} (after Casler, 2001).

MARKER-ASSISTED SELECTION

Marker-assisted selection in grasses

Of the emerging genetic technologies, marker-assisted selection (MAS) is the most likely to have an immediate impact on plant breeding. It relies on the detection of DNA variation among individuals, using a variety of techniques including restriction fragment length polymorphism (RFLP), random-amplified polymorphic DNA (RAPD), simple sequence repeats (SSRs – microsatellites), expressed sequence tags (ESTs) and amplified fragment length polymorphism (AFLP) (Henry, 2001).

Successful breeding using MAS depends on understanding the genetic architecture of relevant traits. Traditionally, major gene and polygenic variation have been analysed in different ways, but the techniques of Quantitative Trait Locus (QTL) analysis now allow a more integrated approach in dissecting complex traits and assessing gene effects. For example, genetic fingerprinting of *Hordeum spontaneum* has revealed genetic marker associations with ecogeographic factors and experimentally imposed stresses (Forster *et al.*, 2000). Examples of QTLs associated with valuable traits are increasing in a wide range of crops, including rice, wheat, maize, sorghum, barley, forage and turf grasses, brassicas, soybean, pea, alfalfa, beans, tomato, beet and potato (McCouch and Doerge, 1995; Mohan *et al.*, 1997; Duncan and Carrow, 1999). QTL analysis can be used to determine the location or locations of variation for a complex trait directly on the genetic map of a particular species. At the same time, the density of markers on these maps has increased dramatically, associated with the use of the DNA markers described above.

Modern conventional plant breeding is using both these new marker technologies and the increased understanding of the mechanisms and inheritance of complex traits. Terms such as "designer breeding" and "breeder's toolkits" have been used to describe new genotypes that contain genes of known function whose presence confer adaptations to specific stress conditions (Thomas, Humphreys and King, 2001). Heterozygous molecular markers that are tightly linked to traits of interest can be used to speed up selection in breeding programmes. For example, to improve abiotic stress tolerance, molecular markers linked to root traits and osmotic adjustment were identified in rice (Nguyen, Babu and Blum, 1997). In sorghum, molecular markers have been used to identify regions of the genome specifically associated with pre-flowering drought tolerance (Tuinstra *et al.*, 1996). Dormancy is a drought-avoidance mechanism. Many genes determine its expression and in a few cases these have been mapped to specific chromosome regions. Genes that are differentially expressed between dormant and non-dormant seeds have been isolated, although their identity and function remain obscure (Li and Foley, 1997). Other genes and QTLs linked to drought-tolerance traits are listed in Duncan and Carrow (1999).

Although research associated with the identification of QTLs is expanding rapidly, examples of where MAS methods have been applied in breeding programmes are still relatively limited. Resistance to abiotic and biotic stresses are the most popular targets for MAS so far. However, despite the promising potential of MAS, poor precision in locating QTLs, bias of individual gene effects and risks of identifying "false-positive" associations have generated disappointing responses (Kearsey and Farquhar, 1998). More work is required on the practical use of MAS in breeding programmes and the value of specific markers. MAS should be more efficient than conventional phenotypic selection when traits are difficult or expensive to measure. MAS is superior to phenotypic selection (if phenotypic selection is at all possible) if the cost ratio of obtaining phenotypic measurements compared to scoring marker loci is greater than 1 and the heritability of the traits is less than 30 percent (Xie and Xu, 1998).

Work is progressing on genetic mapping and QTL analysis in ryegrasses and fescues, based on strong international collaboration (Jones *et al.*, 2002). As a result of the International *Lolium* Genome Initiative (ILGI; Jones *et al.*, 2002) involving the Institute of Grassland and Environmental Research (IGER) and groups in Australia, Japan and France, a reference linkage map of perennial ryegrass has been produced. It comprises 240 loci covering 811 cM on seven linkage groups. The map contains 124 co-dominant markers, of which 109 are heterologous anchor RFLP probes from wheat, barley, oat and rice, allowing comparative relationships to be investigated between perennial ryegrass and other Poaceae. The linkage groups of perennial ryegrass are numbered to correspond to the homologous groups of the Triticeae cereals. The genetic maps of perennial ryegrass and the Triticeae cereals are highly conserved in terms of synteny and colinearity (Jones *et al.*, 2002). There is also general agreement in the syntenic relationships between perennial ryegrass, meadow fescue, oat and rice, and those between the Triticeae and these species.

An IGER linkage map of perennial ryegrass based on an F_2 mapping population was produced initially for the genetic analysis of water-soluble carbohydrate (WSC) accumulation (Figure 12.1). This map identified 7 linkage groups covering a total of 515 cM and was compared with the ILGI map produced from a first backcross (BC_1)-type population (Armstead *et al.*, 2002). The maps could be aligned using 38 common loci and a marker order for all mapped loci in either population was identified in an integrated map. Using this map, QTLs associated with WSC accumulation in the leaves and leaf sheaths of ryegrass have been identified (Figure 12.2; Humphreys and Turner, 2001; Turner *et al.*, 2001). Under the prevailing cultivation conditions in the UK, WSC is an

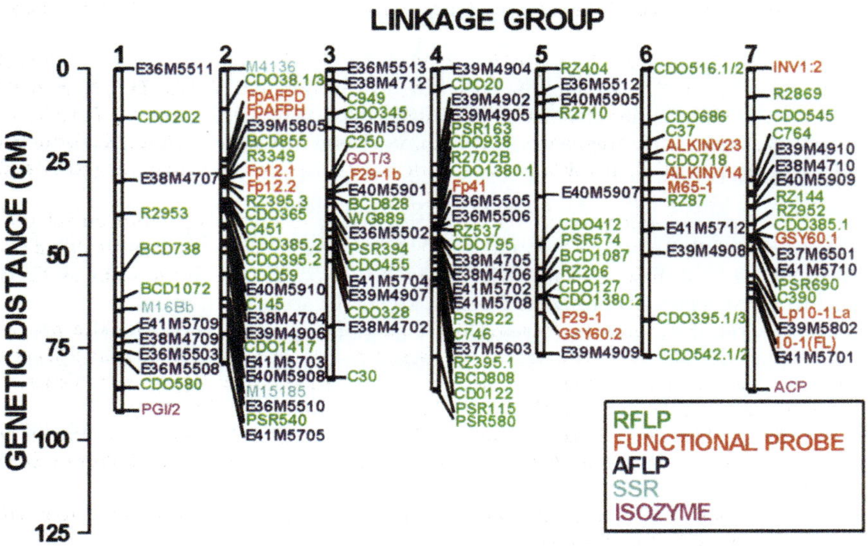

Figure 12.1 Linkage map of perennial ryegrass (*Lolium perenne*) derived from an F_2 mapping population.

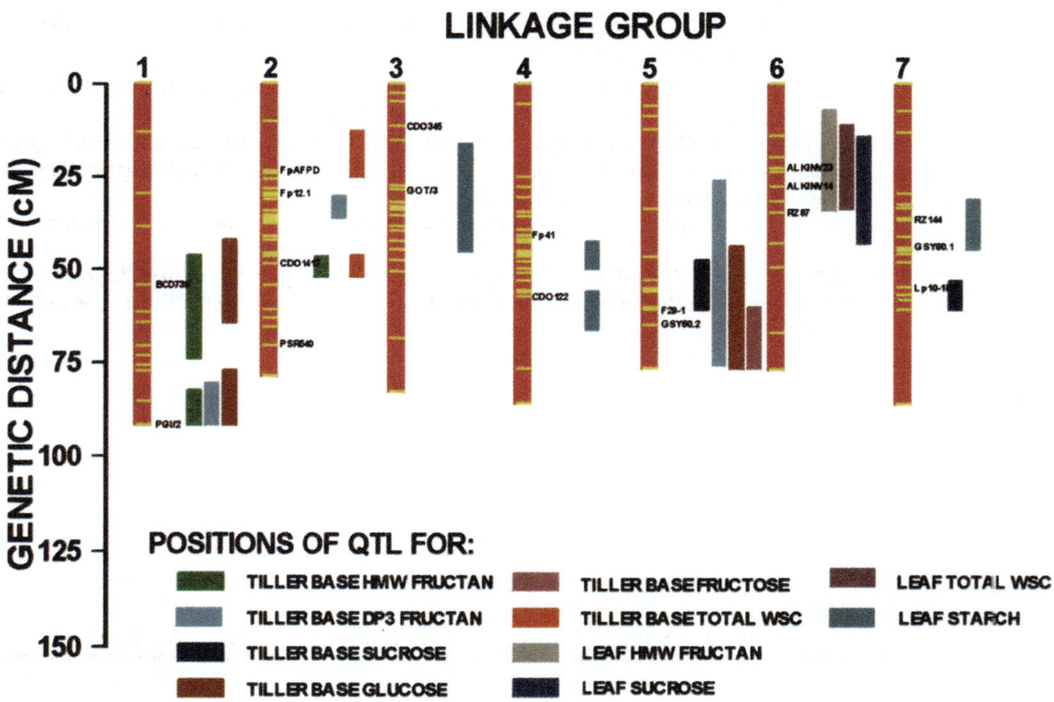

Figure 12.2 WSC NIMGRASS map of perennial ryegrass showing carbohydrate QTL.

important determinant both of forage quality in grazed herbage (where it promotes rumen metabolism and reduces the catabolism of amino acids) and of rapid acid accumulation during ensiling. The WSC F_2 mapping family has a 27 percent variation in total WSC of bulk herbage with a normal distribution.

Other traits identified during physiological studies and scored on the mapping family include regrowth (reserve-driven growth), constitutive leaf extension rate (photosynthate-driven growth), leaf and tiller base (mainly sheath) carbohydrate content and leaf chlorophyll content. Additionally plant size (as a measure of vigour) and heading date were scored. QTLs for these traits of interest were spread across the genome, although significant groups of QTLs occurred together in some regions.

Interestingly, some QTLs overlie or fall close to the location of genes with known function. A major tiller base sucrose QTL overlies the location of an alkaline invertase gene. A major heading date QTL lies close to a gene highly homologous to the *Arabidopsis* flowering time gene *CONSTANS* (*CO*) (Ian Donnison, pers. comm.). Allelic variation in the *CO-like gene* correlates with differences in heading date and may account for up to 25 percent of the total variation. Based on the WSC QTL information obtained so far, experimental selections have been made based on the QTL for tiller base high molecular weight fructans accumulation (Figure 12.3, Table 12.2). These are currently being evaluated in the field and glasshouse.

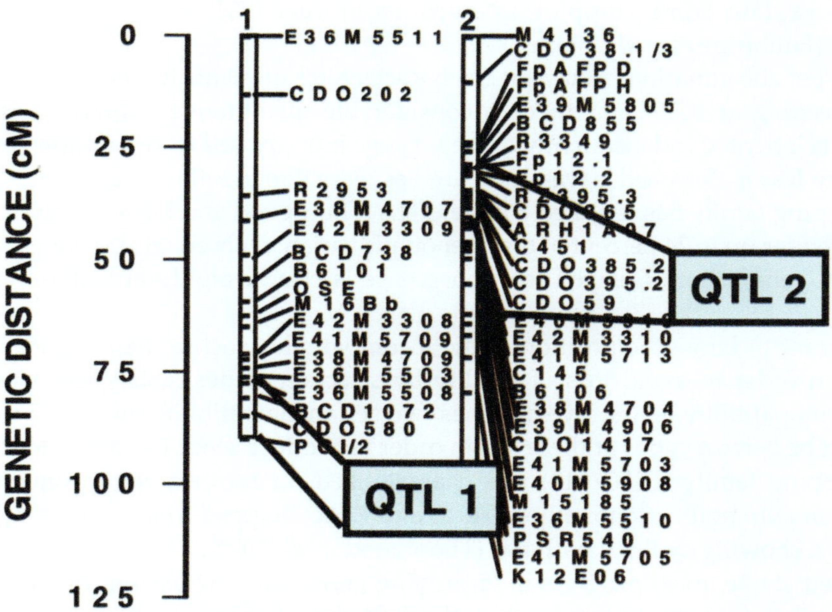

Figure 12.3 Perennial ryegrass linkage groups 1 and 2 showing regions flanked by markers used in selections for tiller base fructan accumulation

Table 12.2 Average trait expression of parent plants selected according to markers for tiller base fructan accumulation

Linkage group	QTL allele type	Fructan (percent)	Total WSC (percent)	Heading date (days after 1 April)
LG1	aa	21.9	27.0	46.4
LG1	bb	15.8	22.3	46.6
LG2	aa	22.1	27.8	46.7
LG2	bb	15.2	19.7	42.2

Resistances to pests and diseases have provided well-defined targets for improvement using MAS in a wide range of species, including forage crops and turf grasses (Humphreys *et al.*, 2001a). Examples of QTL mapping for pest and disease resistance include leaf rust (*Puccinia hordei*) in barley (Qi *et al.*, 1998), and barley leaf stripe (*Pyrenophora graminea*) (Pecchioni *et al.*, 1999). Identification of QTL and closely linked molecular markers for specific resistant genes can facilitate the pyramiding of resistance genes into breeding lines in order to increase durability against pathogen attack in cultivars. Pests and pathogens often overcome in a relatively short time resistance due to single genes. Durability of resistance can be increased by bringing together different resistance genes from diverse sources. Attempts to pyramid resistance genes using MAS have been reported in rice for bacterial blight resistance (Yoshimura *et al.*, 1995; Huang *et al.*, 1997). The pyramid lines in rice showed a wider spectrum and a higher level of resistance than lines with only a single gene. Using the ryegrass map in Figure 12.1 and associated mapping family plants, QTLs for crown rust resistance have been identified (Thorogood *et al.*, 1999). Some significant rust QTLs are associated with QTLs for flowering time and WSC accumulation, which may help to explain some complex interactions of rust QTL with temperature and plant development (Humphreys *et al.*, 2001a).

Exploiting genetic variation in stay-green characteristics of plants has been a major objective of turf grass breeding at IGER and there is considerable quantitative variation within perennial ryegrass that is inherited independently of the *sid* gene introgressed from meadow fescue. Analysis of leaf pigment loss in detached leaves, measured as reduction in reflected hue angle, in a perennial ryegrass mapping family has revealed 6 large QTLs (Thorogood and Laroche, 2001). Individually, these accounted for up to 24 percent of the phenotypic variation observed and were either dominant or additive in gene action. Delayed leaf senescence is clearly highly heritable in this mapping family.

It is important to know the position of the incompatibility loci in the ryegrass genome when using MAS, in order to avoid loss of fertility by selection of desirable genes closely linked to particular incompatibility genes. Ryegrass has two self-incompatibility loci (the S and Z loci) and varieties must be heterozygous at these loci in order to produce seed. The reference ILGI perennial ryegrass mapping family (P150/112) was characterized for the two self-incompatibility loci (S and Z) by an *in vitro* pollination method. The S and Z loci mapped to linkage groups (LGs) 1 and 2, respectively, showing synteny with rye (Thorogood *et al.*, 2002).

As indicated above, most progress in improving perennial ryegrass for forage has come from recurrent selection for plot performance within elite breeding populations. The value of using MAS to increase the efficiency of this process is being investigated. Mother plants of a breeding population were genotyped for 25 RFLP markers that had previously been mapped to single locations in the ryegrass genome, covering all 7 LGs. Markers at 8 loci were associated significantly with various aspects of the performance of their half-sib progeny when growing as plots. Traits associated with markers included leaf WSC, leafiness in mid-summer, yield at first silage cut, recovery after first silage cut, yield during vegetative growth, ground cover at the end of the second harvest year, and resistance to *Rhynchosporium* leaf scald (Wilkins *et al.*, 2001). Synthetic varieties constructed on the basis of marker information are being compared with those constructed on the usual basis of progeny means.

Marker-assisted selection in forage legumes

Introduction

The majority of work in this area has been carried out in alfalfa (*Medicago sativa*), primarily in the United States of America, but with some effort in Europe. In both alfalfa and other forage legumes, current effort is directed mainly towards mapping, with little directed MAS except under experimental conditions. As in other crops, genomic analysis will directly affect both mapping and MAS, and there are significant genomics efforts currently directed towards *Medicago truncatula* and *Lotus japonicus* in the United States of America, Japan and Europe. Major programmes on both these species are under way in Europe (partly through the International Legume Genome Initiative), in the United States of America (mainly on *Medicago*) and in Japan (predominantly on *Lotus*). Already there are initiatives being set up within the European Union to integrate breeding programmes for both grain and forage legumes with these large genomics studies, and this will lead to an acceleration of progress.

Alfalfa

Medicago sativa is the most important temperate forage legume. It is an outbreeder, tetraploid with ($2n = 4x = 32$) and with tetrasomic inheritance. Current cultivars are heterogeneous synthetics. In North America, all of these are derived from intercrossing between nine original accessions. Tetrasomic inheritance causes difficulties for breeders; accordingly there is considerable focus on haploid production, leading to development of cultivated alfalfa at the diploid level. From the genetic analysis viewpoint, accessions and breeding populations have been characterized in terms of genetic diversity by RAPDS and RFLPS, and linkage maps have been produced at the diploid level, based on different populations (Brummer, Bouton and Kochert, 1993; Echt *et al.*, 1993; Kiss *et al.*, 1993). Linkage and other studies are, however, hampered by inbreeding depression (cf white clover, below) and markers have been used to investigate heterosis, its link with performance and the approach to homozygosity (Osborn, Brouwer, and McCoy, 1997). Additionally, polymerase chain reaction (PCR)-based sequence-characterized amplified regions (SCAR) markers (Larsen *et al.*, 2001) and AFLPs (Obert, Skinner and Stuteville, 2000) have been studied in order to identify markers for disease resistance. More recently, SSR (microsatellite) markers have been used and trait mapping carried out on tetraploid alfalfa (Diwan *et al.*, 1997, 2000; Skinner, Loughin and Obert, 2000).

White clover

White clover (*Trifolium repens*) is an important perennial forage legume in Northern Europe, New Zealand and parts of Australia. The first white clover map has been developed using self-compatible inbred lines through a collaboration between IGER and The Plant Biotechnology Centre, Victoria (Abberton, Michaelson-Yeates and Marshall, 2000). This utilizes AFLPs and microsatellites. Previous work had used RAPDS to assess the extent of genetic diversity between inbred lines and compare this with the degree of heterosis in crosses between them (Joyce *et al.*, 1999). Targets for MAS using the white clover map include disease resistance (Humphreys *et al.*, 2001a) and agronomic, physiological and quality traits (Webb and Abberton, 1999; Webb, Abberton and Young, 2002). In parallel with the mapping studies, development of EST collections is being undertaken in a joint programme between the Plant Biotechnology Centre (Victoria, Australia) (PBC) and AgResearch (New Zealand) (Forster *et al.*, 2001).

Red clover

To date, there has been very little use of molecular markers in red clover improvement programmes (Taylor and Quesenberry, 1996). Isozymes and RAPDs have been used for genetic diversity studies (Kongkiatnam *et al.*, 1995), RAPD markers have been developed for disease resistance (Page *et al.*, 1997) and isozyme heterozygosity linked to morphological characteristics (Xie and Mosjidis, 2001). There have also been some studies of gene expression (Nelke *et al.*, 1999). The authors are not aware of a current red clover map in the public domain, although a number of groups in Russia, Japan and the UK are exchanging information with the intention of producing such a map.

Marker-assisted introgression

Marker-assisted introgression in grasses

At IGER, the main targets for introgression have been the introduction of biotic and abiotic stress tolerance from *Festuca* spp. into *Lolium multiflorum* or *L. perenne* (Humphreys *et al.*, 1998). Introgression into *Lolium* is necessary following the initial generation of the hybrids in order to ensure that the high nutritive quality of *Lolium* spp. is retained together with the improved tolerance derived from *Festuca*. Marker-assisted introgression allows a greater degree of precision in this process, whereby the desired *Festuca* genes that govern stress tolerance are targeted and possible associated deleterious gene combinations excluded.

A range of mapping families have been produced to facilitate marker-assisted introgression:

- Triploid hybrids between either *L. perenne* or *L. multiflorum* (4x) and *F. pratensis* (2x) are the starting points for transfer into *Lolium* (2x) of *Festuca* genes for stay-green (Thomas et al., 1994); enhanced winter hardiness via tolerance to either freezing or ice-encasement (Canter, 2001); and crown rust resistance (Roderick, Thorogood and Adomako, 2000). For example, introgression from *F. pratensis* into *L. perenne* with the aid of DNA markers and genomic *in situ* hybridization (GISH) has resulted in an improvement in high temperature rust resistance (Roderick, Thorogood and Adomako, 2000). The introgressed *L. perenne* selection has equivalent rust resistance to the non-recurrent *F. pratensis* parent.
- Pentaploid hybrids between *L. multiflorum* (4x) and *F. arundinacea* (6x) (Humphreys and Ghesquière, 1994) are the starting points for transfer into *Lolium* (2x) of *Festuca* genes for drought resistance and stable protein.
- Tetraploid hybrids between *L. multiflorum* (4x) and *F. glaucescens* (4x) (Ghesquière *et al.*, 1996) are the starting points for transfer into *Lolium* (2x) of *Festuca* genes for drought resistance, chromosome-pairing control and rhizomatous habit.
- Tetraploid hybrids between *L. multiflorum* (4x) and *F. mairei* (4x) (M.W. Humphreys, unpublished observations) are the starting points for transfer into *Lolium* (2x) of *Festuca* genes for drought resistance and heat tolerance.
- Androgenic lines derived from *Lolium–Festuca* (4x) hybrids (Humphreys *et al.*, 2001b) are also used to facilitate assembly of novel gene combinations and expression of traits otherwise difficult to recover by conventional breeding.

Having identified plants that express introgressed *Festuca*-derived traits, they are analysed using GISH. Plants that have just one *Festuca* chromosome segment and that express the trait are selected, ensuring that the genes controlling the trait are on this chromosome segment. The close relationship between *L. perenne* and *F. pratensis* has allowed work on introgression mapping and alignment of the genetic map with a physical map (Armstead *et al.*, 2001; King *et al.*, 2001).

Genomic *in situ* DNA hybridization can be used to determine the location of alien *Festuca* genes, and genetic markers ascribed to the targeted sequence (Humphreys *et al.*, 1998). *Lolium* lines carrying different *Festuca* genes that convey a range of adaptations to abiotic stresses are currently being developed alongside "breeder's toolkits" to aid their commercial exploitation. By introgressing from *Festuca* into *Lolium* different suites of genes concerned with stress resistance, the complex physiological mechanisms involved may be "dissected" and thus be better understood. This work is partly funded under a European Union programme where *Festuca* genes governing adaptations to a range of abiotic stresses encountered throughout Europe are transferred to *Lolium*. The most advanced introgression line is a *L. multiflorum* genotype with genes for drought resistance derived from *F. arundinacea* (Humphreys and Thomas, 1993; Humphreys and Pašakinskienė, 1996).

An introgression genotype with one *Festuca* chromosome segment has been used to allocate molecular markers specifically to that segment. Approximately 200 AFLP markers that would map to that segment were selected by screening the genotype alongside the *Lolium* and *Festuca* parents. When a marker "allele" that was present in the *Festuca* parent but absent in the *Lolium* parents was observed in the hybrid, it would map to the introgressed *Festuca* segment. At meiosis in the introgression genotype, the *Festuca* segment recombined with its *Lolium* homologous chromosome arm. Therefore, by crossing the introgression genotype to the recurrent *Lolium* parent, recombinant series of that one segment was generated (Figure 12.4).

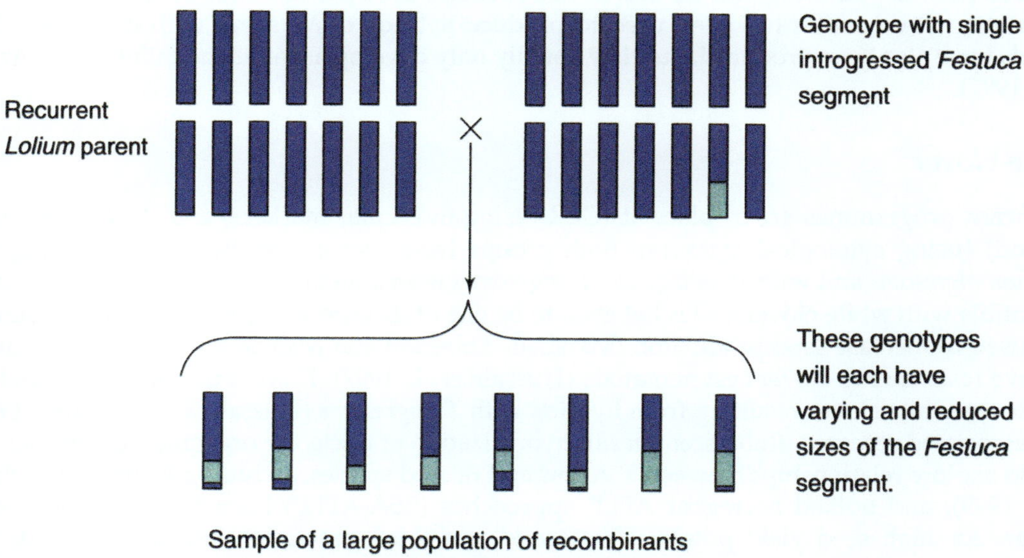

Figure 12.4 Generating a recombinant series of the introgressed *Festuca* segment.

Screening the backcross population for the presence or absence of each marker that was specific to the segment allowed the development of a localized genetic map of the segment. The authors are now using this approach to isolate markers to a single recessive allele responsible for the stay-green or non-senescence phenotype. This recessive mutant, designated *sid*, was identified in *F. pratensis* but transferred to *L. multiflorum* by backcrossing. A BC_1 genotype heterozygous for *sid* and with a single *F. pratensis* chromosome segment was identified by test crossing and by GISH. AFLPs and the *sid* locus are being mapped in the BC_2 population. A small but dense map of the *F. pratensis* segment is now being produced. This approach will be applied to *Festuca* chromosome segments shown to carry genes for drought resistance and winter survival as they are identified.

Marker-assisted introgression in forage legumes

Introduction

Introgression studies have been carried out on a range of forage legumes, with effort concentrated on alfalfa and white clover. Some of these studies are beginning to employ marker-assisted introgression using both molecular and cytological markers. Interspecific hybrids have been produced between red clover and some of its relatives, but with little impact.

Alfalfa

Progress in this species has been limited by polyploidy, inbreeding depression and lack of an easily characterized karyotype. A range of hybrids has, however, been generated, mainly by ovule or embryo culture (Osborn, Brouwer and McCoy, 1997). The most successful introgression has been with *Medicago dzhawakhetica* (2n = 4x = 32), aimed at the introduction of disease resistance and improved winter hardiness. This programme has been constrained by low fertility of the triploid hybrids, low levels of recombination between genomes, and lack of discrimination between chromosomes. In an attempt to address these issues, *M. dzhawakhetica* genome-specific markers have been developed using RAPDs and RFLPs (Osborn, Brouwer and McCoy, 1997).

Somatic hybridization has been used to produce hybrids of *M. sativa* with *M. falcata*. Other hybrids have also been produced, but they usually only develop as far as the callus stage (Arcioni *et al.*, 1997).

White clover

Significant programmes are in place at IGER (using molecular markers) and AgResearch (New Zealand) (using cytological markers). Both groups have focused on two introgressions, with *Trifolium nigrescens* and with *T. ambiguum*. *T. nigrescens* is an annual diploid species that is sexually compatible with white clover and is believed to be one of its ancestral genomes. Introgression has been used to increase seed production (Marshall, Abberton and Michaelson-Yeates, 1999) and to improve resistance to clover cyst nematode (Hussain *et al.*, 1997). There is evidence for unreduced gametes in white clover resulting from hybrids with *T. nigrescens* (Hussain and Williams, 1997).

In associated research, fluorescent *in situ* hybridization of nucleolar organizer regions has been used to explore relationships between *T. repens* and related species, including *T. nigrescens* (Ansari *et al.*, 1999), and bulked segregant AFLP approaches (BSA-AFLP) have been used to identify markers for high seed yield potential in second and third generation backcross hybrids with white clover as the recurrent parent (Abberton *et al.*, 2000).

Hybrids between white clover and the rhizomatous, very persistent, *Trifolium ambiguum* (Caucasian clover) require ovule culture or embryo rescue. This has been used at IGER to produce drought-resistant material with both stolons and rhizomes (Abberton *et al.*, 2000) and used in New Zealand primarily for virus resistance (Woodfield and Brummer, 2001). BSA-AFLP has been used to identify markers for the rhizomatous habit in plants of backcross families (Abberton, Michaelson-Yeates and Marshall, 2000). This material is currently being assessed for cold tolerance at several sites in northern Europe.

THE ROLE OF GENETIC MANIPULATION IN FORAGE IMPROVEMENT

The application of biotechnology to breeding objectives may be an alternative to conventional breeding or an additional element in the programme. The decision-making process is outlined in Figure 12.5 and shows some of the basic questions that breeders must address, together with some of the possible alternative biotechnological solutions. Biotechnology can aid breeding by accessing or creating novel sources of specific plant variation. This can be achieved either by the exchange of genetic material between remotely related species (somatic hybridization), through cytological instabilities produced by *in vitro* culture (somaclonal variation), or by the introduction

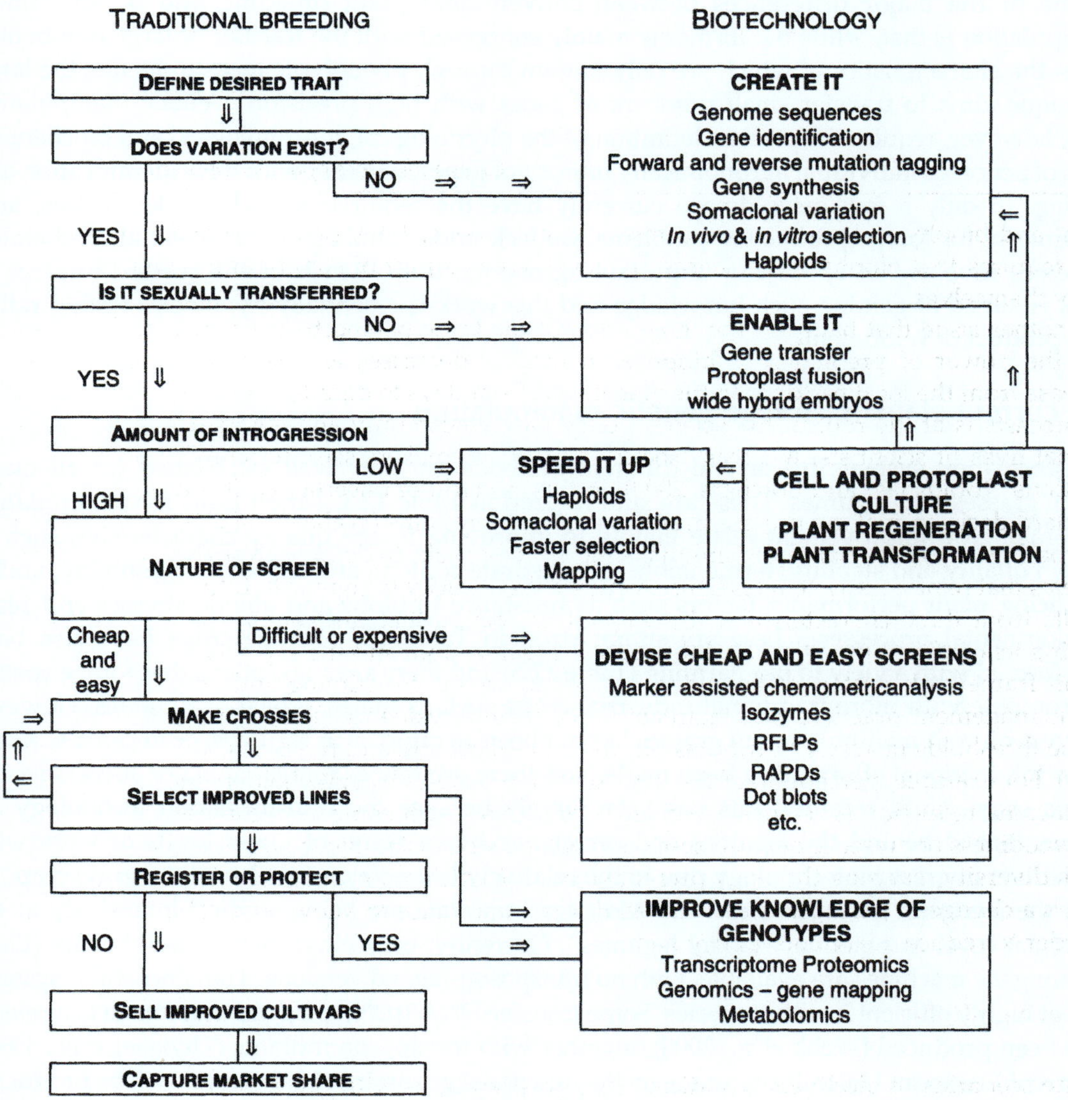

Figure 12.5 Decision tree for plant breeders and biotechnologists.

of cloned genes into the genome using DNA manipulative techniques (genetic manipulation). These approaches complement the processes described above, where DNA-marker techniques and the construction of comprehensive genetic maps of plant genomes facilitate conventional breeding. Use of genetically manipulated organisms (mainly bacteria) is, of course, an essential part of delivering these methods. Genetic manipulation also provides tools to understand basic plant biology and help to identify genes controlling the physiological and biochemical processes underlying traits of agronomic importance. Thus there are at least two ways in which genetic manipulation technology can contribute to the precision breeding of conventional varieties as well as by the direct production of novel genotypes.

One of the major differences between conventional plant breeding and direct genetic manipulation is that, while the former is mainly concerned with the transfer of large numbers of genes the characteristics of which are only known through phenotypic measurements, the latter technique aims to transfer small numbers of genes with high precision. Genetic manipulation does, however, require the precise definition of the physiological, morphological and biochemical basis of action of individual agronomically important genes, together with their identification and cloning. In only a few cases do we currently have the required knowledge to do this, and, particularly for forage production problems, we lack understanding of the essential biochemical processes involved in specifying important agronomic traits. Much of the current research at IGER is aimed at gaining such knowledge and this work is essentially hypothesis testing rather than product development.

The current status of forage genetic manipulation

Transformation systems by several different methods are now well established for all major forage grasses and legumes. These are summarized in Table 12.3. Target traits for manipulation for forage crop improvement relate mainly to improving forage quality characteristics such as protein quality and stability, fermentable carbohydrate content, and cell wall digestibility, and to enhancing plant performance factors such as resistance to biotic and abiotic stresses and plant developmental processes. These are summarized in Table 12.4. Forage crops have also been transformed with a view to Bio-Farming (Table 12.5) and there are suggestions that forage species can compete with more traditional industrial crops such as maize, wheat or potatoes. However progress on trait manipulation in grasses has not been as rapid as in legumes, or in cereals, where major international efforts have been made, and there are few examples of major success stories at this stage. Until recently, this has been largely because the transformation technology for grasses lagged behind the legumes and cereals, and as a result of lower levels of worldwide investment in grass biotechnology due to the relatively low economic value of the grass crop.

In the legumes, *Lotus japonicus* and *Medicago truncatula* are being studied intensively at the molecular level as potential "model legumes". Currently, only ethyl methanesulphonate (EMS) mutant gene machines are available, with no transposon-tagged versions. Transformation systems are not highly efficient in either species. Some transfer-DNA (tDNA) tagged mutants in *L. japonicus* have been produced (Webb *et al.*, 2000), together with transposon mutants (Thykjaer *et al.*, 1995). A current European Union Programme on the functional genomics of *L. japonicus* will be producing more experimental resources and a large number of ESTs are already available in both species.

Table 12.3 Transformation systems developed for forage species

Species	Transformation method				
				Agrobacterium	
	Protoplasts	Silicon carbide fibres	Bombardment	*tumefaciens*	*rhizogenes*
		Grasses			
Lolium perenne	✓	✓	✓		
Lolium multiflorum	✓	✓	✓	✓	
Lolium temulentum		✓	✓		
Festuca arundinacea	✓	✓	✓	✓	
Festuca rubra	✓		✓		
Agrostis stolonifera		✓	✓		
Agrostis palustris				✓	
Agrostis alba	✓				
Dactylis glomerata	✓				
Dichanthium annulatum			✓		
Cenchrus ciliaris			✓		
Zoysia japonica	✓				
Zea mays	✓	✓	✓	✓	
		Legumes			
Medicago sativa			✓	✓	✓
Trifolium repens				✓	✓
Trifolium pratensis				✓	✓
Trifolium subterraneum				✓	
Lotus corniculatus				✓	✓
Onobrychis viciifolia					✓
Stylosanthes guianensis				✓	
Medicago truncatula				✓	
Lotus japonicus				✓	✓
Lotus tenuis					✓
Lotus angustissimus					✓
Coronilla varia					✓

Table 12.4 Examples of progress in trait manipulation in forage species

Target trait	Species	Gene	Ref.
Fructans	*Lolium multiflorum*	Levan sucrase (SacB)	[1]
	Festuca arundinacea	Levan sucrase (SacB)	[2]
	Trifolium repens	Levan sucrase (SacB)	[3]
	Medicago sativa	Levan sucrase (SacB)	[3]
Digestibility – Cell wall phenolics	*Lolium multiflorum* & *Festuca arundinacea*	Ferulic acid esterase	[4]
Digestibility – Phenolics/lignin	*Festuca arundinacea*	Hydroxycinnamoyl Co A hydratase/lyase (HCHL)	[5]
Digestibility – lignins	*Lolium multiflorum*	Lignin biosynthetic genes (CAD, COMT, 4CL CCR)	[6]
	Medicago sativa	Cinnamyl alcohol dehydrogenase	[7]
	Festuca arundinacea	Cinnamyl alcohol dehydrogenase	[38]

(*Contd.*)

(*Contd.*)

Target trait	Species	Gene	Ref.
Pollen allergy	*Lolium multiflorum*	Lol PII	[8]
Leaf senescence	*Lolium multiflorum*	*Agrobacterium tumefaciens* ipt	[9]
	Trifolium repens	*Agrobacterium tumefaciens* ipt	[10]
Nodulation/N-fixation	*Medicago sativa*	Aspartate amino transferase	[11]
	Lotus japonicus	T-DNA tagging	[12]
	Lotus corniculatus	Leghaemoglobin	[13]
		Nodulins (ENOD2, sucrose synthase)	[14]
		Glutamine synthetase	[15]
Fungal resistance	*Medicago sativa*	ß-1-3 glucanase	[16]
		Rice class 1 chitinase (RCC2)	[17]
		Superoxide dismutase	[18]
	Trifolium subterraneum	Antifungal proteins	[19]
Insect resistance	*Trifolium repens*	Bt toxin (Cry Ba)	[20]
	Medicago sativa	Bt toxin (CryIC)	[21]
Viral resistance	*Trifolium repens*	Alfalfa mosaic virus coat protein	[22]
	Trifolium repens	Clover yellow vein virus coat protein	[23]
	Medicago sativa	Alfalfa mosaic virus coat protein	[24]
	Lolium perenne	Ryegrass mosaic virus coat protein	[25]
Rumen stable protein	*Festuca arundinacea*	Sunflower albumin	[26]
	Trifolium subterraneum	Sunflower albumin	[27]
	Medicago sativa	β-phaseolin	[28]
		Chicken ovalbumin	[29]
	Trifolium repens	Pea albumin I	[30]
Phosphate use efficiency	*Medicago sativa*	Malate dehydrogenase	[31]
	Trifolium repens	Citrate synthase	[32]
Tannins – reduced protein degradation/bloat	*Lotus corniculatus*	Tannin genes (DFR, CHS)	[33] [39]
		Anthocyanin Myb (Sn)	[34]
	Medicago sativa, Lotus angustissimus, L. tenuis	Anthocyanin Myb (Sn)	[35] [40]
	Trifolium repens	Anthocyanin regulatory genes	[36]
	Onobrychis viciifolia	Myb transcription factor	[37]

Sources: [1] Ye *et al.*, 2001. [2] Perret, Cairns and Morris, 2002. [3] Jenkins *et al.*, 2000. [3] Jenkins *et al.*, 2000. [4] Dunn-Coleman, Langdon and Morris, 2002. [5] Bettany *et al.*, 2002. [6] Spangenberg *et al.*, 2000. [7] Baucher *et al.*, 1999. [8] Bhalla, Swoboda and Singh, 1999. [9] Qiang, 2000. [10] Ludlow *et al.*, 2000. [11] Mett *et al.*, 1996. [12] Webb *et al.*, 2000. [13] Stougaard *et al.*, 1986. [14] Skøt *et al.*, 1996. [15] Forde *et al.*, 1989. [16] Masoud *et al.*, 1996. [17] Mizukami *et al.*, 2000. [18] McKersie *et al.*, 1996. [19] Aldo *et al.*, 2000. [20] Voisey *et al.*, 2000. [21] Strizhov *et al.*, 1996. [22] Garrett and Chu, 1997. [23] Chu *et al.*, 2000. [24] Hill *et al.*, 1991. [25] Alpeter *et al.*, 2000. [26] Wang *et al.*, 2001. [27] Khan *et al.*, 1996. [28] Bagga *et al.*, 1992. [29] Schroeder *et al.*, 1991. [30] Ealing, Hancock and White, 1994. [31] Tesfaye *et al.*, 2001. [32] Lyons *et al.*, 2000. [33] Morris and Robbins, 1997. [34] Damiani *et al.*, 1999. [35] Damiani *et al.*, 1998. [36] de Majnik *et al.*, 2000. [37] Gruber, Ray and Blathut-Beatty, 2000. [38] Chen *et al.*, 2003. [39] Carron, Robbins and Morris, 1994. [40] Robbins *et al.*, 2002.

Table 12.5 Progress in bio-farming with forage crops

Commodity	Species	Gene	Reference
Industrial enzymes	*Medicago sativa*	Cellulases	[1]
		Lignin peroxidase, α-amylase	[2] [3]
		Phytase	[4]
		Wall degrading enzymes	[5]
	Lolium multiflorum & *Festuca arundinacea*	Ferulic acid esterase from *Aspergillus niger*	[6]
Plastics	*Medicago sativa*	PHB genes from *Ralstonia eutropha*	[7]
Antibodies	*Medicago sativa*	Foot-and-Mouth viral protein VP1	[8]
		Diagnostic monoclonals	[9]

Sources: [1] Ziegelhoffer, Will and Austin-Phillips, 1999. [2] Austin-Phillips *et al.*, 1995. [3] Austin-Phillips and Ziegelhoffer, 2000. [4] Austin-Phillips *et al.*, 1999a. [5] Austin-Phillips *et al.*, 1999b. [6] Dunn-Coleman, Langdon and Morris, 2002. [7] Purev, Somers and Samac, 2000. [8] Wigdorovitz *et al.*, 1999. [9] Khoudi *et al.*, 1999.

FUTURE PROSPECTS

The pace of future advances will depend significantly on the further development of the technology for transformation. *Agrobacterium* transformation of grasses would provide efficient single copy gene transfer with segregation of unlinked tDNA sequences, but this requires development of new vectors and higher efficiency. Development of transformation systems that will handle agronomic traits requires the delivery and integration of large DNA fragments (bacterial artificial chromosome – BAC) clones fine mapped to agronomic traits that would be functionally equivalent to mini-introgression segments). This would be further aided by the availability of heritable, self-replicating plant artificial chromosomes based on self-replicating centromeric sequences. Similar systems are available in animals but require building and testing in plants.

In parallel, gene discovery and function testing would be speeded up by the use of mutation gene machines for forward and reverse genetics based on transposon tagging (as in maize). This requires the introduction of reliable transposon systems into forage species (see above), although there are good prospects for the future use of maize Mutator transposon in other Gramineae. Activation tagging in grasses and legumes would also assist function testing, although this requires development of transformation systems for grasses and legumes that are as efficient as those in *Arabidopsis*.

Precision genetic manipulation requires accurate targeting of transgene expression, either to different cell types or to specific intercellular compartments (endoplasmic reticulum, vacuole, apoplast or golgi). Some progress has been made using cell-specific promoters in grasses and legumes (but many more cell-specific promoters are needed) and for compartment targeting using 5′ and 3′ signal sequences in grasses associated with specific wall degrading enzymes (Dunn-Coleman, Langdon and Morris, 2002). Plastid transformation gives higher expression of prokaryotic genes and may facilitate "containment" of transgenes within the transformed plants, but there are marked species-dependent differences in the level of paternal inheritance of plastid DNA.

FUTURE TARGETS FOR TRAIT MANIPULATION

The polygenic nature of inheritance of many desirable characteristics in forages will inevitably require the manipulation of multiple gene pathways. In principle this can be achieved by the manipulation of higher-order regulatory genes. In practice, this will require the cloning and functional testing of master genes controlling whole metabolic pathways, cellular differentiation, tissue developmental programmes and plant architecture.

Control of reproductive development can lead to improvements in leafiness (and hence quality), but is also highly desirable to prevent transgene spread where there are abundant wild relatives that will intercross. Currently this is a major barrier to the introduction of transgenic forages. Control also requires the development of gene-switch technology to allow normal seed production. The ability to manipulate both apomixis and male sterility would help to fix genetic variation in outbreeding species and prevent gene spread via pollen, and there is a current European Union programme aimed at cloning the genes that control apomixis. Such control would also facilitate the incorporation of novel variation in tropical forages that are currently apomictic in nature.

As our understanding of the basic biochemistry of the processes improves, it is likely that there will be a range of alternative transformation targets aimed at improving quality traits via protein stability or enhanced carbohydrate accumulation (Table 12.4). Increasingly sophisticated strategies to improve biotic and edaphic stress tolerance through transformation are already being employed in major crops and are likely to influence forage research priorities.

The fact that forage is both a direct and major component of the ruminant diet means that forages also offer the potential for direct delivery of therapeutic proteins (antibodies, vaccines, viral inhibitors, etc.) to improve animal health. The incorporation of such agents into forages using genetic manipulation is theoretically possible, scientifically attractive but currently untested. There is considerable generic interest in the expression of pharmacologically active compounds in crops for human consumption ("pharming"), but there has been opposition on both ethical and safety grounds. However, improvement of animal feeds by the incorporation of bioactive molecules may offer opportunities to test the overall validity of the approach whilst avoiding direct human ingestion.

CONCLUSIONS

The development of a range of biotechnological approaches to forage improvement is both feasible and demonstrable. Progress has been slow relative to key arable crops because of the global level of investment, the complexity of the key traits and, to some extent, the innate recalcitrance of the species involved. In principle, the application of biotechnology to forage improvement offers a number of distinct opportunities. First, it offers the ability to create novel traits within specific crops. Second, it offers greater precision, in terms both of transgenic and marker-based approaches. Finally, it offers the potential for an increase in the rate of introduction of novel traits into varieties. To date, this increase in rate has not been realized, due to a mixture of regulatory and scientific challenges. However, the massive increase in genomics resources worldwide, coupled to complete genome sequencing for *Arabidopsis*, rice and, in due course, model legumes, will speed the rate of progress. Key targets for improvement will vary between species and production systems, but will emphasize quality, persistence and efficiency of resource acquisition and utilization. The key factors that will militate against progress are the relatively low economic value of the crop and the fact that, in many cases, perenniality reduces the need for seed of new varieties. There is also considerable opposition to the cultivation of genetically manipulated forage varieties in locations where gene flow to wild relatives will be ubiquitous. Under such circumstances, it is important to utilize the investment in forage biotechnology effectively. In the present circumstances, the benefits of genetic manipulation will come more through precise and more rapid conventional breeding, subsequent to being able to assign specific gene functions to QTLs, rather than from novel transgenic varieties. This may mask the impact of the technology, and avoid the exaggerated claims for the benefits of genetic manipulation and biotechnology that have largely contributed to its public demise, but should provide breeders with powerful new tools if they choose to use them.

Much of the work described in this chapter involves the development of platform technologies with broad applicability, whereas varieties are often intended for fairly narrow agronomic niches. Collaboration on platform technology is highly desirable. IGER has agreements on sharing basic technologies and genetic resources, markers, etc., with five other countries that have interests in the Institute's mandate species, with a wish to extend these arrangements further if possible. If a common resource base can be established, then the next generation of forages in many countries will accrue the many benefits that appropriate application of biotechnology could bring, with improvements to both the overall efficiency of animal production and indeed the quality of the animal product. Inevitably, biotechnological improvement of forage crops will be linked to advances in the arable sector, for both genetic and technical reasons. Nevertheless, forages do offer unique opportunities for the application of such technologies, based upon the distinctive nature of ruminant intake and digestion and the intimate relationship between livestock agriculture and landscape. The authors would, therefore, expect to see developments proceeding in parallel with those elsewhere, but with some characteristic and distinctive applications emerging.

REFERENCES

Abberton, M.T., Michaelson-Yeates, T.P.T. & Marshall, A.H. 2000. Molecular marker analysis in white clover. pp. 192–195, *in:* N.A. Provorov, I.A. Tichonovich and F. Veronesi (eds). *New Approaches and Techniques in Breeding Sustainable Fodder Crops and Amenity Grasses.* Proceedings of the 22nd Eucarpia Fodder Crops and Amenity Grasses Section Meeting, St Petersburg, Russia, 17–21 October 1999. All-Russia Institute for Agricultural Microbiology, St Petersburg.

Abberton, M.T., Marshall, A.H., Michaelson-Yeates, T.P.T., Williams, T.A., Thornley, W., Prewer, W., White, C. & Rhodes, I. 2000. An integrated approach to introgression breeding for stress tolerance in white clover. pp. 311–314, *in:* G. Parente and J. Frame (eds). *Crop Development for the Cool and Wet Regions of Europe: Achievements and Future Prospects.* Proceedings of the Final COST Action 814 Conference, Pordenone, Italy, 10–13 May 2000 (EUR 19683). Office for Official Publications of the European Communities.

Aldo, G., Drayton, M., Kalla, R., Cammue, B. & Spangenberg, G. , 2000. Development of transgenic subterranean clover expressing different chimeric AFP genes for enhanced resistance to fungal diseases. p. 110, *in:* Abstracts of the *2nd International Symposium on Molecular Breeding of Forage Crops.* Lorne and Hamilton, Victoria, Australia, 19–24 November 2000.

Alpeter, F., Xu, J.P., Salahuddin, Posselt, U.K. & Scubert, J. 2000. RNA-mediated virus resistance in fertile transgenic perennial ryegrass (*Lolium perenne* L.) plants. p. 103, *in:* Abstracts of the *2nd International Symposium on Molecular Breeding of Forage Crops.* Lorne and Hamilton, Victoria, Australia, 19–24 November 2000.

Ansari, H., Ellison, N.W., Reader, S.M., Badaeva, E.D., Friebes, B., Miller, T.E. & Williams, W.M. 1999. Molecular cytogenetic organisation of 5S and 18S-26S rDNA loci in white clover (*Trifolium repens* L.) and related species. *Annals of Botany,* **83**: 199–206.

Arcioni, S., Damiani, F., Mariani, A. & Pupilli, F. 1997. Somatic hybridisation and embryo rescue for the introduction of wild germplasm. pp. 61–89, *in:* B.D McKersie and D.C.W. Brown (eds). *Biotechnology and the Improvement of Forage Legumes.* (Biotechnology in Agriculture Series, No. 17.) Wallingford, UK: CAB International.

Armstead, I.P., Bollard, A., Morgan, W.G., Harper, J.A., King, I.P. & Jones, R.N. 2001. Genetic and physical analysis of a single *Festuca pratensis* chromosome segment substitution in *Lolium perenne*. *Chromosoma*, **110**: 52–57.

Armstead, I.P., Turner, L.B., King, I.P., Cairns, A.J. & Humphreys, M.O. 2002. Comparison and integration of genetic maps generated from F_2 and BC_1-type mapping populations in perennial ryegrass (*Lolium perenne* L.). *Plant Breeding* (In Press)

Austin-Phillips, S. & Ziegelhoffer, T., 2000. The production of value added proteins in transgenic alfalfa. pp. 285–302, *in:* G. Spangenberg (ed). *Molecular Breeding of Forage Crops*. Dordrecht, The Netherlands: Kluwer Academic Publishers.

Austin-Phillips, S., Bingham, E.T., Mathews, D.E., Shahan, M.N., Will, J. & Burgess, R.R. 1995. Production and field performance of transgenic alfalfa (*Medicago sativa*) expressing alpha-amylase and manganese-dependent lignin peroxidase. *Euphytica*, **85**: 381–393.

Austin-Phillips, S., Burgess, R., Ziegelhoffer, T. & German, T.L. 1999a. Transgenic plants as an alternative source of lignocellulosic degrading enzymes. US patent #5981835.

Austin-Phillips, S., Koegel, R., Straub, R. & Cook, M. 1999b. Animal feed composition containing phytase derived from transgenic alfalfa and methods of use thereof. US patent # 5900525.

Bagga, S., Sutton, D., Kemp, J.D. & Sengupta-Gopalan, C. 1992. Constitutive expression of the ß-phaseolin gene in different tissues of transgenic alfalfa does not ensure phaseolin accumulation in non-seed tissue. *Plant Molecular Biology*, **10**: 9–12.

Baucher, M., Bernard-Vaihe, M.A., Chabbert, B., Besle, J.M., Opsomer, C., Van Montagu, M. & Botterman, J. 1999. Down-regulation of cinnamyl alcohol dehydrogenase in transgenic alfalfa (*Medicago sativa* L.) and effects on lignin composition and digestibility. *Plant Molecular Biology*, **39**: 437–447.

Bettany, A.J.E., Dalton, S.J., Timms, E., Hauck, B., Michael, T.J., Walton, N.J. & Morris, P. 2002. Modifying lignin and cell wall phenolics in forage grasses by altering precursor supply. p. 62, *in:* J.-L. Durand, J.-C. Emile, C. Huyghe and G. Lemaire (eds). *Multi-functional grasslands*. Proceedings of the 19th European Grassland Federation Meeting. La Rochelle, France, 27–30 May 2002.

Bhalla, P.L., Swoboda, S. & Singh, M.B. 1999. Antisense-mediated silencing of a gene encoding a major ryegrass pollen allergen. *Proceedings of the National Academy of Sciences, USA*, **96**: 11676–11680.

Brummer, E.C., Bouton, J.H. & Kochert, G. 1993. Development of an RFLP map in diploid alfalfa. *Theoretical and Applied Genetics*, **83**: 329–332.

Canter, P.H. 2001. Winter hardiness in *Festuca pratensis*. PhD thesis, University of Wales, Aberystwyth.

Carron, T.R., Robbins, M.P. & Morris, P. 1994. Genetic modification of condensed tannin biosynthesis in *Lotus corniculatus*. 1. Heterologous antisense dihydroflavonol reductase down-regulates tannin accumulation in "hairy-root" cultures. *Theoretical and Applied Genetics*, **87**: 1006–1015.

Casler, M.D. 2001. Breeding forage crops for increased nutritional value. *Advances in Agronomy*, **71**: 51–107.

Chen, L., Auh C-K., Dowling P., Bell, J., Chen, F., Hopkins, A., Dixon, R.A., & Wang Z-Y. 2003. Improved forage digestibility of tall fescue (*Festuca arundinacea*) by transgenic down-regulation of cinnamyl alcohol dehydrogenase. *Plant Biotechnology Journal*, **1**: 437–449 .

Chu, P., Holloway, B., Venables, I., Lane, L., Gibson, J., Larkin, P. & Higgins, T.J. 2000. Development of transgenic white clover (*Trifolium repens*) with resistance to clover yellow vein virus. p. 145, *in:* Abstracts of the *2nd International Symposium on Molecular Breeding of Forage Crops*. Lorne and Hamilton, Victoria, Australia, 19–24 November 2000.

Damiani, F., Paolocci, F., Consonni, G., Crea, F., Tonelli, C. & Arcioni, S. 1998. A maize anthocyanin transactivator induces pigmentation in hairy roots of dicotyledonous species. *Plant Cell Reports*, **17**: 339–344.

Damiani, F., Paolocci, F., Cluster, P.D., Arcioni, S., Tanner, G.J., Joseph, R.J., Li, Y.G., de Majnik, J. & Larkin, P.J. 1999. The maize transcription factor *Sn* alters proanthocyanidin synthesis in transgenic *Lotus corniculatus* plants. *Australian Journal of Plant Physiology*, **26**: 159–169.

Diwan, N., Bhagwat, A., Bauchan, G.B. & Cregan, P.B. 1997. Simple sequence repeat DNA markers in alfalfa and perennial and annual *Medicago* species. *Genome*, **40**: 887–895.

Diwan, N., Bouton, J.H., Kochert, G. & Cregan, P.B. 2000. Mapping of simple sequence repeat (SSR) DNA markers in diploid and tetraploid alfalfa. *Theoretical and Applied Genetics*, **101**: 165–172.

Duncan, R.R. & Carrow, R.N. 1999. Turfgrass molecular genetic improvement for abiotic/edaphic stress resistance. *Advances in Agronomy*, **67**: 233–305.

Dunn-Coleman, N., Langdon, T. & Morris, P. 2002. Manipulation of the phenolic acid content and digestibility of plant walls by targeted expression of genes encoding cell wall degrading enzymes. *US Patent Application*, 20030024009.

Ealing, P.M., Hancock, K.R. & White, D.W.R. 1994. Expression of the pea albumin 1 gene in transgenic white clover and tobacco. *Transgenic Research*, **3**: 344–354.

Echt, C.C., Kidwell, K.K., Knapp, S.J., Osborn, T.C. & McCoy, T.J. 1993. Linkage mapping in diploid alfalfa (*Medicago sativa*). *Genome,* **37**: 61–71.

Forde, B.G., Day, H.M., Turton, J.F., Wen-Jun, S., Cullimore, J.V. & Oliver, J.E. 1989. Two glutamine synthetase genes from *Phaseolus vulgaris* L. display contrasting developmental and spatial patterns of expression in transgenic *Lotus corniculatus* plants. *The Plant Cell,* **1**: 391–401.

Forster, B.P., Ellis, R.P., Thomas, W.T.B., Newton, A.C., Tuberosa, R., This, D., El-Enein, R.A., Bahri, M.H. & Ben Salem, M. 2000. The development and application of molecular markers for abiotic stress tolerance in barley. *Journal of Experimental Botany,* **51**: 19–27.

Forster, J.W., Jones, E.S., Kolliker, R., Drayton, M.C., Dumsday, J.L., Dupal, M.P., Guthridge, K.M., Mahoney, N.L., van Zijl de Jong, E. & Smith, K.F. 2001. Development and implementation of molecular markers for forage crop improvements. pp. 101–134, *in:* G. Spangenberg (ed). *Molecular Breeding of Forage Crops.* Proceedings of the 2nd International Symposium on Molecular Breeding of Forage Crops. Lorne and Hamilton, Victoria, Australia, 19–24 November, 2000. Dordrecht, The Netherlands: Kluwer Academic Publishers.

Garrett, R.G. & Chu, P.W.G. 1997. White clover expressing the seed coat protein of alfalfa mosaic virus: field trial issues. pp. 125–136, *in:* G.D. McLean, P.M. Waterhouse, G. Evans and M.J. Gibbs (eds). *Commercialisation of transgenic crops: risks, benefits and trade considerations.* Canberra: Australian Government Publishing Service.

Ghesquière, M., Emile, J.C., Jadas-Hécart, J., Mousset, C., Traineau, R. & Poisson, C. 1996. First *in vivo* assessment of feeding value of *Festulolium* hybrids derived from *Festuca arundinacea* var. *glaucescens* and selection for palatability. *Plant Breeding,* **115**: 238–244.

Gruber, M.Y., Ray, H. & Blathut-Beatty, L. 2000. Genetic manipulation of condensed tannin synthesis in forage crops. pp. 189–202, *in:* G. Spangenberg (ed). *Molecular Breeding of Forage Crops.* Dordrecht, The Netherlands: Kluwer Academic Publishers.

Henry, R.J. (ed). 2001. *Plant Genotyping – The DNA Fingerprinting of Plants.* Wallingford, UK: CABI Publishing. 325p.

Hill, K.K., Jarvis-Eagan, N., Halk, E.L., Krahn, K.J., Liao, L.W., Mathewson, R.S., Merlo, D.J., Nelson, S.E., Rashka, K.E. & Loesch-Fries, L.S. 1991. The development of virus-resistant alfalfa, *Medicago sativa* L. *Biotechnology,* **9**: 373–377.

Huang, N., Angeles, E.R., Domingo, J., Magpantay, G., Singh, S., Zhang, G., Kumaravadivel, N., Bennett, J. & Khush, G.S. 1997. Pyramiding of bacterial blight resistance genes in rice: Marker-assisted selection using RFLP and PCR. *Theoretical and Applied Genetics,* **95**: 313–320.

Humphreys, M.O. 1999. The contribution of conventional plant breeding to forage crop improvement. pp. 71–78, Vol. 3, *in:* J.G. Buchanan-Smith, L.D. Bailey and P. McCaughey (eds). *Proceedings of the XVIII International Grassland Congress.* Winnipeg, Manitoba, Saskatoon, Saskatchewan, Canada, 1997. Canadian Forage Council, Canadian Society of Agronomy and Canadian Society of Animal Science.

Humphreys, M.O., Abberton, M.T., Cook, R., Turner, L.B., Roderick, H.W. & Thorogood, D. 2001a. pp. 77–87, *in:* V.H. Paul and P.D. Dapprich (eds). *Molecular breeding in pasture species and its application to improving pest and disease resistance.* Proceedings 3rd International Conference on Harmful and Beneficial Micro-organisms in Grassland, Pastures and Turf. Paderborn, Germany, 26–29 September, 2000. Universiteit Paderborn, Germany.

Humphreys, M.O. & Turner, L.B., 2001. Molecular markers for improving nutritive value in perennial ryegrass. Abstract P1.17, *in: Plant Breeding: Sustaining the Future.* Abstracts 16th Eucarpia Congress, Edinburgh, Scotland, UK. 10–14 September, 2001.

Humphreys, M.W. & Thomas, H. 1993. Improved drought resistance in introgression lines derived from *Lolium multiflorum* × *Festuca arundinacea* hybrids. *Plant Breeding,* **111**: 155–161.

Humphreys, M.W. & Ghesquière, M. 1994. Assessing success in gene transfer between *Lolium multiflorum* and *Festuca arundinacea. Euphytica,* **77**: 283–289.

Humphreys, M.W. & Pašakinskienė, I. 1996. Chromosome painting to locate genes for drought resistance transferred from *Festuca arundinacea* into *Lolium multiflorum. Heredity,* **77**: 530–534.

Humphreys, M.W., Pašakinskienė , I., James, A.R. & Thomas, H. 1998. Physically mapping quantitative traits for stress-resistance in the forage grasses. *Journal of Experimental Botany,* **49**: 1611–1618.

Humphreys, M.W., Zwierzykowski, Z., Collin, H.A., Rogers, W.J., Zare, A.G. & Lesniewska, A., 2001b. Androgenesis in grasses – Methods and aspects for future breeding. pp. 5–13, *in:* Proceedings COST 824, *Biotechnological Approaches for Utilization of Gametic Cells.* Bled, Slovenia, 1–5 July, 2000.

Hussain, S.W. & Williams, W.M. 1997. Evidence of functional unreduced gametes in *Trifolium repens* L. *Euphytica,* **97**: 21–24.

Hussain, S.W., Williams, W.M., Mercer, C.F. & White, D.W.R. 1997. Transfer of clover cyst nematode resistance from *Trifolium nigrescens* Viv. to *T. repens* L. by interspecific hybridisation. *Theoretical and Applied Genetics,* **95**: 1274–1281.

Jenkins, C., Simpson, R., Higgins, T.J. & Larkin, P. 2000. Transgenic white clover and lucerne containing bacterial fructan as a novel carbohydrate for grazing animals. p. 32, *in:* Abstracts of the *2nd International Symposium on Molecular Breeding of Forage Crops.* Lorne and Hamilton, Victoria, Australia, 19–24 November 2000. Dordrecht, The Netherlands: Kluwer Academic Publishers.

Jones, E.S., Mahoney, M.L., Hayward, M.D., Armstead, I.P., Jones, J.G., Humphreys, M.O., King, I.P., Kishida, T., Yamada, T., Balfourier, F., Charmet, G. & Forster, J.W. 2002. An enhanced molecular marker-based genetic map of perennial ryegrass (*Lolium perenne* L.) reveals comparative relationships with other Poaceae genomes. *Genome,* **45**(2): 282–295.

Joyce, T.A., Abberton, M.T., Michaelson-Yeates, T.P.T. & Forster, W.J. 1999. Relationships between genetic distance measured by RAPD-PCR and heterosis in inbred lines of white clover (*Trifolium repens*). *Euphytica,* **107**: 159–165.

Kearsey, M.J. & Farquhar, A.G.L. 1998. QTL analysis in plants; where are we now? *Heredity,* **80**: 137–142.

Khan, M.R.I., Ceriotti, A., Tabe, L., Aryan, A., McNabb, W., Moore, A., Craig, S., Spencer, D. & Higgins, T.J.V. 1996. Accumulation of a sulphur-rich seed albumin from sunflower in leaves of subterranean clover (*Trifolium subterraneum* L.). *Transgenic Research,* **5**: 179–185.

Khoudi, H., Laberge, S., Ferullo, J.-M., Bazin, R., Darveau, A., Castonguay, Y., Allard, G., Lemieux, R. & Vezina, L.-P. 1999. Production of a diagnostic monoclonal antibody in perennial alfalfa plants. *Biotechnology and Bioengineering,* **64**: 135–143.

King, I., Thomas, H.M., Armstead, I.P., Roberts, L.A., Morgan, W.G. & Kearsey, M. 2001. Introgression mapping elucidates the relationship between chiasma frequency and genetic distance and how genetic distance relates to physical distance. *Journal of Experimental Botany,* **52**(Supplement) C2.13.

Kiss, G.B., Csanadi, G., Kalman, K., Kalo, P. & Okresz, L. 1993. Construction of a basic genetic map for alfalfa using RFLP, RAPD and isozyme and morphological markers. *Molecular and General Genetics,* **238**: 129–137.

Kongkiatnam, P., Waterway, M., Fortin, M.G. & Coulman, B.E. 1995. Genetic variation within and between two cultivars of red clover (*Trifolium pratense* L.): Comparison of morphological, isozyme and RAPD markers. *Euphytica,* **84**: 237–246.

Larsen, R.C., Vandemark, G.J., Gritsenko, M.A., Hollingsworth, C.R. & Gray, F.A. 2001. PCR-based SCAR markers for the detection and identification of *Phoma sclerotiodes,* the cause of brown root rot disease of alfalfa. *Phytopathology,* **91**(supplement) S53.

Li, B.L. & Foley, M.E. 1997. Genetic and molecular control of seed dormancy. *Trends in Plant Science,* **2**: 384–389.

Ludlow, E., Lin, Y.H., Chalmers, J., Kalla, R., Pallaghy, C. & Spangenberg, G. 2000. Development of transgenic white clover with delayed leaf senescence. p. 84, *in:* Abstracts of the *2nd International Symposium on Molecular Breeding of Forage Crops.* Lorne and Hamilton, Victoria, Australia, 19–24 November 2000. Dordrecht, The Netherlands: Kluwer Academic Publishers.

Lyons, R., Mahoney, M., Yonglin, D., Lidgett, A., Pallaghy, C. & Spangenberg, G. 2000. Development of transgenic white clover expressing a chimeric bacterial citrate synthase gene for enhanced phosphorus use efficiency and aluminium tolerance. p. 79, *in:* Abstracts of the *2nd International Symposium on Molecular Breeding of Forage Crops.* Lorne and Hamilton, Victoria, Australia, 19–24 November 2000. Dordrecht, The Netherlands: Kluwer Academic Publishers.

Marshall, A.H., Abberton, M.T., Michaelson-Yeates, T.P.T. 1999. New approaches to the genetic improvement of *Trifolium repens.* p. 20, *in: Proceedings of the XVth Trifolium Conference.* Madison, Wisconsin, USA, 10–12 June 1998.

de Majnik, J., Weinman, J.J., Djordjevic, M.A., Rolfe, B.G., Tanner, R.J., Joseph, R.G. & Larkin, P.J. 2000. Anthocyanin regulatory gene expression in transgenic white clover can result in an altered pattern of pigmentation. *Australian Journal of Plant Physiology,* **27**: 659–667.

Masoud, S.A., Zhu, Q., Lamb, C. & Dixon, R.A. 1996. Constitutive expression of an inducible ß-1,3-glucanase in alfalfa reduces disease severity caused by the oomycete pathogen *Phytophthora megasperma* f.sp. *medicaginis,* but does not reduce disease severity of chitin-containing fungi. *Transgenic Research,* **5**: 313–323.

McCouch, S.R. & Doerge, R.W. 1995. QTL Mapping in Rice. *Trends in Genetics,* **11**: 482–487.

McKersie, B.D., Bowley, S.R., Harjanto, E. & Leprince, O. 1996. Water-deficit tolerance and field performance of transgenic alfalfa overexpressing superoxide dismutase. *Plant Physiology,* **111**: 1177–1181.

Mett, V.L., Podivinsky, E., Tennant, A.M., Lochhead, L.P., Jones, W.T. & Reynolds, P.H.S. 1996. A system for tissue-specific copper-controllable gene expression in transgenic plants: nodule-specific antisense of aspartate aminotransferase-P2. *Transgenic Research,* **5**: 105–113.

Mizukami, Y., Houmura, I., Takamizo, T. & Nishizawa, Y. , 2000. Production of transgenic alfalfa with chitinase gene (RCC2). p. 105, *in:* Abstracts of the *2nd International Symposium on Molecular Breeding of Forage Crops.* Lorne and Hamilton, Victoria, Australia, 19–24 November 2000. Dordrecht, The Netherlands: Kluwer Academic Publishers.

Mohan, M., Nair, S., Bhagwat, A., Krishna, T.G., Yano, M., Bhatia, C.R. & Sasaki, T. 1997. Genome mapping, molecular markers and marker-assisted selection in crop plants. *Molecular Breeding,* **3**: 87–103.

Morris, P. & Robbins, M.P. 1997. Manipulating condensed tannins in forage legumes. pp. 147–173, *in:* B.D McKersie and D.C.W. Brown (eds). *Biotechnology and the Improvement of Forage Legumes.* (Biotechnology in Agriculture Series, No. 17.) Wallingford, UK: CAB International.

Nelke, M., Nowak, J., Wright, J.M., McLean, N.L., Laberge, S., Castonguay, Y. & Vezina, L.P. 1999. Enhanced expression of a cold-induced gene coding for a glycine-rich protein in regenerative somaclonal variants of red clover (*Trifolium pratense* L.). *Euphytica,* **105**: 211–217.

Nguyen, H.T., Babu, R.C. & Blum, A. 1997. Breeding for drought resistance in rice: Physiology and molecular genetics considerations. *Crop Science,* **37**: 1426–1434.

Obert, D.E., Skinner, D.Z. & Stuteville, D.L. 2000. Association of AFLP markers with downy mildew resistance in autotetraploid alfalfa. *Molecular Breeding,* **6**: 287–294.

Osborn, T.C., Brouwer, D. & McCoy, T.J. 1997. Molecular marker analysis in Alfalfa. pp. 91–109, *in:* B.D McKersie and D.C.W. Brown (eds). *Biotechnology and the Improvement of Forage Legumes.* (Biotechnology in Agriculture Series, No. 17.) Wallingford, UK: CAB International.

Page, D., Delclos, D., Aubert, G., Bonavent, J.F. & Mousset-Declas, C. 1997. Sclerotinia rot resistance in red clover: Identification of RAPD markers using bulked segregant analysis. *Plant Breeding,* **116**: 73–78.

Pecchioni, N., Vale, G., Toubia-Rahme, H., Faccioli, P., Terzi, V. & Delogu, G. 1999. Barley-*Pyrenophora graminae* interaction: QTL analysis and gene mapping. *Plant Breeding,* **118**: 29–35.

Perret, S.J., Cairns, A.J. & Morris, P. 2002. Improving feed quality of grasses by manipulation of water-soluble carbohydrate content. *J. Experimental Botany Suppl.* **53:** P2.21.

Purev, S., Somers, D.A. & Samac, D.A. 2000. Synthesis of biodegradable plastics in alfalfa plants. p. 100, *in:* Abstracts of the *2nd International Symposium on Molecular Breeding of Forage Crops.* Lorne and Hamilton, Victoria, Australia, 19–24 November 2000. Dordrecht, The Netherlands: Kluwer Academic Publishers.

Qi, X., Niks, R.E., Stam, P. & Lindhout, P. 1998. Identification of QTLs for partial resistance to leaf rust (*Puccinia hordei*) in barley. *Theoretical and Applied Genetics,* **96**: 1205–1215.

Qiang, L. 2000. Characterisation and genetic manipulation of leaf senescence in *Lolium multiflorum* Lam. PhD Thesis. University of Wales, Aberystwyth.

Robbins, M.P., Paolocci, F., Hughes, J.-W., Turchetti, V., Arcioni, S., Morris, P. & Damiani, F. 2002. *Sn,* a maize bHLH gene, transactivates anthocyanin and condensed tannin pathways in *Lotus corniculatus. Journal of Experimental Botany,* **54**: 239–248.

Roderick, H.W., Thorogood, D. & Adomako, B. 2000. The expression of resistance to crown rust infection in perennial ryegrass (*Lolium perenne* L.). *Plant Breeding,* **119**: 93–95.

Schroeder, H.E., Khan, M.R.I., Knibb, W.R., Spencer, D. & Higgins, T.J.V. 1991. Expression of a chicken ovalbumin gene in three lucerne cultivars. *Australian Journal of Plant Physiology,* **18**: 495–505.

Skinner, D.Z., Loughin, T. & Obert, D.E. 2000. Segregation and conditional probability association of molecular markers with traits in autotetraploid alfalfa. *Molecular Breeding,* **6**: 295–306.

Skøt, L., Minchin, F.R., Timms, E., Fortune, M.T., Webb, K.J. & Gordon, AJ. 1996. Analysis of the two nodulins, sucrose synthase and ENOD2, in transgenic *Lotus* plants. *Plant and Soil,* **186**: 99–106.

Spangenberg, G., Baera, K., Bartkowski, A., Huxley, H., Lidgett, A., Lynch, D., McInnes, R. & Nagel, J. 2000. The manipulation of lignin biosynthesis in pasture grasses. p. 41, *in:* Abstracts of the *2nd International Symposium on Molecular Breeding of Forage Crops.* Lorne and Hamilton, Victoria, Australia, 19–24 November 2000. Dordrecht, The Netherlands: Kluwer Academic Publishers.

Stougaard, J., Marcker, K.A., Otten, L. & Schell, J. 1986. Nodule-specific expression of a chimaeric soybean leghaemoglobin gene in transgenic *Lotus corniculatus. Nature,* **321**: 669–674.

Strizhov, N., Keller, M., Mathur, J., Konczkalman, Z., Boch, D., Prudovsky, E., Schell, J., Sneh, B., Koncz, C. & Zilberstein, A. 1996. A synthetic cryIC gene, encoding a *Bacillus thuringiensis* delta-endotoxin, confers *Spodoptera* resistance in alfalfa and tobacco. *Proceedings of the National Academy of Sciences (USA),* **93**: 15012–15017.

Taylor, N.L. & Quesenberry, K.H. 1996. Red Clover Science. (Current Plant Science and Biotechnology in Agriculture, Vol. 28.) Dordrecht, The Netherlands: Kluwer Academic Publishers. 226p.

Tesfaye, M., Temple, S.J., Allan, D.L., Vance, C.P. & Samac, D.A. 2001. Overexpression of malate dehydrogenase in transgenic alfalfa enhances organic acid synthesis and confers tolerance to aluminium. *Plant Physiology,* **127**: 1836–1844.

Thomas, H.M., Morgan, W.G., Meredith, M.R., Humphreys, M.W., Thomas, H. & Leggett, J.M. 1994. Identification of parental and recombined chromosomes in hybrid derivatives of *Lolium multiflorum* × *Festuca pratensis* by genomic *in situ* hybridisation. *Theoretical and Applied Genetics,* **88**: 909–913.

Thomas, H.M., Humphreys, M.W. & King, I.P. 2001. Molecular markers and marker-assisted selection in forage grasses Abstract O1.3, *in: Plant Breeding: Sustaining the Future.* Abstracts 16th Eucarpia Congress, Edinburgh, Scotland, UK, 10–14 September, 2001.

Thorogood, D., Paget, M.F., Humphreys, M.O., Turner, L.B. & Roderick, H.W. 1999. A QTL analysis of crown rust resistance in perennial ryegrass. pp. 200–207, *in: New Approaches and Techniques in Breeding Sustainable Fodder Crops and Amenity Grasses.* 22nd Eucarpia Fodder Crops and Amenity Grasses Section Meeting. St Petersburg, Russia, 17–21 October 1999. All-Russia Institute for Agricultural Microbiology, St Petersburg.

Thorogood, D. & Laroche, S. 2001. QTL analysis of chlorophyll retention during leaf senescence in perennial ryegrass. Abstract P1.27, *in: Plant Breeding: Sustaining the Future.* Abstracts 16th Eucarpia Congress, Edinburgh, Scotland, UK, 10–14 September, 2001.

Thorogood, D., Kaiser, W.J., Jones, J.G. & Armstead, I.P. 2002. Self-incompatibility in ryegrass 12. Genotyping and mapping the S and Z loci of *Lolium perenne* L. *Heredity,* **88**(5): 385–390.

Thykjaer, T., Stiller, J., Handberg, K., Jones, J. & Stougaard, J. 1995. The maize transposable element Ac is mobile in the legume *Lotus japonicus. Plant Molecular Biology,* **27**: 981–993.

Tuinstra, M.R., Grote, E.M., Goldsbrough, P.B. & Ejeta, G. 1996. Identification of quantitative trait loci associated with pre-flowering drought tolerance in sorghum. *Crop Science,* **36**: 1337–1344.

Turner, L.B., Humphreys, M.O., Cairns, A.J. & Pollock, C.J. 2001. Comparison of growth and carbohydrate accumulation in seedlings of two varieties of *Lolium perenne. Journal of Plant Physiology,* **158**: 891–897.

Voisey, C.R., Dudas, B., Briggs, R., Burgess, E.P.J., Wigley, P.J., McGregor, P.G., Lough, T.J., Beck, D.L., Forster, R.L.S. & White, D.W.R. 2000. Transgenic pest and disease resistant white clover plants. pp. 239-250, *in:* G. Spangenberg (ed). *Molecular breeding of forage crops.* Dordrecht, The Netherlands: Kluwer Academic Publishers.

Wang, Z.Y., Ye, X.D., Nagel, J., Potrykus, I. & Spangenberg, G. 2001. Expression of a sulphur-rich sunflower albumin gene in transgenic tall fescue (*Festuca arundinacea* Schreb) plants. *Plant Cell Reports,* **20**: 213–219.

Webb, K.J. & Abberton, M.T. 1999. Molecular genetics of white clover. pp. 53–63, *in:* N. Sanchez-Diaz, J.J. Irigoyen, J. Aguirreolea and K. Pithan (eds). *Crop Development for Cool and Wet Climate of Europe,* COST 814 Workshop, held at the University of Navarra, Pamplona, Spain, 19–21 October 1998. OPEC Sales Agents, Luxembourg.

Webb, K.J., Skot, L., Nicholson, M.M., Jorgensen, B. & Mizen, S. 2000. *Mesorhizobium loti* increases root-specific expression of a calcium-binding protein homologue identified by promoter tagging in *Lotus japonicus. Molecular Plant-Microbe Interactions,* **13**: 606–616.

Webb, K.J., Abberton, M.T. & Young, S.R. 2002. Molecular genetics of white clover. pp. 239–253, *in:* J.K. Jaiwal and R.P. Singh (eds). *Applied Genetics of Leguminosae Biotechnology.* Great Britain: Kluwer Academic Publishers.

Wigdorovitz, A., Carrillo, C., Dus-Santos, M.J., Trono, K., Peralta, A., Gomez, M.C., Rios, R.D., Frazone, P.M., Sadir, A.M. & Escibano, J.M. 1999. Induction of a protective antibody response to foot and mouth disease virus in mice following oral or paternal immunisation with alfalfa transgenic plants expressing the viral structural protein VP1. *Virology,* **255**: 347–353.

Wilkins, P.W., Allen, D.K., Mytton, L.R. & Lister, S.J. 2001. Improvement of perennial ryegrass in nitrogen recovery and water-soluble carbohydrate content. Abstract P2.74, *in: Plant Breeding: Sustaining the Future.* 16th Eucarpia Congress, Edinburgh, Scotland, UK, 10–14 September, 2001.

Woodfield, D.R. & Brummer, E.C. , 2001. Integrating molecular techniques to maximise the genetic potential of forage legumes. pp. 51–65, *in:* G. Spangenberg (ed). *Molecular Breeding of Forage Crops.* Proceedings of the 2nd International Symposium on Molecular Breeding of Forage Crops. Lorne and Hamilton, Victoria, Australia, 19–24 November 2000. Dordrecht, The Netherlands: Kluwer Academic Publishers.

Xie, C. & Xu, S. 1998. Efficiency of multistage marker-assisted selection in the improvement of multiple quantitative traits. *Heredity,* **80:** 489–498.

Xie, C. & Mosjidis, J.A. 2001. Inheritance and linkage study of isozyme loci and morphological traits in red clover. *Euphytica,* **119**: 253–257.

Ye, X.D., Wu, X.L., Zhao, H., Frehner, M., Nösberger, J., Potrykus, I. & Spangenberg, G. 2001. Altered fructan accumulation in transgenic *Lolium multiflorum* plants expressing a *Bacillus subtilis* SacB gene. *Plant Cell Reports,* **20**: 205–212.

Yoshimura, S., Yoshimura, A., Iwata, N., McCouch, S.R., Abenes, M.L., Baraoidan, M.R., Mew, T.W. & Nelson, R.J. 1995. Tagging and combining bacterial blight resistance genes in rice using RAPD and RFLP markers. *Molecular Breeding,* **1**: 375–387.

Ziegelhoffer, T., Will, J. & Austin-Phillips, S. 1999. Expression of bacterial cellulase genes in transgenic alfalfa (*Medicago sativa* L.), potato (*Solanum tuberosum* L.) and tobacco (*Nicotiana tabacum* L.). *Molecular Breeding,* **5**: 309–318.

Section V
GEOGRAPHICAL INFORMATION SYSTEMS (GIS)

CHAPTER 13

GIS-based Forage Species Adaptation Mapping

D.B. Hannaway[1], C. Daly[2], L. Coop[3], D. Chapman[4] and Y. Wei[5]

ABSTRACT

Selecting forage crops adapted to the climatic and edaphic conditions of specific locations is essential for economic sustainability and environmental protection. This is particularly true for large countries and regions with diverse climates and soils and intended uses. It is also important for deciding what could be grown instead of what has been grown in areas undergoing substantial changes. Better matching of plants to locations would increase economic returns and reduce environmental hazards associated with suboptimal performance. Historically, deciding what plant to grow in an area was based on tradition or adaptation zones created from qualitative measurements and generalizations. They were not site specific and did not consider more than a few factors, usually annual temperature and precipitation. Currently, species selection could be more precise and accurate because the technology is available but remains difficult to apply due to the absence of user-friendly computer-based selection tools. Climate and soil geographical information system (GIS) layers, matched with a matrix of forage characteristics wrapped in an easy-to-use tool would greatly improve the selection process. GIS-based climate and soils maps are being developed and reviewed. A matrix of species characteristics is being developed for the major forage crops in Australia, People's Republic of China and United States of America. Base layer climate and soils maps and species adaptation maps could be combined on a CD-ROM to help decision-makers match their conditions to suitable forage crop species. World Wide Web segments could provide a source of current information and links to original data sources and supplementary materials. The technology is ripe, but integration of various components is now needed. The future holds the promise of many more improvements for putting each plant in the best location.

[1]Department of Crop and Soil Science, Oregon State University, Corvallis, OR 97331, USA.
[2]Spatial Climate Analysis Service, Oregon State University, Corvallis, OR 97331, USA.
[3]Department of Entomology, Oregon State University, Corvallis, OR 97331, USA.
[4]Institute of Land and Food Resources, University of Melbourne, Victoria, Australia.
[5]Meteorological Institute of Inner Mongolia, Hohhot 010051, People's Republic of China.

INTRODUCTION

A major focus in agriculture today is sustainability – seeking to utilize plant, animal, soil and water resources in the most productive yet least damaging manner. Consequently, land managers frequently request recommendations on which species and varieties would best fit into their production system (Bolte *et al.*, 1991; Hannaway *et al.*, 1992). Current crop selection tools are too generalized and do not utilize the wealth of knowledge currently available from different disciplines. Forage crop selection tools are needed that, first, better capture and disseminate the collective knowledge of forage agronomists, grassland ecologists, geographical information system (GIS) specialists, climatologists, soil scientists, information science specialists, farmers and ranchers, and, second, facilitate individually tailored, on-farm or on-ranch decision-making. With interagency and interdisciplinary teamwork, and collaboration with the private sector, it is technically possible to improve the forage crop selection process, thereby reducing the economic risks and environmental hazards implicit in the selection of less appropriate or inappropriate crops.

HISTORICALLY

Before current sophisticated computer tools became available for generating maps, most textbooks, seed catalogues, crop fact sheets, and technical guides included an adaptation zone description or a map that showed general zones where the species being discussed could be grown. Typically, these maps were produced by a graphic artist working with a crop specialist or agrometeorologist, and used general agricultural concepts and broad groupings of precipitation, temperatures and soils. Thus, the best that could be done was to use a few factors and develop a general idea of where a crop might be successful (Figure 13.1) (Sleper and Buckner, 1995).

These maps were (are) of minimal value in decision-making. They are inadequate since they do not give specific locations for successful or optimal yield. Most maps do not reflect the critical factors (minimum, maximum, and optimal ranges for precipitation, temperature, photoperiod, soil pH and drainage, elevation, slope, aspect, etc.). Broad zone maps are too general, too limited

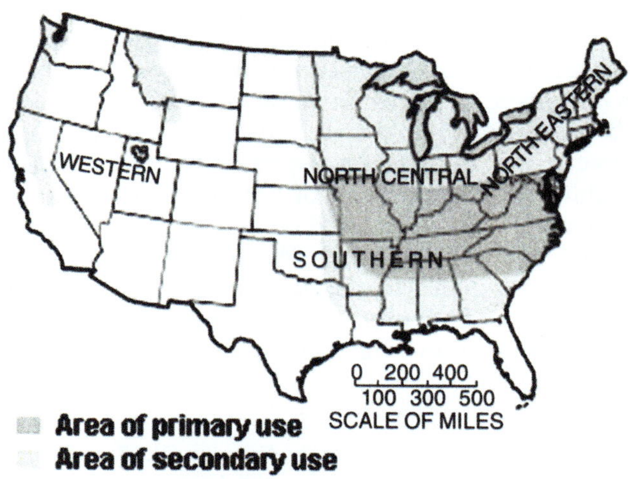

Figure 13.1 Generalized tall fescue adaptation map for the United States of America.

in representing various climate and soil factors contributing to plant growth, and are not capable of depicting changes in elevation and other topographical measurements. There has been a definite need for maps that are more specific, consider more information, and can be changed to reflect anticipated changes in factors.

CURRENTLY

With many advances in science and technology, there is now an abundance of tools that could be combined to develop better plant-to-site matching. Current computer-based tools include GIS, expert systems (ES), decision-support systems (DSS), and Web-based delivery systems. Each of these has value individually, but together they offer the potential to present huge amounts of information in a very user-friendly form.

THE TOOLS

GIS

A GIS is a family of powerful and dynamic computer software systems that manipulate and display layers of spatially variable data (Figure 13.2).

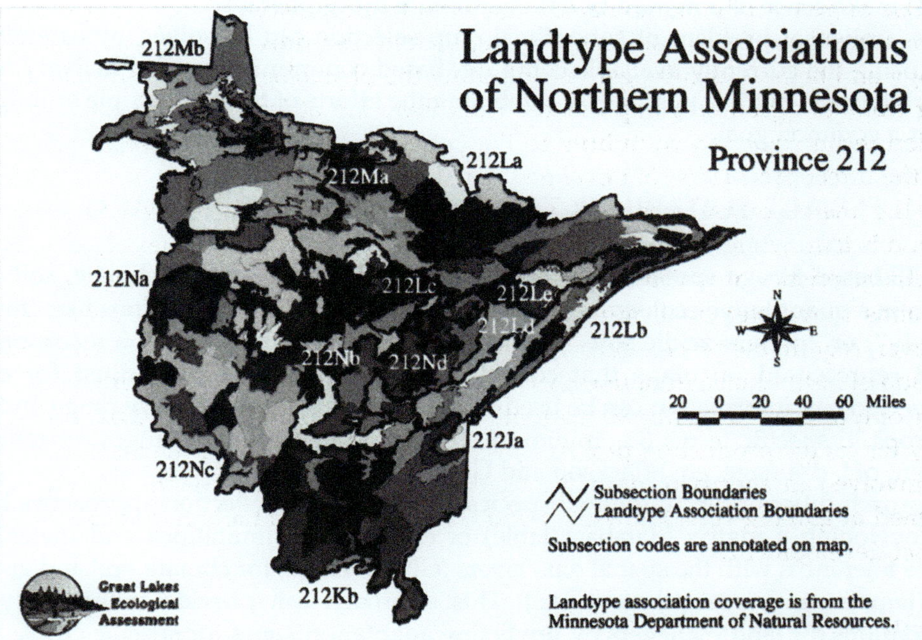

Figure 13.2 Graphic representation of GIS layered information.
Map Key: "212" identifies the area as an ecological subregion of the United States of America, as defined in US Forest Service Publication WO-WSA-5 (See: http://www.fs.fed.us/land/pubs/ecoregions/index.html), and subdivided as Ja = Glacial Lake Superior Plain; La = Border Lakes; Lb = Northern Shore Highlands; Lc = Nashwauk Uplands; Ld = Toimi Uplands; Le = Laurentian Uplands; Na = Chippewa Plains; Nb = St. Louis Moraines; Nc = Pine Moraines and Outwash Plains; Nd = Tamarack Lowlands; Ma = Littlefork Vermillion Uplands; Mb = Big Woods.

A variety of data types are used in GIS systems. Layers may include climate spatial data (precipitation, temperature, radiation), geophysical features (topography, mineral and soil traits, ground and surface water), biological characteristics (plant and animal information and tolerances), socio-economic factors (demographics, values for inputs and products), and geopolitical information (political boundaries, urban centres, transport infrastructure).

These factors all need careful consideration when formulating strategies for activities affecting natural and human ecosystems. By integrating these individual spatial data layers in a GIS, it is possible to better understand their interrelationships and respond accordingly.

Expert systems and decision-support systems

The integrated GIS spatial layers are even more useful when embedded in a DSS or ES that permits agriculture and natural resource management decision-makers to "see the big picture": protect fragile ecozones, minimize inputs of fertilizers and pesticides, optimize the use of land for agricultural production, and improve economic returns.

Web-based delivery

Delivering this information 24 hours a day, 7 days a week through Web-based information systems can now provide the entire package of information, integration tools and expert-based interpretation strategies to make sense of a myriad of data heretofore indecipherable.

Thus, the universal problem of suboptimal crop selection can be solved by assembling and effectively using the currently available computer-based communication and delivery tools. It is still not a trivial task, but it is possible by combining the efforts of talented people willing to work together for a common goal.

CREATING FORAGE SUITABILITY ZONE MAPS: A NEW APPROACH

Creating GIS-based forage species adaptation maps involves integrating climate, soil and crop information in a quantitative ecology approach. Instead of using generalizations like "moderately winter hardy", or "prefers acidic soils", more specific measurements can be incorporated into layers and represented on maps that consider many factors and can adjust for elevation. Quantitative ecology information can be used to define the highly productive range and survival limits of crops in specific terms (i.e. minimum, maximum and optimal ranges for temperature, precipitation, pH, drainage, etc.) (Jackson and Gaston, 1992).

This approach differs from traditional species adaptation and selection approaches because it involves developing a matrix (database table) of crop growth limitations and matching these quantitative tolerances with the spatial data layers (GIS coverage) for climate, soil and geophysical elements (temperature, precipitation, etc.). This approach can provide individually tailored recommendations for land managers by predicting adaptation zones for specific crops.

THE CHALLENGES

Developing integrated systems for the delivery of huge amounts of data on many topics in a way that is understandable and easy to use presents three fundamental challenges:

- assembling and organizing the data,
- obtaining missing data through research or use of surrogates for unavailable data, and
- working together in ways that are efficiently collaborative.

Scientists are often trained to work in a particular subject matter specialty. Typically, scientists are not well trained to integrate information with related or potentially related subject matter areas. In addition, scientists are most often rewarded for individual accomplishments rather than joint efforts. Thus, developing Web-based GIS Expert Systems for natural resource management decision-making is for scientists both a technical challenge and a social challenge.

STEPS NEEDED

Assembling and organizing the data

The crop ecophysiology information currently required is scattered throughout various research papers and isolated crop species literature, so profitable use is impeded. Assembling and making these data readily usable is an immediate need in developing a more ecologically based selection process for identifying crop adaptation zones (Hannaway *et al.*, 2000).

Obtaining missing data

Missing data must be identified and obtained through cooperative sharing of existing data and by conducting applied research projects to obtain data currently non-existent. Developing surrogates for missing data (developing relationships that can be used to reasonably predict the needed information) requires an understanding of the fundamental relationships and what can be used creatively to replace that currently unavailable.

Working together

It is inefficient for scientists and teachers to duplicate work when sharing information, and working cooperatively could provide much more information. Collaborative networks also provide a wonderful checks-and-balances system for research findings. It would be advantageous for scientists in specific disciplines to collaborate, and many new applications could result when experts from numerous fields and disciplines mingle their expertise.

TWO CURRENT EFFORTS

China and Australia – both huge landmasses with diverse climates –are currently making strides in utilizing collaborative applications of technology.

PEOPLE'S REPUBLIC OF CHINA

For the past several years, Oregon State University has been collaborating with several Chinese organizations to produce climate, soil and species suitability maps of China. The result has been the most detailed climate, soil and species adaptation maps currently available.

Climate modelling

This project utilizes the uniquely capable state-of-the-science climate modelling software called Parameter-elevation Regressions on Independent Slopes Model (PRISM) to produce GIS climate maps, including monthly and annual precipitation and minimum and maximum temperature (Figures 13.3–13.5). The data used to create the surfaces in the figures were mean monthly values from 1961–1990. The colours provide a way to make sense of a range of mean values.

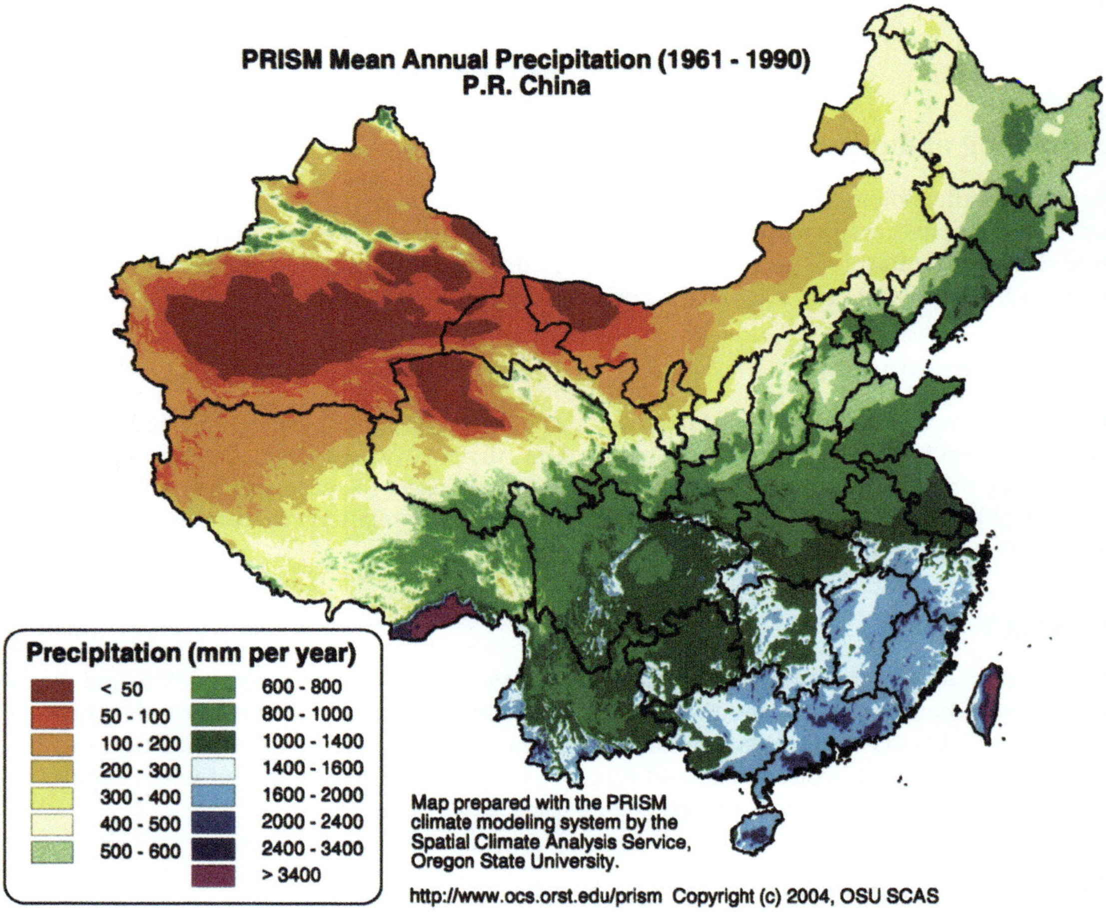

Figure 13.3 Annual precipitation of China as modelled by PRISM.

PRISM (Daly, Nielson and Phillips, 1994; Daly, Taylor and Gibson, 1997; Daly *et al*., 2002) is especially suited to mapping climate in complex landscapes such as those of China. The regression-based PRISM uses point data, a digital elevation model (DEM), other spatial data sets, a spatial climate knowledge base and expert interaction to generate repeatable estimates of annual, monthly, daily and event-based climatic elements. These estimates are interpolated to a regular grid, making them GIS-compatible. Recent mapping efforts include peer-reviewed, official USDA precipitation and temperature maps for all 50 states and Pacific Islands of the United States of America (USDA-NRCS, 1998; Daly and Johnson, 1999; Daly *et al*., 2001); a new official climate atlas for the United

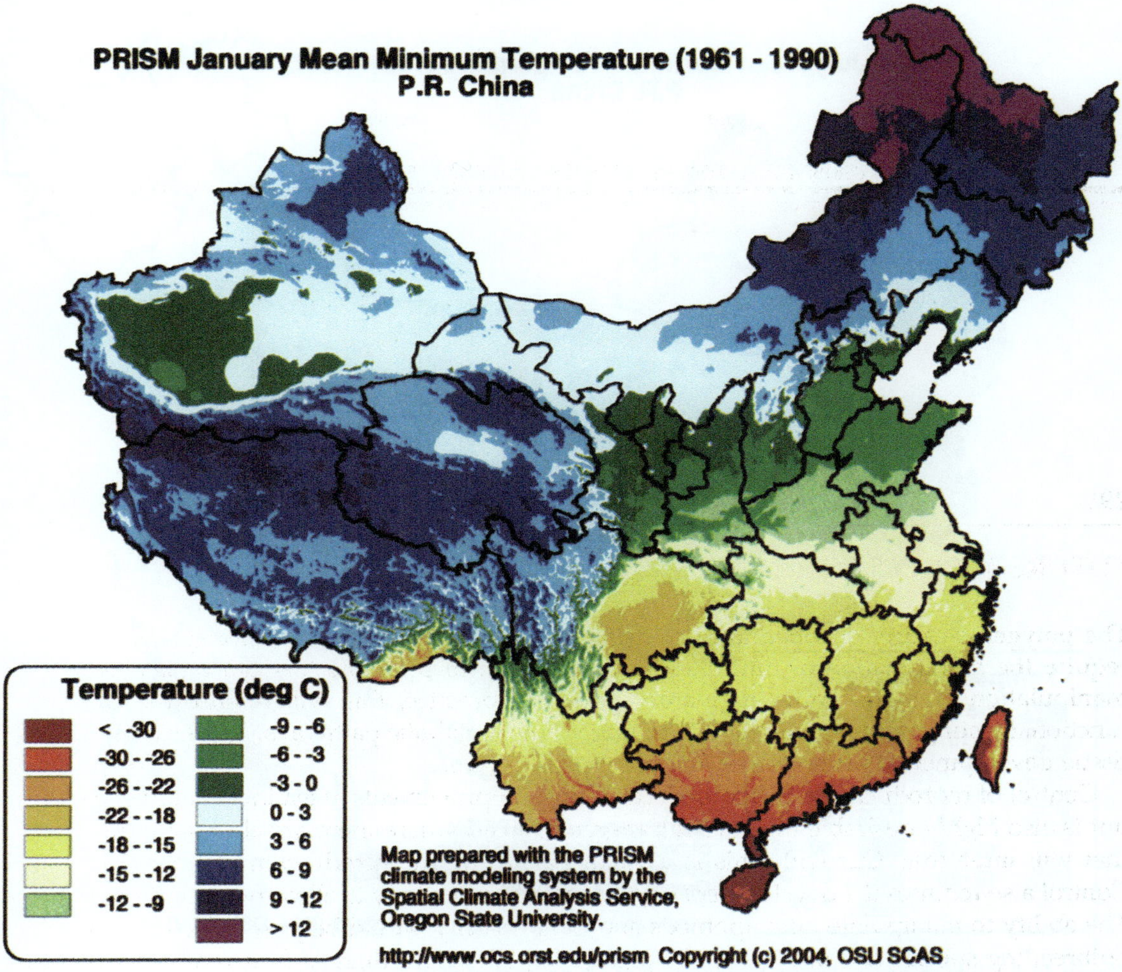

Figure 13.4 January mean minimum temperature of China as modelled by PRISM.

States of America (Plantico *et al.*, 2000); a 103-year series of monthly temperature and precipitation maps for the conterminous 48 states of the United States of America (Daly *et al.*, 1999, Daly *et al.*, 2000b); precipitation and temperature maps for Canada, China and Mongolia (Daly *et al.*, 2000a); and the first comprehensive precipitation maps for the European Alps region (Schwarb *et al.*, 2001a,b).

PRISM knowledge base for spatial climate

Successful interpolation of a climate variable requires that the major physical forcing functions be identified and accounted for in the interpolation procedure. A useful way to present these factors is to review the spatial climate knowledge base incorporated into the PRISM modelling system (Daly *et al.*, 2002). The knowledge base encoded into PRISM contains key concepts that describe the spatial patterns of climate. The knowledge base draws upon over a century of observations

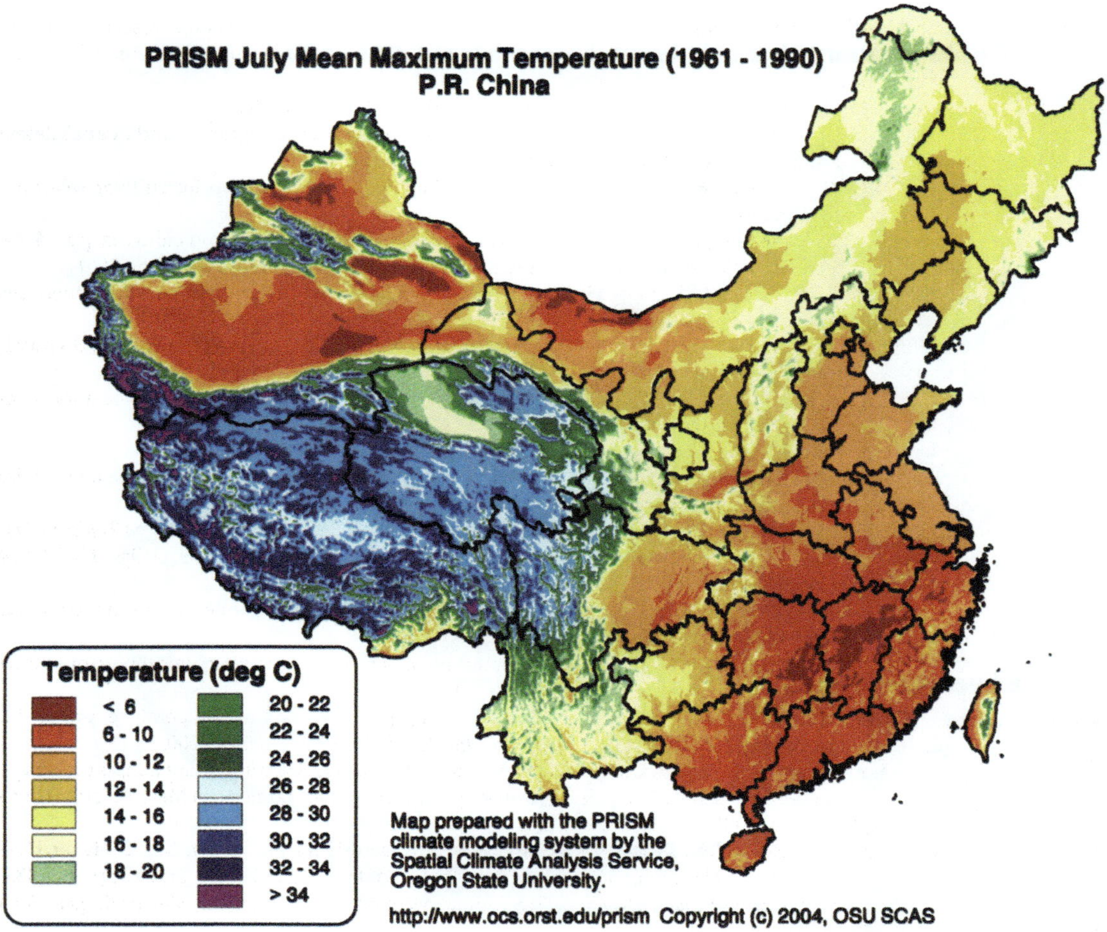

Figure 13.5 July mean maximum temperature of China as modelled by PRISM.

and research in a discipline that can be termed geospatial climatology – the study of the spatial patterns of climate and their relationships with geographical features. The key concepts included to date are discussed briefly below.

- *Elevational influence on climate* Climate varies strongly with elevation. Temperature typically decreases with altitude, and precipitation generally increases (Oke, 1978; Barry and Chorley, 1987). Elevation is an excellent statistical predictor variable, because it is usually sampled at a far greater spatial density than climate variables and is often estimated on a regular grid (i.e. as a DEM).
- *Terrain-induced climate transitions* In complex terrain, climatic patterns are defined and delineated by topographic slopes and barriers, creating a mosaic of hill slopes, or facets, each potentially experiencing a different climatic regime (Daly, Nielson and Phillips, 1994; Gibson, Daly and Taylor, 1997). Topographic facets can be delineated at a variety of scales. The major leeward and windward sides of large mountain ranges occur at relatively large scales,

while north- and south-facing hill slopes with different radiation regimes exist at small scales.

- *Coastal effects* Proximity to a large water body can be a major determinant of climate regime. For example, gradients in summer maximum temperature can exceed 20°C in 5 to 20 km, and precipitation patterns are often delineated by proximity to coastal moisture sources.
- *Two-layer atmosphere* While climate usually varies with elevation monotonically, some cases arise for which a monotonic change is not realistic. Examples are mid-slope precipitation maxima where the moist boundary layer is shallow relative to the terrain height (Giambelluca and Nullet, 1991; Juvik *et al.*, 1993); and wintertime temperature inversions in sheltered valleys, where temperature increases of 2.5 to 3.0°C/100 m are not uncommon.
- *Orographic effectiveness of terrain* Terrain features produce varying precipitation-elevation gradients, depending partly on their effectiveness in blocking and uplifting moisture-bearing air. Steep, bulky features oriented normal to the flow can generally be expected to produce steeper precipitation-elevation slopes than low, gently rising features oriented parallel to the flow.

PRISM formulation

PRISM assumes that, for a localized region, elevation is the most important factor in the distribution of temperature and precipitation (Daly *et al.*, 2002). PRISM calculates a linear climate-elevation relationship for each DEM grid cell, but the slope of this line changes locally with elevation as dictated by the data points. Beyond the lowest or highest station, the function can be extrapolated linearly as far as needed. A simple, rather than multiple, regression model was chosen because controlling and interpreting the complex relationships between multiple independent variables and climate is difficult. Instead, weighting the data points (discussed later) controls the effects of variables other than elevation.

The climate-elevation regression is developed from x,y pairs of elevation and climate observations supplied by station data. A moving-window procedure is used to calculate a unique climate-elevation regression function for each grid cell. Upon entering the regression function, each station is assigned a weight that is based on several factors. The combined weight of a station is a function of distance, elevation, cluster, vertical layer, topographic facet, coastal proximity and effective terrain weights. Distance, elevation and cluster weighting are relatively straightforward in concept. A station is down-weighted when it is relatively distant or has significantly different elevation from the target grid cell, or when it is clustered with other stations (which leads to over-representation). Coastal proximity weighting is used to define gradients in precipitation or temperature that may occur due to proximity to large water bodies (Daly, Taylor and Gibson, 1997; Daly and Johnson, 1999; Daly *et al.*, 2002; Daly, Helmer and Quinones, 2003). Facet weighting effectively groups stations into individual hill slopes (or facets), at a variety of scales, to account for sharp changes in climate regime that can occur across facet boundaries. Vertical layer weighting is used to simulate situations where rapid changes, or even reversals, in the relationship between climate and elevation are possible (i.e. temperature inversions). Effective terrain weighting accounts for differences in the ability of terrain features to enhance precipitation through mechanical uplift of moisture-bearing air (Daly, 2002).

Evaluation

PRISM was compared to other popular statistical interpolation methods in the Willamette River Basin, Oregon, United States of America (Daly, Neilson and Phillips, 1994). In a jackknife cross-validation exercise, PRISM exhibited the lowest overall bias and mean absolute error. PRISM was also applied to northern Oregon and to the entire western United States of America. PRISM's cross-validation bias and absolute error in northern Oregon increased a small to moderate amount compared to those in the Willamette River Basin; errors in the western United States of America showed little further increase. PRISM has since been applied to the entire United States of America, with excellent results, even in regions where complex terrain does not dominate climate patterns.

By relying on pixel-specific relationships between climate and elevation rather than on a single, domain-wide, relationship, PRISM continually adjusts its frame of reference to accommodate local and regional changes in climate regime, with minimal loss of predictive capability. The PRISM methodology and output products underwent extensive evaluation early in a project with the US National Resources Conservation Service to develop state-of-the-science isohyetal maps of monthly and annual precipitation for all 50 states of the United States of America. A panel of State Climatologists from several western states, plus additional experts, critically reviewed PRISM methods and maps of precipitation in their areas of interest. It was concluded that PRISM produced precipitation maps that equalled or exceeded the quality of the best manually-prepared maps available (Daly and Johnson, 1999).

Cooperators

In the People's Republic of China, assistance with climate modelling is being provided by a large group of cooperators, working together in a novel mode of collaboration. Cooperators include: China's National Meteorological Centre; CAAS Institute of Agrometeorology; CAAS Institute of Remote Sensing; China Agricultural University – Agrometeorology Department; Nanjing Agricultural University – Physiological Ecology Department; Inner Mongolia Institute of Meteorology; CAS Institute of Geography and Natural Resources; Institute of Remote Sensing Applications; Jiangsu Academy of Agricultural Sciences – Agrometeorology Department; Wuhan University (formerly Wuhan Technical University of Surveying and Mapping); Hubei Province Soil and Water Resources Bureau; Yunnan Province Institute of Geography; Yunnan Province Pasture Research Center; and Yunnan Agricultural University.

Soils mapping

Soil characteristics have been mapped with the help of the CAAS Soil and Fertilizer Institute; Yunnan Province Institute of Geography; and Jiangsu Academy of Agricultural Sciences. These maps include national and provincial maps for soil type, texture, drainage, pH, salinity and alkalinity (Figures 13.6–13.8).

Species tolerances

Initial draft quantitative species tolerances for climate factors have been defined for three example species (Table 13.1). These values are being refined and soil tolerances are being added with the help of global cooperators using Web-based mapping tools.

Figure 13.6 Soil types in China.

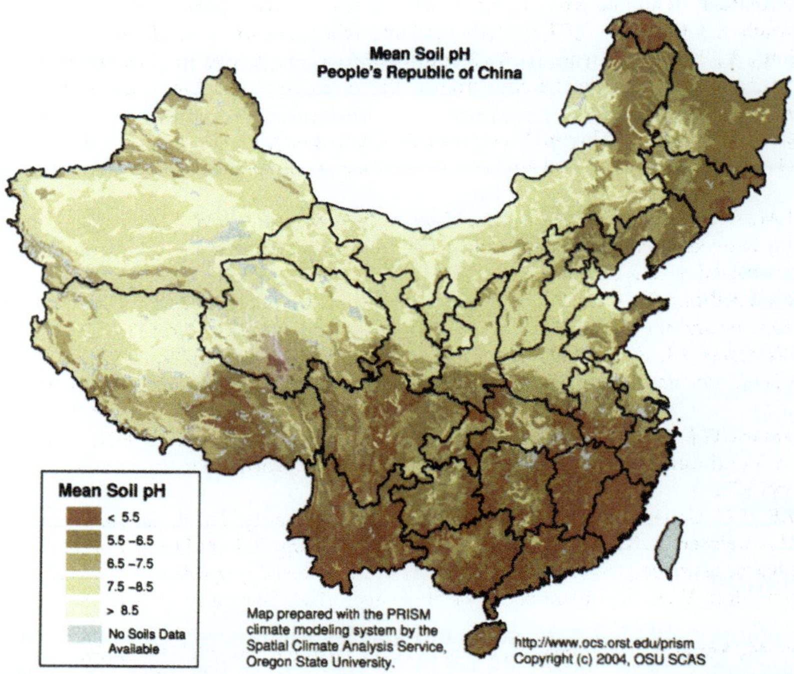

Figure 13.7 Soil pH values in China.

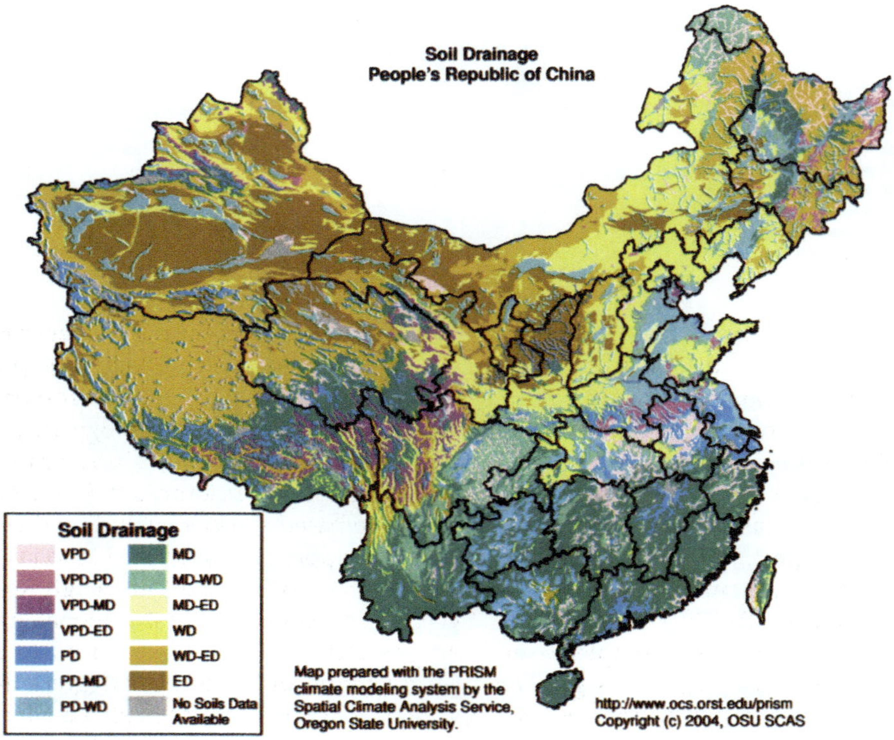

Figure 13.8 Soil drainage in China.

Table 13.1 Climatic tolerances for tall fescue, cocksfoot and perennial ryegrass

Species	Maximum temperature (°C)	Minimum temperature (°C)	Annual precipitation (mm)
Well adapted			
Tall fescue	22–32	≥ –10	≥ 625
Cocksfoot	22–31	≥ –7.5	≥ 625
Perennial ryegrass	22–30	≥ –5	≥ 625
Moderately adapted			
Tall fescue	20–34	≥ –15	≥ 450
Cocksfoot	20–33	≥ –12.5	≥ 490
Perennial ryegrass	20–32	≥ –10	≥ 525
Marginally adapted			
Tall fescue	18–36	≥ –20	≥ 300
Cocksfoot	18–35	≥ –17.5	≥ 375
Perennial ryegrass	18–34	≥ –15	≥ 450

Species adaptation

Using the three components (climate, soils and species tolerances), GIS-based maps can be made for species adaptation. Project design allows for refinement of the maps by subject experts via an on-line dynamic map server, as well as for the creation of additional layers for other environmental and economic data (Figures 13.9–13.11).

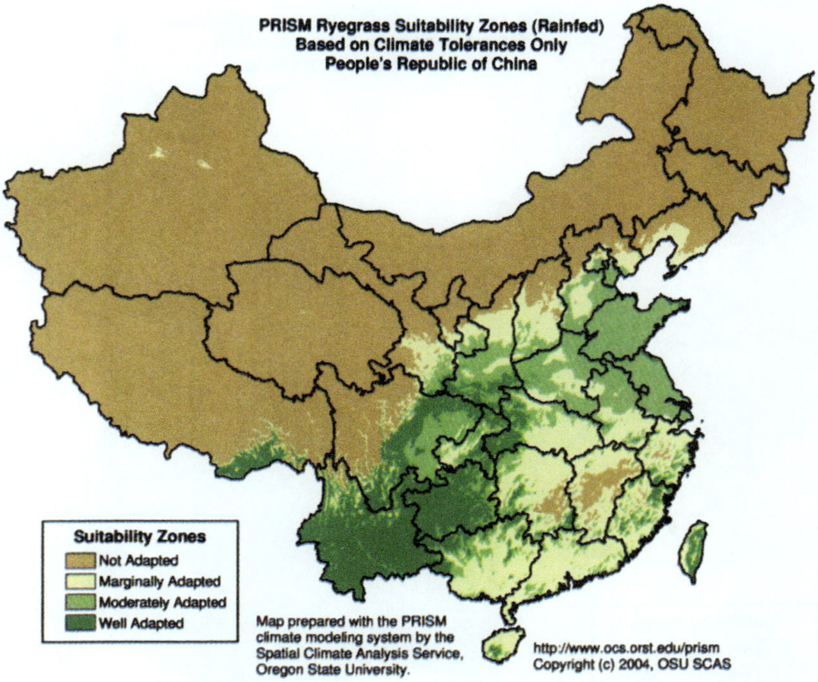

Figure 13.9 Perennial ryegrass adaptation map.

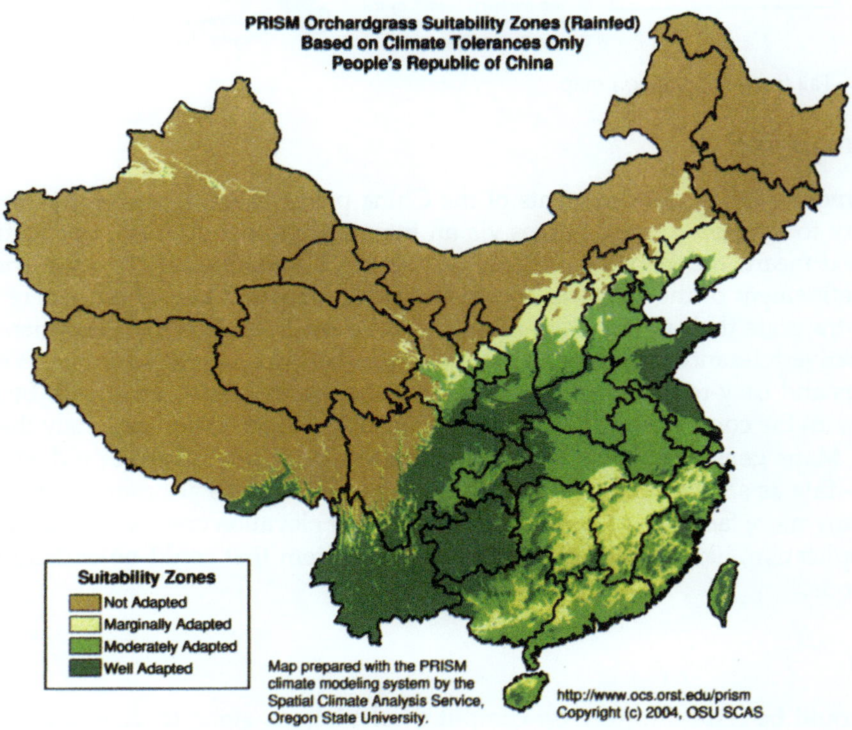

Figure 13.10 Cocksfoot adaptation map.

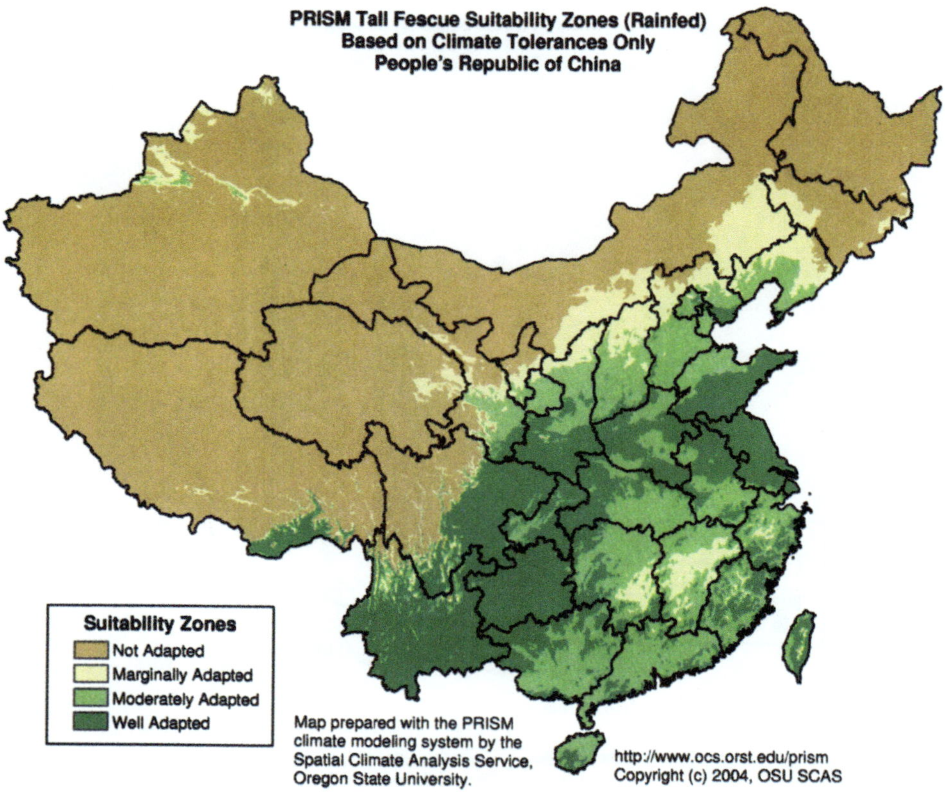

Figure 13.11 Tall fescue adaptation map.

Dynamic mapping

One of the most exciting developments of the China project work has been the development of the capability to develop dynamic maps via an Internet Map Server (IMS) application. Dynamic in this context means the maps can be changed almost instantaneously to reflect new data. This allows for refinement of the adaptation maps by subject experts located anywhere in the world. Experts can translate their mental map of what they know to be true from experience in the field into quantitative tolerance-based digital maps. With the IMS presented on the Web with drop-down menus and easy-to-use forms, an expert can adjust the measurements to be reflected in a map. Then with the computer making thousands of calculations instantaneously the revised map can appear. Maps generated in this manner are completely different from old, static maps that were out-of-date as soon as they were published. Dynamic maps are more current but can also depict so many more factors and handle the changes that elevation creates. This dynamic mapping software application, like PRISM, is a one-of-a-kind system that could potentially be applied to similar projects.

Validation

Scientists would be remiss to rely on computer calculations alone to determine the validity of maps when there are many experts and techniques that can add to the picture. Ground truthing can be strengthened with the help of collaborators collecting data from applied research trials, including those developed specifically for this project, and from trials in other projects adapted to include the data needed for this validation effort. With the maps tested on the ground through the collaborative efforts of many experts, adjustments can be made and measurements refined, resulting in more accurate and useful maps.

Summary

For China, initial countrywide climate maps have been developed for precipitation, maximum temperature and minimum temperature. These maps are based on 1961–1990 monthly means from approximately 2500 stations. National soil maps have been developed from the most current soil survey data. Draft crop characteristics data have been developed and applied to the climate maps. The soils data may now be added to the tolerances for species through the IMS application. A Web site is being developed to provide links to the climate, soil and crop characteristics maps. Maps are being developed using ArcInfo (commercial GIS software), GRASS (public-sector software), and PRISM (Daly, 2000; Taylor, 2000). Maps will be posted to the Web site as developed and verified. Economists and marketing and transportation specialists could provide additional overlays to help determine potential market availability, probable profitability and feasible transportation options. Other social factors are being considered and could be added as the concepts are developed with collaborators with other expertise.

AUSTRALIA

A pasture species database has been developed jointly by the Grassland Society of Victoria and the University of Melbourne (Chapman, Clements and Roberts, 2001; www.meu.unimelb.edu.au/grasslands). Its main purpose is to provide a comprehensive, up-to-date source of information on all species and cultivars currently available for use in pastures in southeast Australia. Currently it is restricted to temperate species. Adding a Web-based GIS ES interface is planned through cooperative cooperation with scientists in New Zealand, the United States of America, People's Republic of China, and the European Union.

Why is a pasture species and cultivars database needed?

There are scores of species and hundreds of cultivars of pasture plants available commercially in Victoria, Australia. This list has grown rapidly in the past 15 years, and will continue to grow as plant breeders turn their focus to new and alternative species of grasses, legumes and forage herbs. These will complement the tried-and-true species that have been the mainstay of agricultural development since early last century. Recently, Craig (2001) described developments in the breeding of alternative legumes. Similarly, the range of cultivars among the traditional species continues to expand, as described by Stewart (2001), Mitchell (2001) and Milne (2001).

These trends in Australia mirror trends in other countries. Harris (2000) noted that, in New Zealand, farmers had a limited range of cultivars of 17 species and hybrids to select from in 1968, whereas thirty years later this had grown to 32 species and at least 113 cultivars. Similar cultivar diversification has occurred in North America and other areas (e.g. Frame, Baker and Henderson, 1996). The great diversity of climatic condition, soil types and conditions, and animal production systems in Australia explains the need for species and cultivars with varied environmental and edaphic tolerances.

Expanded adaptation range

Enlarging the pool of available genetic resources has clearly helped expand the adaptive range of the traditional species, and introduced new species with particular adaptations to help overcome limitations to development of productive pastures. For instance, Ayres *et al.* (2000) estimated that the potential area across southern Australia where subterranean clover could be grown successfully more than doubled in three decades, from about 34 million hectares in 1960 to 71 million hectares in 1990, as a result of the improved adaptability of cultivars released from the National Sub Clover Breeding Programme over this time. Recent developments in tall fescue, described by Milne (2001), mean that farmers can now select winter-active fescue cultivars for use in medium-rainfall areas in preference to the traditional continental (summer-active) types. Therefore, suitable tall fescue material with improved nutritive value is now available for the whole of lowland Victoria, excluding only the low-rainfall NW region of the State. This is a significant step forward from the days of the original, slow-establishing, poorly digestible, winter-dormant fescue cultivars.

Spatial data layers: GIS-based mapping of adaptation zones

Hill (1996) used climate spatial data layers for eastern and southern Australia along with a set of simple rules defining the adaptability of nine pasture species to climatic conditions (rainfall, evaporation and maximum and minimum temperature) to identify adaptation zones for each species. His analysis revealed that, collectively, these nine species could be used on nearly 90 million hectares of freehold and leasehold land stretching from southern Queensland across to SW Western Australia. Alfalfa was the most widely adapted species, estimated to be suited to 96% of the land in the survey area. At the other extreme, perennial ryegrass and white clover were estimated to be suited to only about 22% of the survey area, or 19 million hectares.

One of the conclusions that emerged from this analysis was that the area where these species are *actually* used is much less than the potential. For instance, perennial ryegrass is used on about 6 million hectares (Cunningham *et al.*, 1994), while it is potentially suited to 19 million hectares.

Combining technology, applied ecology, and agricultural production systems

Combining Hill's (1996) quantitative ecology approach with computer mapping techniques offers hope for suitable solutions to current competing demands of land managers. Australian agricultural industries and scientific organizations would benefit enormously by optimal matching of climate, soil and species tolerances for developing sustainable systems. Identifying fragile ecosystems for preservation and more robust environments for agricultural production is possible with current and emerging GIS and remote sensing (RS) techniques and spatial data set resources.

To develop effective decision-aid tools, there is a great need for expanding and updating the climate, soil and species characteristics databases. This addresses the first challenge in making the most of new technology: gathering data. New techniques are being developed for creating spatial data layers and new cultivars are being released at a fast pace. All of the species reported by Hill in 1996 now have new cultivars, with other characteristics. This, no doubt, will have altered their potential areas of adaptation.

Improving the management of grasslands in northern China

There is a tremendous need for improving the management of grasslands in northern and western China. These lands are under extreme overgrazing pressure, and the grasslands are deteriorating and desertification is an ever-increasing problem. A significant component of the effort to improve these grasslands involves the use of GIS-based mapping and the development of a Web-delivered DSS.

Currently

Estimates of pasture lands in China vary between 400 and 541 million hectares. This represents more than half of the total land area of the country. Usable grasslands represent nearly a quarter of the total land area. This enormous, invaluable resource presents both opportunities and responsibilities. The opportunity is to develop a strong animal industry leading to lessening of rural poverty and increasing the nutrition of millions of Chinese. The responsibility is to manage this resource in a way that increases its multifunctional capabilities, including reduced soil erosion, improved water quality, and prevention and reversal of desertification. Balancing the ecological, social and economic needs requires scientific management.

Historical Background

The disturbance of natural vegetation has caused repeated disasters in China (Hu, Hannaway and Youngberg, 1992). The removal of vegetation and improper land management in the Loess Plateau has caused serious soil erosion and resulted in floods and droughts in north China. The effects of excessive logging of forests during previous decades has led to extensive flooding in western Sichuan Province and other portions of northern and western China.

Loss of soil and degradation of water quality through wind and water erosion has resulted from improper deforestation and cultivation of hilly, mountainous lands and grasslands. In the grassland areas, overgrazing and improper reclamation have resulted in deterioration and desertification in arid, semi-arid and dry sub-humid areas. People in regions affected by desertification suffer directly from shortage of food and degraded environmental quality. Loss of productivity in arid regions creates a major environmental constraint for sustainable development. Even in recent years, serious flooding and dust storms continue to emphasize the need for re-vegetation and improved management of China's extensive grassland areas.

High-level recognition of the problem

At present, the conservation of water, soil, forests and grassland areas is considered one of the most important projects in China. The government has directed that resources be focused on improving the conditions of the grasslands of north China and the lives of the rural population of west China. Recent policy changes mandate cessation of annual cropping of highly erodable lands (slopes greater than 25°) and a concentration of research and management applications in northern and northwestern China.

Current technology can develop tools that could lead to sustainable economic development while exercising environmental stewardship. Optimal land use and management, which consider both economic and environmental issues, is the goal. Joint teams, involving expertise from China and the United States of America, are needed in the areas of advanced computer technologies (RS, GIS, global positioning systems (GPS), DSS and Web-based information systems), grassland ecology, forage–livestock systems, soil erosion and desertification control, and education. These teams, through their joint work, will ensure improved management for the grasslands of north China and, as a result, improve the condition of the natural resources and the quality of life for the people living in this region.

The overall goal is to improve the management of China's grasslands, leading to improvement in the condition of this vast resource and enhancing the quality of life of the millions of people dependent upon it. More specifically, four major computer-based systems need to be developed:

- a spatial data analysis system (using RS, GIS and GPS) for assessing and monitoring changes in grassland conditions,
- a standardized system for quantitative characterization of grassland species,
- an optimal grassland species selection system, and
- a grassland and livestock management system for sustainable utilization.

PROPOSED STEPS

Monitoring, modelling, and managing grasslands

- Develop an RS-based integrated monitoring network for north China grasslands.
- Create detailed, GIS-based climate and soil maps for north China.
- Create a database of forage and range species characteristics.
- Develop a system to deliver dynamic maps on demand (dynamic map server).
- Develop a DSS for species and cultivar selection linked to vendors.
- Develop seasonal production profiles (forage models) of species linked to grazing capacity (sustainable resource-based forage–livestock system).

Field-based validation

- Coordinate georeferencing of RS information using GPS technology.
- Review and revise climate and soils maps using local and provincial experts.
- Review and revise species characteristics database in collaboration with grassland ecologists.
- Evaluate usability of dynamic mapping server.
- Review and revise DSS species and cultivar recommendations in collaboration with provincial experts and local farmers and ranchers.
- Evaluate seasonal productivity model estimates using local data.

Using an Internet Map Server and GIS applications for teaching species adaptation

GIS information is not often incorporated and used in current teaching activities. At the same time, teaching which plants will grow productively in various locations can be difficult with traditional methods. Students are unaware of, and consequently not using, the dynamic maps now possible that could be successfully used to facilitate plant selection. Students need to become familiar with new computer mapping technologies and plant selection DSS. This familiarity will allow them to better select plants for specific sites using GIS data and an IMS.

With Oregon State University and USDA research project funding, climate, soil and plant information adaptation maps for the United States of America and The People's Republic of China are being created and made available at http://forages.orst.edu/projects/ofsss/enmain.cfm?pageid=110

Additionally, a dynamic mapping application is being developed as a password-protected Web segment (http://blitzen.oce.orst.edu/arcims/adapt/) to allow research scientists to refine the growth characteristics of various plant species to be quantitative instead of qualitative (e.g. pH of 6.2–6.4 instead of "prefers a near-neutral pH"). Quantitative data result in more detailed and accurate mapping. With these materials beginning to appear at the fingertips of Internet users, students can now be taught to make better-informed decisions about selection of plants for a location. Students can assemble, organize and evaluate quantitative tolerances for a specific species, integrate climate and soil information, use the map server to generate an adaptation map, compare species adaptation maps, past and present, and evaluate maps and make inferences regarding their applications and use.

REFERENCES

Ayres, J.F., Davies, H.L., Farquharson, R.J. & Murison, R.D. 2000. The contribution of pasture research for animal production from legume-based pastures in temperate Australia. *Asian-Australian Journal of Animal Science*, **13** (Supplement July 2000): B1–4.

Barry, R.G. & Chorley, R.J. 1987. *Atmosphere, Weather and Climate*. 5th Ed. London: Routledge.

Baylor University. 2000. GRASS GIS Homepage. Baylor University Center for Applied Geographic and Spatial Research, Waco, Texas, USA. See: Website at http://www3.baylor.edu/grass/index2.html.

Bolte, J.P., Hannaway, D.B., Shuler, P.E. & Ballerstedt, P.J. 1991. An intelligent frame system for cultivar selection. *AI Applications in Natural Resource Management*, **5**: 21–31.

Chapman, D.F., Clements, R. & Robertson, D. 2001. An interactive database of pasture species and cultivars. pp. 34–44, *in:* Proc. 42nd Grassland Society of Victoria Conference. Mount Gambier, South Australia, 14–15 June 2001.

Coop, L. 2002. GRASSLinks Public Access GIS at OSU IPPC. Oregon State University Integrated Plant Protection Center Homepage at http://ippc2.orst.edu/glinks.

Craig, A. 2001. Developments in publicly bred sub clovers and alternative legumes. pp. 70–74, *in:* Proc. 42nd Grassland Society of Victoria Conference. Mount Gambier, South Australia, 14–15 June 2001.

Cunningham, P.J., Blumenthal, M.J., Anderson, M.W., Prakash, K.S. & Leonforte, A. 1994. Perennial ryegrass improvement in Australia. *New Zealand Journal of Agricultural Research*, **37**: 295–310.

Daly, C. 2000. Spatial Climate Analysis Service. See: http://www.ocs.orst.edu/PRISM/

Daly, C. 2002. Variable influence of terrain on precipitation patterns: Delineation and use of effective terrain height in PRISM. World Wide Web document. See: www.ocs.orst.edu/prism/effter.pdf.

Daly, C. & Johnson, G.L. 1999. PRISM spatial climate layers: their development and use. *in: Short Course on Topics in Applied Climatology*. 79th Annual Meeting of the American Meteorological Society, 10-15 January 1999. Dallas, TX, USA. 49p. See: www.ocs.orst.edu/prism/prisguid.pdf.

Daly, C., Helmer, E.H. & Quinones, M. 2003. Mapping the climate of Puerto Rico, Vieques and Culebra. International Journal of Climatology, **23:** 1359–1381.

Daly, C., Neilson, R.P. & Phillips, D.L. 1994. A statistical-topographic model for mapping climatological precipitation over mountainous terrain. *Journal of Applied Meteorology*, **33**: 140–158.

Daly, C., Taylor, G. & Gibson W. 1997. The PRISM approach to mapping precipitation and temperature. pp. 10–12, *in:* American Meteorological Society *10th Conference on Applied Climatology*, Reno, NV, USA, 2000.

Daly, C., Gibson, W.P., Hannaway, D. & Taylor, G.H. 2000a. Development of new climate and plant adaptation maps for China. pp. J62–65, *in:* Proceedings 12th AMS Conference on Applied Climatology, American Meteorological Society, Asheville NC, USA, 8–11 May 2000.

Daly, C., Kittel, T.G.F., McNab, A., Royle, J.A., Gibson, W.P., Parzybok, T., Rosenbloom, N., Taylor, G.H. & Fisher, H. 1999. Development of a 102-year high-resolution climate data set for the conterminous United States. pp. 480–483, *in: Proceedings 10th Symposium on Global Change Studies.* 79th Annual Meeting of the American Meteorological Society. Dallas, Texas, 10–15 January 1999.

Daly, C., Kittel, T.G.F., McNab, A., Gibson, W.P., Royle, J.A., Nychka, D., Parzybok, T., Rosenbloom, N. & Taylor, G.H. 2000b. Development of a 103-year high-resolution climate data set for the conterminous United States. pp. 249–252, *in:* Proceedings 12th AMS Conference on Applied Climatology, American Meteorological Society, Asheville NC, USA, 8–11 May 2000.

Daly, C., Taylor, G.H., Gibson, W.P., Parzybok, T.W., Johnson, G.L. & Pasteris, P. 2001. High-quality spatial climate data sets for the United States and beyond. *Transactions of the American Society of Agricultural Engineers,* **43**: 1957–1962.

Daly, C., Gibson, W.P., Taylor, G.H., Johnson, G.L. & Pasteris, P. 2002. A knowledge-based approach to the statistical mapping of climate. *Climate Research,* **22**: 99–113.

Frame, J., Baker, R.D. & Henderson, A.R. 1996. Advances in grassland technology over the past fifty years. pp. 31–63, *in:* G. Pollett (ed). *Grasslands into the 21st Century: Challenges and Opportunities.* Occasional Symposium of the British Grassland Society, No. 19.

FSF [Free Software Foundation]. 1991. GNU General Public License. Free Software Foundation, Inc. Boston, USA, Website at http://www.gnu.org/copyleft/gpl.html.

Giambelluca, T.W. & Nullet, D. 1991. Influence of the trade-wind inversion on the climate of a leeward mountain slope in Hawaii. *Climate Research,* **1**: 207–216.

Gibson, W.P., Daly, C. & Taylor, G.H. 1997. Derivation of facet grids for use with the PRISM model. pp. 208–209, *in: Proceedings of the 10th AMS Conference on Applied Climatology,* American Meteorological Society, Reno, NV, 20–23 October 1997.

Hannaway, D.B., Bolte, J.P., Shuler, P.E. & Ballerstedt, P.J. 1992. ACE: Alfalfa cultivar expert. *Journal of Production Agriculture,* **5**: 85–88.

Hannaway, D.B., Daly, C., Gibson, W., Taylor, G., Johnson, D., Wan, H., Luo, W., Liu, L., Han, X., Feng, L., Wei, Y., Hu, Y., Gao, L., Li, X., Zhang, W., Yang, X., Xu, Z. & Gu, A. 2000. GIS-based forage species adaptation mapping for China. International Symposium on Intelligent Agricultural Information Technology. Beijing, China, 1–4 December 2000.

Hannaway, D.B., Daly, C., Gibson, W., Taylor, G., Bolte, J.P., Sriprisan, I. & Griggs, T. 2001. GIS-based forage species adaptation mapping. pp. 1080–1081, *in:* Proceedings XIX International Grassland Congress. São Pedro, Brazil, 10–21 February 2001.

Harris, W. 2000. Formulation of pasture seed mixtures with reference to competition and succession in pastures. pp. 149–174, *in:* P.G. Tow and A. Lazenby (eds). *Competition and Succession in Pastures.* Wallingford, UK: CABI Publishing.

Henry, D. 2002. Measuring pastures from space. pp. 91–94, *in:* Proceedings 43rd Grassland Society of Victoria Conference. Morwell, Victoria, Australia, 13–14 June 2002.

Hill, M.J. 1996. Potential adaptation zones for temperate pasture species as constrained by climate: a knowledge-based logical modelling approach. *Australian Journal of Agricultural Research,* **47**: 1095–1117.

Hu, S.T., Hannaway, D.B. & Youngberg, H.W. 1992. *Forage Resources of China.* Wageningen, The Netherlands: PUDOC. 327p.

Huse, S. 1998. REGIS GRASSLinks 3.1: Public Access GIS. University of California, Berkeley, Research Program in Environmental Planning and GIS. Homepage at http://www.regis.berkeley.edu/grasslinks.

Jackson, P.L. & Gaston, G.G. 1992. Maximum extent of potential non-irrigated crop growing areas in the conterminous United States: Maps of the relative geographic range of crop production based on inherent environmental factors. Contract EPA-OBO201NAEX, U.S. Environmental Protection Agency, Global Climate Research Program, Corvallis, Oregon.

Juvik, J.O., Nullet, D., Banko, P. & Hughes, K. 1993. Forest climatology near the tree line in Hawai'i. Agricultural and Forestry Meteorology, **66**: 159–172.

Milne, G. 2001. Tall fescue dramatic changes. pp. 60–69, *in:* Proceedings of the 42nd Conference of the Grassland Society of Victoria. Mount Gambier, South Australia, 14–15 June 2001.

Mitchell, G. 2001. Phalaris and cocksfoot variety and management update. pp. 51–59, *in:* Proceedings of the 42nd Conference of the Grassland Society of Victoria. Mount Gambier, South Australia, 14–15 June 2001.

Murlikrishna, I.V. 2001. Web-based EO/GIS education and training opportunities and challenges with reference to [the] Indian scenario. GIS Development, U.P., India, Web site article: http://www.gisdevelopment.net/education/papers/edpa0005.htm.

Oke, T.R. 1978. *Boundary Layer Climates.* London: Methuen. 372p.

Plantico, M.S., Goss, L.A., Daly, C. & Taylor, G.H. 2000. A new U.S. climate atlas. pp. 247–248, *in:* Proceedings 12th American Meteorological Society (AMS) Conference on Applied Climatology. Asheville, NC, 8–11 May 2000.

Schwarb, M., Daly, C., Frei, C. & Schar, C. 2001a. Mean seasonal precipitation throughout the European Alps, 1971–1990. Hydrologic Atlas of Switzerland, National Hydrologic Service, Bern, Switzerland.

Schwarb, M., Daly, C., Frei, C. & Schar, C. 2001b. Mean annual precipitation throughout the European Alps, 1971–1990. Hydrologic Atlas of Switzerland, National Hydrologic Service, Bern, Switzerland.

Sleper, D.A. & Buckner, R.C. 1995. The Fescues. p. 346, *in:* R.F. Barnes, D.A. Miller and C.J. Nelson (eds). *Forages: An Introduction to Grassland Agriculture*. Ames, Iowa: Iowa State University Press.

Stewart, A. 2001. A brief overview of ryegrasses. pp. 46–50, *in:* Proceedings 42nd Grassland Society of Victoria Conference. Mount Gambier, South Australia, 14–15 June 2001.

Taylor, G.H. 2000. Oregon Climate Service. See: www.ocs.orst.edu/

USACERL. 1993. GRASS Version 4.1. Geographic Resources Analyses Support System. User's Reference Manual. US Army Corps of Engineers Construction Engineering Research Laboratories, Champaign, IL. 563p.

USDA-NRCS [United States Department of Agriculture – Natural Resources Conservation Service]. 1998. PRISM Climate Mapping Project – Precipitation. Mean monthly and annual precipitation digital files for the continental U.S. USDA-NRCS National Cartography and Geospatial Center, Ft. Worth TX. CD-ROM.

CHAPTER 14

The Application of Remote Sensing Technology to Grassland Resource Management in the People's Republic of China

Z.B. Nan[1]

ABSTRACT

Application of remote sensing (RS) technology to grassland resource management and livestock production in China is outlined. The role that the technology has played in developing sustainable grassland farming systems in western China is presented, and opportunities for further development are discussed. Use of RS technology in grassland surveys started in China in 1979. The technology has been used extensively at various levels since then in order to achieve better management of grassland resources. RS information is mainly obtained from aerial photography and satellite-based sensors, including the Landsat satellites' Multispectral Scanner (MSS) and Thematic mapper (TM) and the National Oceanic and Atmospheric Administration (NOAA) Polar Orbiter's Advanced Very High Resolution Radiometer (AVHRR). Supported in part by RS technology, the first national grassland survey and inventory was completed in 1995. The technology has been used to monitor changes in grassland use; assess grassland productivity and grassland degradation; and evaluate conservation and recreation sites. Special techniques have been also developed for monitoring grassland fires; detecting snow cover and depth on grasslands; evaluating drought; and determining areas damaged by rodents.

Recently, the techniques have been used in western China development to facilitate reversion of steep sloping arable land into pasture and forest. RS technology has become a

---

[1]College of Pastoral Agriculture Science and Technology, Lanzhou University; and Gansu Grassland Ecological Research Institute; P.O. Box 61, Lanzhou 730021, PR China.

powerful tool for grassland scientists, extension officers and policy-makers. Based on RS technology with Geographical Information System (GIS) and Global Positioning System (GPS) technology, an *Expert System for Grassland Husbandry and Ecological Construction* at provincal scale has been developed, and a national grassland fire monitoring system is being constructed based on information obtained from the Moderate-resolution Imaging Spectroradiometer (MODIS).

As RS technology develops and there is improvement of training and education programmes at higher educational levels, the technology will become an integral part of precision grassland agriculture in China. An era of digital grassland is coming and the science and technology of grassland RS as an integrated, independent subject will gain formal recognition by the public both in China and the world in the near future.

INTRODUCTION

The area of grassland in China is about 393 million hectares, and covers 41 percent of the total land area. Grassland is the largest biogeographical unit in China, its area being two and three times bigger than cropland and the forestland, respectively (Ren, 1985). Due to population pressure and the dominance of cereal monoculture on the most fertile land, most of China's grassland is characterized by aridity with extremes of heat or cold, or is too rocky, steep or saline for growing cereal crops or tree crops; thus, it is mainly in marginal land and isolated regions. Grassland supports 25.6 percent of all cattle and 34.7 percent of sheep in China. It produces 26.1 percent of the total yield of beef and mutton, and 29.4 percent of the milk. Detailed and comprehensive information on grassland in China has been given by Liao and Jia (1996) and Ren (1999).

However, compared with developed countries, the productivity of grassland in China has been very low. According to Ren and Huang (1978), the grassland productivity of Australia, the USA and New Zealand is 9, 26 and 81 times higher, respectively (Table 14.1). Although their study was carried out more than 20 years ago, the situation indicated then can still be considered typical of the huge differences between China and those countries.

Table 14.1 Grassland livestock productivity in China compared with some other countries

Country	Productivity (APU ha^{-1})$^{(1)}$	Other country/China
PR China	0.58	1
Australia	5.6	9.7
USA	15.8	27.2
New Zealand	47.9	82.6

Note: (1) APU = Animal Product Unit (as defined by Ren, 1985).
Source: Ren and Huang, 1978

Due to the extensive nature of the grasslands and the urgent need to manage them properly and effectively at low cost, RS technology has significant promise, and has been used to support grassland resources development and management for the last two decades in China. However, its history goes back to the 1950s, when black-and-white aerial photography was used in natural resource surveys (Niu, Hu and Lu, 1998). Aerial photography was used in grassland vegetation surveys during 1962–1964 for Heilongjiang province, in northeastern China. The area surveyed was about 60 000 km^2 each year, and grassland vegetation was mapped at a scale of 1 : 25 000

(Jiang, Wang and Yu, 1964). This was probably the earliest work in China using RS technology for large-scale grassland surveys.

The launch of Landsat 1 in 1972 ushered in a new era, extending RS beyond aerial photogrammetric interpretation into the realm of digital analysis of multispectral and multitemporal data worldwide (Tuller, 1989). Following technical progress in developed countries, application of satellite imaging to grassland resource management developed rapidly in China. The techniques were introduced into China in the late 1970s. Supported jointly by the Food and Agriculture Organization (FAO) of the United Nations, the United Nations Development Programme (UNDP), the Chinese Ministry of Science and Technology, and the Chinese Ministry of Agriculture, a National Training and Application Centre for Agricultural Remote Sensing was established at Beijing Agricultural University in 1979 (Xia, Chia and Chang, 2002).

Some pilot trials on a small scale were carried out for grasslands of Hebei and Gansu provinces and the Inner Mongolia Autonomous Region (Jia, Xia and Chen, 1981; Ren, 1984). The techniques were being used extensively in grassland resource surveying by the middle of 1980s, with the information obtained mainly from the Landsat Thematic Mapper (TM), Landsat Multispectral Scanner (MSS) and the National Oceanic and Atmospheric Administration (NOAA) polar orbiter's Advanced Very High Resolution Radiometer (AVHRR). The herbage dry matter (DM) yields for various types of grassland were monitored and predicted from the early 1990s using Landsat TM and Landsat MSS data as information resources, together with NOAA-AVHRR data.

In the mid-1990s, the application of RS technology to grassland resource management in China was extended into a new era, with Geographical Information System (GIS), Global Positioning System (GPS) technology coupled with RS techniques being used to develop a system of information technology for comprehensive monitoring of grassland production (Li, 1994; Li, Liang and Chen, 1998).

This chapter summarizes the major developments in the application of RS technology in grassland resource management in China, and considers possible developments in this area in the next decade.

GRASSLANDS INVENTORY AND LAND USE CHANGE

Grasslands inventory

Application of RS technology

Survey and inventory of grassland resources with the aid of RS technology blossomed during the 1980s following the introduction of RS technology into China in the late 1970s. Surveys on small areas were undertaken at an early stage to learn and master the techniques. Those surveys included the work carried out on the Inner Mongolian steppe jointly by Beijing Agricultural University and the Grassland Research Institute of the Chinese Academy of Agricultural Sciences (Xia, Jia and Chang, 1986), by Beijing Agricultural University on the arid grasslands of Weichang county, Heibei province (Jia, Xia and Chen, 1981) and by Gansu Grassland Ecological Research Institute on the alpine grasslands of Tianzhu county, Gansu province (Ren, 1984). All those works used a combination of aerial photographs and Landsat MSS image as the information resources. Grassland classification maps at scales of 1 : 45 000 to 1 : 100 000 for various areas were one of the outcomes of those surveys. Experience accumulated and technical staff were trained through those projects. A solid basis was laid for later work on larger areas.

During the late part of the 6th (1980–1985) and the entire the 7th (1986–1990) national five-year economic development plan periods, application of RS technology to grassland resources survey developed rapidly and the technique was used directly to promote development of the national economy. Several projects involving the application of RS technology were launched under the aegis of the Ministry of Science and Technology and the related provincial governments. Those included the Inner Mongolian grassland resources survey, a natural resource survey for establishing protective forest areas in northern, northeastern and northwestern China, and the grassland resources survey for the Naqiu prefecture of Tibet. Landsat TM and Landsat MSS images were used as the information resources for those projects (Table 14.2).

Table 14.2 Grassland survey projects using remote sensing technology in China during the 1980s

Region	Location	Area surveyed (km^2)	Information resources	Period covered	Ref.
Weichang county, HB	41°51′N 111°50′E	100	Aerial photography and Landsat MSS	1980	[1]
Balinyouqi, IM	43°12′–44°27′N 118°10′–120°5′E	1.03×10^4	Aerial photography and Landsat MSS	1984	[2]
Tianzhu county, GS		7.15×10^3	Aerial photography and Landsat MSS	1983	[3]
Beijing; Chende and Zhangjiaguo, HB	39°30′–42°40′N 113°50′–119°15′E	8.55×10^4	Aerial photography, Landsat TM and Landsat MSS	1987	[4]
Inner Mongolia		1.157×10^6	Landsat TM and Landsat MSS	1983-1987	[5]
Naqiu Tibet	29°55′–36°30′N 83°55″–95°5′E	3.94×10^5	Landsat MSS and Landsat TM	1985–1987	[6]
Yulin and Yanan, Shx	36°10′–39°34′N 107°28′–111°15′E	5.6×10^4	Landsat TM	1986–1989	[7]
Grasslands in forest areas in northern China	36°10′–39°34′N 107°28′–111°15′E	4.01×10^5	Landsat TM	1986–1989	[8]

Key to provinces: HB = Heibei province; IM = Inner Mongolia Autonomous Region; GS = Gansu Province; Shx = Shaanxi Province

Sources: [1] Jia *et al.*, 1981. [2] Xia, Jia and Chang, 1986. [3] Ren, 1984. [4] Liu *et al.*, 1990. [5] Li, 1987. [6] Chen, 1991. [7] Ren *et al.*, 1990. [8] Ren, Liu and Li, 1994.

Books on the survey results and grassland classification maps for each project have been published. For detailed information on each project, see Li (1987) for Inner Mongolian grasslands; Xu (1994) for northern, northeastern and northwestern China forests and grasslands; Chen (1991) for Tibetan Niaqu grasslands; and Gong and Xu (1991) for Loess Plateau natural resources, including grasslands. The grassland survey for the whole Inner Mongolian Autonomous Region is typical of similar works carried out in the country. The project was conducted by a multidisciplinary team consisting of more than 100 scientists from 12 institutions, including Beijing University and Inner Mongolian University. During 1983–1987, they completed a survey of 1 157 000 km^2 of grasslands using Landsat TM and Landsat MSS as the main information resources. At the end of the project, a grassland classification map at 1 : 350 000 or 1 : 500 000 and at 1 : 1 000 000 scale was produced

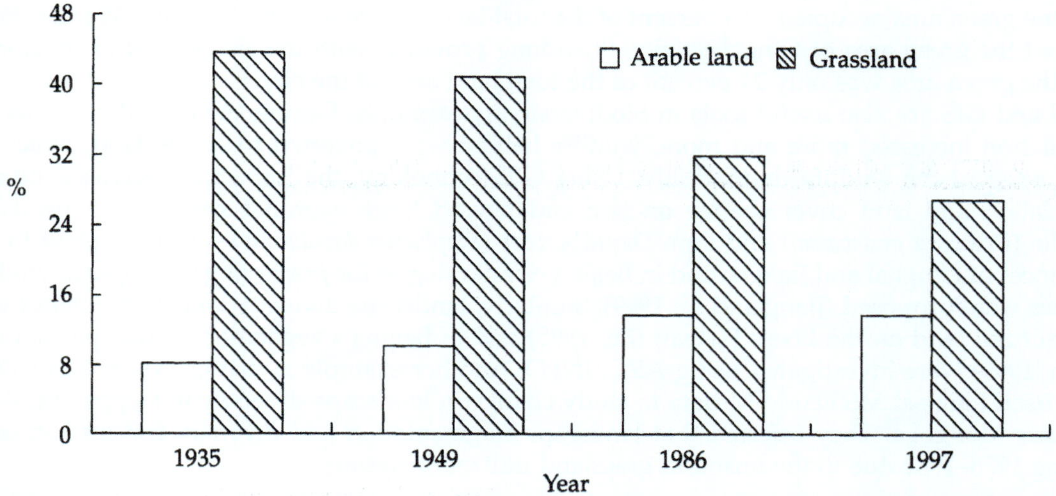

Figure 14.1 Area of arable land and grassland as percentage of the total land area of China from 1935 to 1997
Source: Ge, Zhao and Zheng, 2000.

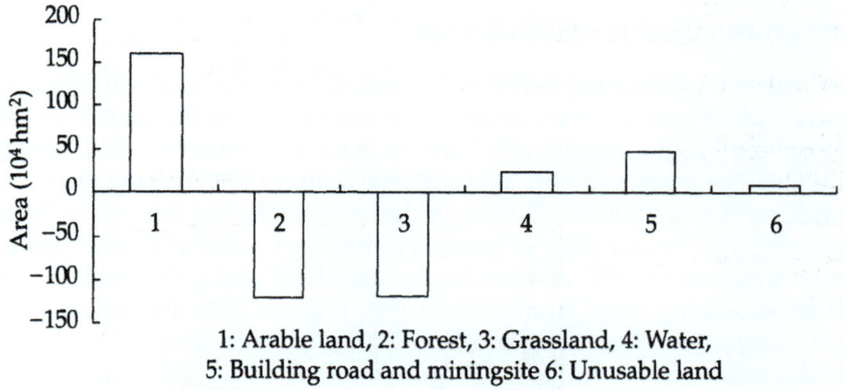

Figure 14.2 Changes in land use (area) in the period 1996 to 2000 in China.
Source: Wang *et al.*, 2001.

CONSERVATION AND RECREATION SITE VALUING

As humans enter the twenty-first century, the concept of multi-functionality for grasslands is accepted by more and more people. Grassland is used not only for livestock production, but also for landscape construction, biodiversity preservation, tourism and recreation (Durand *et al.*, 2002). In China, a modern grassland agricultural production system that consists of four production levels has been proposed (Ren, 1985) and this has been accepted well by scientists, policy-makers and the public. The system includes pre-plant production (conservation, tourism and recreation); plant production; animal production; and post-biological production (preliminary processing of plant and animal products, and marketing). RS technology coupled with GIS and GPS becomes more apparent in Ren's pre-plant production level, including biodiversity conservation and recreation. Yu *et al.* (1994), for instance, used aerial photographs as the information resource to study the coverage of turf and street trees in Hefei city, the capital of Anhui province. They found that the green area occupied 33.5 percent of the total land area of the city. Similarly Wang (1994a) studied the green area of Qingzhou city, Shandong province, with the aid of RS and concluded that the green area was only 24 percent of the total land area of the city.

RS and GIS are also useful tools in biodiversity assessments. During the last 50 years, as the population increased more and more, wildlife habitat was converted to arable land, which, in turn, affected the wildlife dramatically. Using GIS technology, the impact of increased human population and land cover change on two endangered large mammal species – Przewalski's gazelle (*Procapra przewalski*) and Pere David's deer (*Elaphurus davidianus*) – were studied in the provinces of Qinghai and Jiangsu and in Beijing city. Strategies for preserving endangered wildlife species were proposed (Jiang and Li, 1999). Similarly, landscape diversity and spatial patterns in agricultural land on the Loess Plateau (Fu, 1995) and in Beijing's vegetation landscape (Liu and Chen, 1999) were investigated using ARC/INFO. Another example is that of Wu and Li (2000), who used Landsat MSS and TM data to study change in landscape diversity in steppe grassland in Inner Mongolia. They inferred that landscape fragmentation and diversity indices increased during 1978–1982 due to the unsound grassland utilization system.

GRASSLAND PRODUCTIVITY ASSESSMENT

Net primary production at national scale

Grassland net primary production (NPP) is the central carbon-related variable summarizing the interface between plant and other processes in an ecosystem. At the launch of the International Geosphere-Biosphere Programme (IGBP), Chinese scientists initiated global change studies that included assessment of grassland NPP at a national scale in 1985 (Zhang *et al.*, 1997). However, the most studies involving RS and GIS were undertaken in the last few years. For studies carried out internationally, NOAA-AVHRR is currently the most reliable option (Nan, 2001). Zhang (2001) reviewed progress in NPP assessment for grassland and crop yield in China. Grassland NPP showed an increasing trend from 1982 to 1999 (Figure 14.3). Rainfall is the major limiting factor for NPP distribution (Piao, Fang and Guo, 2001a; Sun and Zhu, 2000, 2001), resulting in an NPP gradient decreasing from southeast to northwest (Piao, Fang and Guo, 2001b). The total NPP of terrestrial vegetation in China was 6.13×10^9 tonnes of carbon in 1990, of which 24.5 percent was from grasslands (Chen, Liu and Feng, 2001). The NPP of meadow grassland is about 6.3×10^3 kg DM ha^{-1} annually, while the annual NPP for dry steppe is only 0.75×10^3 kg DM ha^{-1} (Xiao *et al.*, 1996).

GRASSLAND HERBAGE DRY MATTER YIELDS

Livestock in the pastoral regions of northern China graze on grassland throughout the year, so animal nutrition is entirely dependent on the forage supply from grasslands. Herbage DM yields have great seasonal fluctuations, while the needs of grazing animals are relatively constant. There are thus large seasonal gaps between forage supply from grassland and animal needs. The livestock virtually starves for several months every year due to shortage of feed (Ren *et al.*, 1978). Therefore, herbage DM yields are of great interest to grassland scientists, animal scientists and policy-makers. RS technology has been employed in herbage yield assessments and monitoring since the mid-1980s in China, mostly based on NOAA-AVHRR as the information resource. Similarly to the

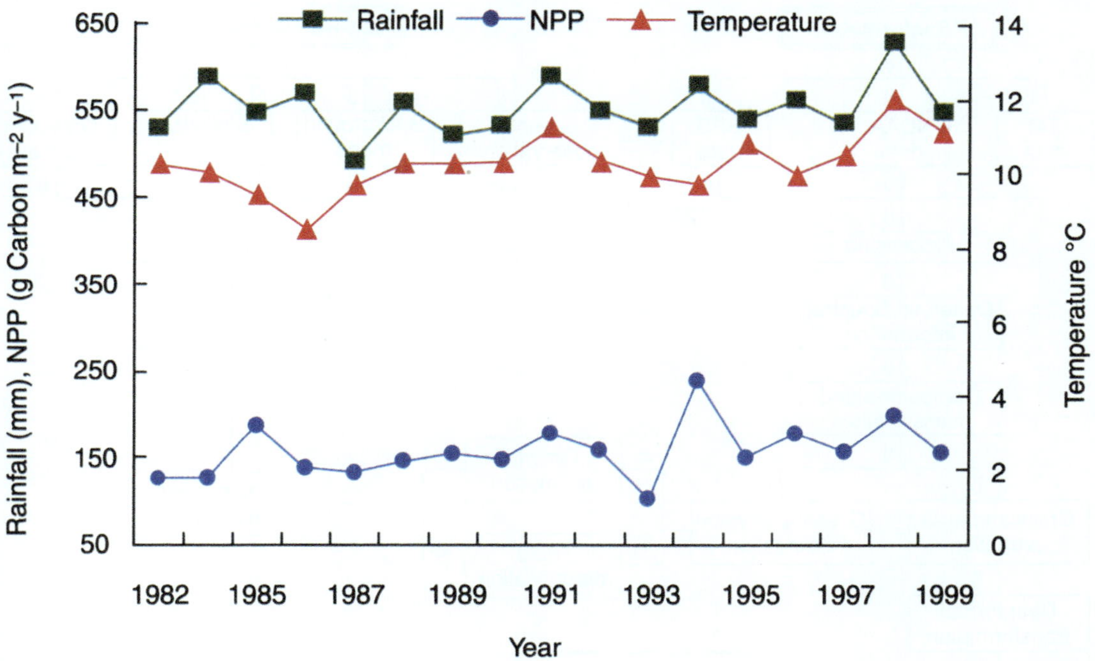

Figure 14.3 Mean daily temperature (■), annual rainfall (●) and grassland NPP (▲) in China during 1982–1999 (excluding 1994 because the data were missing).
Source: Redrawn from Piao, Fang and Guo, 2001a.

progressive development seen internationally (Nan , 2001), the studies undertaken in China focused on grassland productivity assessment. Herbage DM yields have been measured for steppe grasslands in Inner Mongolia (Li, 1993; Liu, 1994), alpine grasslands in Tibet (Wei and Chen, 2001), Qinghai (Xu *et al.*, 2001; Zhou and Wang, 1996) and Gansu province (Wang *et al.*, 2001), and mountainous grasslands in Xijiang Autonomous Region (Chen, Wei and Liang, 1994b; Li, Liang and Chen, 1998).

A strong positive correlation ($r = 0.62$; $p<0.01$) was found between steppe aboveground biomass and NDVI for arid grassland on the Loess Plateau (Ren, Liu and Li, 1994) and steppe grassland in Inner Mongolia (Kawamura *et al.*, 2003). Among those, the largest extent of grassland productivity monitoring was achieved by a team led by the late Professor Li Bo. They developed a technical system to monitor the herbage DM yield for the entire Inner Mongolian Autonomous Region (Li, 1993).

Based on their experiences in Xijiang, Li, Jiang and Liang (1998) summarized the procedure for RS yield estimation over large area grasslands in China (Figure 14.4). To improve the accuracy of yield measurement, various vegetation indices have been developed to establish the correlation between vegetation condition and herbage yields (Chen, Wei and Liang, 1994a; Xu *et al.*, 1985; Zhang and Li, 1991; Tong *et al.*, 1986; Li and Jiang, 1996; Huang and Shang, 1993; Huang *et al.*, 2001; Li, Jiang and Dai, 1998). Liang, Chen and Wei (1996), for instance, compared two vegetation indices and seven existing yield prediction models, and found that the model that could be used for herbage yield prediction of flat grassland was not suitable for rolling country. Chen *et al.* (1998) developed the so-called Average Normalized Difference Vegetation Index (ANDVI) and

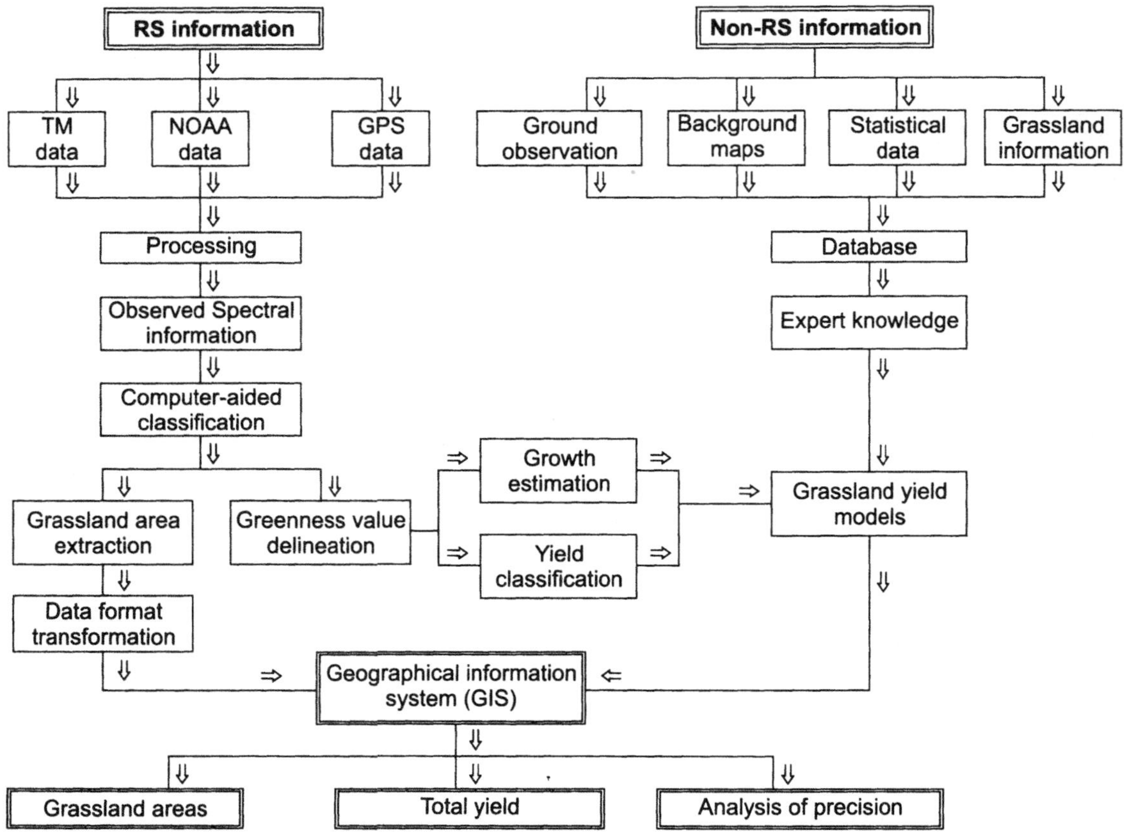

Figure 14.4 Procedure for yield estimation using remote sensing over large areas of grasslands
Source: Redrawn from Li, Jiang and Liang, 1998.

believed that ANDVI was more appropriate when using NOAA-AVHRR data to measure grassland productivity in China. Using their system, they monitored herbage yield and seasonal carrying capacity of Changji Prefecture, Xinjiang Autonomous Region. At the same time, Li, Jiang and Liang (1998) found that both ratio vegetation index (RVI) and NDVI correlated well with fresh herbage yields in four grassland types in Fukang county, Xinjiang Autonomous Region. Similar results have been reported by Ren, Liu and Li (1994) for estimation of northern grassland herbage yield.

The relationships between multi-temporal NOAA-AVHRR NDVI images and rainfall and temperature were studied at national (Li and Shi, 2000), regional (Xiao *et al.*, 1997) and local (Huang and Wang, 1994; Li, 1995) scales. Zhu, An and Long (1998) indicated that a spatial structure method could well be used to analyse the Landsat TM image and estimate grassland herbage yield. Liu *et al.* (2000) modelled the spatial pattern of NDVI in a post-nomadic sedentary grazing system in the Xilin river region, Inner Mongolia. Wang and Yong (1990) established the correlation between spectral reflectance and grassland herbage yield at Xilingol, Inner Mongolia. To improve accuracy of herbage yield estimation, relationships between contents of leaf chlorophyll and NDVI were established (Zhang *et al.*, 1997), and a method was developed of measuring with NOAA-AVHRR data standing litter on grassland (Pei, Li and Fan, 1995).

GRASSLAND DEGRADATION AND DESERTIFICATION ASSESSMENT

Grassland degradation is the major problem challenging agricultural production, rural development and environmental improvement in China. About 90 percent of total usable grassland in the country has been degraded to various extents, with about 1.35×10^9 ha of medium-degraded grassland. The degradation has mainly occurred in northern and alpine grasslands, the area degraded being about 41.5 percent of the total area of these two categories of grassland. More importantly, the degraded grassland is expanding, at a rate of about 2 million hectares annually. The area degraded annually is equivalent to 0.5 percent of the total grassland area in China, while the area improved annually for the whole country is a mere 0.3 percent of all grassland in the country (X.L. Li, pers. comm.). This shows clearly that currently the rate of grassland degradation is greater than the rate of grassland improvement, despite Chinese governments at all levels making huge efforts to combat grassland degradation. Degraded grassland produces much less herbage biomass and of poorer quality than non-degraded grassland (Table 14.6). Consequently there are serious problems for livestock production and environmental protection, not least among which are sand storms.

Table 14.6 Plant community changes in the degraded grasslands of Zhenglan county, Inner Mongolia, PR China

Measurement	Degree of degradation		
	Light	Medium	Severe
Dominant species	Lc+Ac+Cs+Af	Af+Cs+Pa+Thm	Cha+As+Ks+Pa
Sward height (cm)	29	14	3
Coverage (percent)	26.5	18	11
DM yield (kg ha^{-1})	1 287	576	410

Key to species: Lc = *Leymus chinensis*. Af = *Artemisia frigida*. Ac = *Agropyron cristatum*. Cs = *Cleistogenes squarrosa*. Pa = *Potentilla acaulis*. Thm = *Thymus mongolicus*. Cha = *Chenopodium acuminatum*. As = *Artemisia sieversiana*. Ks = *Kochia scoparia*.
Source: From Ren and Shen,1990.

Grassland degradation control and environmental improvement have been given top priority by the Chinese central government in development policy for western China. However, due to the extensive and diverse nature of the land, and isolation from the centres of society, it is extremely difficult to achieve the goal. For this reason, RS technology has enhanced grassland degradation assessment. Satellite data from Landsat TM, Landsat MSS and NOAA-AVHRR have been used in most of the studies carried out in China.

At the county and regional levels, Pei and Pan (1993) and Pei, Ao and Pan (1994) used NOAA-AVHRR information to assess the area and degree of steppe grassland degradation in Xilingol Prefecture, Inner Mongolia. They concluded that the percentage of grassland area degraded was 23.83 percent for meadow grassland, 45.62 percent for typical steppe grassland and 73.85 percent for desert grassland.

Chen, Wei and Liang (1998a,b) studied alpine grassland productivity in Dari county, Qinghai province. They compared the data from the ground survey taken in 1985 with the information obtained from both NOAA-AVHRR and Landsat TM in 1997, using ERDAS and ARC/INFO, and found that the area of severe degraded grassland had increased from 0.9 percent in 1985 to 5.52 percent of the total grassland area in 1997, a rate of 4000 ha year^{-1}. Tu and Shi (1998) employed

Landsat TM data to survey degraded alpine grassland on the Qinghai-Tibetan plateau and concluded that, among the grassland types studied, alpine meadow was degraded the most severely.

Landsat TM has also been used to survey water erosion in grasslands in western China (Liu *et al.*, 2001), and Landsat TM, Landsat MSS and aerial photos have been used to assess soil and water erosion in Inner Mongolia (Shi *et al.*, 1987). The results indicated that medium and severe water erosion occurred on 63.19 percent of total grassland areas in 12 provinces located in western China. Among the 12 provinces, Xinjiang, Sichuan, Gansu, Inner Mongolia, Shaanxi, Yunnan and Qinghai were the provinces with the most severe water erosion (Liu *et al.*, 2001). About 74.1 percent of total land area in Inner Mongolia showed erosion problems, of which 75.4 percent was attributed to wind erosion of grasslands (Shi *et al.*, 1987). RS technology has been used for monitoring grassland desertification routinely. Wang, Zhao and Xiao (1999) systematically monitored desertification progress in northern China (Figure 14.5). They reported that the area of grassland desertification had increased by 312 000 ha year^{-1} in northern China, based on comparison of aerial photographs taken in the 1950s and 1970s. With the additional support of GIS and Landsat TM images from 2000, they found that the desertified area had increased continuously, and extended over an area of 38.58 million hectares by 2000, having expanded at an annual rate of 350 000 ha year^{-1}, from a total desertified area of 33.9 million hectares in 1987 (Wang *et al.*, 2003). A multi-disciplinary research effort indicated that grassland degradation and desertification were the major causes of sandstorm weather occurring in northern China (Ye *et al.*, 2000).

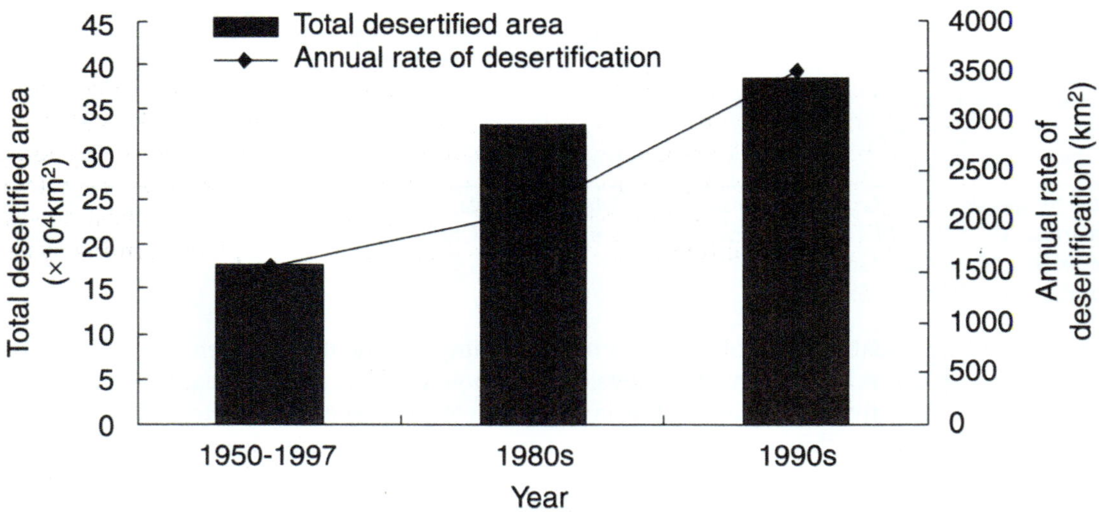

Figure 14.5 Desertification in northern China from the 1950s to the 1990s
Source: Based on data from Wang *et al.*, 2003.

STRESS DETECTION AND MONITORING

RS technology combined with GIS has been used in detecting and monitoring both biotic and abiotic stresses in grassland ecosystems (Nan, 2001). During the last decade in China the technology has mainly been used in fire monitoring, snow detection, drought evaluation and rodent damage assessment.

FIRE DETECTION AND MONITORING

Fire is a frequent disaster on China's grasslands. It has been estimated that more than 50 000 fire disasters have occurred since 1949. The accumulated area damaged by fire amounts to over 200 million hectares of grassland, and has caused losses exceeding US$ 7 500 million. A Fire Control Office was established under the aegis of the Ministry of Agriculture in 1991, and RS technology has been employed in fire detection and monitoring since then (Zhang, 1998). The pilot studies for use of NOAA-AVHRR data in fire detection were carried out in the 1980s (Hu, Pei and Wu, 1993) and a technical system developed and used routinely in grassland fire detection (Su and Liu, 1995, 1998; Dong and Su, 2002). From 1994 to 1998, 2 287 grassland fires were detected in the provinces of Inner Mongolia, Heilongjiang, Xingjiang, Jilin and elsewhere in northern China (Su and Liu, 1998). About 55 percent of the detected fires occurred in Inner Mongolian grassland. March to June each year is the most likely season for fire (Li *et al.*, 1997; Liu, Su and Li, 1999). This information has greatly helped fire prevention and detection, and consequently annual economic losses due to fires have fallen by about US$ 100 million (Zhang, 1998).

DETECTION OF SNOW COVER AND DEPTH

Snow, the so-called "white disaster" in China's pastoral regions, is another problem for grassland livestock production. It has been estimated that more than 70 snow disasters have occurred in China's grassland during the last 50 years, causing economic losses of nearly US$ 12.5 million (Zhou, Cheng and Lu, 1999). Efforts have been made to detect snow coverage and depth in grasslands using RS technology, and Bai and Feng (1997) have summarized progress. During 1990–1995, a project called *Snow disaster monitoring and assessment using remote sensing in main pastoral areas of China* was implemented by Lanzhou Institute of Glaciology and Geocryology (LIGG) of the Chinese Academy of Sciences, with support from the Chinese Ministry of Science and Technology. Naqu Prefecture of Tibetan Autonomous Regions was selected as the target area for study. An Operational System for Snow Disaster Monitoring and Assessment (SDMAOS) was developed, combining NOAA-AVHRR data and ground survey data. The results of this project were published in the Institute's Memoirs series in 1995 by Science Press, China. The majority of the works reported have used NOAA-AVHRR as the information resource. However, the Chinese climate satellite, FY–1C, has also been used to study the snow cover area in Inner Mongolia and Xinjiang in 2000–2001 (Liu, Li and Zhang, 2001).

Areas covered by snow at national and region scales have been detected with the support of NOAA-AVHRR data (Liu, Yuan and Zhang, 1992; Xu and Tian, 2000). The snow depth is more critical for livestock production than simply snow coverage. Technically more complicated systems were needed for detecting snow depth using RS technology. Progress has been made in developing techniques and methods for detecting snow depth, although none have been used routinely at the practical level (Li and Zhang, 1995; Lu, Feng and Zeng, 1996; Lin and Fan, 2003). Yong, Tong and Yong (1994) developed a snow depth index (SDI) based on Channel 1, 2 and 4 information from NOAA-AVHRR, the correlation between NOAA-AVHRR data and ground survey data was established, and the grassland vegetation map was then overlain with the snow depth map to detect the snow thickness in Xilingol grassland, Inner Mongolia. The results indicated that 86.32 percent of the grassland area monitored was covered by more than 50 cm of snow in the winter of 1992. A similar study was also carried out in Inner Mongolia by Shi and Chen (1996). They concluded that information from NOAA-AVHRR can be used to determine the area covered by snow. The snow depth in the area concerned could be determined from the satellite information in combination with the snow information obtained from ground climate stations and topographic information of the target area, collated using GIS technology.

DROUGHT EVALUATION

Drought is one of the most important limiting factors in crop and grassland production in China since most areas in northern China are classed as arid and semi-arid regions. Studies of soil water content using RS technology in China occurred much later than in other developed countries in the world. Wang (1994b) and Li, Xu and Tang (2000) summarized progress in this field in China. Several reports were made on determining grassland soil moisture contents with NOAA-AVHRR data and GIS. Examples include Pei and Ao (1996), who generated a soil drought map for the Inner Mongolia grasslands, and Ai *et al.* (1996) found that temporal and spatial characteristics of soil moisture content in the Changji grassland, Xinjiang Autonomous Region, could be effectively monitored by combining Landsat TM and NOAA-AVHRR data. From comparison of soil moisture content data obtained from ground measurement and from NOAA-AVHRR, Liu and Li (1996) concluded that the accuracy using RS technology to determine the soil moisture content of grassland in Xinjiang could reach 75–97 percent, depending on the vegetation type. Liang *et al.* (1998) and Liang, Wang and Dai (2000) simulated the annual rainfall distribution and runoff on the Loess Plateau, using ARC/INFO GIS, and reported that the runoff in both large and small streams could be simulated by a digital elevation model (DEM) and by one of the runoff coefficient methods.

BIOTIC STRESS DETECTION

Grasslands are subjected to biotic stresses such as plant disease, insect pests, rodents and noxious plants, all of which militate against grassland productivity reaching its potential. It has been estimated that the annual yield losses of forages caused by plant diseases in China is about US$ 317 million (Nan, 2000). In 2000, the area damaged by insect pests and rodents in 12 provinces in northern China was about 17.4 percent of all the grasslands (Yang, 2001). Significant inputs were made by governments at all levels to control grassland pests, but few studies have used RS technology to detect grassland pests, although the technology has been used successfully in other countries (Jackson, 1986; Everitt, Escobar and Davis, 1995; Nelson, Orum and Ramon, 1999). The few examples found in the Chinese literature included Wang, Liang and Chao (2001), who attempted to develop an expert system for alfalfa diseases diagnosis, and Zhou *et al.* (1999), who identified alpine grassland areas damaged by *Myospalax baileyi*, *Ochotona curzoniae* and *Marmeta himalayana*, using NOAA-AVHRR data, knowledge of which will improve the efficiency of rodent control.

THE FUTURE

As humankind enters the twenty-first century, facilitating sustainable development of grassland is becoming a new target for application of RS technology. Information technology has developed rapidly, including diverse earth observing satellite networks with coarse, fine and very high resolutions; and the integration of RS, GIS and GPS. A new era of Digital Earth is approaching, based on high-speed computers and the Internet. All these recent developments will enhance the use of RS technology in all sectors of agriculture. One can confidently predict that RS technology will play a more important role in grassland resource management than before, although it will take some time for the technology to be fully used in daily management and production in the country. From a scientific point of view, the following section discusses some aspects of use of RS technology in grassland resources management expected in the near future in China, illustrated by recent achievements in this area in China.

DEVELOPING PRECISION GRASSLAND AGRICULTURE

Precision agriculture is one of the frontiers in world agriculture. Based on the information obtained using RS technology, GIS and GPS, plant and soil management practices can be adjusted according to conditions (Jin, 1998). During the last 15 years the central government has continuously supported research and development projects in information technology. Significant progress has been made in development of hardware, software and implementation of techniques in resources survey and management. Some of those achievements will be extended nationwide to speed up the application of information technology to grassland management and to develop precision grassland agriculture in the near future.

By integrating RS, GIS and GPS, an *Expert System for Grassland Husbandry and Ecological Construction*, for instance, has been developed for Gansu province by Ren, Liang and Chen (2001). The System comprises text, graphics, image and video information, and includes 19 databases covering soil, topography, climate, grassland resources, 15 knowledgebases to help in the identification and control of pasture plant diseases, insect pests, rodents, weeds, forage cultivation and grassland classification, and 5 model databases (Liang, Chen and Su, 2001). The System has three main functions: (i) to provide information on any site within Gansu province, and its administrative status; (ii) to indicate the natural conditions of the site concerned; and (iii) to suggest suitable forage species for the particular area, and proper management for forage cultivation and pest control. This work is seen as a good example of application of information technology in grassland production in China, and ranks as one of China's outstanding scientific achievements. This System now is available in the form of a CD-ROM so that anyone, including farmers who have their own computer, could use the System to help grassland production in the province. The project indicates the future use of RS technology in China and has provided a good basis for developing precision grassland agriculture. In 2003, the Ministry of Science and Technology launched a national programme in China called *Reducing Digital Gaps Between East and West*, to promote the use of information technology in western China. With support from this programme, the System is being extended to include livestock production, and the coverage is being expanded, to include seven provinces in western China (Ren, Liang and Chen, 2001).

ENTERING A NEW DIGITAL GRASSLAND ERA

About 15 years ago, the Chinese Ministry of Agriculture indicated that future application of RS technology in grassland production should be able to report grassland productivity at monthly intervals on the national television and radio stations, analogous to routine weather forecasting, so that policy-makers, grassland managers and farmers at different levels can adjust their management practices according to changing circumstances. Recently, Guangdong province, one of the most developed provinces in China, has set up a GIS network integrating ten GPS receiving stations and a central facility. The network will combine RS, GPS and GIS, and will be utilized for sectors of agriculture, meteorology, communication and security of the province.

Promoting reversion of steep sloped arable land to pastures and forests is a major task in western China development policy, and has been given top priority by central government. RS technology has been used to facilitate the process. Chen, Liang and Xu (2001), for instance, use Landsat TM and SPOT (System pour l'Observation de la Terre) data with GIS to complete the plan for reverting arable land to pastures in Dingxi county, Gansu province. Similarly, Yang *et al.* (2001) developed a decision-support system, including RS and GIS, for Yunnan province. The system was specially designed for reverting arable land to pasture and forest.

At regional scale, a grassland information system based on RS, GIS and GPS for western China has been completed (Wu *et al.*, 2001). A national grassland monitoring and warning system has been under construction since 2000, using modern information technology (Tang, Miao and Li, 2001). A national grassland fire monitoring system based on MODIS (Moderate-resolution Imaging Spectroradiometer) data is also under development with the support of the Chinese Ministry of Science and Technology. All of these represent a dawn for digital grassland activities, and indicate a bright future for application of RS technology, GIS and GPS in grassland production and resource management in China.

DEVELOPING THE SCIENCE AND TECHNOLOGY OF GRASSLAND REMOTE SENSING

Application of RS technology to grassland management is not an art, it is an interdisciplinary science. The two major limiting factors to grassland RS application in China are shortage of skilled personnel and lack of multidisciplinary teamwork. Currently, there are not enough experienced and skilled personnel in China with knowledge in both RS and grassland science. Very few grassland scientists and extension officers have been taught to apply grassland RS. Recently, RS technology as a subject has been taught to students majoring in grassland science at undergraduate level in some Chinese universities. Postgraduate students with a grassland science background have studied RS, or *vice versa*, at both Master and PhD levels.

At the same time, grassland is perhaps the most complicated ecosystem, with much variability within and between ecosystems globally, differing in species present, forage productivity, soils, climate, species of grazing animal and human activities and management. Past experience shows that the most efficient and successful application of RS technology has derived from multidisciplinary approaches. As the use of the technology spreads, inter-disciplinary barriers are breaking down, and the traditional Chinese philosophy of self-support is changing, making multidisciplinary teamwork a necessity. New theories and technical systems can be expected to be developed, and the science and technology of grassland RS as a branch of grassland science will gain formal recognition in both China and the world.

RS technology, together with GIS and GPS, will become an integral part of precision grassland agriculture in the near future. As the largest terrestrial ecosystem in China, grasslands have played a key role in the development of civilization. Grasslands not only provide forages to feed livestock, but also play a critical role in alleviating many of the most challenging environmental and ecological problems that humankind is facing. These include desertification, land degradation, climate change, and loss of biodiversity. As more grassland scientists obtain more knowledge of

RS systems, they will become more cost effective, and digital grassland will come of age for grassland resource management.

ACKNOWLEDGEMENTS

This paper was supported in part by the National Key Basic Research Foundation Project of China (G2000018602). The author thanks Professor Liu Zhong-Ling, Professor Chen Quan-Guang, Professor Han Jian-Guo, Dr Liang Tian-Gang and Mr Nie Bin for providing valuable references, and Ms Ma Yan-Jing for typing tables and generating the figures.

REFERENCES

Ai, L.X., Liu, P.J., Zhang, L. & Chang, P. 1996. Monitoring soil water distribution at Changji prefecture, Xinjiang, with remote sensing technology [in Chinese]. *Xinjiang Meteorology*, **19**(1): 18–21.

Bai, Y.C. & Feng, X.Z. 1997. Introduction to some research work on snow remote sensing [in Chinese, with English abstract]. *Remote Sensing Technology and Application*, **12**(2): 59–65.

Chen, L.J., Liu, G.H. & Feng, X.F. 2001. Estimation of net primary productivity of terrestrial vegetation in China by remote sensing. *Acta Botanica Sinica*, **43**(11): 1191–1198.

Chen, Q.G. (ed). 1991. *Grassland Resource and Animal production in the Naqu Region of Tibet* [in Chinese]., Lanzhou, PR China: Gansu Science and Technology Press. 210p.

Chen, Q.G., Wei, Y.X. & Liang, T.G. 1994a. The monitoring of grassland resources using NOAA data [in Chinese, with English abstract]. *Pratacultural Science*, **11**(1): 56–63.

Chen, Q.G., Wei, Y.X. & Liang, T.G. 1994b. Researches on monitoring method of grassland biomass yield and carrying capacity by using NOAA-AVHRR information [in Chinese, with English abstract]. *Acta Prataculturae Sinica*, **3**(4): 50–60.

Chen, Q.G., Wei, Y.X. & Liang, T.G. 1998a. The investigation of grassland degradation in Dari county of Qinghai province I: Remote sensing and mapping [in Chinese, with English abstract]. *Acta Prataculturae Sinica*, **7**(2): 58–63.

Chen, Q.G., Wei, Y.X. & Liang, T.G. 1998b. Causes of grassland degradation in Dari county of Qinghai province [in Chinese, with English abstract]. *Acta Prataculturae Sinica*, **7**(4): 44–48.

Chen, Q.G., Liang, T.G., Mao, Y.L. & Wei, Y.X. 1998. Applications of NOAA-AVHRR data on grassland monitoring. pp. 383–392, *in:* U.N. Singh, H.L. Hu and G.C. Wang (eds). *Optical Remote Sensing for Industry and Environmental Monitoring*.Proceedings of a conference, . Beijing, China 15-17 September 1998. SPIE – The International Society for Optical Engineering, Vol. 3504.

Chen, Q.G., Liang, T.G. & Xu, Z.B. 2001. Research based on 3S technology in monitoring of ecological environment and restoring native vegetation to hill slopes in Dingxi experimental area of Gansu province [in Chinese, with English abstract]. *Journal of Lanzhou University*, **37**: 68–71.

Dong, Y.P. & Su, H. 2002. Information extraction for grassland fire using NOAA-AVHRR image of channel 1, 2, 4 and 5 [in Chinese, with English abstract]. *Acta Agrostia Sinica*, **10**(4): 298–301.

Durand, J.L., Emile, J.C., Huyghe, C. & Lemaire, G. (eds). 2002. *Multi-function grasslands: quality forages, animal products and landscapes*. Proceedings of the 19th General Meeting of the European Grassland Federation. La Rochelle, France, 27–30 May 2002, 1124p.

Everitt, J.H., Escobar, D.E. & Davis, M.R. 1995. Using remote sensing for detecting and mapping noxious plants. *Weed Abstracts*, **44**(12): 639–649.

Fu, B.J. 1995. The spatial pattern analysis of agricultural landscape in the Loess area [in Chinese, with English abstract]. *Acta Ecologica Sinica*, **15**(2): 114–120.

Ge, C.S., Zhao, M.C. & Zheng, J.Y. 2000. Land use change of China during the 20th century [in Chinese, with English abstract]. *Acta Geographica Sinica*, **55**(6): 698–706.

Gong, Z.T. & Xu, S.S. (eds). 1991. *Application of Remote Sensing to Loess Plateau Resource Survey* [in Chinese]. Nanjing, PR China: Nanjing Press. 243p.

Hu, T., Pei, H. & Wu, R. 1993. A study of using satellite data to monitor grassland fire [in Chinese]. pp. 208–215, *in:* Li Bo (ed). *Research on Dynamics Monitoring of Grazing Ecosystems in the North of China*, Beijing: China Agricultural Science and Technology Press.

Hu, Z.Z., Zhang, P.J., Nan, Z.B., Fan, X.R. & Zheng, W.L. 1978. Grassland classification of Gansu province [in Chinese, with English abstract]. *Journal of Gansu Agricultural University*, **1**: 1–26.

Hu, Z.Z. (ed). 1997. *Introduction to Grassland Taxonomy* [in Chinese]. Beijing: China Agricultural Press. 261p.

Huang, F., Liu, X.N., Zhang, Y.Z. & Zhao, H.Y. 2003. Study on spatial-temporal patterns of Songnen sandy land use/cover based on RS and GIS technology [in Chinese, with English abstract]. *Journal of Desert Research*, **23**(2): 136–141.

Huang, J.F. & Shang, C.Q. 1993. The remote sensing dynamic monitoring model of the grass yield of natural grassland in the middle section of the Tianshan north slope [in Chinese, with English abstract]. *Journal of Natural Resources*, **8**(1): 10–17.

Huang, J.F. & Wang, X.Z. 1994. Application of weather satellite data to Xingjiang natural grassland monitoring [in Chinese, with English abstract]. *Prataculturál Science*, **11**(4): 7–72.

Huang, J.F., Wang, X.Z., Wang, R.C. & Hu, X.B. 2001. A study on monitoring and predicting models of grass yield in natural grassland using remote sensing data and meteorological data [in Chinese, with English abstract]. *Journal of Remote Sensing*, **5**(1): 69–74.

Jackson, R.D. 1986. Remote sensing of biotic and abiotic plant stress. *Annual Review of Phytopathology*, **24**: 265–287.

Jiang, D.F., Wang, G.H. & Yu, T.G. 1964. Developing grassland vegetation map with aerial photography [in Chinese]. pp. 83–89, *in:* Proceedings of Grassland Conference for the western region of Northeastern China, The Botany Society for Northeastern China. Harbin, Heilongjiang, China, 15–23 October, 1964,

Jia, S.X., Xia, J.X. & Chen, M.X. 1981. Application of remote sensing to grassland resource survey [in Chinese]. *Chinese Journal of Animal Science*, **5**: 25–31.

Jiang, Z.G. & Li, D.Q. 1999. Land cover change and conservation of Przewalski's gazelle and Pere David's deer [in Chinese, with English abstract]. *Journal of Natural Resources*, **14**(4): 334–338.

Jiao, F. 2000. Use of information system for land resource monitoring in Loess Plateau [in Chinese]. *Natural Resources and Ecological Environmental Study Network*, **11**(1): 31–35.

Jin, J.Y. 1998. Precision agriculture and its perspective in China [in Chinese, with English abstract]. *Plant Nutrition and Fertilizer Science*, **4**(1): 1–7.

Kawamura Kensuke, Tsuyoshi Akiyama, Osamu Watanabe, Hisahito Hasegawa, Zhang, F.P., Hiro-omi Yokota & Wang, S.P. 2003. Estimation of aboveground biomass in Xilingol steppe, Inner Mongolia, using NOAA/NDVI. *Grassland Science*, **49**(1): 1–9.

Li, B. (ed). 1987. *Application of Remote Sensing to the Inner Mongolian Grassland Resource Survey.* Huhhot, PR China: Inner Mongolian University Press. 545p.

Li, B. (ed). 1993. *Research on Dynamics Monitoring of Grazing Ecosystem in the North of China* [in Chinese]. Beijing: China Agricultural Science and Technology Press. 225p.

Li, B. 1994. Ecology and grassland management [in Chinese, with English abstract]. *Grassland of China*, **1**: 1–8.

Li, C.Y., Dai, Y.J., Zheng, Z.Y. & Zhang, M. 1997. Climatic analysis and prediction of fire for Zhelimu grassland [in Chinese]. *China Agricultural Meteorology*, **18**(3): 30–32.

Li, J.L. 1995. Application of remote sensing environment comprehensive technology system in grassland resources monitor and management [in Chinese, with English abstract]. *Journal of Xinjiang Agricultural University*, **18**(3): 59–64.

Li, J.L. & Jiang, P. 1996. Studies on grassland large area yield estimation using remote sensing and environment data [in Chinese, with English abstract]. *Prataculturál Science*, **13**(2): 14–20.

Li, J.L., Jiang, P. & Dai, R.L. 1998. Advances in study on the remote sensing technology and GPS and GIS integration systems in estimating grassland yield applications in the north of Xinjiang, China [in Chinese, with English abstract]. *Acta Ecologica Sinica*, **18**(5): 504–510.

Li, J.L., Jiang, P. & Liang, T.G. 1998. The developing progress, contents and prospects of grassland remote sensing science in China [in Chinese, with English abstract]. *Grassland of China*, **3**: 51–56.

Li, J.L., Liang, T.G. & Chen, Q.G. 1998. Estimating grassland yields using remote sensing and GIS technology in China. *New Zealand Journal of Agricultural Research*, **41**: 31–38.

Li, Y.C., Xu, M. & Tang, Y. 2000. The status and advances of thermal inertia models in measuring soil moisture by remote sensing in China [in Chinese, with English abstract]. *China Agricultural Meteorology*, **21**(2): 40–43.

Li, X.B. & Shi, P.J. 2000. Sensitivity analysis of variation in NDVI, temperature and precipitation in typical vegetation types across China [in Chinese, with English abstract]. *Acta Phytoecologica Sinica*, **24**(3): 379–382.

Li, Z. & Zhang, W.Y. 1995. Extracting the information of snow-cover from NOAA-AVHRR data and overlaying with vector data [in Chinese, with English abstract]. *Remote Sensing Technology and Application*, **12**(4): 19–24.

Liang T.G., Chen, Q.G. & Su, Q.Z. 2001. The advances of database, knowledge base and model base of Gansu expert systems for pratacultural development [in Chinese, with English abstract]. *Journal of Lanzhou University*, **37**: 26–40.

Liang, T.G., Chen, Q.G. & Wei, Y.X. 1996. Research on the dynamic monitoring model of grassland resources yield in Fukang county of Xinjiang [in Chinese, with English abstract]. *Remote Sensing Technology and Application*, **11**(1): 27–32.

Liang, T.G., Wei, Y.X. & Chen, Q.G. 1994. Research on the series mapping methods of grassland resources using PC ARC/INFO [in Chinese, with English abstract]. *Pratacultural Science*, **6**: 55–60.

Liang, T.G., Wang, X.Z. & Dai, R.L. 2000. Simulating methods of average annual rainfall on spatial changes [in Chinese, with English abstract]. *Acta Botanica Boreali-Occidentalia Sinica*, **20**(5): 856–862.

Liang, T.G., Zhang, S.L., Dai, R.L. & Xu, Y.Q. 1998. Modelling of surface runoff based on raster system in GIS for rain-harvesting agriculture [in Chinese, with English abstract]. *Journal of Hydrology*, **7**: 26–29.

Liao, G.F. & Jia, Y.L. 1996. *Rangeland Resources of China* [in Chinese]. Beijing: China Science and Technology Press. 607p.

Lin, J. & Fan, H.J. 2003. A method to monitor heavy snow-related disasters in the Inner Mongolia area [in Chinese, with English abstract]. *Meteorology*, **29**: 27–32.

Liu, C.R. & Chen, L.Z. 1999. Distribution characteristics of patch size in the vegetation landscape in Beijing [in Chinese, with English abstract]. *Acta Botanica Sinica*, **41**(2): 199–205.

Liu, C., Li, Y.J. & Zhang, H.P. 2001. Monitoring snow disaster with FY-1C satellite [in Chinese]. *Remote Sensing Information*, **1**: 42–44.

Liu, F.Y. & Jia, S.X. 1986. Spectral reflection features of grassland vegetation [in Chinese]. *Grassland of China*, **6**: 26–30.

Liu, F.Y., Li, Z.Y., Han, J.G. & Xia, J.X. 1990. Application of remote sensing to grassland resource survey at stony mountainous region in northern China [in Chinese]. pp. 94–103, *in:* X.W. Zhao (ed). *Renewable Resources Remote Sensing – Areas for Establishing Protective Forests in Stony Mountainous Region in Northern China*. Beijing: China Forest Press.

Liu, G.X. 1994. Approach of NOAA-AVHRR data applied in dynamic monitoring of grassland resources [in Chinese, with English abstract]. *Pratacultural Science*, **11**(4): 43–45.

Liu, G.X., Su, H. & Li, S.L. 1999. The summarization on the fire accident of grassland in Inner Mongolia [in Chinese, with English abstract]. *Grassland of China*, **4**: 76–78.

Liu, J.J. & Pan, F. 2001. Dynamic and analysis of land use in the lower region of Talimu river [in Chinese, with English abstract]. *Arid Environmental Monitoring*, **15**(3): 142–146.

Liu, P.J. & Li, L.X. 1996. Accuracy of monitoring soil water contents with remote sensing technology [in Chinese]. *Xinjiang Meteorology*, **19**(1): 25–27.

Liu, X.H., Chen, Z.Z., Tsuyoshi Akiyama, Mo, W.H. & Ayumi Fukuo. 2000. Modelling the spatial pattern of normalized difference vegetation index (NDVI) under a post-nomadic sedentary grazing system [in Chinese, with English abstract]. *Acta Phytoecologica Sinica*, **24**(6): 662–666.

Liu, Y.J., Yuan, X.Y. & Zhang, H. 1992. Snow cover monitoring using meteorological satellite data [in Chinese, with English abstract]. *Remote Sensing of Environment*, **7**(1): 24–31.

Liu, Y.P., Wang, Q., Wang, W.J. & Shen, W.M. 2001. Present situation of water erosion in west China [in Chinese, with English abstract]. *Journal of Soil and Water Conservation*, **15**(6): 17–19.

Lu, A.X., Feng, X.Z. & Zeng, Q.Z. 1996. Study on the application of GIS in the research of pastureland snow disaster [in Chinese, with English abstract]. *Journal of Catastrophology*, **11**(1): 2–29.

Ma, M.G., Wang, X.M., Jiao, Y.M. & Chen, X.Z. 2003. Study on the change of landscape pattern of oasis in arid region based on RS and GIS methods [in Chinese, with English abstract]. *Journal of Desert Research*, **23**(1): 53–58.

Mu, X.D., Long, R.J., Chen, G. & Kang, T.F. 1993. A study on spectral reflection character in alpine grassland [in Chinese, with English abstract]. *Acta Prataculturae Sinica*, **2**(2): 8–10.

Nan, Z.B. 2000. Establishing sustainable management system for diseases of pasture crops in China [in Chinese, with English abstract]. *Acta Prataculturae Sinica*, **9**(2): 1–9.

Nan, Z.B. 2001 Remote sensing application to grassland monitoring. pp. 1037–1042, *in:* Proceedings of the XIX International Grassland Congress. Sao Pedro, Sao Paulo, Brazil, 11–21 February 2001.

Nelson, M.R., Orum, T.V. & Ramon, J.G. 1999. Applications of geographic information systems and geostatistics in plant disease epidemiology and management. *Plant Disease*, **83**(4): 308–319.

Niu, J.M., Hu, H. & Lu, G.F. 1998. Trends of remote sensing and geographical information system (GIS) applications on vegetation ecology [in Chinese, with English abstract]. *Acta Scientiarum Naturalium Universitatis NeiMongol*, **29**(4): 569–573.

Pei, H. & Ao, Y.Q. 1996. The research on the measures of monitoring soil drought by polar orbit meteorology satellite [in Chinese, with English abstract]. *Grassland of China*, **5**: 40–45.

Pei, H., Ao, Y.H. & Pan, Y.Z. 1994. Study on degraded grassland [in Chinese, with English abstract]. *Grassland of China*, **1**: 58–61.

Pei, H. & Pan, Y.Z. 1993. Monitoring grassland degradation at Xilingol, Inner Mongolia using satellite data [in Chinese]. pp. 203–207, *in:* Li Bo (ed). *Research on Dynamics Monitoring of Grazing Ecosystems in the North of China*. Beijing: China Agricultural Science and Technology Press.

Pei, H., Li, Y.P. & Fan, Y.D. 1995. Monitoring herbage dry mass in temperate grassland by NOAA-AVHRR meteorological satellite [in Chinese, with English abstract]. *Grassland of China*, **6**: 44–47.

Piao, S.L., Fang, J.Y. & Guo, Q.H. 2001a. Terrestrial net primary production and its spatio-temporal patterns in China during 1982-1999 [in Chinese, with English abstract]. *Acta Scientiarum Naturalium Universitatis Pekinensis*, **37**(4): 563–569.

Piao, S.L., Fang, J.Y. & Guo, Q.H. 2001b. Application of CASA model to the estimation on Chinese terrestrial net primary productivity [in Chinese, with English abstract]. *Acta Phytoecologica Sinica*, **25**(5): 603–608.

Ren, J.Z. (ed). 1984. Survey of grassland resources with remote sensing technology for Tianzhu county of Gansu province [in Chinese]. Special Issue of the Chinese Journal of Grassland and Forage Science, 146p.

Ren, J.Z. (ed). 1985. *Grassland Agro-ecology* [in Chinese]. Beijing: China Agricultural Press. 185p.

Ren, J.Z. 1999. A preliminary discussion on grassland ecological-economic regions in China [in Chinese, with English abstract]. *Acta Prataculturae Sinica*, **8**(Suppl.): 12–22.

Ren, J.Z. & Huang, W.H. 1978. Comparison of grassland science and technology between China and other developed countries [in Chinese]. Special Report to the Ministry of Agriculture. Published by the Chinese Academy of Agricultural Sciences, Beijing. 24p.

Ren J.Z., Hu Z.Z. & Mu, X.D. 1979. A study on developing animal product unit [in Chinese]. *Chinese Journal of Animal Science*, **2**: 21–26.

Ren, J.Z., Hu, Z.Z., Mu, X.A. & Zhang, P.J. 1980. Integrated orderly grassland classification system and its significance in grassland development [in Chinese, with English abstract]. *Grassland of China*, **1**: 1–8.

Ren, J.Z., Liang, T.G. & Chen, Q.G. 2001. Advances of ecological construction and expert systems for pratacultural development in Gansu province [in Chinese, with English abstract]. *Journal of Lanzhou University*, **37**: 1–8.

Ren, J.Z. & Shen, Y.Y. 1990. The ecological crisis faced by the China's grasslands and the strategies for control the degradation [in Chinese]. *Research of Agricultural Modernization*, **11**(3): 9–12.

Ren, J.Z., Wang, Q., Mu, X.D., Hu, Z.Z., Fu, Y.K. & Sun, J.X. 1978. Grassland production process and seasonal grassland livestock production system [in Chinese, with English abstract]. *Acta Agricultura Sinica*, **2**: 87–92.

Ren, Z.B., Se, Y.B.T., Shi, Y.H., Wu, F.S. & Zhang, M.H. 1990. Remote sensing interpreting and charting on grassland of north Shaanxi province [in Chinese, with English abstract]. *Grassland of China*, **6**: 40–43.

Ren, Z.B., Liu, F.Y. & Li, Z.K. 1994. Grassland resource survey with remote sensing technology [in Chinese]. pp. 196–216, *in:* Xu, G.H. (ed). *Theory and Application of Remote Sensing Technology to Renewable Resource Survey in Areas for Establishing Protective Forests in Northern China*. Beijing: China Forest Press.

Se, Y.B.T., Ren, Z.B., Li, W. & Hu, H.J.L.T. 1999. The application of TM image to the investigation and evaluation of the exploitation and utilization of the land resources in the Northeast Inner Mongolia [in Chinese]. *Grassland of China*, **6**: 37–41.

Shi, P.J. & Chen, J. 1996. Study on monitoring snow disaster in large areas supported by GIS and RS [in Chinese, with English abstract]. *Acta Geographica Sinica*, **51**(4): 296–304.

Shi, P.J., Gong, P., Li, X.B., Chen, J., Qi, H. & Pan, Y.X. 2000. *Methods and Practice of Land Use and Land Cover Change Studies* [in Chinese]. Beijing: Science Press. 154p.

Shi, P.J., Jing, Z.P., Zhao, Y. & Hao, Y.C. 1987. Application of remote sensing technology to soil erosion survey in Inner Mongolia [in Chinese, with English abstract]. pp. 127–137, *in:* Inner Mongolian Grassland Remote Sensing Survey Team (eds). *Application of Remote Sensing to Inner Mongolian Grassland Survey*. Huhhot, PR China: Inner Mongolian University Press.

Su, D.X. 1996. The compilation and study of the grassland resource map of China on the scale of 1:1 000 000 [in Chinese, with English abstract]. *Journal of Natural Resources*, **11**(1): 75–83.

Su, H. & Liu, G.X. 1995. Application of NOAA satellite data to monitor fire disaster in Inner Mongolia grassland area [in Chinese, with English abstract]. *Grassland of China*, **2**: 12–14.

Su, H. & Liu, G.X. 1998. Research on real-time monitoring of grassland fire [in Chinese, with English abstract]. *Grassland of China*, **6**: 47–49.

Sun, R. & Zhu, Q.J. 2000. Distribution and seasonal change of net primary productivity in China from April 1992 to March 1993 [in Chinese, with English abstract]. *Acta Geographica Sinica*, **55**(1): 36–45.

Sun, R. & Zhu, Q.J. 2001. Effect of climate change of terrestrial net primary productivity in China [in Chinese, with English abstract]. *Journal of Remote Sensing*, **5**(1): 58–61.

Tang, H.J., Miao, J.M. & Li, W.F. 2001. Application of remote sensing to agriculture [in Chinese]. pp. 121–125, *in:* Q.X. Tong and L.Z. Zheng (eds). *Proceedings of Symposium on the 20 Years Development of Remote Sensing Technology Application in China*. Beijing: Meteorological Press. Beijing, China, 15–16 July, 2000.

Tian, G.J., Zhuang, D.F. & Liu, M.L. 2003. The spatial-temporal dynamic change of cultivated land in China in 1990s [in Chinese, with English abstract]. *Advance in Earth Sciences*, **18**(1): 30–36.

Tong, Q.X., Ding, Z., Zheng, L.F., Wang, E.H., Xiao, Q.G., Chen, W.Y. & Zhou, Z.S. 1986. A preliminary study on NOAA images for the estimation of pasture biomass [in Chinese, with English abstract]. *Journal of Natural Resources*, **1**(1): 87–95.

Tuller, P.T. 1989. Remote sensing technology for rangeland management applications. *Journal of Range Management*, **42**(6): 442–453.

Tu, J. & Shi, C.C. 1998. Study on the degeneration of alpine meadow grassland in Qingzhang plateau with remote sensing techniques [in Chinese, with English abstract]. *Acta Agrostia Sinica*, **6**(3): 226–233.

Wang, J.F. 1994a. Application of remote sensing technology to green area survey of Qingzhou city [in Chinese, with English abstract]. *Engineer Prospecting*, **3**: 65–66.

Wang, R.H. 1994b. Advance in monitoring soil water contents with remote sensing technology [in Chinese]. *Remote Sensing Technology and Application*, **6**: 1–5.

Wang, S.Y., Liang, T.G. & Chao, Z.H. 2001. Preliminary study on alfalfa disease diagnosis system in Gansu [in Chinese, with English abstract]. *Journal of Lanzhou University (Natural Sciences)*, **37**: 120–126.

Wang, S.Y., Liu, J.Y., Zhang, Z.X., Zhou, Q.B. & Zhao, X.L. 2001. Analysis on spatial-temporal features of land use in China. *Acta Geographica Sinica*, **56**(6): 631–639.

Wang, T., Wu, W., Xue, X., Zhang, W.M., Han, Z.W. & Sun, Q.W. 2003. Time-space evolution of desertification land in northern China [in Chinese, with English abstract]. *Journal of Desert Research*, **23**(3): 230–235.

Wang, T., Zhao, H.L. & Xiao, H.L. 1999. Advances in desertification research of China [in Chinese, with English abstract]. *Journal of Desert Research*, **19**(4): 299–311.

Wang, X.L. 1999. A research on the change of agricultural land use and impact on rural sustainable agricultural development in Hulunbeier distract based on remote sensed data [in Chinese, with English abstract]. *Progress in Geography*, **18**(4): 322–329.

Wang, X.Z., Du, G.Z., Liang, T.G., Dai, R.L. & Wang, G. 2001. The establishment of grassland productive assessment model and the determination of precipitation resource distribution pattern of Gannan Tibetan Autonomous Prefecture based on RS and GIS [in Chinese, with English abstract]. *Acta Prataculturae Sinica*, **10**(2): 95–102.

Wang, Y.R. & Yong, S.P. 1990. A correlative analysis between spectral reflectance and yield of grassland on Xilingol, Inner Mongolia [in Chinese, with English abstract]. *Acta Phytoecologica et Geobotanica Sinica*, **14**(3): 258–265.

Wang, Y.R. & Yong, W.Y. 1998. An analysis of spectral reflectance diurnal change for *Leymus chinensis* steppe in different seasons [in Chinese, with English abstract]. *Acta Scientiarum Naturalium Universitatis NeiMongol*, **29**(2): 214–222.

Wei, Y.X. & Chen, Q.G. 2001. Grassland classification and evaluation of grazing capacity in Naqu prefecture, Tibet Autonomous Region, China. *New Zealand Journal of Agricultural Research*, **44**: 253–258.

Wu, B. & Ci, L.J. 2001. Temporal and spatial patterns of landscape in the Mu Us Sandy land, Northern China [in Chinese, with English abstract]. *Acta Ecologica Sinica*, **21**(2): 191–196.

Wu, Q., Yang, B.J., Zhang, S.L. & Pei, Z.Y. 2001. 3S-based grassland information system for western China [in Chinese, with English abstract]. *Transactions of the CSAE*, **17**(5): 142–145.

Wu, Y.N. & Li, Z.H. 2000. Changing landscape diversity with time in Xilinguole steppe [in Chinese, with English abstract]. *Acta Phytoecologica Sinica*, **24**(10): 58–63.

Xia, J.X., Chia, S.S. & Chang, X.G. 2002. A study on application of Landsat imagery to grassland resource surveys. pp. 236–240, *in:* S.X. Jia. *Jia Shen-xiu Works*. Beijing: China Agricultural University Press.

Xia, J.X., Jia, S.X. & Chang, X.G. 1986. Application of remote sensing to grassland resource survey at Balinyouqi, Inner Mongolia [in Chinese, with English abstract]. *Grassland of China*, **6**: 11–16.

Xiao, Q.G., Chen, W.Y., Du, J. & Guo, L. 1997. Monitoring the ecological transect in east Asia monsoon region by meteorological satellite remote sensing [in Chinese, with English abstract]. *Acta Botanica Sinica*, **39**(9): 826–830.

Xiao, Q.G., Chen, W.Y., Sheng, Y.W. & Guo, L. 1996. Estimating the net primary productivity in China using meteorological satellite data [in Chinese, with English abstract]. *Acta Botanica Sinica*, **38**(1): 35–39.

Xu, G.H. (ed). 1994. *Theory and Application of Remote Sensing Technology to Renewable Resource Survey in Areas for Establishing Protective Forests in Northern China* [in Chinese]. Beijing: China Forest Press. 212p.

Xu, P., Zhu, J.Z. & An, S.Z. 1988. Study on the interpretation of grassland type by means of remote sensing image [in Chinese, with English abstract]. *Grassland of China*, **6**: 61–66.

Xu, W.X., Yang, Y.L., Jia, S.H., Li, F.X., Hu, L. & Yin, Q.J. 2001. Monitoring grassland productivity of the last 20 years at southeastern Qinghai province with remote sensing technology [in Chinese]. pp. 358–363, *in:* Q.X. Tong and L.Z. Zheng (eds). *Proceedings of Symposium on the 20 Years Development of Remote Sensing Technology Application in China*. Beijing: Meteorological Press. Beijing, China , 15–16 July, 2000,

Xu, X.R., Jing, L.F., Dian, C.G., Zhou, C.S. & Zhang, M. 1985. The principle and method for measurement of grass output of Inner Mongolian grassland using NOAA-CCT [in Chinese, with English abstract]. *Acta Geographica Sinica*, **40**(4): 333–346.

Xu, X.K. & Tian, G.L. 2000. Dynamic distribution and change of snowing in China [in Chinese, with English abstract]. *Journal of Remote Sensing*, **4**(3): 178–182.

Yan, C.Z., Wang, Y.M., Feng, Y.S., Wang, J.H. & Wu, W. 2000a. Macro-scale study on grassland cover of Ningxia Region by remote sensing in the digital way [in Chinese, with English abstract]. *Journal of Desert Research*, **20**(3): 298–300.

Yan, C.Z., Wang, Y.M., Feng, Y.S., Wang, J.H. & Wu, W. 2000b. Macroscopic study on Gansu grassland coverage by remote sensing in the digital way. *Grassland of China*, **2**: 1–7.

Yang, A.L. 2001. Grassland insect pests and rodents in 2000 [in Chinese]. *Prataculturai Science*, **18**(5): 23–26.

Yang, C.J., Liu, J.Y., Zhang, Z.X. & Dang, C.L. 2001. Decision on reforesting and regrassing some of the dry sloping cultivated land with support of GIS and remote sensing in Yunnan [in Chinese, with English abstract]. *Acta Geographica Sinica*, **56**(2): 181–188.

Ye, D.Z., Chou, J.F., Liu, J.Y. & Zhang, Z.X. 2000. Causes of sand-stormy weather in northern China and control measures [in Chinese, with English abstract]. *Acta Geographica Sinica*, **55**(5): 513–521.

Yong, S.P., Tong, C. & Yong, W.Y. 1994. A study on the spatial overlaying relationship between winter snow cover and vegetal cover in Xilin Gole steppe in Inner Mongolia [in Chinese, with English abstract]. *Journal of Glaciology and Geocryology*, **16**(3): 238–244.

Yu, S.F., Li, X., Deng, J.H. & Fang, Y. 1994. Investigation on current urban afforestation situation in Hefei by aerial remote sensing [in Chinese, with English abstract]. *Journal of Anhui Agricultural University*, **21**(1): 63–66.

Zhang, F., Zhou, B., Chen, X.Z., Chen, J.Q. & Li, Z.C. 2002. Remote sensing background study on land use and land cover of Loess Plateau hill and ravine region in central part of Gansu province [in Chinese, with English abstract]. *Remote Sensing Information*, **3**: 29–31.

Zhang, J.H. 2001. Study on models for evaluating primary productivity of natural vegetation and crop yield [in Chinese, with English abstract]. *Acta Agriculturae Shanghai*, **17**(3): 83–89.

Zhang, R.H., Sun, X.M. & Zhu, Z.L. 1997. A remote sensing model for determining chlorophyll content and its distribution using Landsat images. *Acta Botanica Sinica*, **39**(9): 821–825.

Zhang, X.P., Liu, J., Li, R., Wang, Q.K. & Liu, Y.F. 2001. Dynamic monitoring and analysis of land use based on RS and GIS in Zhifanggou watershed [in Chinese, with English abstract]. *Agricultural Research in the Arid Areas*, **19**(4): 115–121.

Zhang, X.S., Zhou, G.S., Gao, Q., Ni, J. & Tang, H.P. 1997. Study of global change and terrestrial ecosystem in China [in Chinese, with English abstract]. *Earth Science Frontiers* (China University of Geosciences, Beijing), **4**(1–2): 131–144.

Zhang, X.Y., Zhao, H.K., Fan, G.Y., Liu, Y. & Zhou, J.S. 1991. Application of TM image to the land use survey of pastoral country [in Chinese]. pp. 200-201, *in:* D.F. Chen (ed). *Proceedings of Agricultural Remote Sensing*. Beijing: China Survey Press. Beijing, China, 21–23 January 1991,

Zhang, Y.X. & Li, J.D. 1991. A study on non-destructive methods for estimating above ground biomass of grassland plant communities [in Chinese, with English abstract]. *Acta Phytoecologica et Geobotanica Sinica*, **15**(2): 177–182.

Zhang, Z.S. 1998. Current situation of grassland fire prevention in China and thought on its development [in Chinese, with English abstract]. *Grassland of China*, **1**: 54–56.

Zhou, H., Cheng, Z.Z. & Lu, X.S. 1999. The natural calamities in rangeland of China and its preventive strategy [in Chinese, with English abstract]. *Grassland of China*, **2**: 1–3.

Zhou, S., Xie, Y.L., Guo, W.X., Yu, B.L., Xue, Q., Luo, X.L., Zhang, C.H. & Yun, Z.L. 1999. The application of NOAA-AVHRR data for identifying rat damaged areas in grassland [in Chinese, with English abstract]. *Acta Prataculturae Sinica*, **8**(3): 16–21.

Zhou, Y.M. & Wang, J.S. 1996. A method of monitoring grassland resource in Qinghai province using satellite data [in Chinese, with English abstract]. *Quarterly Journal of Applied Meteorology*, **7**(4): 507–510.

Zhou, Y.M., Wang, J.H., Ma, A.Q., Qi, Y. & Ba, Y. 2003. Study on land use dynamics based on RS and GIS in Linze county [in Chinese, with English abstract]. *Journal of Desert Research*, **23**(2): 142–146.

Zhu, J.Z., An, S.Z. & Long, J. 1998. The study of estimating forage yield in natural grassland by spatial structure analyse of TM image [in Chinese, with English abstract]. *Grassland of China*, **2**: 21–24.

Section VI
FARMER AND PASTORALIST PARTICIPATION

CHAPTER 15

New Forages for Smallholder Livestock Systems in Southeast Asia: Recent Developments, Impacts and Opportunities

P.M. Horne[1], W.W. Stür[2], P. Phengsavanh[1], F. Gabunada Jr.[3] and R. Roothaert[4]

ABSTRACT

Planting forages is a new concept for smallholder farmers in Southeast Asia, who traditionally had access to ample natural feed resources for their cattle, buffalo and goats. This is no longer the case. Traditional feed resources are disappearing and degrading, and farmers are having difficulty in finding sufficient feed for their animals. They need to invest more and more time in collecting natural feed resources. At the same time, the demand for livestock products is increasing rapidly in Asia and raising ruminant livestock is becoming an increasingly attractive option for farmers.

Recent experiences show that smallholder farmers are willing to embrace this new concept of planting feed for their animals. In Southeast Asia, there are now thousands of farmers who have integrated forages into their farms and many are deriving substantial benefits from doing so. Forages are used mainly as a supplement to other available feed resources, such as natural grasses and crop residues. Farmers grow forages in a variety of ways such as intensive cut-and-carry plots near houses, or contour hedgerows in cropping areas, and these forages are fed to a variety of animals, ranging from large ruminants such as cattle, to fish, pigs and rabbits. While many farmers were initially attracted to planting forages to save time on days when they were busy with other things – they could quickly cut enough feed for their animals from a forage plot near the house – many saw opportunities

[1]Lao-CIAT Forages and Livestock Systems Project, P.O. Box 6766, Vientiane, Lao PDR.
[2]CIAT Livelihood and Livestock Systems Project, 22 Seventh Avenue, Windsor, Qld. 4030, Australia.
[3]CIAT, Livelihood and Livestock Systems Project, Leyte State University, 6521-A Baybay, Leyte, the Philippines.
[4]CIAT, ILRI-Ethiopia, P.O. Box 5689, Addis Ababa, Ethiopia.

to increase their livestock production and developed livestock production into an income-generating component of their farming system.

This chapter describes the process of forage technology development that has led to the acceptance of planted forages by smallholder farmers in Southeast Asia. It describes the forage technologies developed and used by farmers, discusses lessons learnt in working with farmers and analyses the impact of planted forages on livelihoods of smallholder farm households. It concludes with a discussion of future needs and priorities for forage research and development in the region.

INTRODUCTION

It comes as a surprise to many that new forage varieties and systems can be of significant interest to smallholder farmers. Forages (especially grasses) are perceived to be everywhere, being the most basic and freely available of feed resources. What could possibly be achieved by trying to convince farmers to invest their time and labour in establishing plants to feed animals when grass has always been freely available to all? This chapter presents some recent experiences from Southeast Asia to illustrate why many farmers are doing just this, even in places dominated by large areas of grasslands.

About 95 percent of domesticated livestock in Southeast Asia are kept by resource-poor smallholder farmers, providing significant benefits, beyond simply being a source of capital accumulation.

- *Livestock can be sold at any time* in a market, which has a constant demand and relatively stable prices. By comparison, most fruit and vegetable crops must be picked when ripe and often sold in a market where prices are depressed by oversupply.
- *Larger livestock (cattle, buffalo and goats) can be walked long distances to market*. The marketing flexibility this provides is especially important for remote communities. In Laos, for example, the nearest road for many farmers is one day's walk or more from their village. This severely limits their cash-crop options, as most crop products are bulky and would have to be carried long distances to market for meagre returns per unit weight.
- *Livestock provide manure to sustain crop yields*. In many areas of Southeast Asia, smallholder farmers have either no cash to buy fertilizers or no access to reasonably priced fertilizer supplies. These farmers often rely on manure as an essential input for maintaining the productivity of small areas of rice fields and home-gardens.
- *Livestock provide a relatively high profit for very low labour input.*
- *Larger livestock (cattle, buffalo and goats) use feed resources that cannot be utilized for any other purpose*. In the past, these livestock were commonly grazed in forests, grasslands or in fallow fields, where they survive by foraging on grasses, shrubs, tree leaves and crop residues. The communally owned feed resources provided farmers with substantial returns and livelihood security for little or no management input.

To most smallholder livestock producers in Southeast Asia, planted forages are much more than simply a new variety of a crop with which they are already familiar. "Planting forages" is an entirely new concept, requiring major changes in the way farmers think about and manage their crop–livestock systems. Such significant changes do not occur easily unless there are strong pressures for change – and these pressures are coming fast in Southeast Asia. They include the disappearance and degradation of traditional grazing resources, and rapidly increasing demand for livestock products. The total area of native forages in Southeast Asia is declining as the more favourable areas are converted into arable land, leaving only the poorest land for grazing and gathering of feed for animals. Between 1984 and 1994, there was an average 13 percent increase in land area under annual and permanent crops in Southeast Asia (Stür *et al*., 2002). The end result is that farmers have to spend more time finding sufficient feed for their animals.

Coupled with this declining traditional feed resource is a projected explosion in demand for livestock products (e.g. Vercoe *et al*., 1997; Steinfeld, 1998; Delgado, *et al*., 1999). This demand is partly due to the rapidly expanding populations of the region, but also in part due to the astonishing rate of urbanization of these populations. In China, for example, the percentage of the population in urban areas is predicted to double between 1975 and 2025, while the total population would increase by 60 percent in the same period (Table 15.1). Urbanization on this scale will be associated inevitably with changes in food consumption. Diets in rural communities are typically higher in calories and less diversified than urban diets, which are characterized by higher consumption of livestock products (Ghirotti, 1999). Consequently, domestic demand for ruminant meat per capita will increase by something like 40–60 percent across the region by 2010 compared with 2000. It is expected that non-ruminant meat production will be largely satisfied by domestic production, but many countries will be unable to meet the increasing demand for ruminant meat domestically (Table 15.2). This is already occurring in Thailand, which imports large numbers of cattle and buffalo from Myanmar, Laos and Cambodia each year. Vercoe *et al*. (1997) conclude that the major constraint to satisfying this increasing regional demand for meat products is limited feed supplies and feed quality.

Table 15.1 Projected human populations (× 10^6) and percentage of population that is rural for selected Southeast Asian countries

	1975		2000		2025	
Country	Total	Percent rural	Total	Percent rural	Total	Percent rural
Cambodia	7	90	13	83	22	67
PR China	932	82	1282	64	1480	43
Indonesia	135	81	212	59	273	39
Lao PDR	3	89	5	81	9	67
Malaysia	12	62	22	43	31	29
Myanmar	30	76	48	72	60	57
Philippines	42	64	76	41	107	27
Thailand	41	85	63	80	77	70
Viet Nam	48	81	78	76	105	62

Source: FAOSTAT, data as of March 2003.

TRADITIONAL FORAGE RESOURCES IN SOUTHEAST ASIA

The diversity of smallholder crop–livestock systems in Southeast Asia has been described in detail by other authors (e.g. Devendra *et al*., 1997) and is illustrated in Figure 15.1. Within these crop–livestock systems, livestock are usually maintained on a mix of naturally occurring forages and locally available crop and tree by-products. Access to naturally occurring forages is commonly the main factor influencing the livestock feeding and management systems. The best access to natural forages is typically in shifting cultivation areas and in extensive upland farming systems

Table 15.2 Projected ruminant and non-ruminant meat production and demand (kg per capita) for selected Southeast Asian countries

Country	Ruminant meat				Non-ruminant meat			
	2000		2010		2000		2010	
	Demand	Prod.	Demand	Prod.	Demand	Prod.	Demand	Prod.
Cambodia	3.6–3.9	3.3	5.2–5.9	3.3	9.7–10.5	9.1	14.2–16.3	9.1
PR China	4.1–5.1	5.4	5.8–7.4	8.5	38.4–48.5	38.5	54.5–71.3	49.8
Indonesia	3.1–3.6	2.3	4.6–6.0	2.5	8.3–9.7	12.3	12.3–16.7	26.2
Lao PDR	3.2–3.5	3.0	4.8–5.4	3.0	8.4–9.0	7.8	12.4–14.2	7.8
Malaysia	4.4–4.9	0.8	6.3–7.9	0.6	49.6–56.5	71.0	73.7–93.9	139.6
Philippines	3.4–3.6	2.9	4.4–4.9	2.7	21.5–23.0	24.0	27.8–31.7	29.8
Thailand	6.0–7.4	5.7	10.1–14.8	6.3	20.4–25.6	25.0	35.5–53.5	42.3
Viet Nam	2.7–2.9	2.7	3.6–4.2	3.0	15.8–17.2	16.0	21.5–25.2	18.9

Note: Prod. = production.
Source: Vercoe *et al.*, 1997

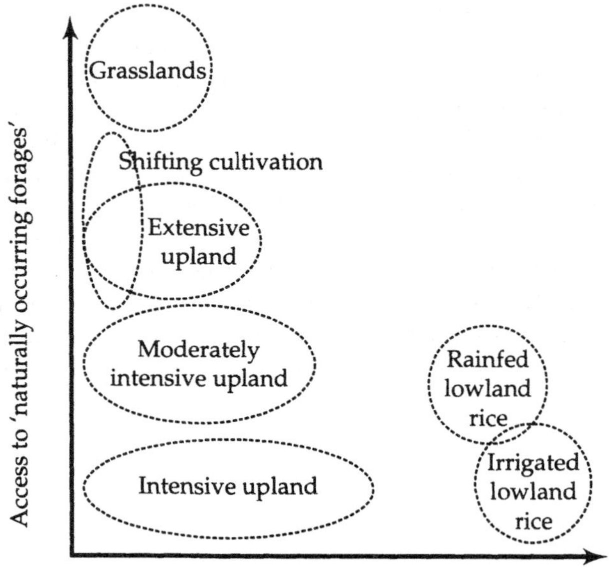

Figure 15.1 Land use intensity of mixed crop-livestock farms in Southeast Asia
Source: Stür *et al.*, 2002.

(e.g. grasslands, forests and forest margins), which encourages farmers to raise cattle to take advantage of the freely available feed resource. Access to grazing lands is more limited in the moderately intensive and intensive upland systems. Small ruminants (goats and sheep) supplemented with cut feed (forages, crop residues and agro-industrial by-products) are playing an increasingly important role in these systems as grazing land becomes ever more limited. Livestock in lowland areas have little or no access to grazing land apart from crop residues and fallows. In some lowland areas of Southeast Asia, however, the densities of livestock (especially

(Contd.)

Forage species	Country			
	Indonesia (6 sites)	Lao PDR (5 sites)	Philippines (13 sites)	Viet Nam (4 sites)
	Legumes			
Aeschynomene sp.	2	8	1	8
Alysicarpus sp.	0	4	0	3
Arachis glabrata	1	0	4	0
Arachis pintoi	2	4	6	5
Calliandra calothyrsus	1	10	2	1
Calopogonium mucunoides	1	1	4	1
Centrosema sp.	1	3	2	3
Centrosema macrocarpum	4	1	2	2
Centrosema pascuorum	1	1	1	1
Centrosema pubescens	2	2	4	3
Chamaecrista rotundifolia	1	4	0	3
Clitoria ternatea	0	1	1	2
Codariocalyx gyroides	1	0	0	0
Cratylia argentea	1	1	1	0
Desmanthus virgatus	0	1	6	3
Desmodium cinerea	1	0	3	0
Desmodium heterophyllum	1	1	1	1
Desmodium ovalifolium	2	2	4	2
Desmodium sp.	1	5	1	4
Flemingia macrophylla	1	0	1	1
Gliricidia sepium	4	3	4	3
Lablab purpureus	0	0	2	0
Leucaena leucocephala	2	8	3	8
Leucaena sp.	1	5	2	5
Macroptilium sp.	0	4	2	3
Macroptilium gracile	1	4	1	4
Pueraria phaseoloides	0	0	5	0
Sesbania grandiflora	1	0	1	0
Stylosanthes guianensis	3	6	8	2
Stylosanthes hamata	1	2	0	2
Stylosanthes scabra	2	1	0	2
Teramnus uncinatum	0	1	0	1
Vigna sp.	0	4	0	1
Zornia latifolia	0	1	0	1
Subtotal legume accessions	39	88	72	75
Total accessions	64	113	98	104

Source: Stür, 2002.

Regional evaluations

Regional evaluations usually involved fewer than 20 accessions that were selected from the nursery evaluations and were sown at many locations to provide more information on how broadly adapted the narrowing suite of varieties really was. The regional evaluations were controlled by researchers, but were often managed by farmers. The regional evaluation sites also provided a venue for early feedback on the kinds of forages that farmers preferred and were a valuable source of vegetative planting material for on-farm evaluations.

On-farm evaluations

On-farm evaluations were both formal and informal, and usually involved very few accessions (<6) and were conducted on more than 1500 farms at 19 sites in four countries in the region (see Figure 15.2; Table 15.5). Sites were selected to give a broad range of soil and climatic conditions and a broad range of farming systems (more detail about each site can be obtained from Stür, 2002). Sites were also selected on the basis of a set of criteria that were developed to identify

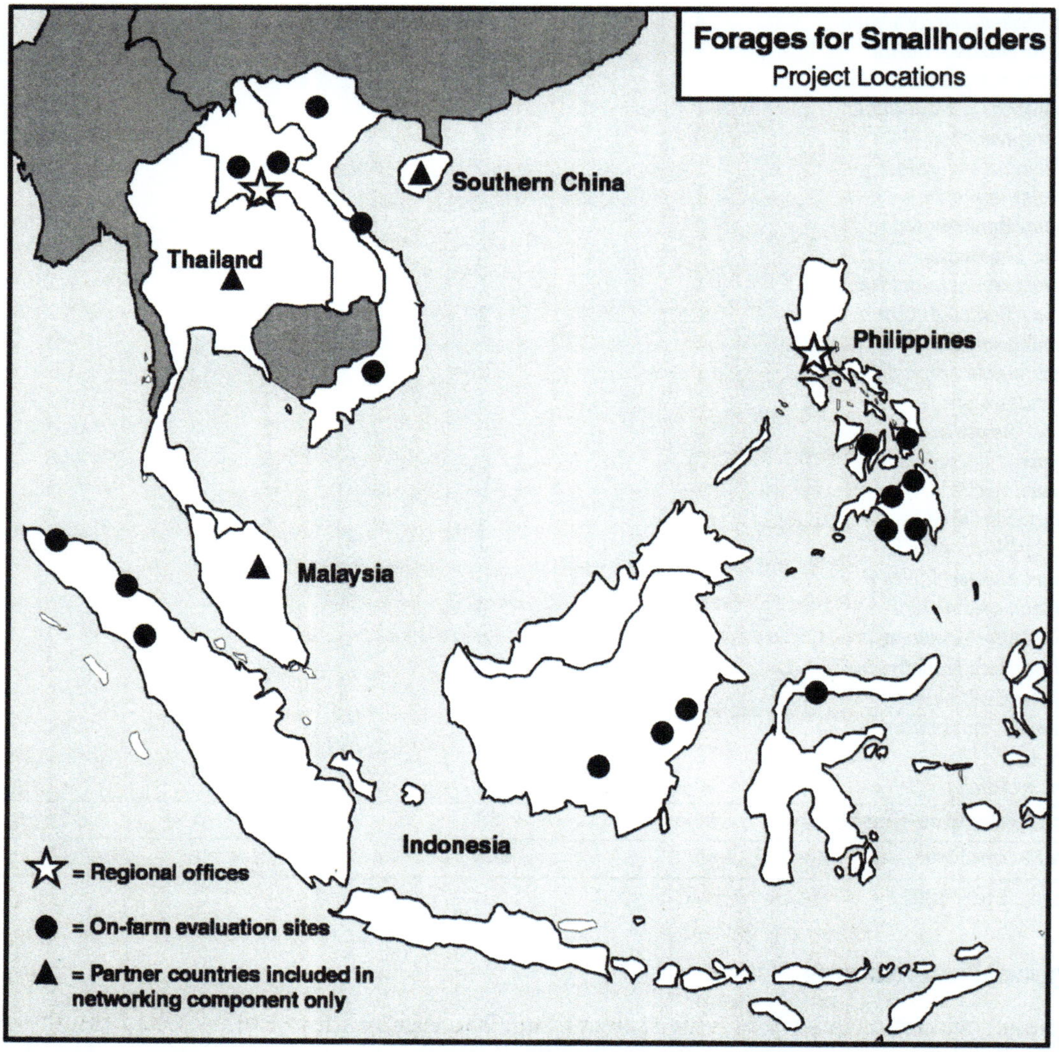

Figure 15.2 Location of on-farm forage evaluation sites throughout Southeast Asia

Table 15.5 Farming systems in which the on-farm evaluations of forages occurred

Farming systems	Country			
	Indonesia	Lao PDR	Philippines	Viet Nam
Short-rotation slash and burn	–	3	–	–
Grasslands	2	–	–	1
Extensive uplands	2	–	1	–
Moderately intensive uplands	1	–	3	1
Intensive uplands	1	–	1	1
Rainfed lowlands	1	–	1	–

Source: Gabunada *et al.*, 2000.

locations that had the greatest potential to benefit from the forage options that were being offered (Table 15.6). These criteria proved to be touchstones for the potential for forage development in any community.

The participatory approaches that were used in the on-farm evaluations have been described in detail by Horne *et al.* (2000) and Horne and Stür (2003).

Table 15.6 Key questions for selecting areas and communities for participatory forage technology development

Questions	For successful sites ...
1 Is there a genuine (livestock feeding) problem? 2 Do farmers think that this problem is important enough to commit their time in working towards a solution? 3 Are there many farmers and other communities who have the same problem? 4 Do we have potential solutions to offer that can provide substantial benefits? 5 Can you achieve YOUR goals by working with these farmers? 6 Are there active local individuals or groups who are committed to working with farmers to solve this problem?	'YES' is essential
7 Are there farmers who are already trying to solve this problem? 8 Is there a local enthusiast who may 'champion' technology development and expansion of the resulting technologies?	'YES' is desirable but not essential

Source: Horne *et al.*, 2000.

This evaluation strategy conducted over the five years between 1995 and 1999 resulted in the identification of a small range of forage varieties for Southeast Asia, many with broad adaptation (from humid to sub-humid and cool tropics, and from fertile to extremely infertile, acid soils) (Table 15.7). Each of these varieties is described in more detail in Horne and Stür (1999). There are many grasses that can be suitable for many soils and climates, but the range of legumes is much more limited, particularly for acid, infertile soils and for cooler, higher altitude areas.

This list of varieties is not exhaustive, since the goal of the evaluation process was to come up with a small suite of broadly adapted, robust varieties. With time, other varieties with specific adaptation to particular areas or environmental niches are likely to emerge. Three legumes in this category are:

Table 15.7 Adaptation of selected forage varieties to climate, soils and use on smallholder farms in Southeast Asia

Species and cultivar or variety	Climate			Soil fertility and acidity			Ways of growing forages							
	Wet tropics with no or short dry season	Wet/dry tropics with long dry season	Cooler tropics (e.g. high elevation)	Fertile (neutral to moderately acid soils)	Moderately fertile (neutral to moderately acid soils)	Infertile (extremely acid soils)	Cut and carry plots or rows	Grazed plots	Living fences	Contour hedgerows	Improved fallows	Cover crops in annual crops	Cover crops under trees	Ground covers for erosion control
Grasses														
Andropogon gayanus 'Gamba'	•	••		•	•	•	••	•		•				
Brachiaria brizantha 'Marandu', 'Karanga' 'Serengeti'	•	••	••	•	••	•	••	•		•				
Brachiaria decumbens 'Basilisk'	•	••	••	•	••	•	•	••						•
Brachiaria humidicola 'Tully', 'Yanero'	••	•	•	•	•	••	•	••						••
Brachiaria ruziziensis 'Ruzi'	••		•	••	•		•	••						•
Panicum maximum 'Simuang'	••	•	•	••	•		••	•		•				
Paspalum atratum 'Terenos'	••		•	•	••	•	••	••		••				
Pennisetum purpureum and *Pennisetum* hybrids 'Napier', 'Mott', 'King'	••		•	••	•		••			•				
Setaria sphacelata 'Lampung', 'Solander'	••	•	••	••	•		••	•		••				
Legumes														
Arachis pintoi 'Itacambira', 'Amarillo'	••		•	••	••			•					••	••
Calliandra calothyrsus 'Besakih'	•		••	•	••		••		•	•				
Centrosema macrocarpum 'Ucayali'	••	•		••	•		•				••	••	•	•
Centrosema pubescens 'Barinas'	••	•	•	••	•		•				••	••	•	•
Desmanthus virgatus 'Chaland'	••		••	••	•		••			•				
Desmodium cinerea 'Las Delicias'	•	•		•	•		••			••				

(*Contd.*)

(Contd.)

Species and cultivar or variety	Climate: Wet tropics with no or short dry season	Climate: Wet/dry tropics with long dry season	Climate: Cooler tropics (e.g. high elevation)	Soil fertility and acidity: Fertile (neutral to moderately acid soils)	Soil fertility and acidity: Moderately fertile (neutral to moderately acid soils)	Soil fertility and acidity: Infertile (extremely acid soils)	Ways of growing forages: Cut and carry plots or rows	Ways of growing forages: Grazed plots	Ways of growing forages: Living fences	Ways of growing forages: Contour hedgerows	Ways of growing forages: Improved fallows	Ways of growing forages: Cover crops in annual crops	Ways of growing forages: Cover crops under trees	Ways of growing forages: Ground covers for erosion control
Gliricidia sepium 'Retalhuleu', 'Belen Rivas'	••	••		•	••		••		••	•				
Leucaena leucocephala 'K636', 'K584'	••	••	•	••	•		••	•	•	•				
Stylosanthes guianensis 'Stylo 184'	••	••	•	••	••	••	••	•		•	••	••	•	

Key to symbols: •• = highly suitable; • = possible; no dot = not suitable.
Source: Stür *et al.*, 2002.

- *Centrosema pascuorum* cv. "Cavalcade", an annual twining legume that is finding a niche in Thailand for making hay (although *Rhizoctonia* susceptibility problems still need to be resolved).
- *Macroptilium gracile* cv. "Maldonado", a short-lived twining legume that is an excellent short-term cover crop and is well adapted to poorer soils and waterlogged areas.
- *Sesbania grandiflora* cv. "Turi", a fast growing, short-lived tree legume that has very high feeding value and is well adapted to areas with a long dry season. It is being widely used as a component of mixed upland farming systems in eastern Indonesia.

Additionally, new promising varieties continue to emerge from basic research. Two that may have potential uses in the near future in the Southeast Asian region are:

- *Brachiaria brizantha* × *Brachiaria ruzizensis* hybrid "Mulato". This hybrid Brachiaria (developed by John Miles at CIAT) has many of the high feed quality characteristics of *B. ruziziensis* but with the drought tolerance and persistence of *B. brizantha*. It is currently under trial in Thailand for seed production potential and in several other countries for performance in smallholder livestock systems.
- *Cratylia argentea* cv. Veranera is a drought tolerant shrubby legume native to South America. It has been evaluated previously in the humid areas of Southeast Asia without much success but will be much better adapted to the drier areas of the region.

The active involvement of farmers at so many locations in the evaluation process provided an enormous amount of feedback on environmental adaptation of forages, and on the ways that farmers would grow and utilize these varieties and on their criteria for selecting the varieties with most potential. However, most of the information was in the form of farmers' experiences and not 'hard' experimental data, making objective analysis difficult. To obtain a quantitative assessment of the performance of the selected forage varieties, it was decided to take the evaluation process one step further and conduct a formal, researcher-managed genotype × environment interaction experiment at 12 sites in the region.

All of the herbaceous varieties in Table 15.7 were included, plus several other varieties that might have niche uses. These were the grasses *Brachiaria brizantha* CIAT 26110 "Toledo"; *Digitaria milanjiana* cv. Jarra; *Panicum maximum* CIAT 6299 "Tobiata"; and *Urochloa mozambicensis* CPI 60147; and the legumes *Aeschynomene histrix* CIAT 9690; *Calopogonium caeruleum* CIAT 7304; *Chamaecrista rotundifolia* cv. Wynn; *Desmodium ovalifolium* CIAT 13305; *Flemingia macrophylla* CIAT 17403 "Chumphon"; *Macroptilium gracile* cv. Maldonado; *Pueraria phaseoloides* CIAT 7182; and *Zornia latifolia* CIAT 728. The characteristics of the 12 sites where the experiment was replicated are shown in Table 15.8. Final results were expected to be available from CIAT in 2004.

Table 15.8 Characteristics of sites where the genotype × environment experiment was conducted

Location	Temperature		Dry season length		Soil pH	Soil fertility		
	cool	warm	long	short		infertile	moderate	fertile
Nam Suang, Lao PDR		✓	✓		4.8	✓		
Los Baños, Philippines		✓		✓	6.5			✓
Hainan, PR China	✓	✓		✓	5.5		✓	
Sei Putih, Indonesia		✓		✓	5.0		✓	
Pakchong, Thailand		✓	✓		5.5			✓
BaVi, Viet Nam	✓	✓	✓		4.8	✓		
Alor Setar, Malaysia		✓		✓	4.7		✓	
Sepaku, Indonesia		✓		✓	4.6	✓		
Timor, Indonesia		✓	✓		7.0		✓	
Bali, Indonesia		✓			6.8		✓	
Binh Thuan, Viet Nam		✓			6.5		✓	
Gen. Tino, Philippines			✓		5-6		✓	

OUTCOMES OF FARMERS' EVALUATIONS OF FORAGES

The rate of adoption of forage varieties and the ways of using them varied enormously between sites. This reflected the variation and complexity of factors and "enabling mechanisms" affecting farmers' decisions. At one site, for example, farmers found that *Brachiaria humidicola* grew well in an area that was not suitable for crops because of very low fertility. This provided them with an opportunity to utilize the poor soils more productively for livestock. At some sites, despite initial optimism, there ended up being very few opportunities for adopting forages, mainly because the criteria described in Table 15.6 were not strongly satisfied at those sites.

At all sites farmers were initially cautious, often having never considered the concept of planting forages to feed their animals, let alone having never seen examples of forage varieties. Consequently, they inevitably chose to start with small plots near their houses, to get a feel for the varieties. Subsequently, as their confidence in some forage varieties grew, they started to expand their areas and looked for ways to achieve significant impacts with these varieties. This generally happened in two ways. First, farmers were able to solve immediate problems (such as dry-season feed shortages and the large amount of labour required to cut native forages). Second, some changed their farming systems to take advantage of new opportunities presented by the forage varieties they selected (such as feeding of introduced grasses to ponded fish in Tuyen Quang). In most cases, this expansion of area was based on intensive permanent or semi permanent forage systems, especially cut-and-carry feeding systems (Figure 15.3 and Table 15.9). Within two years, these systems ran into nutrient rundown problems. At some sites, the more experienced farmers started to recognize the value of maintaining the productivity of their forage areas and have started to apply nutrients in form of manure or fertilizer to their cut-and-carry forage plots, such as in Tuyen Quang and Sepaku. Others have planted frequently cut hedgerows in the middle of cropping land (e.g. rows of grasses between strips of crops, or contour hedgerows in sloping land) that catch nutrients from the crops, or they have started to develop less intensively managed (relatively more sustainable) systems, such as grazed plots in East Kalimantan that require less labour and result in some automatic nutrient return to the forages. Given a little time and confidence building, farmers could develop either intensive fodder systems with input of nutrients, intensive multipurpose systems that optimize the use of resources and minimize risk, or less intensive grazed systems (Figure 15.3).

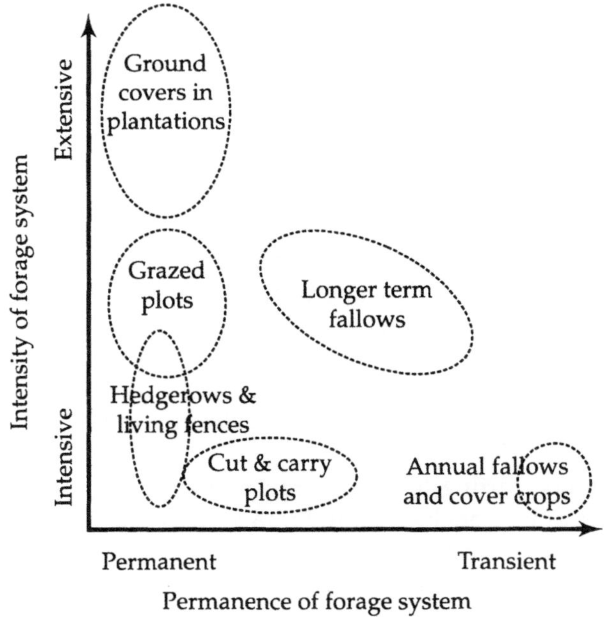

Figure 15.3 Intensity of forage systems on farms in Southeast Asia.
Source: Based on the concept of Perkins *et al.*, 1986.

The outcomes of the field evaluations summarized in Table 15.9 illustrate several things.

- There were rarely opportunities for forages in lowland rice systems, neither irrigated nor rainfed rice areas. These areas have become too valuable for cropping. This situation may change if the price of meat continues to increase more rapidly than the price for lowland crops such as rice.
- Despite early optimism, forage systems with significant long-term potential for natural resource management (such as long-term fallows in shifting cultivation areas) did not emerge quickly. However, in areas where such systems were developed successfully by a small

Table 15.9 Forage systems being adopted by farmers

Farming systems	Forage Systems: Cut-and-carry plots	Grazed plots	Living fences	Hedgerows	Improved fallows	Cover crops in annual crops	Cover crops under trees	Ground covers for erosion control
Short-rotation slash and burn	✓ (100)	✓ (1)	✓ (2)					
Grasslands	✓ (88)	✓ (9)	✓ (29)	✓ (21)				
Extensive uplands	✓ (99)	✓ (1)		✓ (13)				
Moderately intensive uplands	✓ (73)	✓ (4)	✓ (27)	✓ (16)		✓ (6)	✓ (6)	✓ (6)
Intensive uplands	✓ (63)		✓ (6)	✓ (35)				
Rainfed lowlands								

Key: Shaded cells indicate that before the trials it was expected that farmers would adopt this option. ✓ = farmers did adopt this option. Numbers in brackets are the percentage of expanding farmers who adopted each forage system.
Source: Adapted from Gabunada *et al.*, 2000.

group of farmers (such as contour hedgerows controlling runoff and providing feed for livestock in Malitbog, the Philippines), new farmers have subsequently adopted these systems quickly.

- At no site were farmers interested in replacing large areas of extensive pastures of native species with introduced forages. The strong message from this work was that introduced forages are used as a supplement for existing feed resources (native grasses, crop residues and agro-industrial by-products). However, after several years of developing forage technologies in these areas, some farmers (e.g. in Sepaku, Indonesia) have expanded their animal numbers and are slowly replacing *Imperata cylindrica* grasslands on their own land with more nutritious forages.
- The number of farmers starting to grow forages in an area (e.g. district) increases rapidly once a small number of farmers have tested forages and started to integrate forages on their farms. This generates a demand for forage planting material, which provides additional income for early adopters (see Plate 15.1).

LESSONS FROM ON-FARM EVALUATION OF FORAGES

Some key lessons have emerged from the process of selecting and evaluating forages with farmers described earlier (Horne and Stür, 2003):

- ***Start evaluating forage varieties with farmers as soon as possible***
  There is no need for researchers and development workers to wait until they fully understand the complexity of the farming system before offering forage options to farmers. With their in-depth knowledge of the farming system, farmers can quickly decide which varieties are likely to provide substantial benefits. Farmers are keen experimenters and will want quick action once they have decided which forage varieties to test.

Plate 15.1 Smallholder farmer in Laos feeding forages to his buffalo. [Photograph: P.M. Horne]

- ***Where possible offer a broad range of varieties, not just the 'favourite' variety***
  Each farmer and each farm is different and no single forage variety or system will be appropriate for all farms. Researchers and development workers need to offer a broad range of varieties and ideas about forage systems that are realistic within the resources available to the village. Farmers may see other problems they can solve with the forages or changes they can make to their farming practices to take advantage of new opportunities presented by the new forages.
- ***Offer basic ingredients of technologies, not 'fully developed' technologies***
  Often researchers and development workers feel that they cannot offer technologies to farmers until they are "fully developed". They may, for instance, feel reluctant to offer farmers forage varieties without also telling them exactly how these should be grown, managed and fed to animals. In practice, this is not necessary. Farmers need to develop management systems to fit their own circumstances. It is better to provide them with ideas (e.g. examples of what other farmers are doing in similar situations) and principles of forage agronomy and utilization (Stür and Horne, 2001) rather than specific recommendations. This will help them to make informed choices about each option.

- ***Identify technology "entry points" when you start working with new farmers***
  Often, it is helpful to identify entry points that provide early benefits to farmers, building trust and enthusiasm. Fast-growing grasses for cut-and-carry systems, or a vigorous legume for supplementing livestock, can give the quick early impacts that generate enthusiasm among farmers to further experiment with forage systems that provide fewer immediate benefits.
- ***Start small and keep evaluations relatively simple***
  Typically, farmers will first want to test forage varieties on a small scale as this minimizes risk and gives them the opportunity to experiment with the varieties. Small plots near the house allow them to answer questions like "How well does each variety grow in the dry season?" "How easy are they to cut?" and "Do my animals like to eat these varieties?" A large numbers of varieties (typically more than six) are difficult to manage and compare at one time.

Three misconceptions about forages were common among farmers and extension workers:

- ***There are no miracle forage varieties***
  No forage varieties can produce high yields in adverse conditions (such as in heavy shade or on very poor soils). For example, a common request is for forages that produce high yields in the heavy shade of tree crops, such as densely planted coffee or bananas.
- ***Forages need careful early management***
  Like all cultivated crops, forages need management to survive and be productive. The most common causes of problems in early farmer trials with forages are sowing seeds too deep, heavy competition from weeds, and grazing of young forage seedlings by wandering stock.
- ***Communal grazing land is not ideal for forage development***
  Communal grazing land cannot easily be improved with introduced forages unless free grazing of livestock is controlled (either by fencing or regulation).

WHAT EVIDENCE EXISTS OF IMPACTS FROM FORAGES IN SOUTHEAST ASIA?

Having identified broadly adapted, productive and palatable forage varieties was not sufficient to guarantee improvements in the feed supply for smallholder livestock systems. In Southeast Asia, there are many examples of forage and pasture programmes that have foundered because of the expectation that "the more seed we distribute the more likely we are to be successful". Similarly, implementing strong community-based research and extension processes can be a waste of time if they are not matched with the best technical options available. Within any forage species there can be huge variation in morphology, adaptation and feeding value. Working with farmers to evaluate whichever variety of *Panicum maximum* grass happened to be available, for example, rather than working with the best variety for that region, is likely to have disappointing outcomes. In our experience, *the most successful way to develop significant impacts from forages in smallholder livestock systems comes from matching the best varieties available with participatory approaches to research and extension*. The information that follows illustrates some of the impacts that are possible from such an approach, both on livestock production and on poverty (for further details see Peters *et al.*, 2001). The impacts from forages (and the systems that farmers develop for utilizing them) can often be unexpected and novel. The participatory approaches described earlier encourage the emergence of this diversity rather than the spread of an average (or single) innovation.

In Sepaku, both the net yearly income per household from livestock and labour productivity of livestock production was higher for forage adopters (from a mean forage area of 0.3 ha) compared with non-adopters (Table 15.10). This was due to, first, an increase in the value of cattle and buffalo when fed new forages, as traders are paying a higher price for cattle fed on new forages because of better carcass quality, and, second, a reduction in the time required for herding. Some forage adopters used the time saving to capitalize on the opportunity offered by the forages to increase their herd size. Others put the saved time to good use in off-farm activities to increase household income. Manure, used mostly for upland crops and sale, accounted for almost 40 percent of the total income from livestock.

In Samboja and Makroman the main benefit of adopting forages was increased labour productivity (Table 15.10). This was due mainly to a significant reduction in the amount of time needed for herding in Samboja and for cutting and carrying forages in Makroman. The availability of manure from goats to maintain soil fertility was mentioned as an important benefit for farmers in Makroman.

The impact of growing forages on labour distribution varied (Table 15.12). At Makroman, no major changes were reported by farmers. In Samboja, women were more implicated in cutting and carrying feed after forage adoption, but it was reported that this freed up time for the men to access lucrative off-farm income opportunities. In Sepaku, before forage adoption, men women and children had to be involved in cutting and transport of grass. After forage adoption, children spent less time in this activity. The total time needed for cutting forage decreased and, when herd sizes remained the same, the workload of women in livestock management decreased.

Table 15.12 Impacts of forage adoption on labour distribution for cutting forage and feeding animals at three sites in East Kalimantan, Indonesia

Location	Before adoption (percent)			After adoption (percent)		
	young	men	women	young	men	women
Sepaku	33	33	35	19	44	36
Samboja	24	74	2	25	59	15
Makroman	19	56	25	16	63	22

Source: Bosma, Roothaert and Ibrahim, 2001

The cases reported above are all examples of positive impacts from forage adoption and imply (from comparison between early and late adopters) that impacts increase with time (as a farmer's forage area, experience, livestock capital and investment capacity all increase). Not all farmers, however, experienced positive impacts, and, in a very small number of cases, there were negative impacts. These negative impacts were mostly either that farmers provided labour to establish forages but received no benefits (due to the failure of seed to emerge or to other external effects such as wandering livestock destroying the forage plots), or fallow land was in short supply and the forages provided less benefit than the alternative uses of the land available to the farmer. The negative impacts in these few cases were limited by two factors: first, these problems usually occurred very early in the forage development process, when forages were a completely new concept to farmers and, by design, the areas planted were very small to reduce the extent of any possible risks (see Horne and Stür, 2003), and second technical issues such as poor forage establishment, were resolved within the first two years of evaluation.

PRIORITY ISSUES FOR FORAGE DEVELOPMENT IN SOUTHEAST ASIA

Given that there is now a small but strong suite of forage varieties being progressively adapted, adopted and integrated into smallholder farming throughout Southeast Asia, the following issues need to be considered:

- The scale of integration of forages onto smallholder farms is still small – possibly 5000 farms in the region – and many of these farmers are still in a testing phase, evaluating the potential benefits forages can provide on their farms before expanding forages to larger areas. There is a need to document the many impacts of these forage systems and to use the knowledge to provide ideas to farmers in other areas for starting forage evaluations on their own.
- As farmers increase livestock production, taking advantage of the increased forage supply, they are also changing the way they raise animals. Frequently, animals are kept longer or completely in stalls, or are tethered near the house and forages are cut-and-carried to the animals. Farmers need help with matching feed availability with the needs of animals to maximize livestock productivity. Often, this means lack of protein in the diet of productive animals, which can be overcome by providing some concentrates or increasing the proportion of legumes in the cut feed. Farmers may also encounter other production constraints, such as animal diseases, nutritional deficiencies because of the complete reliance on cut-and-carry feed (and thus selection by the farmer rather than by the animals), availability of appropriate concentrate feeds, and marketing of livestock products. These constraints need to be addressed to ensure that improved forage production translates into improved livestock production and improved livelihoods.
- Intensification of livestock systems will inevitably occur in some places, but the smallholder livestock sector, which dominates the region, will not disappear in the foreseeable future. What prospects are there for intensification of forage production systems to meet the demands of those peri-urban areas where livestock production will be intensified?
- We have only just begun to understand the high potential for forages for non-ruminants, especially pigs, poultry and fish, in the smallholder sector of some countries (especially Indochina). These livestock systems are often important in poor villages and among poor people within villages, so there is good potential for a positive impact on poverty through improvements in forages for these livestock systems.
- Most of the smallholder farmers growing forages are keeping livestock for multiple reasons, such as wealth generation and draught power; few reasons are clearly market oriented. There is an opportunity for farmers to improve their income from animal production by producing animals for the market, rather than to sell animals only when they are in need of cash. Farmers need assistance to learn how to manage risk associated with increased livestock production, and ways of maximizing productivity and return from their investment.
- Forages will continue to be just one component of the feed resources available to smallholder farmers in Southeast Asia. Researchers can work with farmers who already have forages integrated on their farms to identify ways to combine forages with other locally available feed resources (e.g. "Stylo 184" combined with sweet potato and rice bran) in ways that can deliver maximum benefits to farmers (whether it be reduced labour inputs, increased growth rates of animals, or reduction of feed shortages at particular times of year).
- Most countries are dependent on imports for their supplies of seed of the varieties evaluated. This in turn means that most communities within these countries are dependent on national organizations for supplies of seed. This inevitably means there are constraints to

expansion. Now that farmers are starting to ask for seed of these particular forage varieties, there is an opportunity to develop local seed or vegetative planting material production and delivery systems.

- There is a need for research to identify forage varieties for particular niches (e.g. legumes for less fertile soils) and adaptive research on new forage varieties emerging from basic research, such as the *Brachiaria* hybrid cv. "Mulato" and anthracnose-resistant *Stylosanthes guianensis*.

ACKNOWLEDGEMENTS

Much of the work that is reported here is the result of a network of committed forage researchers and extension workers in countries throughout Southeast Asia. The authors wish to thank them all for their ongoing commitment.

REFERENCES

Blair, G.J., Ivory, D.A. & Evans, T.R. 1986. Forages in Southeast Asian and South Pacific Agriculture. *in:* Proceedings of an international workshop held at Cisarua, Indonesia, 19–23 August 1985. ACIAR Proceedings Series, No. 12. 202p.

Bosma, R.H., Roothaert, R.L. & Ibrahim 2001. Economic and social benefits of new forage technologies in East Kalimantan, Indonesia. CIAT Working Document, No. 190. Centro Internacional de Agricoltura Tropical, Los Baños, Philippines. 59p.

Bosma, R.H., Roothaert, R.L., Asis, P., Saguinhon, J., Binh, L.H. & Yen, V.H. 2003. Economic and social benefits of new forage technologies in Mindanao, Philippines, and Tuyen Quang, Viet Nam. CIAT Working Document No. 191. Centro Internacional de Agricoltura Tropical, Los Baños, Philippines. 92 .

Delgado, C., Rosegrant, M., Steinfeld, H., Ehui, S. & Courois, C. 1999. Livestock to 2020: the next food revolution. Food, Agriculture and the Environment Paper, No. 28. IFPRI, Washington, DC, USA. 72p.

Devendra, C., Thomas, D., Jabbar, M.A. & Kudo, H. 1997. Improvement of livestock production in crop-animal systems in rainfed agro-ecological zones of South-East Asia. International Livestock Research Institute, Nairobi, Kenya. 116p.

FFTC. 1984. Asian Pastures: recent advances in pasture research and development in Southeast Asia. Food and Fertiliser Technology Center, Taiwan.

Gabunada Jr., F., Heriyanto, Phengsavanh, P., Phimphachanhvongsod, V., Khanh, T.T., Nacalaban, W., Asis, P., Yen, V.T.H., Tugiman, Ibrahim & Stür W.W. 2000. Integration of adapted forages on farms in Southeast Asia – Experiences from the Forages for Smallholders Project. pp. 221–226, *in:* Stür *et al.*, 2000, q.v.

Ghirotti, M. 1999. Making better use of animal resources in a rapidly urbanizing world: a professional challenge. *World Animal Review*, **92**(1): 1–14.

Halim, R. (ed). 1989. *Grasslands and Forage Production in South-East Asia.* Proceedings of the first meeting of the Regional Working Group on Grazing and Feed Resources of South-East Asia, 27 February–3 March 1989, Serdang, Malaysia. FAO, Rome, 216p.

Horne, P.M. & Stür, W.W. 1999. *Developing forages with smallholder farmers – how to select the best varieties to offer farmers.* ACIAR Monograph, No. 62. 80p.

Horne, P.M. & Stür, W.W. 2003. *Developing agricultural solutions with smallholder farmers – how to get started with participatory approaches.* ACIAR Monograph, No. 99. 19p.

Horne, P.M., Tajuddin, I. & Chong, D.T. 1994. Agroforestry plantation systems: sustainable forage and animal production in rubber and oil-palm plantations. pp. 89–98, *in: Agroforestry and Animal Production for Human Welfare.* Proceedings of a conference held at the 7th Animal Science Congress of the Australian-Asian Animal Production Societies (AAAP), Bali, Indonesia, 11–16 July 1994. ACIAR Proceedings, No. 55. Canberra: ACIAR.

Horne, P.M., Magboo, E., Kerridge, P.C., Tuhulele, M., Phimphachanhvongsod, V., Gabunada Jr., F., Binh, L.H. & Stür, W.W. (2000). Participatory approaches to forage technology development with smallholders in Southeast Asia. pp. 23–31, *in:* Stür *et al.*, 2000, q.v.

Ibrahim, Khanh T.T. & Heriyanto. 2000. Forage technologies for smallholders in grassland areas. pp. 277–280, *in:* Stür *et al.*, 2000, q.v.

Nacalaban, W., An, L.V., Asis, P., Moneva, L. & Balbarino, E. 2000. How do forages fit into smallholder farms in mixed upland cropping systems? pp. 273–276, *in:* Stür *et al.*, 2000, q.v.

Perkins, J.M., Petheram, R.J., Rachman, R. & Semali, A. 1986. Introduction and management prospects for forages in Southeast Asia and South Pacific. pp. 15–23, *in:* G.J. Blair, D.A. Ivory and T.R. Evans (eds). *Forages in Southeast Asian and South Pacific agriculture*. Proceedings of an international workshop, Cisarua, Indonesia, 19–23 August 1985. ACIAR Proceedings, No. 12.

Peters, M., Horne, P., Schmidt, A., Holmann, F., Kerridge, P.C., Tarawali, S.A., Schultze-Kraft, R., Lascano, C.E., Argel, P., Stür, W.W., Fujisaka, S., Müller-Sämann, K. & Wortmann, C. 2001. The role of forages in reducing poverty and degradation of natural resources in tropical production systems. Agren Network Paper, No. 117. 16p. (Agriculture and Research Extension Network, DfID).

Phaikaew, C. & Hare, M. 1998. Thailand's experiences with forage seed supply systems. pp. 2–14, *in:* P.M. Horne, C. Phaikaew and W.W. Stür (eds). *Forage Seed Supply Systems*. Proceedings of a workshop, Animal Nutrition Research Centre, Tha Pra, Khon Kaen, Thailand, 31 October–1 November 1996. CIAT Working Document, No. 175.

Purcell, T., Nacalaban, W. Gabunada Jr., F. & Cramb, R. 2000. Assessing the impact of agricultural technologies in smallholder farming systems – results from a participatory monitoring and evaluation study on forages in Malitbog, northern Mindanao, the Philippines. pp. 297–308, *in:* Stür *et al.*, 2000, q.v.

Reynolds, S.G. 1995. *Pasture–cattle–coconut systems*. FAO-RAPA Publication, Bangkok, Thailand.

Shelton, H.M. & Stür, W.W. (eds). 1991. Forages for Plantation Crops. pp. 25–31, *in:* Proceedings of a workshop, Sanur Beach, Bali, Indonesia, 27–29 June 1990. ACIAR Proceedings, No. 32.

Shelton, H.M., Gutteridge, R.C., Mullen, B.F. & Bray, R.A. (eds). 1998. Leucaena – *adaptation, quality and farming systems*. ACIAR Proceedings, No. 86.

Shelton, H.M., Piggin, C.M., Acasio, R., Castillo, A., Mullen, B.F., Rika, I.K., Nulik, J. & Gutteridge, R.C. 2000. Case studies of locally-successful forage tree systems. pp. 120–131, *in:* Stür *et al.*, 2000, q.v.

Steinfeld, H. 1998. Livestock production in the Asia and Pacific region – current status, issues and trends. *World Animal Review*, **90**(1): 14–21.

Stür, W.W. (ed). 2002. Proceedings of the third regional meeting of the Forages for Smallholders Project, Samarinda, East Kalimantan, Indonesia, 23–26 March 1998. CIAT Working Document, No. 188. 220p.

Stür, W.W. & Horne, P.M. 2001. Developing forages with smallholder farmers – how to grow, manage and use forages. ACIAR Monograph No. 88. 96 pages.

Stür, W.W. & Shelton, H.M. 1991. Review of forage resources in plantation crops of Southeast Asia and the Pacific. pp. 25–31, *in:* H.M. Shelton and W.W. Stür (eds). *Forages for Plantation Crops*. Proceedings of a workshop, Sanur Beach, Bali, Indonesia, 27–29 June 1990. ACIAR Proceedings, No. 32.

Stür, W.W., Cameron, A.G. & Hacker, J.B. 1995. Forages for Smallholders. Proceedings of the Third Meeting of the Southeast Asian Forage Seeds Project, Samarinda, Indonesia, 23–28 October 1994. CIAT Working Document, No. 143. CIAT, Cali, Colombia. 149p.

Stür, W.W., Horne, P.M., Hacker J.B. & Kerridge, P.C. (eds). 2000. *Working with farmers: the key to adoption of forage technologies*. Proceedings of an international workshop, Cagayan de Oro City, Mindanao, the Philippines, 12–15 October 1999. ACIAR Proceedings, No. 95.

Stür, W.W., Horne, P.M., Gabunada Jr., F., Phengsavanh, P. & Kerridge, P.C. 2002. Forage options for smallholder crop-animal systems in Southeast Asia – working with farmers to find solutions. *Agricultural Systems*, **71**: 75–98.

Tajuddin, I., & Chong, D.T. 1995. Ruminants in plantation systems in Malaysia. pp. 97–99, *in:* B.F. Mullen and H.M. Shelton (eds). *Integration of ruminants into plantation systems in Southeast Asia*. Proceedings of a workshop, Lake Toba, Indonesia, 09–13 September 1994. ACIAR Proceedings, No. 64.

Vercoe, J., Coffey, S., Farrell, D.J., Rutherford, A. & Winter, W.H. 1997. ILRI in Asia: an assessment of priorities for Asian livestock research and development. International Livestock Research Institute, Nairobi, Kenya. 62p.

Yen, V.T.H. & Binh, L.H. 2000. Forage options for fish and pigs in Viet Nam. 281-283, *in:* Stür *et al.*, 2000, q.v.

CHAPTER 16

Pastoralists, Present and Future: Rangeland Development in West Asia and North Africa

Jon Rae[1]

ABSTRACT

For 50 years, governments across West Asia and North Africa have sought to raise the productivity of steppe rangelands through a series of centralized rehabilitation and grazing schemes. Technically-led and top-down, these interventions have left their legacy in state and customary institutions, but with little noticeable impact on the steppe pastures. Indeed, by most accounts, the degradation of biodiversity and the resource base has accelerated. The inadequacy of a centralist approach and the need to involve pastoralists and their institutions has been recognized for some time as fundamental to rangeland management and conservation. Nevertheless, some of the factors behind why they were excluded in the past still echo today. Largely political in nature, these factors constrain the nature of the participatory process and increase its complexity, although considerable opportunity remains for a sustainable future.

INTRODUCTION

The West Asia and North Africa (WANA) area faces unprecedented threats to its management, conservation and sustainable use of expansive dry scrublands (steppe) and desert areas. At the same time, valuable local knowledge in these areas is rapidly disappearing. Common factors lie behind both these trends, not the least of which are increasing human populations, unsustainable consumption patterns (Plate 16.1), climate change and economic forces beyond the reach of local

[1]School of the Environment, Brighton University, East Sussex, United Kingdom.

Plate 16.1 Rangelands are being damaged through widespread cutting and uprooting of fodder shrubs as a source of firewood. [Photograph by Marzio Marzot]

people. Furthermore, these trends bring into relief the inadequacy of long-standing and costly state-centric policies in natural resource management (NRM). For two decades now, participatory approaches to NRM have been explored in response, though, as one recent review suggests, more often in word than in substance (Shackleton *et al.*, 2002). This said, in south Asia and Sub-Saharan Africa, devolution of power and community-based NRM are increasingly familiar currency in forestry, wildlife and rangeland management policy. In the WANA region, such moves are less evident, despite broad recognition amongst most governments of the economic and administrative gains derived from wider participation. The Islamic Republic of Iran has, for instance, made participation of local communities in the development process a cornerstone of its constitution. Nevertheless, here as elsewhere in the WANA region, the extent of participation in steppe management remains restricted.

Most WANA countries became independent in the mid-twentieth century and without exception the subsequent process of nation building involved policies to circumscribe both nomadic activity and the legal autonomy enjoyed by many pastoral groups. Apart from the political imperatives, the dissipation or abolition of traditional authority among the migratory tribes was driven by widespread perceptions concerning the inadequacies, indeed damaging impact, of customary practices on steppe productivity, and by extension on the fledgling national economy. This awareness was partially inherited from the departing colonial powers, but it also reflected long-held, widely ambivalent perceptions amongst many settled populations of the region towards their migratory brethren. The ascendant paradigm during this formative period, and one that echoes to this day, advocated state-led development and the introduction of "scientific" range management to integrate the national economy and boost overall productivity. With the exception of Yemen, all countries in the region nationalized their rangelands (FAO, 2002a) as a prerequisite to centralised management. Half a century later, the question of range management remains as vexed as ever. Central control over resource access amongst a sparse and migratory population utilizing a wide and poorly productive region has universally proved logistically impossible and prohibitively expensive.

Most WANA states have since retreated from systematic management of all rangeland resources, preferring instead to focus rehabilitation and conservation efforts on discrete, more easily controlled enclosures. To date, these have largely been developed and managed without the participation of the pastoralists, even though they currently represent "the main approach [to range management] in the WANA region[s]" (Ngaido *et al.*, 1998). Despite this trend, there are parallel moves, often in the same countries, in search of suitable mechanisms to facilitate participation in rangeland management. Syria is a particularly good example of this, with 30 percent and 40 percent of the steppe targeted for participatory and non-participatory reserve establishment respectively (FAO, 2002b). Like many other countries in the region, Syria is apparently in a state of transition between development paradigms, a transition that is slow and largely defined by institutional parameters (inertia, vested interests, dogma, organizational structure, etc.). The discussion will return to Syria later in the chapter to examine more closely the participatory approach adopted and the associated limitations.

Partnering the shift towards a participatory paradigm, and indeed laying the groundwork for it, have been parallel shifts in ecological and institutional theory. Traditional ecological thinking on rangelands focused on carrying capacity to regulate the interaction between plant and livestock populations and so avert resource degradation. While there is little doubt that this thinking remains in part valid, new ecological research in rangeland areas now suggests that rainfall variability and other episodic events, rather than grazing *per se*, control plant and animal production (Behnke and Scoones, 1991). Records frequently show consecutive periods in which rainfall is consistently above or below the long-term average. Such extremes of drought or exceptional rainfall, sometimes coupled at the margins with pastoral activities, are quite likely to trigger radical changes in vegetation composition. This evolving understanding helps explain why past state interventions have so rarely been able to make a lasting impact. Through settlement and western-style grazing systems, most states across the region sought to regulate carrying capacity and raise rangeland productivity. However, the management implications of the new theory places less emphasis on carrying capacity and more on drought-coping mechanisms, such as mobility and livelihood diversity maintenance.

This recognition of mobility has raised the profile of the much-maligned customary systems of resource tenure as the foundation for sustainable resource management. Though weakened by state-centric polices, many customary institutions have nevertheless adapted and survived, and patterns of mobility today widely remain a product of customary resource tenure (FAO, 1991). This reflects not only the deep cultural roots of the customary system but the necessity of it for facilitating and regulating mobility, and reducing conflict in the absence of a credible and sustainable alternative. Supporting customary structures and institutions where they exist, and building their capacity to better evolve to meet rapidly changing circumstances, represents a key opportunity for rangeland management in the WANA region (FAO, 1991). It follows that to achieve sustainable participation requires a tenure system that is both familiar to the pastoralists and returns the benefits of participation (better long-term grazing and wildlife conservation) to those involved.

An incorporation of customary tenure within an integrated management strategy would reduce the costs and raise the efficiency of the intervention. Past centralized schemes largely succumbed to the crippling transaction costs of introducing, monitoring and enforcing a new institutional regime. Since many of these transaction costs are readily translated into financial costs, such schemes have often proved burdensome to the state. Their impact has been compounded over the past two decades by the rising overall debt incurred by WANA states. Subsequent economic restructuring in many countries, much of it donor-driven, has focused on the roll-back of state involvement in the economy and the decentralization of decision-making to facilitate administrative responsiveness and efficiency. Discussion of participatory approaches in NRM in the region neatly coincides with these liberalizing shifts and often reflects donor aspirations. Nevertheless, with the ineffectiveness of past state interventions, coupled with the fiscal crises of states throughout the region, it is argued here that local institutions often make the devolution of management responsibility to local organizations a feasible and attractive option for policy-makers, rangeland scientists and conservationists alike (McCulloch, Meinzen-Dick and Hazell, 1998).

The first part of this chapter examines current changes in the broader socio-economic context in which pastoralists in the region operate, and speculates in general terms how this might shape pastoralism in the twenty-first century. The discussion then turns to the history of participation, or lack of it, in rangeland management, and broader issues of conservation in the WANA region, in order to identify the parameters affecting whether decentralization is or is likely to take place. Given that land tenure is crucial, both in terms of range management and of participation, the flexibility and adaptability of the customary system is examined. However, as tenure reform is a politically charged issue, the final section turns to a case study of Syria to highlight how the state is reconciling entrenched institutional positions – and indeed ideology and geopolitics – with the need for a pragmatic and decentralized approach to unlocking the benefits of genuine participation.

EXTENSIVE PASTORALISM IN THE TWENTY-FIRST CENTURY

The WANA region is home to 440 million inhabitants, and while this figure is increasing more slowly than in the recent past, it is still expanding by 2.4 percent per annum (FAO, various dates). This trend is paralleled by rapid urbanization, with the rural population becoming a minority in most countries during the 1990s. Poverty affects one quarter of the population, with concentrations in rural and, especially, low-rainfall areas (LRAs) (regions with under 300 mm rainfall per annum) (FAO, various dates). The growing affluence to be found in the urban and peri-urban zones is reflected in the rising demand for animal protein. Much of the latter comes from small ruminants (sheep and goats), an economy that contributes a substantial proportion of the income of farmers and nomadic or semi-nomadic herders of LRAs (see Plate 16.2). Livestock numbers have reflected demand and, though growing more slowly than the human population, have nevertheless doubled in size over the past 40 years (see Figure 16.1). However, most of this growth and indeed most of the region's human and sheep populations (94 percent and 94.5 percent, respectively) are concentrated in the lower-middle and low income countries of the region, with 82 percent of the total flock of 225 million sheep being raised in Afghanistan, Algeria, Islamic Republic of Iran, Morocco, Pakistan, Sudan, Syria and Turkey (FAO, various dates). Two of these countries, Iran and Pakistan, are now the world's largest producers of sheep and goat meat.

Further facilitating the rise in sheep numbers has been the favourable price ratios between animal products and barley – the feed that has come to dominate animal diets across the region since the 1960s. Indeed, this "new" barley cultivation has come at the cost of the better rangelands,

Plate 16.2 Sheep grazing on the rangelands. [Photograph by Marzio Marzot.]

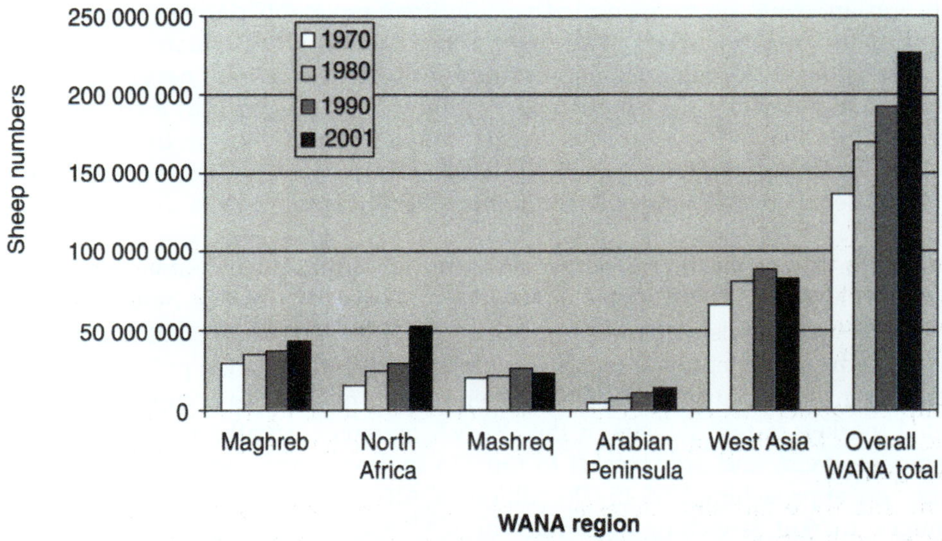

Figure 16.1 Growth in sheep numbers over the past generation across the WANA regions.
Source: FAO, various dates.

reducing the areas of pasture and increasing grazing and wood collection pressures on what remains. Rangeland contribution to animal diet in the region has correspondingly declined from an estimated 70 percent in the 1960s to between 15 and 60 percent in Tunisia and Algeria currently (Norblom and Shomo, 1995). Many of those undertaking this "new" cultivation have been pastoralists themselves, who have sought to diversify livelihood strategies and raise the reliability of their production system and, by extension, their livelihood strategy, with the introduction of barley-based supplementary feed (Wachholtz, 1996). However, a significant proportion of barley derives from the government, sourced from farmers at a set price, or more often from overseas, and transferred with the aid of subsidies to pastoralists. Though in part this practice features in standard drought relief policy, the scale of the subsidies in question reflects both an effort to bolster and satisfy urban demand for meat as well as to secure the tribes and desert areas.

Such subsidies and price supports are now increasingly threatened as state development policy across the region shifts from state capitalism towards greater reliance on market forces. Some countries such as Egypt have now phased out agricultural subsidies significantly. Morocco and Tunisia have removed fertilizer subsidies and are reducing livestock feed subsidies. Tariffs across the region are also being re-examined. Despite substantial protection of home production, net imports of cheap meat across the region stand at 20 percent of volume consumed and are forecast to rise as economies open to the market and demand continues its long ascendancy.

The reduction or removal of subsidies, tariffs and price supports and the removal of constraints more generally on factor mobility (the ease with which productive factors, such as capital, labour and natural resources, can be reallocated across sectors within the domestic or regional economy) are forecast to disadvantage sheep herders and barley production in the LRAs and have negative effects on household income (Bachta and Abassi, 1997; Bendaoud *et al.*, 1997). In Jordan, for instance, overall pastoral incomes are expected to fall by 21 percent, falling hardest on the poorer sections of local society. Jabarin and Chaherli (1997), however, suggest that such (agro-)pastoralists, at least in Jordan, could be compensated with indirect payments derived from savings of twice the amount made from reducing government intervention; in this instance, on the reduction or removal of subsidized feed. Political ramifications of such actions may nevertheless forestall such moves. Indeed, Syria, which elsewhere in its economy is liberalizing, has increased the amount of subsidized feed available to pastoralists, nearly doubling the amount in the past five years, from 27 kg in 1997 to 52 kg per sheep in 2002 (FAO, 2002b). Such subsidies are absent in Oman, but across the border in the United Arab Emirates, US$ 300 each year is available for each camel (PDO, 2001).

Liberalization is forecast to reduce pressures for agricultural expansion and, indeed, to send expansion into retreat at the margins, as well as to reduce or reverse herd growth (Sidahmed, 1996). Whether such trends materialize remains to be seen. Two important caveats exist:

- The first, driven by national security concerns, is the entrenched policy on food security in many countries in support of self-sufficiency over reliance on comparative advantage and global markets. Grain deficits in the WANA regions currently average around 50 million tonnes per year, and are forecast to top 114 million tonnes per year by 2020, at a cost to importers approaching US$ 15 000 million (ICARDA, 2000). Current discussions of an Arab common market, stretching over much of WANA and including the supposed breadbasket Sudan, may well shift this emphasis, though movement in this direction is slow.
- The second is the focus being put on increasing the drought tolerance of barley cultivars, given the advent of biotechnology.

With an emphasis on self-sufficiency as the foundation of food security, such developments may well encourage further expansion of cropland and force pastoralists into ever more inhospitable terrain.

With or without new technology, many of the better pastures that do remain are increasingly coming into focus as reserves of biodiversity. The remoteness of these areas has enabled the survival of some species eliminated in high-density agricultural areas (Blench, 1999). Both internal and external pressures are manoeuvring states into declaring substantial areas as protected, often to the full exclusion of pastoralists. It is this and wider issues of conservation and participation to which the discussion now turns.

PARTICIPATION IN A HISTORICAL PERSPECTIVE

The institutional history for conservation and rangeland management in the region, which for all practical purposes begins after the Second World War, has explicit non-participatory origins. Reflecting long-entrenched perceptions amongst those of the settled and urban areas, customary law and practices amongst pastoralists, as well as their tribal social structures, were and remain widely considered to be anarchic and barriers to progress and modernization. Ibn Khaldun, a famous fourteenth century Arab geographer, divided society between that part based on tribalism (*badawah*), with its heartland amongst the nomads of the steppe, and that, by contrast, focused on urban and settled areas, which is civilised (*hadarah*). The former, in crude terms, is destructive; the latter civilising or, as Ibn Khaldun suggested, the desert tribes

> "are a ... nation ... they enjoy it, because it means freedom from authority and subservience to leadership ... [s]uch a natural disposition is the negation and anti-thesis of civilization"
> (Ibn Khaldun's *Al-Muqaddimah* translated by Rosenthal, 1958).

The pan-Arab Ba'th Party, in power in Iraq and Syria from the 1960s, has taken a similar view of the regressive nature of tribalism and the damaging impact of custom on rangeland management. The 1966 Ba'thist Constitution depicts "nomadism" as "a primitive social state" and goes on to describes how:

> "It decreases the national output and makes an important part of the nation a paralysed member and an obstacle to its development and progress. The party struggles for the sedentarization of nomads by the grants of land to them [and] for the abolition of tribal custom." (Article 43)

Such sentiments were ascendant across much of the region and remain so, if somewhat diluted by participatory credos, in various institutional forms to this day. Partly underlying the longevity of this and related positions on the desert tribes is the geopolitical imperative of protecting the desert and border areas from subversive activities. Mobile tribes operating in these areas are often seen as unstable political actors. While settlement would bring them under greater administrative control, it has proved unpopular and often impractical. What has lasted is the formal abolition of custom and tribal authority to subdue local powers, while co-opting them through client-patron channels and the maintenance of subsidies. If this is indeed the case, there are clear limitations to the extent and depth of participatory programmes.

Scientists and international development agencies challenged the pastoralists from a slightly different angle, seeing a rapidly degrading rangeland and viewing overgrazing as the chief catalyst. An FAO synthesis of range reports on the region through to the mid-1950s concluded

> "that without exception, range management specialists and ecologists have stressed that the range grazing of the steppe and semi desert regions ... is progressively deteriorating as a result of overgrazing" (Shier, 1956).

The implication of this was that the inadequacies of the customary system were either inherent or the result of the inevitable decline of tribal structures and institutions in the face of the modern administration of a nation state. Indeed, that the latter was to come about was paramount for the United Nations, who saw it a prerequisite to modernization that "traditional chiefs ... be brought under close administrative control, [and that] customary law and native courts ... be abolished" (ILO, 1964). Settlement, social transformation and the introduction of western ranching practices were widely adopted as panaceas to the problem of steppe management and productivity.

The inadequacy of such an approach has since been reflected in the clear failure of enforced settlement and the collapse of ranching schemes across the region and further afield. Perhaps the most infamous involuntary settlement scheme was undertaken in pre-revolutionary Iran amongst the Qashqaie and other groupings of the steppe. Seen as an element of imperial rule, such settlement schemes were abolished following the 1979 revolution and the steppe witnessed a widespread return to migratory pastoralism (Beck, 1981). Only in some of the oil-rich countries where significant funds were available did settlement programmes eventually produce results, such as the *hejar* system in Saudi Arabia (Hajrah, 1982). Nevertheless, here, as in other Gulf States, Bedouin livestock have been maintained on the steppe often supported by subsidies and, with few exceptions, herded by expatriate workers from the sub-continent.

The ranching schemes much trumpeted in the 1960s and 1970s suffered a similar fate to those of enforced settlement, to a point that it deterred further international investment in the rangelands for the next two decades (de Haan, 1994). The case of Syria provides a useful illustration for the broader region, since it was here that the famous *hema* system for grazing management was eventually developed and widely adopted elsewhere. The hema (derived from the Arabic word for "protection") system was formulated by Omar Draz, who distilled research on customary grazing systems in Saudi Arabia and neighbouring countries, and coupled them together with modern ranching principles and the desired objective of greater nomadic settlement (Draz, 1971; 1981). The termination report for the highly centralized precursor to the hema system in Syria (as elsewhere), underlined the need for a fresh approach that sought greater participation of the range users:

> It is now clearly recognized that any programme for improvement both of pastures and sheep will require the full co-operation of the Bedouins and that this co-operation is more likely to be achieved through a well designed extension programme linked to a programme for the establishment of co-operatives rather *than through force and intimidation* [italics by author]. FAO (1969:3).

In response, FAO backed the hema system. Following pilot trials in Syria, it was then extended across the region (including Egypt, Jordan and Sudan) by FAO as a participatory model for range management. The system, however, was centralist in its conception and application, enforcing cooperative structures upon, and often across, existing tribal groupings as well as their supposed territories. This constituted "participation". The rules remained draconian and unfamiliar to the pastoralists, and included the transfer of responsibility for herd movement within and between cooperative territories from the households to government-appointed cooperative committees. Other rules included a static carrying capacity, herd-size limit, as well as a ban on fuelwood collection (without an alternative offered). The hema system no longer functions in any country; indeed, it is uncertain if it anywhere became effectively operational in the first instance. Cooperatives were soon abandoned in the Sudan (M. Zaroug, pers. comm., 2002), while in Syria and Algeria they remain, although their role in grazing management has long since been substituted by government largesse, often in the form of subsidized feed and "development" funds (Bedrani, 1991; FAO, 1999).

The failures of the hema system in Syria and elsewhere in the region during the 1970s led both international agencies and governments alike to withdraw from broad-coverage rangeland management. Favoured instead from the mid-1980s onwards, and increasingly supported by agencies through the 1990s, has been a focus on smaller, centrally-managed grazing reserves. As noted above, these reserves constitute the main approach to rangeland management in the region (Ngaido *et al.*, 1998). In Algeria, such arrangements cover some 2 percent of the steppe area, and 3 percent in Iran, while in Syria the figure currently stands at 8 percent, with plans to expand to over 40 percent of the steppe by 2010 (FAO, 2002b). Similar growth in exclusionary enclosures is forecast for other WANA countries (World Bank, 1995).

The grazing reserves represent the effective privatization of rangelands by the state. In an era of participation, the reserve concept nevertheless continues top-down management, with an unknown impact on pastoralists. Just how the concept may fare is not wholly clear, though the prospects are not promising. Indeed, the reasons for past failures of centralized schemes are likely to plague grazing reserves, a notable problem being excessive transaction costs. The only study examining this aspect of centrally-managed grazing reserves was conducted by the International Centre for Agricultural Research in Dry Areas (ICARDA) in 1995 and 1996 (Nordblom *et al.*, 1995; Rae *et al.*, 1999). The studies took in four reserves, each of between 8 000 and 12 000 ha, and were based on extensive interviews with pastoralists, reserve guards and government officials. Pastoralists were not consulted in the establishment of the reserves and consequently the borders often cut across existing customary territories. The displacement of pastoralists from tribal territories subsumed by the reserves and the impact on their household economy was not a consideration in the establishment of the reserves. For discussion on this aspect see Rae, Arab and Nordblom, 2002. This not only caused problems for the local pastoral population, but also led to disputes once the reserves were opened under contract some five years after establishment. Recorded trespass over all four reserves numbered 410 in the previous year. At one reserve, such records were kept for the preceding five years and tallied 386. Furthermore, overlap between formal and customary tenure coupled with the rule that reserve contracts be open to all Syrians, precipitated often difficult and dangerous disputes as they regularly involved tribal groupings (Ngaido *et al.*, 1998).

As this brief history of state intervention in rangeland management suggests, most states across the region continue in their reluctance to deal directly with tribal structures and institutions as a basis for rangeland management. In Tunisia and Morocco, the state does recognize tribal territories and structures, but considers their institutions for management as redundant. In Oman, Saudi Arabia and elsewhere in the Gulf, the state recognizes tribal structures but stops short of formal acknowledgement of their territories. The formal non-recognition of tribal structures and territories, coupled with institutions regulating access, is common to Algeria, Egypt, Iraq and Syria, a position broadly based on ideology, geopolitical concerns and latent notions of modernity. Only in the Sudan, where the hema cooperatives were briefly tried but soon abandoned, does the state recognize both tribal structures and institutions, although the expansion of commercial cultivation by city merchants and government officials continues to raise concerns (FAO, 2002a).

A FUTURE FOR PARTICIPATION

Some 50 years after the birth of the Nation State, the predicted demise of traditional social formations and the ascendancy of "western" concepts of civil society have yet to be realized. While a great deal has changed, the role of family, clan, patron-client networks and customary dispute resolution mechanisms remain some of the strongest institutions in the region, and constitute an important element of an evolving homegrown civil society. It is the resilience of these, however, that has been overlooked in the development process and no more so than in projects for management of the semi-arid and arid environments in the region. That almost all of such projects have failed in part reflects this neglect. One does not have to look far in the Arabian Peninsula to find resonant tribal tradition expediting the state-building process. Even in one of the most democratic Arab states, Jordan, most party affiliation is either an expression of family or tribal ties. Despite this, when it comes to rangelands, most aid agencies have consistently judged, as they have in Jordan, that "centralized power and growing individualism has broken allegiances within tribal groupings ... and ... broken down traditional [land] tenure" (IFAD, 1998).

Many state governments in the region are now openly recognizing that many traditional formations are resilient, and many of the region's ruling elite have been willing or obliged to accommodate them through consultative groups with advisory roles (Kamrava, 1998). The Islamic government in Iran, for instance, has established an advocacy group, Organization for the Nomadic Pastoralists of Iran, for consultation on policy decisions. Rarely, however, have Arab countries extended such recognition to the specific task of managing rangelands. This said, aware of past policy failures towards the moving tribes, governments are now shying away from forced settlement of nomads and adopting, in word at least, participatory practices.

A round-table discussion on pastoralism for the region, conducted by FAO in Iran in 1992, with representatives from various governments and organizations, concluded that without tenure reform there was little chance of securing participation and halting rangeland degradation (FAO, 1992). The participants of the round table went on to recommend that in most instances it was appropriate to start from existing tribal or other customary forms of institutions as a basis for building new institutions for range management. The reasoning for this is threefold:

- in the first instance, customary tenure is understood and broadly followed by the pastoralists;
- second, it provides a mechanism to allocate real decision-making power to established and organised groups; and
- third, customary tenure enables flexibility of range use that is able to cope with wide environmental fluctuations.

More recently, institutions including International Union for the Conservation of Nature (IUCN), UNESCO, World Commission on Protected Areas (WCPA), Royal Society for the Conservation of Nature (Jordan), WWF, GTZ, Government of the Islamic Republic of Iran, University of Oxford and European University Institute convened a meeting in Jordan to consider a comprehensive approach to mobile peoples and conservation. The product was the Dana Declaration, a collection of principles to promote better conservation practices in areas dominated by mobile peoples. The first principle is that conservation approaches must seek effective empowerment of mobile groups through recognition of mobile peoples' rights, management responsibilities and capacities. Second, partnership between conservation interests and mobile peoples should be based on mutual trust and respect and address the long-standing issue of discrimination against mobile peoples. Third, in planning conservation programmes there is a need to respect and incorporate local social institutions and management practices. Fourth, there should be broad adoption of adaptive management practices that build on existing cultural models and incorporate mobile peoples' aspirations and customary law. Lastly, conservation structures must reflect collaborative management based on the concept of equitable sharing of decision-making and management responsibility.

As the signatories to the Dana Declaration and others recognize, the basic building block for participation of migratory groups is land tenure.

Customary land tenure

Customary tenure systems in the WANA regions have only recently become the focus of study within development circles and, in many instances, they remain poorly understood and appreciated. The assumption has been that such systems were either inherently inefficient or lost, given expanding state power, or both. Nevertheless, tenure systems have long been an interest to anthropologists and what investigations have been undertaken suggest often complex and adaptable regimes, as found by Stewart (1986) amongst the Bedouin of Southern Sinai.

What seems common to customary systems studied in the WANA regions is that, from the densely farmed areas along the steppe margins all the way to hyper-arid zones, territory and exclusive rights are built around whatever in the particular case is the most limiting factor (Dresch, 2000; Wilkinson, 1983). In the steppe and desert, this factor has predominately been water. Establishing a right in water brings with it a parallel usufruct right in the surrounding pastures, since the former is essential for the exploitation of the latter. Size of area and fluidity of border tend to increase or become irrelevant with rising aridity or where competition over resources is limited. Conversely, in higher rainfall areas of the steppe where land scarcity is perhaps more of a factor or where land is partially settled, there are pressures to more clearly define borders and access between clans or villages.

The last 40 years have witnessed a dramatic increase in livestock numbers, particularly sheep, as well as in mobility of both herd and water. Within customary tenure, the emphasis in many instances has now shifted away from water *per se* to pastures as the limiting factor. Largely in keeping with Boserup's hypothesis that suggests a strong correlation between rising populations on the one hand and increased regulation and resource management on the other, many customary regimes have responded to such pressures by clarifying borders and access rules in the more favourable and pressured areas. Along Egypt's northwest coast, for instance, increasing pressure from developers has precipitated clear divisions among the Awlad Ali clans along a 30-km-deep strip from Alexandria to the Libyan border, as well as greater levels of grazing restrictions on outside groups (Rabenau, 1994). A similar pattern is being followed in Morocco, though it is not directly recognized as such by the authors of that study (Nassif, Boughala and Ngaido, 1997). More detailed studies of customary tenure have been conducted in the Syrian steppe through ICARDA and FAO (Chatty, 1998; FAO, 1999; Rae *et al.*, 2002). These confirm a gradient of tenure across the aridity scales and demonstrate the evolution of custom toward greater clarification of borders and rights under pressure from rising sheep populations. Similar to developments in Egypt, a belt of clear clan territories, often in excess of 100 km in depth, has evolved in the higher rainfall areas along the margins of the Syrian steppe. The definition of such borders usually follows a dispute, typically over investment rather than grazing *per se*, and is settled by each party, often with the arbitration of a third party. In Syria, as elsewhere, that third party is increasingly the state, which alone has the power to guarantee solutions. This said, in Syria at least, third-party tribal elders facilitate resolution and have often endorsed written agreements (Rae *et al.*, 2002; Rae 2002). Such participation of the state may seem contrary to formal policy but it resolves a potentially damaging dispute in a fashion familiar and legitimate to the disputants (Chatty, 1998). It should be noted that borders between territories mark the extent of the tribe's peace, on which the honour of the whole tribe depends. Customary access to these territories is less contentious and operates through social networks when they are able or allowed to operate.

The virtues of building upon existing customary tenure regimes, in rangeland management and conservation, centre on existing social capital, low rule enforcement costs, and their flexibility in time and space to shifting realities of resource availability. In particular, supporting and building existing socio-political organizations that fulfil management roles is unquestionably easier than introducing new institutions, let alone structures. For wildlife conservation, the precarious nature of "release schemes" such as those for oryx in Oman and Saudi Arabia, or those being contemplated in Syria and Jordan, could benefit substantially from greater recognition of customary tenure and appropriate dispute resolution mechanisms.

THE POLITICS OF PARTICIPATION: CASE STUDY OF SYRIA

Syria provides a good illustration for the region for a variety of reasons. It is the original home of the hema concept, sold as the first participatory grazing system in the region, that was exported through the 1970s and 1980s across WANA. Its failure, described below, precipitated the adoption of the so-called "plantation" concept or grazing reserves, again leading the way in the region. The current large-scale experiment in the country, with state-led and "participatory" grazing reserves, is once more unprecedented in the region and may well be an indicator of future developments.

In Syria, as elsewhere in WANA, there is both continuity and change in traditional and new socio-political forms on the steppe, each influencing the other. As a former Ba'thist leader conceded after long experience implementing Ba'thist policy in Syria, all

> "intellectual, political and ideological currents that spread in … the Arab countryside … are much more affected by tribal tradition than that the latter are influenced the other way round"
>
> (Jalal al-Sayyid, 1973, *Hizb al-Ba'th al 'Arabi*, Beirut, pp. 252-253, in Van Dam, 1996: 22).

The hema system demonstrated this phenomenon with its widespread reliance on tribal structures and territories in the establishment of most cooperatives. Though the hema system failed, the cooperatives continue as a strong link between the ruling Ba'th Party and local steppe communities. This position in the power structure of the country ensures their central role in any attempt to foster the participation of pastoralists. For a long time conduits for subsidized feed to pastoralists, the cooperatives have recently become involved in a major new initiative between the Syrian government and the International Fund for Agricultural Development (IFAD) for the development of a participatory management strategy for the steppe (see FAO, 2002b).

Despite the IFAD project's focus on the steppe regions, it is not working in association with the Syrian Steppe Directorate but operates through a new and separate unit within the Ministry of Agriculture and Agrarian Reform. The reasons for this division of responsibility, and indeed steppe territory, are unclear, though it has permitted the development of two distinct approaches to range management: one described as participatory by IFAD; the other immersed within the centralist traditions of the Steppe Directorate. The former operates on the northern half of the steppe and covers approximately three million hectares, or 30 percent of the steppe area, while the Steppe Directorate remains responsible for the rest, with a fresh programme of reserves planned for four million hectares. An examination of the two programmes will highlight similarities and differences, and underscore the limitations and opportunities placed on institutional form by institutional inertia and politics.

Steppe Directorate

The Steppe Directorate is a technically orientated department with a focus on rangeland rehabilitation and management, principally through the use of "grazing reserves". Such reserves are demarcated with a trench or earth fence and part planted or re-seeded with shrubs. The current ten-year plan (which began in 2000) envisages the expansion of such reserves over four million hectares from a pre-existing base of 720 000 ha (see Table 16.1). If all goes to plan, some 40 percent of the Syrian steppe will be consumed by Steppe Directorate reserves, 15 percent of which will be replanted or re-seeded, with the reminder benefiting from protection only. After a period of establishment, three years in the first instance but perhaps longer depending on conditions, the reserves are opened under restricted contract to pastoralists for periods of between one and two months each year. The cost to pastoralists of access is in the range of SY$ 100–120 per ha (approximately US$ 1) with a maximum of three sheep per hectare, depending on an annual assessment of carrying capacity by Directorate technicians. Once fully up and running, income from the scheme should be around US$ 8 million per annum.

Such reserves are located without reference to customary resource tenure, though purportedly they are located outside cooperative borders despite borders never having been mapped. Nevertheless, the Deputy Minister responsible for the Steppe did confirm that cooperatives did

Table 16.1 Grazing reserve programme of the Steppe Directorate Grazing Reserve Plan for the Period 2000–2010

	2000	2010
Reserve number	50	280 (est.)
Reserve area (ha)	720 000	4 000 000
Planted or seeded (ha)	150 000	600 000
Natural rehabilitation (ha)	670 000	3 400 000
Potential income	US$ 1 440 000	US$ 8 million
Proportion rehabilitated		15 percent
Proportion untouched		85 percent
Annual expansion		200 000 ha $annum^{-1}$
Final coverage (by 2010)		40 percent of steppe
Period of reserve establishment		>3 years
Carrying capacity		1–3 head ha^{-1}
Grazing fees		SY$ 100–120 ha^{-1} programme
Period for grazing		1–2 months $annum^{-1}$
Admission		Under contract, to all Syrians
Level of participation:		None

cover the whole steppe area prior to the formal abolition of borders in 1997 (they were abolished in practice in the mid-1970s) (FAO, 2002b).

Reserve establishment and management is top-down, while access to them, once open to grazing, is available to all Syrian citizens. Though such "inclusiveness" is technically the case, past experience with reserves in Syria suggests that priority of access goes to local tribal groupings (Nordblom and Shomo, 1995). Nevertheless, the grazing reserves, as envisaged by the Steppe Directorate, constitute the effective privatization of a substantial proportion of the Syrian steppe.

The IFAD Badia Rangelands Development Project contrasts markedly with the programme of the Steppe Directorate. In the Badia project, rangeland management is judged the most crucial element and is being pursued through "the development of a replicable participatory methodology" (IFAD, 2002). In the Syrian political context, the starting point for this methodology or any other participatory process is invariably cooperative structures and their territories. The project has five components: rangeland management, infrastructure, livestock development, social and economic activities (including health, illiteracy, income generation with particular focus on women), and staff management (including training).

Operating through existing cooperatives and their leaderships, the Badia project seeks the allocation of land for project activities. Meetings of all members in each cooperative involved in the project have not been possible and all practical activities are agreed to and coordinated through the cooperative leadership. In many instances this leadership often reflects tribal structures, though even where they do, notions of "equity" cannot be assumed.

In the same way as the channels of participation were pre-determined prior to the participatory process, so too were the technical measures and the method of their execution. Once allocated by the cooperatives, these lands, though left unmarked, are off-limits to grazing or fuel collection, and are either replanted or re-seeded, with remaining areas benefiting from protection. It is envisaged that, by the end of the project (2007), all three million hectares will come under protection. In Homs province, for example, 20 of the 40 eligible cooperatives are so far participating in the programme, with the remainder expected to be incorporated in the near future. Amongst those already involved, six sites totalling 16 800 ha have so far been allocated for protection, of which 1 600 ha have been replanted, 5 200 ha re-seeded, and the remaining 10 000 ha left untouched. What the project hopes to achieve in Homs province by 2007 is the protection of all 772 200 ha, 90 percent of which will remain untouched though protected, 7 percent re-seeded and the remainder replanted.

After a two-year period, the protected lands are returned to the cooperatives and placed under the control of grazing committees composed of cooperative members, of which at least one must be a woman. Technical decisions, such as carrying capacity and period of grazing, remain with the project staff and are re-assessed each year, while the grazing committee is responsible for organizing grazing as well as monitoring and enforcing compliance. In addition, it will be left to the grazing committee to decide on the level of the grazing fee to be charged, though it estimated that it will fall between 7 and 10 US cents per head. The fees will go to the cooperative treasury to fund future development, as well as to pay the salary of guards (1 guard for every 5 000 ha, at approximately US$ 840 per year per guard). Access to the protective areas by neighbouring cooperatives will be by decision of the grazing committee.

As of June 2002, no protected areas had yet been returned to the cooperatives in Homs Province or elsewhere in the IFAD area. During the two years of planting and re-seeding of protected areas, salaries for the guards would be paid for by the project, and once these areas are "handed back" this responsibility falls on the cooperative itself. The project judges that the there is need for 1 guard for every 5 000 ha, at an initial annual salary of SY$ 42 000 annually.

The participatory methodology adopted by the project begins with the use of top-down structures in the form of cooperatives, a political though necessary compromise. Despite this, the participatory approach did not seek the active participation of pastoralists in identifying problem areas or their solutions; these were predetermined by technicians and shaped the cooperative organization. Participation is left to the management of the protective areas once they are returned to the cooperatives. Elections amongst cooperative members for seats on the grazing board were not conducted; rather, the grazing board reflects the original long-standing cooperative board with the addition of a female representative. With considerable power vested in the grazing boards, including the right to levy and collect fees, mechanisms to resolve disputes between grazers and grazing boards as well as oversee overall functioning is needed, though remains unplanned.

The IFAD project, nevertheless, is in marked contrast to similar activities being conducted by the Steppe Directorate. Both suffer from being unable to deal directly with tribal realities on the ground. As a compromise, the IFAD project has turned to the cooperative structures, although it seeks participation only for implementation of work and methods predetermined as necessary. The Steppe Directorate forgoes participation and collaboration with cooperatives altogether, preferring a rather more direct approach. Both programmes are planned to end between 2007 and 2010; which one proves more successful will help shape the future of pastoralism in Syria.

CONCLUSION

The rapid socio-economic change that has characterized the working environment of pastoralists in the region will continue in the twenty-first century. Liberalization will gain a firmer foothold on the steppe, though it will be constrained by food security and geopolitical concerns. The current trend from a statist to a participatory approach will further unfold, though the limits to that participation are likely to remain restrictive in most instances. Where these limitations emanate from the state, they will reflect not only institutional inertia and vested interests but also the political equation individual nations formulate to maintain political control and national security. Other limitations to participation stem from development agencies and conservation bodies, and it is these that are the target of the Dana Declaration. The prevailing system for conservation in the region and elsewhere has been based on the exclusionary model, which entails displacement of local populations and envisages for them only limited roles in the management or economic activities of the reserve. This is clearly evident not only in wildlife conservation in the region but also in range management and the current popularity of grazing reserves. The former typically exclude herds most if not all the time, while the latter generally permit access during restricted periods of the year.

Nevertheless, participatory techniques are gaining a higher profile in the region, if coloured by political and institutional realities. The latter constrain what is possible, and indeed where the opportunities lie. In much in the same way that evolution is thought to function in biology, evolution in human organization must, in most instances, use what is already in place together with innovations that are often more incremental than revolutionary. This is not to say that change is absent, just that inertia, established vested interests and political imperatives will continue to measure and direct such change.

REFERENCES

Bachta, M. & Abassi, M. 1997. Property rights analysis and investment in the low rainfall areas of Tunisia. pp. 201–209, *in:* N. Chaaherli and P. Hazell (eds). *Agricultural growth, sustainable resource management, and poverty alleviation in the low rainfall areas of West Asia and North Africa*. Proceedings of an International Conference, Amman, Jordan, 2–6 September 1997.

Beck, L. 1981. Government policy and pastoral land use in southwest Iran. *Journal of Arid Environments*, 4: 253–267.

Bedrani, S. 1991. Legislation for livestock on public lands in Algeria. *Nature and Resources*, **27**(4): 24–30.

Behnke, R. & Scoones, I. 1991. Rethinking range ecology: implications for rangeland management in Africa. *IIED Drylands Network Programme Issue Paper*, No. 33.

Bendaoud, M., Boughlala, M., Chaherli, N. & Moussaoui, M. 1997. A community model for evaluating the impact of policy, technology and property rights changes in low rainfall areas of Morocco. pp. 225–239, *in:* N. Chaaherli and P. Hazell (eds). *Agricultural growth, sustainable resource management, and poverty alleviation in the low rainfall areas of West Asia and North Africa*. Proceedings of an International Conference, Amman, Jordan, 2–6 September 1997.

Blench, R. 1999. *'You can't go home again': Pastoralism in the new millennium*. Overseas Development Institute and the Food and Agriculture Organization of the United Nations.

Chatty, D. 1998. Enclosure and exclusions: conserving wildlife in pastoral areas of the Middle East. *Anthropology Today*, **4**(4): 1–7.

De Haan, C. 1994. An overview of the World Bank's involvement in pastoral development. *Pastoral Development Network Paper*, No. 36b.

Draz, O. 1971. An approach for the settlement of nomads through revival of the ancient Hema system of range reserves in the form of cooperatives within an integrated programme of range improvement in Syria. *FAO Eexpert Consultation of the Settlement of Nomads in Africa and the Near East*, 1971 FAO, Rome, 9 p.

Draz, O. 1981. An integrated programme for range improvement and sheep feeding for the Near East and North African Countries. pp. 1–20, *in: Proceedings of an FAO Workshop on the improved utilization of feed resources for sheep fattening in the Near East*. 2–5 September 1981. Department of Agriculture, Damascus, Syria & FAO, Rome.

Dresch, P. 2000. *A History of Modern Yemen*. Cambridge: Cambridge University Press.

FAO. 1969. *Interim report: Syrian Arab Republic 002*, FAO, Intergovernmental committee 13-18 October 1969, FAO, Rome. 10 p.

FAO. 1991. *Pastoral systems in the Near East: Traditional systems in evolution*, Amman, Jordan 1-5 December 1991. FAO, Rome.

FAO. 1992. Roundtable on Pastoralism, Tehran, Islamic Republic of Iran, 1–3 November 1992. FAO, Rome

FAO. 1999. Land resource tenure: Range rehabilitation and establishment of a Wildlife Reserve, Project GCP/SYR/003, Syria: Palmyra

FAO. 2002a. Land tenure review for the Near East. FAO, Rome

FAO. 2002b. Participatory land management: Range rehabilitation and establishment of a Wildlife Reserve, Project GCP/SYR/009: Palmyra, Syria.

FAO. Various dates. FAOSTAT for various years (data as of 2002).

Hajrah, H. 1982. Public land distribution in Saudi Arabia. London: Longman.

ICARDA [International Centre for Agricultural Research in Dry Areas]. 2000. Introduction by the Director General, *Caravan* (ICARDA's Newsletter), 25 April 2000.

IFAD [International Fund for Agricultural Development]. 1998. Rangeland degradation and socio-economic changes among the Bedu of Jordan: results of the 1995 IFAD Survey (coordinated by R. Blench). pp. 397–423, *in:* R. Squires and V. Sidahmed (eds). *Drylands: Sustainable use of rangelands into the twenty-first century*. IFAD Technical Reports.

IFAD. 2002. *The Syrian "Badia Project"*, www.ifad.org/lrkm/region/pn/sy-1073.htm

ILO. [International Labour Office]. 1964. *Technical meeting on problems of nomadism and sedentarization*. ILO, Geneva.

Jabarin, A. & Chaherli, N. 1997. Assessing major agricultural policies in low rainfall areas of Jordan: "A multimarketing modelling approach", pages 72-89, *in:* N. Chaaherli and P. Hazell (eds). *Agricultural growth, sustainable resource management, and poverty alleviation in the low rainfall areas of West Asia and North Africa*. Proceedings of an International Conference, Amman, Jordan, 2–6 September 1997.

Kamrava, M. 1998. Non-democratic states and political liberalisation in the Middle East: a structural analysis. *Third World Quarterly*, **19**(1): 63–85.

McCulloch, A., Meinzen-Dick, R. & Hazell, P. 1998. Property rights, collective action and technologies for natural resource management: a conceptual framework. *CGIAR System-wide Program on Property Rights and Collective Action Working Paper*, No. 1.

Nassif, F., Boughlala, M. & Ngaido, T. 1997. Land policies, property rights and long-term land improvements in the low rainfall areas of Morocco. pp. 181–195, *in:* N. Chaaherli and P. Hazell (eds). *Agricultural growth, sustainable resource management, and poverty alleviation in the low rainfall areas of West Asia and North Africa*. Proceedings of an International Conference, Amman, Jordan, 2–6 September 1997.

Ngaido, T., Nordblom, T., Gintzburger, G. & Osman, A. 1998. A policy shift towards sustainable resource development. pp. 185-201, *in:* V. Squires and E. Sidahmed, E. (eds). *Drylands: Sustainable use of rangelands into the twenty-first century*. IFAD Technical reports Series, Rome

Nordblom, T. & Shomo, F. 1995. Population, Land, Grain Production and Consumption, Range and Other Feed Sources for Livestock: Past, Present and Projected to 2025 for West Asia and North Africa. Socio-economic Series, No. 1, Aleppo, Syria: ICARDA.

Nordblom, T., Arab, G., Osman, A. & Gintzburger, G. 1995. April 1995 survey of Bedouin groups with contracts to graze the government rangeland plantation at Maragha, Aleppo province, Syria. *Regional symposium on integrated crop-livestock systems in the dry areas of West Asia and North Africa*, 10–13 September 1995, ICARDA, Aleppo.

PDO [Petroleum Development Oman]. 2001. Social Impact Assessment for the Mukhaizna Project. PDO, Muscat, Oman.

Rabenau, B. 1994. Sustainable development strategies for the North-West Coast, Egypt. El Qasr Rural Development Project, World Bank.

Rae, J. 2002. Tribes and territory in modern day Syria. *The Arab Geographer*, **5**(2): 102–113.

Rae, J., Arab, G. & Nordblom, T. 2002. Conservation and displacement in the Syrian Steppe. pp. 105–124, *in*: D. Chatty and M. Colchester (eds). *Conservation and Mobile Indigenous Peoples: Displacement, Forced Settlement and Sustainable Development*. Oxford, UK: Berghahn Books.

Rae, J., Arab, G., Jani, K. & Nordblom, T. 1999. Socio-economics of shrub plantations. pp. 196–201, *in:* N. Chaaherli and P. Hazell (eds). *Agricultural growth, sustainable resource management, and poverty alleviation in the low rainfall areas of West Asia and North Africa*. Proceedings of an International Conference, Amman, Jordan, 2–6 September 1997.

Rae, J., Arab, G., Nordblom, T., Jani, K. & Gintzburger, G. 2002. Tribes, state, and technology adoption in arid land management in Syria, pp. 147–165, *in:* R. Meinzen-Dick, A. Knox, F. Place and B. Swallow (eds). *Innovation in Natural Resource Management: The role of property rights and collective action in developing countries*. Baltimore, USA: John Hopkins.

Rosenthal, F. 1958. *Ibn khaldun's The Muqaddimah: an introduction to history*. Translated from Arabic by F. Rosenthal. New York, NY: Pantheon Books.

Shackleton, S., Campbell, B., Wollenberg, E. & Edmunds, D. 2002. *Devolution and community-based natural resource management: Creating space for local people to participate and benefit*. Rural Policy and Environment Group, Overseas Development Institute.

Shier, F. 1956. *Report to the government of Syria on a program for improvement in sheep breeding and management*, No 571. FAO, Rome.

Sidahmed, A.E. 1996. The rangelands of the arid/semi-arid areas: challenges and hopes for the 2000s. Keynote address to Symposium D: Range Management. *International Conference on Desert Development in the Arabian Gulf Countries*, Kuwait, 23–26 March 1996.

Stewart, F. 1986. *Bedouin boundaries in central Sinai and the southern Negev*. Wiesbaden, Germany: Otto Harrassowitz.

Van Dam, N. 1996. *The Struggle for Power in Syria: Politics and Society Under Asad and the Ba'th Party* London: I.B. Tauris.

Wachholtz, R. 1996. *Socio-economics of Bedouin farming systems in dry areas of northern Syria*, Stuttgart-Hohenhiem, Germany: Wissenschaftsverlag Vauk.

Wilkinson, J. 1983. Traditional concepts of territory in South East Arabia. *The Geographical Journal*, **149**: 301–315.

World Bank. 1995. *North Africa and Iran: Rangelands development in arid and semi-arid areas, strategies and policies*. World Bank, Washington DC.

Section VII
REGIONAL DEVELOPMENTS

CHAPTER 17

Forage and Ruminant Livestock Integration in Tree Crop Plantations of Southeast Asia

C.C. Wong[1], F. Moog[2] and C.P. Chen[3]

ABSTRACT

The current scenario for the development of forage, livestock and tree crop integration in Southeast Asia is presented and discussed in relation to its potential for future development of a sustainable ruminant industry in ASEAN countries. Using the Malaysian model, the R&D on cattle–oil palm integration from its early conception in the seventies to commercialisation of its technology in the late nineties, was highlighted. Various aspects are reviewed, including forage feed availability and its changes with tree-crop maturity, and potential use of crop by-products by ruminants. Scientific findings have helped to dispel negative attitudes of planters towards ruminant rearing in plantations, and also provide a firm basis for a holistic approach in livestock–tree crop integration. Although nutrient availability per tonne of DM forage declines with age of the plantation crop, the major nutritional drawback is its forage yield reduction per unit land area. The use of agricultural by-products from the plantations to compensate for diminishing supply of forages in mature stands is discussed. The cattle component is now regarded as complementary to plantation crop production, and collaboration among research institutions, government agencies and plantation entrepreneurs has made cattle–tree-crops integration a successful and economically viable investment. Future research and development priorities will depend on reorientation to a multidisciplinary and multisectoral systems approach that addresses efficient resource use and management for environmentally sustainable development of the ruminant industry within plantation systems.

[1]Rice and Industrial Crops Research Centre, MARDI, P.O. Box 12301 GPO. 50774 Kuala Lumpur, Malaysia.
[2]Bureau of Animal Industry, Department of Agriculture, Visayas Avenue, Diliman, Quezon City, Philippines.
[3]Wilstar Sdn. Bhd., Wisma Southern, 6th Floor, 26-34 Jalan Dato Hamzah, 41000 Klang, Selangor DE, Malaysia.

INTRODUCTION

The new millennium will witness the inevitable increasing urbanization of Southeast Asia, a steadily rising standard of living as a consequence of economic development, and a rapid increase in population. Despite the recent economic downturn of the Southeast Asian countries, the projected trends in per capita consumption of livestock products are for significant increases in the next years. In the past, total meat consumption per capita increased from 9.4 kg in 1961 to 21.0 kg year^{-1} in 1995, with pigs and poultry being the major meat sources (FAO, 1999). This increase in demand for livestock products has exacerbated competition for agricultural land and labour between animal and crop production. As livestock density increases, particularly in areas of intensive cropping systems, it is only a matter of time until issues of land availability constrain future livestock production, and environmental sustainability will need to be addressed rationally.

Traditionally, cropping agriculture in Southeast Asia is already intensive. Plantation agriculture is the only sector that has the potential for maximizing land use and promoting agricultural diversity and sustainability using environmental friendly cultural practices. It is for these reasons that integration of livestock within the agriculture cropping industry in Southeast Asia has received renewed emphasis in the new millennium (Table 17.1)

Table 17.1 Availability of forages for ruminants from potential grazing areas of Southeast Asia.

Land use type	Area (million ha)	Yield [(1)] (t DM ha^{-1} yr^{-1})	Availability (million t DM yr^{-1})
Permanent pasture	17.1	0.8	13.68
Forest and woodland	229.6	0.2	45.93
Permanent crops	28.5	0.4	11.40
Other land use	98.6	0.4	39.43
Total	373.8		110.44

Notes: (1) Average yields proposed by Roxas, Wanapat and Winugroho, 1997.
Source: Adapted from Pezo *et al.*, 1999.

Over 210 million hectares of agricultural land are estimated to be under perennial tree crops in Southeast Asia (Alexandratos, 1995). The countries in Southeast Asia with huge plantation crops are Indonesia, Malaysia and Philippines, and these three countries account for over 80 percent of the plantation land. Currently oil-palm occupies over 5.2 million hectares of land, coconut 6.6 million hectares and rubber 5.7 million hectares. The potential for integrating such perennial tree crops with large and small ruminants is great. Rubber, oil-palm and coconut plantations in Indonesia, Viet Nam and Malaysia, teak forests in Lao PDR and Myanmar, and coconut and tropical orchards in the Philippines and Thailand have not been fully exploited for ruminant integration purposes. In Malaysia, the preferred options under rubber are small ruminants (Plate 17.1), with large ruminant under oil-palm (Plate 17.2). In coconut plantations, both large (Plate 17.3) and small ruminants are used.

It is not surprising that a number of livestock integration projects with plantation crops, especially coconut, were initiated as early as the 1970s to stabilize, if not increase, the income of smallholders from the dual use of land. One result has been voluminous publications on research and development (R&D) of integration projects, which were extensively reviewed by Reynolds (1995). The 3.16 million hectares of coconuts in the Philippines offer tremendous potential for integration of ruminants and diversification of income (Moog and Faylon, 1991; Deocareza and Diesta, 1993;

Plate 17.1 Sheep grazing under young rubber in Malaysia. [Photograph by Ismail Tajuddin]

Plate 17.2 Beef cattle-oil palm integration on a FELDA [Federal Land Development Authority] plantation in Serting, Negri Sembilan, Malaysia. [Photograph by Dr Stephen Reynolds]

Plate 17.3 Cattle grazing native pastures under old coconuts. [Photograph by Dr Stephen Reynolds].

Castillo, 1995). The present low copra and coconut oil prices and an uncertain future for coconut plantations with declining production due to palm senescence are good reasons for considering livestock integration. Although higher yielding coconut hybrids are being introduced, the adoption of new technologies has been slow for various socio-economic reasons.

In Indonesia, the cattle component was estimated to contribute as much as 75 percent of the total farm income in coconut-cattle enterprises (Iniguez and Sanchez, 1991). In Malaysia, the high import food bill, estimated at over US$ 3 billion per year, has prompted the government to promote a national policy of integration of livestock in plantations as an urgent measure to reduce the food bill in the face of national economic slowdown. At present, livestock-tree cropping integration is still being looked upon as an avenue for the expansion of food crops and the ruminant industry in countries of Southeast Asia with limited land resources.

CURRENT STATUS OF ANIMAL PRODUCTION SYSTEMS IN PLANTATIONS

The concept of integrating livestock under plantation crops such as coconut, rubber and oil-palm is not new (Dalzell, 1983; Ani Arope, 1988; Khusahry *et al.*, 1988; Tajuddin, 1988; Hassan and Sani, 1991; Reynolds, 1995; 1998). Many farmers in countries of Southeast Asia have already implemented such a system, using locally available forages and cattle rearing in plantations to supplement their income and meat supply. The animal species are mainly carabao, beef or dairy cattle, goats, or, more recently, sheep. In addition, various crop by-products (weeds, tree leaves and wayside forages) have been incorporated as animal feed (Devendra, 1979). Generally, the livestock reared are on a low to medium plane of nutrition. The animals are tethered or allowed to graze freely all year round.

In many Southeast Asian countries, ruminant raising is predominately a supplementary enterprise. Livestock are raised for agricultural as well as social reasons, and often not for business investment. Selling of animal takes place only when the need arises, although the pattern is changing slightly in Malaysia, the Philippines and Thailand, where livestock rearing is becoming a business enterprise.

While much R&D in the 1970s concentrated on selection of improved species for integration purposes, the adoption of improved management practices for such technology is insignificant (Chen *et al.*, 1991). Cattle breeding and fattening operations are feasible in coconut plantations, and these can be based on grazing of improved or native pastures, or cut-and-carry feeding systems, depending on the size of the operation. Grazing systems are generally used in more extensive coconut production areas, such as in North Sulawesi, Indonesia, parts of the Philippines and in many South Pacific countries. In Mindanao, the second largest island of the Philippines, approximately 20 percent of the cattle-coconut farms have improved pastures, such as signal grass (*Brachiaria decumbens*) and centro (*Centrosema pubescens*). However, proper management must include proper fertilization and stocking rate control to ensure the sustainability of such undertakings (Castillo, 1995).

On most farms, cattle have been grazed on natural vegetation and in general the investment in forages was less than 1 percent (Arganosa, 1988). In the Philippines, there have been deliberate efforts to develop integrated livestock–tree crop projects. For example, the International Development Research Centre (IDRC) and others have supported tree crops+small ruminants systems, where sociologists, livestock scientists and forage agronomists worked together. Such a structural set-up and system approach has not been effective in promoting the development of the industry.

A new approach is required to increase the awareness of farmers concerning the benefits of using improved practices in integrated livestock-coconut production. A participatory approach involving introduction of R&D with farmers to speed up technology adoption was promoted in Southeast Asia. The success of this approach remains to be seen.

In Malaysia, extension programmes are mainly government-driven through subsidies and incentives. Many of the R&D programmes initiated in the 1970s and 1980s involved government, international aid and development agencies such as the Australian Centre for International Agricultural Research (ACIAR), FAO and others. Verification or demonstration projects were carried out on smallholdings, government-backed land schemes and private estates to evaluate the commercial viability of sheep and cattle under rubber and oil-palm. The adoption of such technology among the farmers has been limited. It was only comparatively recently that cattle production in the rubber (see Plate 17.4) and oil-palm sectors has been identified as a major thrust towards increasing beef production. Consequently, tree crop–livestock integration in the plantation estate sector is taking on a new dimension. There has been an increase in the number of cattle through integration activities by small-scale farmers. However, involvement in livestock production by the estates that occupy about 50 percent and 18 percent of the oil-palm and rubber plantation land, respectively, is still low.

Nevertheless, the positive results generated from R&D in the Malaysian plantation-based livestock integration in the 1980s and 1990s have led to the development of the current national agricultural policy, whereby this sector is expected to contribute significantly towards the overall improvement in red meat production in Malaysia alone. A target of 240 000 tonne has been set, to be achieved by 2010, implying an annual growth rate of 6.9 percent, compared with the average annual growth rate of 0.3 percent in 1985–1995.

The future direction and the expected role of government in encouraging greater involvement of estates in integration activities could be intensified considerably. However, farmers will be constrained by resource endowment, supporting infrastructure, market potential and policies favouring intensification, diversification and specialization (Devendra *et al.*, 1997).

In Thailand, livestock+crop-based production systems are very much concentrated in the southern part of the country, where agriculture is dominated by rubber, oil-palm and tropical orchards. Coconut plantations cover 44 000 ha and para rubber plantations 1 650 000 ha, which could be used for forage integration. The integration of cattle with plantation systems is less practised than cattle in the rice systems. However, some farmers rear cattle and goats as weeders for their plantations. A few have sown improved forages for their animals. Koronivia grass (*Brachiaria humidicola*), guineagrass (*Panicum maximum*), ruzi grass (*Brachiaria ruziziensis*) and stylo (*Stylosanthes guianensis* cvs. Verano, Stylo 184 (= CIAT 184)) and desmanthus (*Desmanthus virgatus*) are some of the improved forages cultivated on coconut plantations. Farmers get income twice a year from 4–10 beef fattening cattle. In most instances, animals rely heavily on native weeds, which thrive in abundance during the rainy season, but become scarce during the dry season (Sophanodora, 1995).

In Indonesia, 70–80 percent of the livestock are found in Java. As a consequence of the limited land area available, small numbers (2–3 animals per farm) are kept in semi-intensive and stall-feeding systems. Major crop–animal systems are associated with both annual food crops and perennial tree crops, mainly in Sumatra and Sulawesi (Kaligis *et al.*, 1995). North Sulawesi has about 300 000 ha of coconut suitable for beef cattle integration. Some efforts are being made to introduce buffalo grass (*Stenotaphrum secundatum*; St Augustine grass) and creeping peanut (*Arachis pintoi*) into coconut stands. In Sumatra, Indonesia, there are several thousand hectares of large plantations planted with rubber and oil-palm. These have been free of animals because they are regarded as pests. The limited research effort conducted at Sei Putih in Sumatra to develop sheep integration in rubber plantations have created growing interest in producing sheep under rubber and oil-palm (Sanchez and Boer, 1989).

CONSTRAINTS IN IMPROVEMENT AND UTILIZATION OF FORAGE SUPPLY IN THE PLANTATION ENVIRONMENT

Environmental factors

The environmental factors affecting forage supply are fairly universal among the three plantation crops in Southeast Asia. The potential for forage supply is greater in coconut plantations, as light transmission increases with tree maturity and thus allows longer-term use of the area below.

Generally, light levels in tree crop plantations are highly variable and dynamic, and are largely determined by the age of the trees, and planting density and pattern. Light levels can be as high as almost full daylight at planting, and then fall as low as less than 10 percent of full daylight when the tree crop canopy closes. Such vast variations in light level produce distinct shade environments to which photosynthetic and morphogenetic adaptations are of ecological significance for plant survival and productivity. The natural ground vegetation normally declines with decreasing light as the tree crop matures.

Natural forages

Luxuriant undergrowth or ground vegetation forms a free source of nutritious feed for ruminant production. Plantation ground vegetation was estimated to comprise as much as 83 percent of the total forage resource available in Malaysia (Devendra, 1981). Survey and sampling of utilization of the ground vegetation in rubber and oil-palm indicated that 60–70 percent of the vegetation could be utilized for ruminant production. The ground vegetation normally comprises grasses, broad-leaved weeds, ferns and others. The most obvious species are cow grass (*Paspalum conjugatum*), carpetgrass (*Axonopus compressus*), mikania (*Mikania cordarta*), lallang or cogongrass (*Imperata cylindrica*), *Cyrtococcum oxyphyllum*, *Nephrolepis bisserata*, *Eupatorium odoratum*, *Glehnia linearis* and many others (Wan Mohamed, 1977; Chen and Bong, 1983; Chee *et al.*, 1997). Not all species are suitable for grazing and selective weeding of non-palatable ones may be necessary. Similar broadly adapted forages species are also found in other countries of Southeast Asia, with the exception of a few site-specific species.

The ground vegetation is affected by several factors, such as quantity and quality of light, soil moisture and nutrients. Growth normally declines with the age of the main crops. Although production of improved forages is good in the juvenile stage of the plantation, their poor persistence and survival could limit their usefulness unless proper management involving lenient defoliation are adopted. The estimated forage dry matter yield of ground vegetation in relation to age of rubber and oil-palm is illustrated in Figure 17.1 and Table 17.2.

Forage production from a number of existing species could amount to 2000–3000 kg DM ha^{-1} or more during the first three years after tree crop planting (Chen *et al.*, 1978; Chen, Wong and Dahlan, 1991; Pillai and Seeveneserajah, 1988). Soon after that it declines sharply until reaching the 6th to 7th year, when production levels off at 400–800 kg DM ha^{-1} until approximately the 20th year. A substantial percentage of the species present are not palatable (7–28 percent) to animals (Dahlan, 1989).

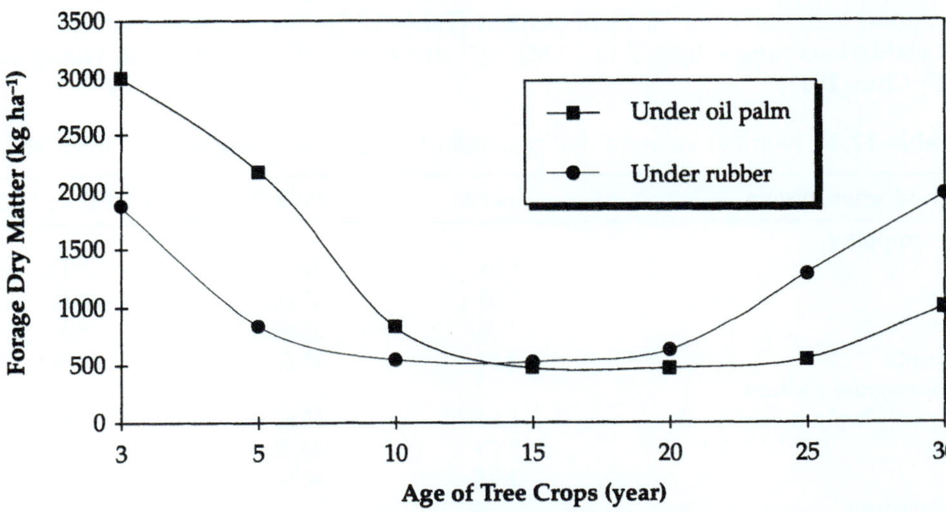

Figure 17.1 Availability of forage dry matter under different ages of oil-palm and rubber.

Table 17.2 Dry matter yield, botanical composition (percentage dry weight) and forage dry matter availability (kg ha^{-1}) under various age groups of oil-palm and rubber canopies

Forage type	Age of plantation trees (years)						
	3	5	10	15	20	25	30
			Under oil-palm				
Non-edible dicot	8.87	3.05	7.26	19.65	24.20	17.05	27.85
Ferns	0.16	3.75	23.60	19.41	26.86	37.79	32.50
Edible dicot	22.54	18.50	13.00	8.08	5.59	17.69	8.44
Legumes	28.65	5.66	0.50	0.40	5.55	2.22	–
Grasses	39.76	59.04	55.64	52.46	37.80	26.25	31.20
DM yield (kg ha^{-1})	2990.00	2164.00	475.00	823.00	467.00	343.00	998.00
			Under rubber				
Non-edible dicot	7.16	4.77	4.06	4.65	17.55	18.16	20.23
Ferns	–	18.21	31.33	62.27	58.00	63.18	55.84
Edible dicot	8.22	8.93	2.52	4.94	1.00	3.66	7.25
Legumes	31.28	15.97	2.07	6.32	4.26	1.43	3.78
Grasses	53.35	52.12	59.99	21.82	19.18	13.57	12.89
DM yield (kg ha^{-1})	18.69	435.00	542.00	520.00	628.00	1282.00	1975.00

Nutritive quality of forages

The nutritive quality of the forages is highly variable and is exacerbated through the complexity and ecological succession of natural forage resources in plantation crops. The feeding value of the ground vegetation is comparable to improved pasture grasses. Crude protein of grasses under trees on most plantations ranges from 8 to 17 percent, broad-leaved weeds from 13 to 22 percent, while plantation legumes from 15 to 18 percent (Table 17.3). The estimated energy content for grasses in plantations ranges from 7 to 10 MJ kg^{-1} dry matter (Wan Mohamed, Hutagalung and Chen, 1987; Chin, 1991).

Table 17.3 Nutritive values of forages under tree crops on plantations in Malaysia

Forage production system	N (%)	DMD (%)	ME (MJ kg^{-1} DM)
Forage–tree cropping			
Grasses	1.58	47.61	7.13
Legumes	3.07	47.80	7.17
Ferns	1.96	35.60	5.34
Edible dicots	2.81	55.6	8.34
Legume–guineagrass pasture			
Guineagrass (*Panicum maximum*)	1.60	52.62	7.80
Legumes	3.55	40.33	6.01
Weeds	1.97	48.04	7.13
N-fertilized pasture			
Mardi digit grass (*Digitaria setivalva*)	2.61	59.40	8.78
Guineagrass (*Panicum maximum*)	2.54	52.62	7.80
Signal grass (*Brachiaria decumbens*)	2.39	56.09	8.31

Key: N = nitrogen; DMD = dry matter digestibility; ME = metabolizable energy.

acceptance among the smallholders in government schemes was obvious during the economic recession of the 1980s. In the early years, R&D was centred on collection of primary data to assess the viability of livestock–tree-cropping systems. Research focused on ecological aspects of livestock-tree-crop integration; and on availability and utilization of agricultural by-products from plantations. These are considered below.

Ecological aspects of livestock–tree-crop integration

Too often, and mistakenly, development of livestock agriculture is seen as harmful to the environment and that livestock are a major factor in deforestation and desertification of the environment. In the 1970s, integration of livestock in plantations was feared, as it might harm estate ecological balance. Alarmed planters claimed that plantations in the country were facing such problems (Tables 17.7 and 17.8). Goats, cattle and buffaloes were a menace as they trampled and chewed up seedling plants, reducing plantation tree growth.

At the same time, R&D on integration indicated beneficial aspects for crops and weed control (Tables 17.9 and 17.10).

Table 17.8 Rubber tree trunk girth increment with different grazing treatments

Treatment	Girth at 2 years (cm)	Girth at 2.5 years (cm)	Increment (cm)
Cover crop ungrazed	12.9	16.4	3.5
Cover crop grazed	12.1	15.3	3.2
Grass-legume grazed	10.5	13.6	3.1

Table 17.9 Effect of weed control on oil-palm fresh fruit bunch (FFB) yield (t ha^{-1})

Treatment year	Spraying			Non-grazing			Grazing		
	1	2		1	2		1	2	
I$^{(1)}$	5.4	1.0		5.7	1.7		5.8	1.4	
II	3.7	6.0	(7.0)	2.9	7.5	(9.2)	2.1	6.9	(8.3)
III$^{(2)}$	1.9	10.6	(17.6)	2.1	13.5	(22.7)	1.7	16.3	(24.6)
IV	–	19.5	(37.1)	1.2	21.7	(44.4)	0.6	23.2	(47.8)
V	–	27.8	(64.9)	0.9	19.6	(64.0)	0.5	25.5	(73.3)
VI	–	25.9	(90.7)	1.5	24.6	(88.6)	0.2	24.5	(97.8)

Key: 1 = Annual weed DM yield; 2 = Annual FFB yield; Parenthesis = Cumulative yield.
Notes: (1) Grazing started 20th September; (2) Sprayed with herbicide from 2nd Half of 3rd grazing year onward.

By their sheer body weight, herds cause soil compaction, which can lead to reduction in yield from rubber and palm oil. Estate planters also reported that the animals were helping to spread an obnoxious weed, *rumput nyonya* (*Asystasia intrusa*), through their seed laden droppings. More often than not, overgrazing by unscrupulous herders also contributes to soil erosion. The above concerns expressed by planters are justified in view of the lack of information and scientific data to support livestock integration in plantations.

Table 17.10 Oil-palm fresh fruit bunch (FFB) yield (t ha^{-1} yr^{-1})

Treatment	Block	Year 11 (1988)	Year 12 (1989)	Year 13 (1990)	Year 14 (1991)	Year 15 (1992)(1)	Overall mean
Grazing (177 ha)	A 10	18.10	21.27	18.49	17.02	11.84	
	A 12	18.62	25.73	18.33	20.49	17.07	
	A 14	17.16	22.52	20.72	23.62	17.04	20.99^{a}
	A 17	27.17	32.71	23.09	23.72	23.79	
	Mean	20.26	25.56	20.41	21.21	17.52	20.00^{a}
Control (171 ha)	B 2	21.66	19.66	19.37	17.09	13.67	
	B 7	21.36	19.78	18.98	18.81	13.72	
	B 8	17.44	22.50	20.30	13.89	15.76	17.98^{b}
	B 9	18.40	21.00	18.27	16.06	11.96	
	Mean	19.71	20.73	19.23	16.46	13.77	17.78^{b}
Overall mean		19.99^{ab}	23.15^{a}	19.82^{ab}	18.84^{bc}	15.65^{c}	

Notes: (1) Estimated yield based on first 9 months. Means with superscripts (a, b, c) are significantly different at $p < 1$ percent.
Source: Harun and Chen, 1995.

Availability and utilization of agricultural by-products from plantations

In areas where the ground vegetation is limited in forage terms, or has deteriorated due to drought, overgrazing or canopy closure, the use of agricultural by-products – oil-palm fronds (OPF), palm press fibre (PPF) and others – may be considered as a roughage supplement source (Alimon, Yusuf and Abu Hassan, 1994). The main by-products useful for livestock production in association with plantation crops are rubber seed meal, copra cake, palm kernel cake (PKC), PPF, palm oil mill effluent (POME) or oil-palm sludge, oil-palm trunk (OPT) and OPF (Alimon and Hair-Bejo, 1996). There is now a wealth of information on availability, nutrient quality, utilization and conservation of these by-products in livestock production systems (Tables 17.11, 17.12 and 17.13). In the Malaysian Agricultural Research and Development Institute (MARDI), interest currently centres on use of OPT and OPF (Abu Hassan, 1995; Abu Hassan *et al.*, 1996).

Table 17.11 By-products of oil-palm and estimated production in Malaysia

Source	By-product	Quantity (million tonne)
On farm	Oil-palm fronds	17.80
	Oil-palm trunks	–
Palm oil mills	Palm kernel cake	1.00
	Palm oil sludge	1.54
	Palm pressed fibre and empty fruit bunches	5.70

The digestibility of PPF is 22.3 percent, with 31.8 percent total digestible nitrogen (TDN). PPF is best fed to animals direct at no more than 20 percent of the total diet. Among the palm residues, PPF is the most nutritious, with a crude protein content of 18.8 percent and TDN of 27.5 percent. It can be fed fresh, or ensiled in drums for longer storage or transportation (Chen, Tajuddin and Chong, 1996). Research results showed that OPT is more amenable to treatment. Its digestibility can be increased from 23.6 to 36.4 percent with steaming. With NaOH treatment, the digestibility

Table 17.12 Chemical composition of oil-palm by-products (as percentage of dry matter)

By-product	CP	CF	NDF	ADF	EE	Ash	ME
Palm kernel cake	17.20	17.10	74.30	52.90	1.50	4.30	11.13
Palm oil sludge	12.50	20.10	63.00	51.80	11.70	19.50	8.37
Palm pressed fibre	5.40	41.20	84.50	69.30	3.50	5.30	4.21
Oil-palm fronds	4.70	38.50	78.70	55.60	2.10	3.20	5.65
Oil-palm trunks	2.80	37.60	79.80	52.40	1.10	2.80	5.95
Empty fruit bunch	3.70	48.80	81.80	61.60	3.20	–	–

Key: CP = crude protein; CF = crude fibre; EE = ether extract; ME = metabolizable energy (MJ k^{-1})
Source: Wong and Wan Zahari, 1992, plus empty fruit bunch data from Dr Mat Rasol Awang (pers. comm.).

Table 17.13 Effect of oil-palm frond level in diets on performance of beef cattle (6 Australian Commercial Cross bulls per treatment)

Parameter		Treatment			
		T1	T2	T3	T4
Live body weight					
Initial weight	(kg)	229.10	226.50	232.90	229.40
Final weight	(kg)	396.30^{a}	366.40^{ab}	333.80^{b}	357.20^{ab}
Daily gain	(kg day^{-1})	0.75^{a}	0.62^{ab}	0.45^{c}	0.57^{bc}
Feed intake					
Total diet	(kg day^{-1} DM)	7.02^{a}	6.10^{ab}	5.48^{b}	5.58^{b}
Feed/Gain					
Total diet	(conversion ratio)	9.65^{b}	10.09^{b}	12.39^{a}	9.92^{b}

Notes: a,b,c = means in row with different letters differ significantly ($p<0.05$). OPF = oil-palm frond. PKC = palm kernel cake. Treatments were: T1 = 10% Urea OPF silage + 90% PKC-based concentrate; T2 = 30% Urea OPF silage + 70% PKC-based concentrate; T3 = 50% Urea OPF silage + 50% PKC-based concentrate; T4 = 50% OPF silage + 50% PKC-based concentrate.
Source: Ishida *et al.*, 1994.

of whole oil palm trunk (WOPT) increases to 54–62 percent, which is equivalent to that of the tropical grasses (Abe *et al.*, 1990; Oshio and Abu Hassan, 1990; Abu Hassan, 1995). Both OPF and OPT can be added at up to 5 percent of the total ration. Thus, oil palm residues are promising roughages for ruminate production in Malaysia (Table 17.14).

The constraints are in using them as economical feeds for livestock. This will depend on transportation and processing costs. Although there is no convincing economic evaluation of OPF and OPT as animal feed, it is presumed that the livestock production should be carried out *in situ* to reduce cost of packaging and transportation. In addition, the use of medicated urea-molasses block may be introduced. The medicated urea-molasses block will improve the assimilation of nutrient from poor quality forage and roughage, and at the same time provide useful non-protein nitrogen, energy, mineral supplements and control of gastrointestinal parasites (Wong, Moh Salleh and Mohd. Najib, 1988; Wan Zahari and Nor Ismail, 1993; Sani *et al.*, 1995; Abdul Rahman, 1996).

Table 17.14 Suggested levels (percentage) of inclusion of oil-palm by-products in ruminant rations

Parameter	Cattle	Sheep and goats	Buffalo
Palm kernel cake	50	30	50
Palm pressed fibre	15	15	20
POME	50	30	50
OPT ensiled	30	30	30
OPF ensiled	50	50	50

Notes: POME = palm-oil mill effluent. OPT = oil-palm trunk. OPF = oil-palm frond.
Source: Wong and Wan Zahari, 1992.

EMERGING ISSUES

With the current low world prices for copra and coconut oil, monocropping of plantation crops is regarded as no longer an economic proposition. Similarly, world palm oil prices have been declining recently. Unless the return to farm labour from monocropping can be sustained or increased, livestock and plantation crop integration as a commercial venture will remain a competitive strategy to be adopted for high productivity of crops and maximization of land use. It is hoped that crop and animal integration in plantation management will in future become common and widely practised in Southeast Asia. However, several major issues have to be resolved.

Plantation crop damage

The perpetual concern of planters over soil compaction has been overrated. Field data from continuous stocking trials and mob-grazing systems indicate absence of significant differences between the grazed and ungrazed paddocks in terms of yield of rubber and oil-palm. If soil compaction is a major factor, depression of crop yield should be evident.

Research carried out in a commercial project rearing cattle under oil-palm indicated higher fresh fruit bunch (FFB) yield in the grazed field than from the ungrazed area (Table 17.9 and 17.10). A depression of FFB production was evident with high stocking rates under continuous grazing. This was due to inadequate forage supply and subsequent damage to young fronds.

Over the years, research has helped to dispel these negative notions regarding integration. Many of the constraints posited by the planters are over-exaggerated and can be avoided through proper management of animals if the concept of environmentally sustainable agriculture for the new millennium is accepted. Where there is adequate forage in the field, animals do little damage to the tree crops. It is under high stocking rates and insufficient forage that damage to tree bark and fronds occur as the animals seek alternative sources of feed.

Feed supply and quality

Contrary to preconceptions, research results indicate that grazing animals have access to a greater diversity of plant species under rubber and oil-palm. The grazing animal performed better, with higher liveweight gains. The slightly lower temperature on plantation is a positive factor in the overall growth performance of ruminants, especially for breeds with some exotic blood line. Changes in botanical composition reflect changes in light intensity, stocking rate, species adaptability, palatability and plantation management.

In the Philippines, fertilizing the existing ground vegetation under coconut resulted in higher daily liveweight gain in cattle and carabao, but the native pasture had deteriorated after two years leaving only unpalatable weed species, a feature also observed in grazed oil-palm plantation. Hence selective weeding has been adopted to promote good growth of the native pastures (Deocareza and Diesta, 1993).

Many studies have been undertaken to demonstrate that improvement of natural pastures through introduction of improved forage species can significantly increase daily weight gain of animals. Leguminous cover crops, usually planted in mixture of centro (*Centrosema pubescens*), puero (*Pueraria phaseoloides*), *C. mucunoides* and *C. caeruleum*, are promoted for integration (Moog, Deocareza and Diesta, 1989). However, these species are generally non-sustainable and declined from 50 percent in total composition to zero in 20 months of grazing by sheep under 3-year-old rubber (Chong, Tajuddin and Abd. Samat, 1991). Only the less palatable species, such as *C. caeruleum*, increased, from 42 to 100 percent, in the period. The dominance of *C. caeruleum* is a great asset to plantation management, but the increasing density of non-palatable woody shrubs and noxious weeds in plantations poses more problems. Thus, a special programme of selective weeding once in two or three years, depending on the situation, is needed. The combination of animal grazing with judicious use of herbicides in the management of forages is preferred in some situations (Rosli and Mohd Nasir, 1997).

However, the issue in integration is the decline in forage resources with increasing age of the plantation crop. Research so far has not solved or overcome this constraint, despite selection of shade-tolerant and persistent forages. Dry matter productivity depends on the availability of radiant energy, and it would be technically short-sighted to expect high regrowth performance from constantly defoliated shade-tolerant species in low light conditions.

Nonetheless, the problems associated with limited feed availability in mature plantations present other opportunities for research into various feed options for ruminant. Recognizing these constraints in plantations, alternatives must be sought, including the use of agricultural by-products from the mature plantation, as in use of discarded oil-palm fronds; thinning of existing stands to allow more light penetration; or adopting new planting patterns, like the recommended double hedgerow avenues. Only time will determine which approach or combination of approaches will be adopted.

Animal nutrient requirement

Another issue is animal nutrition. Undeniably, some of the understorey forages are high in nutritive value. Selective grazing will often result in good growth performance. But to manage forages in plantation where excessive growth of the understorey is under control, hard grazing is often needed. This often leads to decline in the overall nutritive quality of the species. Wong and Chin (1998) reported that the forage quality can only meet the nutrient requirement of beef cattle with a liveweight gain of about 250 g head^{-1} day^{-1}. The question of grazing using improved breeds remains questionable unless special provision is made in the feeding programme. Our experience in Malaysia indicates that plantation managers tend to favour exotic breeds or crossbreds for integration programmes for reasons of greater animal size, as well as for their higher prolificacy. Obtaining feed supplements or concentrates can be another barrier to overcome in integration projects in terms of land, labour and cost.

Availability of feeder animals

The small population of ruminants in these countries is yet another issue. Adequate livestock numbers are a vital prerequisite for the development of improved animal production systems. In the Philippines, Indonesia and Malaysia, there are only limited animal resources available. Hence, large numbers have to be imported for fattening, mainly from Australia, requiring access to foreign exchange. In Thailand, there appears to be significant cattle importation from neighbouring countries, while in the Philippines, the large ranches that formerly supplied smallholders with breeding and fattening stock have been broken up as a consequence of land reform programmes. It is not surprising that the implementation of some integrated tree crop–livestock projects have been affected by the availability of feeder cattle in Malaysia. This shortage can lead to higher purchase cost per animal and this indirectly increases development costs of a project and can also influence the participation of resource-poor small-scale farmers.

Lack of capital

A lack of venture capital is a major constraint for ruminant production as integration projects and operations have a long gestation period before income is generated. This is more so for big plantations where the number of animals involved would be high.

Adapted animal breeds

Apart from the limited population base, animal production is also constrained by feed-related factors and by diseases. There is the tendency when importing exotic breeds to replace rather than complement the limited supply of local feeder cattle. One effect is having to utilize natural resources to meet the high nutrient demand of these exotic breeds. Hence, the importance of using adapted breeds in integration becomes obvious. The authors suggest that the full genetic potential of the indigenous breeds, which are better adapted to the local environment and which have been limited in production potential by limited feed intake, be assessed realistically. A good illustration is that of Malaysia trying to use exotic sheep for integration projects. The lack of adaptation and the high incidence of diseases in sheep can put a strain the overall animal husbandry. It is crucial to match nutrient availability with the nutrient requirement of the animal breed if the venture is to be successful.

Economic aspects

Besides R&D being carried out by research stations, application of integration technology in plantations has also been undertaken. The economic projection for sheep integration in plantation was fairly pessimistic, based on the information existing in 1989 (Pillai and Seeveneserajah, 1988; Eddie Chiew and Zainal Abidin, 1989). However, a successfully implemented cattle–oil-palm project in Estet Pekebun Kecil (ESPEK) of the Rubber Industry Smallholders Development Authority (RISDA) at Terengganu Tengah, Malaysia, proved that livestock could be used not only to control weeds, thus saving 20–50 percent of weeding costs, but also could be managed as an enterprise for revenue generation (Harun and Chen, 1995). An investment return of saving of 41.95 percent was recorded (Table 17.15).

Ahmad, A., Othman, A., Ali, M., Mohd.Sukri, I., Shokri, O. & Shafit, H. 1997. Using French crossbreed cattle in increasing beef production (BM). *Teknol. Ternakan, Jil,* **12:** 45–49.

Alexandratos, N. 1995. *World Agriculture: Towards 2010.* Rome, Italy: FAO and John Wiley & Sons. 488p.

Alimon, A. R. & Hair-Bejo, M. 1996. Feeding systems based on oil palm by-products in Malaysia. pp. 105–113, *in:* Proceedings of the First International Symposium on the Integration of livestock into oil plantation production. Kuala Lumpur, Malaysia, 25–27 May 1997.

Alimon, A.R., Yusuf, S. & Abu Hassan, O. 1994. Improved dairy production through strategic utilisation of pasture, forage conservation and agro-industrial by-products. Paper presented at FAO workshop in MARDI, Selangor D. E. Malaysia, 15–18 August 1994.

Ani Arope. 1988. Status and directions in expanding sheep production in Malaysia. Proceedings of a symposium.

Arganosa, A.S. 1988. R & D status of livestock integration under coconut in Philippines. pp. 90–100, *in:* Proceedings of an International Livestock-Tree cropping Workshop. Serdang, Malaysia, 5–9 December 1988.

Ariff, M.O. & Dahlan, I. 1986. Production characteristics of Kedah-Kelantan crosses. Paper presented at a meeting of the formulation of cattle breeding programme in Malaysia. Department of Veterinary Services, Ministry of Agriculture, Kuala Lumpur.

Ariffin, T. 2000. Crop-livestock integration in the oil palm estate sector: A review of recent developments. pp. 131–132, *in:* Proceedings of the 22nd Meeting of the Malaysian Society of Animal Production. Kota Kinabalu, Malaysia, 29 May–1 June 2000.

Basiron, Y. 1998. The future of livestock and crop integration in oil palm as a commercial venture. Paper presented at a national seminar on livestock and crop integration in oil palm – "Towards sustainability". Kluang, Johor, Malaysia, 12–14 May 1998.

Castillo, A.C. 1995. Ruminants in coconut-based plantation systems. Research and Development in the Philippines. pp. 100–103, *in:* Proceedings of a meeting on Integration of ruminants into plantation systems in Southeast Asia. *ACIAR Proceedings,* No. 64.

Chee, Y.K., Chong, D.T., Ahmad Faizl, Tajuddin Ismail, Stür, W.W. & Shelton, H.M. 1997. Forage resources in rubber plantations in central Malaysia. *Malaysian Journal of Animal Science,* **3:** 28–36.

Chen, C.P. 1985. The research and development of pastures in Peninsular Malaysia. pp. 33–51, *in:* International Symposium on Pastures in the Tropics and Subtropics. Tropical Agriculture Research Centre, Ministry of Agriculture, Forestry and Fisheries, Japan. *Tropical Agricultural Research Series,* No. 18.

Chen, C.P. & Bong J. 1983. Performance of tropical forages under the close canopy of oil palm. I. Grass. *MARDI Research Bulletin,* **11:** 248–263.

Chen, C.P., Chang, K.S., Ajit, S.S. & Hassan, W. 1978. Pasture and animal production under five-year-old oil palm at Serdang. pp. 179–192, *in:* Proceedings of MSAP meeting on Integration of Animal Production with Plantation Crops. Penang, Malaysia. 13–15 April 1978.

Chen, C.P. & Dahlan, I. 1995. Tree spacing and livestock production. pp. 35–49, *in:* Report of the FAO First International Symposium on the integration of livestock to oil palm production. Kuala Lumpur, Malaysia, 25–27 May 1995.

Chen, C.P. & Othman, A. 1983. Effect of cattle production on forage under oil palm – a preliminary report. pp. 214–227, *in:* Proceedings of the 7th MSAP Annual Conference. Port Dickson, Malaysia, 1-2 April 1983.

Chen, C.P. & Shamsudin, A.B. 1991. Limitation of forage availability on cattle productivity under oil palm. pp. 139–147, *in:* Proceedings of the 2nd meeting of the FAO Regional Forage Working Group on Grazing and Feed Resources of Southeast Asia. University of Philippines, Los Banos, Philippines. 27 February–5 March 1997.

Chen, C.P., Wong, H.K. & Dahlan, I. 1991 Herbivores and the Plantations. In: W.Y. Ho et al. (eds). Recent advances on nutrition of herbivores. Proceedings of the 3rd International Symposium on nutrition of herbivores. pp. 71-81.

Chen, C.P., Ahmad Tajuddin, Z., Wan Mohamed, W.E., Tajuddin, I, Ibrahim, C.E. & Moh Salleh, R. 1991. Research and Development on Integration Systems in Livestock, Forages and Tree Crops Production in Malaysia. pp. 55–72, *in:* Proceedings of the FAO International Livestock–Tree-cropping Workshop. MARDI, Serdang, Malaysia, 5–9 December 1988.

Chen, C.P., Tajuddin, I. & Chong, D.T. 1996. Strategies for entrepreneur ship of livestock integration in plantation systems. pp. 101–117, *in:* Proceedings of the Silver Jubilee Conference of MSAP. Kuching, Sarawak, Malaysia, 28–31 May 1996.

Chin, F.Y. 1989. Justification, government policy, measures and strategy for the development of grazing reserves in Peninsular Malaysia. pp. 77–82, *in:* Proceedings of the First Meeting of the Regional Working Group on Grazing and Feed Resources of Southeast Asia. UPM, Serdang, Malaysia, 27 February–3 March 1989.

Chin, F.Y. 1991. Some aspects of management and utilization of ground vegetation under rubber and oil palm animal production. pp. 121–128, *in:* Proceedings of the 2nd Meeting of the FAO Region S.E.A. Forage Working Group on Grazing and Feed Resources. UP, Los Banos, The Philippines, 26 February–5 March 1991.

Chong, D.T., Tajuddin, I. & Abd. Samat, M.S. 1991. Stocking rate effects on sheep and forage productivity under rubber in Malaysia. *In:* H.M. Shelton and W.W. Stur (eds). *Forages for Plantation crops*. Proceedings of a workshop, Sanur Beach, Bali, Indonesia, 27–29 June 1990. *ACIAR Proceedings,* No. 32.

Chong, D.T., Tajuddin, I. & Stür, W.W. 1994. Grazing sheep on improved pasture under double hedgerow rubber planting systems in Malaysia. pp. 68–71, *in:* Proceedings of a workshop on integration of ruminants into plantation systems in Southeast Asia. Lake Toba, Indonesia, 9–13 September 1994. *ACIAR Proceedings,* No. 64.

Dahlan, I. 1989. Modelling and simulation of herbage and cattle production systems under oil palm plantations in Malaysia. Ph.D. Thesis, Kyoto University, Japan. 229p.

Dahlan, I. & Mohd. Ariff, O. 1987. Performance of Kedah-Kelantan (KK) cattle and its crossbreeds under various grazing systems. pp. 304–307, *in:* Proceedings 10th MSAP Annual Conference. Genting Highland, Malaysia.

Dahlan, I. & Shahar, M.Z. 1992. Simulation models for quantitative decision in crop-livestock integrated farming systems. *Journal Asian Farming Systems Association,* **1**: 356–360.

Dalzell, R. 1983. Buffalo production on palm oil plantation. pp. 183–187, *in:* Proceedings of the 7th Annual Conference of MSAP.

Deocareza, A.G.F. & Diesta, H.E. 1993. Animal production on improved pasture crops under coconuts. pp. 189–193, *in:* Proceedings of the 3rd Meeting of the FAO Regional Working Group on Grazing and Feed Resources of Southeast Asia- "Strategies for suitable forage-based livestock production in Southeast Asia."

Devendra, C. 1979. Malaysian Feeding stuff. Malaysian Research and Development Institute, Malaysia.

Devendra, C. 1981. The feed requirements for animal production in Peninsular Malaysia: an assessment of the feed resources, implications and strategies for efficient utilisation. Proceedings of the Conference on Exotic and Crossbred Livestock Performance in Malaysia, September 11-12 1981. Genting Highlands.

Devendra, C. Thomas, D., Jabbar, M.A. & Kudo, H. 1997. *Improvement of livestock production in crop-animal systems in rainfed agro-ecological zones of Southeast Asia.* Nairobi, Kenya: International Livestock Research Institute (ILRI). 116p.

Eddie Chiew, F.C. & Zainal Abidin, M. 1989. Economics of crop-livestock production systems for smallholders. pp. 140–148, *in:* Proceedings of the First Meeting of the FAO Regional Working Group on Grazing and Feed Resources of Southeast Asia. UPM, Serdang, Malaysia, 27 February–3 March 1989.

Eng, P.K., 't Mannetje, L. & Chen, C.P. 1978. Effect of phosphorus and stocking rate on pasture and animal production from a guinea grass–legume pasture in Johore, Malaysia. 11. Animal live weight change. *Tropical Grasslands,* **12**: 198–207.

FAO [Food and Agriculture Organization of the United Nations]. 1999. FAOSTAT Database. Electronic Version.

Harun, O. & Chen, C.P. 1995. Better return from an integrated beef-oil palm enterprise. pp. 28–30, *in:* H. Sharif *et al.* (eds). *Towards corporatising the animal and feed industries.* Proceedings of the 17th MSAP Annual Conference. Penang, Malaysia, 28–30 May 1995.

Hassan, A.W. & Abdullah Sani, R. 1991. Effect of buffalo grazing on changes in native pasture under oil palm – A preliminary report. pp. 149–154, *in:* Proceedings of the 2nd. FAO Regional Working Group Workshop on Utilization of native forages for animal production in South East Asia. Manila, the Philippines, 26 February–5 March 1991.

Iniguez, L.C. & Sanchez, M.D. 1991. Summary of working group sessions. pp. 310–311, *in:* Proceedings of the meeting on Integrated tree cropping and small ruminant production systems. Medan, 9–14 September 1990.

Ishida, M., Abu Hassan, O., Nakui, T. & Terada, F. 1994. Oil palm fronds as ruminant feed. Japan International Research for Agricultural Sciences (JIRCAS), Ministry of Agriculture, Forestry and Fisheries, Japan. *Newsletter for International Collaboration,* **2**(1).

Kaligis, D., Sumolang, C., Stür, W.W. & Shelton, H.M. 1995. Cattle production from grazed pastures in North Sulawesi. pp. 50–52, *in:* Integration of ruminants into plantation systems in Southeast Asia. *ACIAR Proceedings,* No. 64.

Khusahry, M.Y.M., Salleh, R.M., Abdullah, A.F., Wahib, S.A., Rahman, M.Y.A. & Ariff, O.M. 1988. Perspectives on the Integration of sheep under perennial crops. pp. 215–263, *in:* Proceedings of Persidangan Pegawai Kanan. MARDI, Subang Jaya.1988.

Moog, F.A., Deocareza, A.G. & Diesta, H.E. 1989. Liveweight gain of carabao under native and native-centro pastures under coconuts in Bicol Region. Paper presented during the 2nd DA-BAR Anniversary, Quezon City, the Philippines.

Moog, F.A. & Faylon, P.S. 1991. Integrated forage-livestock systems under coconut in the Philippines. pp. 144–146, *in: Forages for plantation crops. ACIAR Proceedings,* No. 32.

Mahadevan, P. & Hickman, C.C. 1982. Requirements for cooperative research to evaluate strains and breeding systems. pp. 265–280, *in:* Working papers of expert consultation on dairy cattle breeding in the humid tropics. Hissar, India, 12–17 February 1979.

Oshio, S. & Abu Hassan, O. 1990. Processing and utilisation of oil palm by-products for ruminants. MARDI-TARC collaborative study, July 1987–February 1990. 110p.

Othman, A., Mohd Sukri, H.I., Wong, C.C., Eng, P.K. & Chen, C.P. 1985. Integrasi lembu daging di ladang-ladang kelapa sawit. *Teknologi Ternakan*, **1**, 105–110.

Pillai, K. R. & Seeveneserajah, K. 1988. Integration of sheep with immature rubber and oil palm-Sime Darby Experience. pp. 109–124, *in:* Proceedings of a symposium on Sheep Production in Malaysia. Kuala Lumpur, Malaysia.

Pezo, D.A., Lanting, E.F., Wong C.C. & Kerridge, P. 1999. Feed resources for ruminants in smallholder farming systems in Southeast Asia. pp. 97–111, *in: ACIAR Proceedings*, No. 95.

Reynolds, S. 1995. Pasture-cattle-coconut systems. FAO, Bangkok, Thailand. 664p.

Reynolds, S. 1998. Feed resources and the integration of animal production in coconut plantations. pp. 3–21, *in:* Proceedings of the 6th Meeting of the Regional Working Group on Grazing and Feed Resources for Southeast Asia. "Integrated Crop-Livestock Production Systems and Fodder Trees". Legaspi City, Philippines, 5–9 October 1998.

Rosli, A. [2001]. Systematic integration of beef cattle in oil palm plantation. Proceedings of a National Seminar on Agroforestry. Melaka, 24–26 April 2001. (in press)

Rosli, A. & Mohd. Nasir, W.I. 1997. Environmental interactions on livestock integration production systems in plantations. pp. 7–12, *in:* Proceedings of the 19th Malaysian Society of Animal Production (MSAP) Annual Conference. Johor Bahru, Johor, 8–7 September 1997.

Roxas, D.B., Wanapat, M. & Winugroho, M. 1997. Dynamics of feed resources in mixed farming systems in Southeast Asia. pp. 101–112, *in:* Proceedings of an International Workshop on crop residues in sustainable mixed crop/livestock farming systems. ICRISAT, Patancheru (India), 22–26 April 1996. Wallingford, UK: CAB International.

Sanchez, M.D. & M. Boer. 1989. Supplementation of local and F_1 local and St. Croix growing sheep grazing under rubber trees in North Sumatra. SR-CRSP Sei Putih, Indonesia, *Working Paper*, No. 104.

Sani, R.A., Wan Zahari, M., Chong, D.T., Mazila, M., Subramaniam, K. & Chandrawathani, P. 1995. Efficacy of medicated blocks for parasite control in sheep. pp. 118–119, *in:* Proceedings of the 17th MSAP Annual Conference. Penang, Malaysia, 28–30 May 1995.

Sophanodora, P. 1995. Forages Research in Southern Thailand. pp. 104–108, *in:* Integration of ruminants into plantation systems in Southeast Asia. *ACIAR Proceedings*, No. 64

Tajuddin, Z.A. 1988. Proceedings of the International Livestock-Tree cropping workshop. Serdang Malaysia, 5–9 December 1988. 133p.

Tajuddin, I., Najib, M.A., Chong, D.T., Abd. Samat, M.S. & Vanaja, V. 1990. Status report on RRIM commercial sheep project 1985–1988. RRIM.

Tajuddin, I. & Chong, D.T. 1994. Ruminants in plantation systems in Malaysia. pp. 97–99, *in:* Proceedings of a workshop on integration of ruminants into plantation systems in Southeast Asia. Lake Toba, Sumatra, Indonesia, 9–13 September 1994.

Vidyadaran, M.K., Zamri, S., Abdul Wahid, S., Azmi, T.I. & Rajion, M. A. 1988. Proceedings of a Symposium on sheep production in Malaysia. Kuala Lumpur, Malaysia, 15–16 November 1988. Published by Ministry of Agriculture, Malaysia (DVS), Kumpulan Guthrie Sdn. Bhd., Universiti Pertanian Malaysia and MSAP. 150p.

Wan Mohamed, W.E. 1977. Utilisation of ground vegetation in rubber plantation for animal rearing. pp. 265–281, *in:* Proceedings of the Rubber Research Institute of Malaysia Planter's Conference.

Wan Mohamed, W.E., Hutagalung, R.I. & Chen, C.P. 1987. Feed availability, utilisation and constraints in plantation-based livestock production system. pp. 81–100, *in:* Proceedings of 10th Malaysian Annual Conference on Advances in Animal Feeds and Feeding in the Tropics. Genting Highlands, Pahang, Malaysia, 2–4 April 1987.

Wan Mohamed, W.E., Mohamad, N. & Abdul Rahman., A. 1988. Perspectives and progress of sheep industry at Kumpulan Guthrie. pp. 98–108, *in:* Proceedings of a symposium on sheep production in Malaysia. Kuala Lumpur, Malaysia. 15–16 November 1988.

Wan Zahari, M. & Nor Ismail, M.S. 1993. Methodology of making urea-molasses mineral blocks for ruminants. pp. 40–41, *in:* Proceedings of the 16th MSAP Annual Conference. Pulau Langkawai, Malaysia, 8–9 June 1993.

Wong, C.C. & Chin, F.Y. 1998. Meeting the nutrient requirement of beef cattle from forages in oil palm plantation. Paper presented at the National seminar on livestock and crop integration in oil palm "Towards sustainability". Kuang, Johor, Malaysia, 12-14 May 1998.

Wong, C.C., Rahim, H. & Mohd. Sharudun, M.A. 1985. Shade tolerance potential of some tropical forages for integration with plantations. 1. Grasses. *MARDI Research Bulletin*, **13**: 225–247.

Wong, H.K., Moh Salleh, R. & Mohd. Najib, M.A. 1988. Mineral status of Malin sheep grazing under coconut. pp. 91–95, *in:* Proceedings of the 11th Annual Conference of the Malaysian Society of Animal Production. University of Malaysia, Kuala Lumpur, Malaysia.

Wong, H.K. & Wan Zahari, M. 1992. Characterization of oil palm by-products as feeds for ruminants. pp. 58–61, *in:* Proceedings of the 15th Malaysian Society of Animal Production Annual Conference. Kuala Trengganu, Trengganu, Malaysia. 26-27 May 1992.

CHAPTER **18**

Intensive Forage Production for Smallholder Dairying in East Africa

Apollo Bwonya Orodho[1]

ABSTRACT

East Africa has many areas suited to dairying, especially in the better-watered zones at medium and high altitudes; it also has well-adapted and proven forages, many of local origin. The technologies are based on over seventy years of sound, well documented research and practical application. Although disease and parasite challenge can be strong, acclimatized stock of exotic dairy breeds are readily available for commercial dairying. Since the 1970s, with changes in farm size, emphasis has been on smallholder dairying, which has developed in the milk-catchment of the larger towns. The larger towns are mostly in upland areas, but two larges cities - Mombasa and Dar-es-Salaam - are coastal. Smallholder dairying relies on stall feeding, using cultivated fodder and crop residues; scavenging on rough grazing is now less common. Forages for the various agro-ecological zones, based on altitude (sea level to about 3000 m) and rainfall are listed, and technology for their cultivation and use described. A study of three levels of intensification shows that exotic, stall-fed stock give the highest margin per litre and have the highest input costs. Dairying provides employment in the small-scale farming sector and improves household incomes. Marketing is largely informal, which may make assurance of standards of hygiene problematical. Farmers who adopt improved technology generally get higher yields and profit margins. Fragmentation of holdings is a serious problem, and it is becoming increasingly difficult for farmers to find enough land for both fodders and subsistence crops.

[1]P.O. Box 1667, Kitale, Kenya

INTRODUCTION

East Africa, comprising Kenya, Tanzania and Uganda, has a large range of agro-climatic zones, from sea level to high mountains, with cultivation up to about 3000 metres, and rainfall regimes from arid, through semi-arid to sub-humid. East Africa is wholly equatorial, lying roughly between 4°N and 11°S, and 29° to 42°E. Rainfall is generally bimodal at the coast and to the east of the Great Rift Valley, but in western areas the rainfall pattern is often almost unimodal – as one long season.

There is a great altitude range, from the coast up to mountains, which still retain some permanent ice caps; temperatures are greatly modified by altitude, but the equatorial situation means that there is little or no variation in mean monthly temperatures. Much of the vast lower areas, outside the coastal strip, are semi-arid grazing land, often exploited by traditional pastoralists; these use dairy products for subsistence but are not the subject of this chapter. Crop production and mixed farming is found mainly in the uplands and a narrow coastal strip. The main cities, areas of high population density and intensive farming are in the highlands, at around 1500 metres. There are, however, two important coastal cities: Mombasa, in Kenya, and Dar-es-Salaam, in Tanzania. The relatively high altitude and cool climate of much of the highlands affects forage choice since 1500 metres, under local conditions, is about the upper limit for many "tropical" forages, as well as for crops such as cotton and rice.

Smallholder dairying is found in zones with favourable climate and soils, where human pressure on land has caused farming systems to change from large, to medium to smallholder farms, integrating crops and livestock. Undernutrition and seasonal fluctuations in quantity and quality of feed are major constraints to smallholder dairying. Ways of intensifying forage production are described in terms of production systems, improved and recommend forages and seasonality of production. Forage quality is looked at in terms of the nutritive value, factors influencing quality and ways of improving it. Forages of economic importance are increasingly attacked by diseases and pests, and some of economic importance are discussed, with special emphasis on Napier grass (*Pennisetum purpureum*).

In the past two decades, smallholder dairying has undergone intensification and changes in management. Systems are moving from grazing to stall feeding, and smallholder dairy farmers have developed feed management strategies to cope with these changes. The great importance of crop residues and by-products, their role as feed and associated limitations are also discussed.

Availability of seed of recommended forages is important; the availability of adapted seed production technologies and production of seed of recommended cultivars by the formal and informal seed sectors are discussed. Manure, an important output of dairying, is looked at in the ways farmers manage and use it in a cut-and-carry system and how increasing dairy production may be achieved through proper management of manure. Finally, the importance and future of forage-based smallholder dairying is discussed in the light of the socio-economic factors, opportunities and future prospects.

HISTORICAL SETTING

The smallholder forage programme has had a very solid information and genetic base. Pasture research began early in eastern Africa, and was especially strong in Kenya, where a full-time

Plate 18.1 Small-scale dairy unit with stall-feeding or zero-grazing system.
[Photograph by Dr Stephen Reynolds]

grass (*Cynodon* spp.), Rhodes grass (*Chloris gayana*) and setaria (*Setaria sphacelata*), is common and increasing. Purchase of fodders such as Napier grass (*Pennisetum purpureum*) or hay, some of which is from the roadside or from farmers who do not have livestock, is common in intensive areas such as Kiambu (Staal *et al.*, 1998).

Planted fodders on smallholder farms include sweet potato (*Ipomoea batatas*) vines, various kinds of vetch (*Vicia* spp.) and desmodiums (*Desmodium uncinatum, D. intortum*) or fodder trees such as calliandra (*Calliandra calothyrsus*) and leucaena (*Leucaena leucocephala*). Farmers also buy grain, concentrates and agro-industrial by-products such as bran, wheat pollard and dairy meal. Napier grass (see Plate 18.2) is the major fodder used by smallholders in Kenya (Orodho, 1990).

TANZANIA

In Tanzania, over 80 percent of milk from improved cattle is from smallholder dairies. There are four smallholder dairy production systems (Table 18.2). The smallholder intensive urban dairy production system is found around Dar-es-Salaam and other major urban centres. This system has a cattle population of about 20 000 head. In towns, cattle are few and are fed from forages purchased or gathered from open spaces or public land. Over 70 percent of dairy cattle are in the smallholder intensive rural and semi-intensive dairy-manure cattle production systems. These are in peri-urban areas and in the rural highland areas of Arusha and Kilimanjaro, and Kagera in the Southern Highlands. In the highlands of Kilimanjaro and Mt. Meru, these systems have about 65

Plate 18.2 Napier grass for cut-and-carry and stall-feeding dairy cattle.
[Photograph by Dr Stephen Reynolds]

percent of all dairy cattle in Tanzania. Farms are small and grow cash and food crops besides producing milk. Farmers typically own about 4 cattle on approximately 1.7 ha, and grow coffee (*Coffea arabica*), bananas (*Musa* sp.) and fodder (MOAC/SUA/ILRI, 1998). Livestock are fed on a cut-and-carry system. Farmers may have another 2.2 ha of lowlands where they grow maize (*Zea mays*) and beans (*Phaseolus vulgaris*) (Mdoe and Wiggins, 1997). They often bring crop by-products from the lowlands to feed cattle in the highlands, which has sustainability implications because of transfer of nutrients between landscapes. Zero grazing, using crop by-products and forage, is practised by about 70 percent of farmers. The rest follow semi-intensive systems, especially on larger farms in the lowlands and in the low-population-density areas of the southern highlands. Many farmers use manure (sometimes after it has been used to generate biogas) on their coffee, bananas and fodders.

In ACZs I–IV, smallholders practise mostly subsistence-oriented semi-intensive dairying, with predominantly zebu herds and a few crosses; they have up to 30 head, especially in drier zones (IV), which mostly graze. This zone has approximately 86 percent of dairy cattle in Tanzania. Cattle are kept for meat and draught, and farmers grow cotton, tobacco, sorghum (*Sorghum bicolor*), finger millet (*Eleusine coracana*), rice (*Oryza sativa*), cassava and sweet potato. Crop by-products are fed to cattle and manure is used as fertilizer.

UGANDA

Dairying is concentrated in the southern half of the country, where Kampala, Jinja and Mbarara milk catchment areas each account for one third of national dairy marketing. Improved breeds make up 4 percent of the national 4.2 million head herd, and produce 16 percent of the milk marketed. Smallholder dairies are intensive or semi-intensive. In intensive dairying, the average farm size is about 1.5 ha, significantly smaller than semi-intensive farms. It is generally a high altitude zero-grazing system, with coffee as the main cash crop and bananas, maize, beans, sweet potato, sorghum and vegetables for food. Intercropping is common, and dairy enterprises range from zero-grazing with 1–3 cows to free grazing or fenced paddocks with larger numbers of cattle. Cattle breeds include exotic animals (e.g. Friesian, Ayrshire, Guernsey, Black and White Dane), their grades and crosses. Intensive dairying is prevalent in southwest Uganda (MAAIF, 1993). Dairy stock depend on local pasture, Napier grass, and crop by-products such as sweet potato vines, banana peelings and pseudostems, and damaged crops. Manure is usually put on the banana plots.

Farmers may supplement dairy animals with purchased concentrate (usually 2–3 kg/cow/day at milking). Seasonal variations in quantity and quality of feed strongly influence milk production. Intensive dairying is in tsetse-free areas with more than 1500 mm rain annually. Some farmers grow high value fodders, including Napier grass, Kikuyu grass, star grass and clovers (*Trifolium* spp.). Semi-intensive dairying systems are mainly in medium altitude (1000–1500 m) areas with medium rainfall (1000–1500 mm). The main cash crop is coffee, and food crops include maize, beans, cassava and vegetables. Farms are small, 1–2 ha, where human population density is high, and 4–15 ha in relatively low-population-density areas. Dairy herds in small to medium farms range from 5 to 30 animals. Some stall-feeding of dairy cattle is practised, using local vegetation, Napier grass and crop by-products such as sweet potato vines, bananas peelings and pseudostems and damaged crops. It is common to find animals herded, tethered or grazed on hillsides, valley bottoms, roadsides and fallows. Communal grazing is limited. Natural pastures comprise 80–88 percent of the land used for grazing, and planted pastures occupy the remainder.

MILK PER CAPITA

Milk production and milk availability per capita in eastern Africa is very low (Table 18.3) and does not meet demand. With increasing population, the demand for milk will leave an even bigger gap, which will have to be met either by imports or increased local production. With slow economic growth and inadequate foreign exchange, the most sustainable option is to increase production from smallholder dairy farms.

Table 18.3 Dairying in East Africa: cattle, milk production and per capita milk availability

Parameter		Kenya	Tanzania	Uganda
Cattle ('000 head)				
Zebu		10 400	13 900	5 400
Dairy		3 045	250	150
Percentage dairy cattle		23	2	3
Annual milk production	('000 litres)	3 075	814	485
Annual per capita milk availability (litres LME)[(1)]		85	23	24

Notes: (1) LME = liquid milk equivalent.
Source: Muriuki and Thorpe, 2001.

FORAGES ON SMALLHOLDER FARMS

During the late 1970s and early 1980s there was much emphasis on research into the needs of smallholders, who had become the majority farmers in high and medium potential areas, as well as in the marginal areas of eastern Africa. The sub-region is a centre of genetic variability for many tropical forage grasses, so the richness of the forage germplasm, collected and introduced, formed an excellent basis for further breeding and selection for smallholder systems. The most important species improved to provide better cultivars for smallholder dairy farmers include Rhodes grass (*Chloris gayana*), setaria (*Setaria sphacelata*) and Napier grass (*Pennisetum purpureum*). Rhodes grass is one of the most important ley grasses in eastern Africa, with several cultivars, such as Pokot, Rongai, Mbarara and Masaba – all identified by the original collection sites.

There are basically three systems of forage production in smallholder dairying: (i) from natural plants, (ii) from improved forages, and (iii) from cultivated fodders.

Natural pastures are generally composed of herbaceous species, mostly grasses, although non-gramineous herbs commonly occur and there is usually a storey of trees and shrubs of varying density. Woody species provide forage as browse, while taller trees contribute fruits. Table 18.1 indicates some of the natural vegetation commonly used by smallholders in ACZs I–IV. Since over 1000 grasses have been recorded in eastern Africa, the table shows only the major ones. The potential of dairy production from natural forage is limited by both the quality and quantity of the feed produced, and this is influenced by many factors, including climate, soil, vegetation, animal type and the grazing system. In dairying systems where natural pastures are used, the dairy animals are either freely grazed or semi-zero-grazed on the roadside or communal land on natural forages. Forage produced from natural pastures are usually inadequate or of low quality, or both, especially in overgrazed fields.

Smallholder dairy farmers can increase production by planting improved forage. Sown forage can increase production economically in semi-intensive systems because such forages are of better feeding value and higher yield than natural vegetation. In addition, sown pastures increase soil fertility and improve soil structure for subsequent cropping, and in some cases reduce the incidences of weeds, pests and diseases. In smallholder dairy systems, forages can increase the level of nitrogen in pasture production systems, enhance yields of protein-rich forage. This can be achieved either by use of fertilizer nitrogen on nitrogen-responsive grasses, or through the use of nitrogen-fixing legumes. To choose the most suitable forages for a sown pasture, it is necessary to define accurately both the environment in which they are to be grown and the purpose for which they are to be used.

SEASONALITY IN FORAGE PRODUCTIVITY

Seasonality of forage availability is a major constraint to animal production in smallholder dairying and is determined by climatic conditions. The air temperatures and rainfall are relatively constant during the 7–8 months of rainy season, and a substantial amount of forage is produced during the first quarter, particularly in zones II and III, with their unimodal rainfall pattern. If dairy farmers plant forages at the onset of rain, soil temperatures are high and more nitrogen is released. Variation in soil temperatures exceeds that of air temperature and the peak growth in forages is reached 4 to 6 weeks after the end of the dry season. The quality of the first forage produced and its response to fertilizer nitrogen is high since most forages utilize the released nitrogen to recover from the dry season.

Farmers can increase forage production by using the most suitable and high yielding forages identified for the various agro-climatic zones (Table 18.4), and following recommended agronomic practices. Increase in production can be realized by establishing more new pastures and by applying nitrogenous fertilizer. Farmers need to manipulate early and late sowing of forages to increase production, particularly for the dry season. First year pastures are more vigorous and continue to grow later into the dry season, especially when sown late. While preference is always given to early sowing, followed by a conservation cut, late sown forages are more vigorous in the dry season and more water is saved and stored. Late sown pastures behave in the first dry season like annual fodders, such as oats or grass sorghum. Smallholder dairy farmers can reduce seasonality in forage availability by fodder conservation. Climatic conditions are such that there are usually several months of dry season when feed is scarce.

During the long rains there is often a surplus of forage, which could be conserved; apart from providing forage for the dry season, conservation cuts are also conducive to good forage husbandry. Smallholders could not only save grass that would otherwise be spoilt, but also encourage regrowth that would not be hampered by the old foliage. Forage conservation technologies, such as hay and silage making, have been developed for smallholders, although adoption has been poor. Although hay is more easily cured in dry climates, weather conditions, particularly in ACZ III, are such that there are only a few days of dry weather when hay or silage can be safely made during the long rains.

Forage production can be increased by use of fodder catch or break crops such as oats and sorghums. At the end of the productive life of a pasture, oats or grass sorghum can play a significant role in increasing forage production. Old pastures can be ploughed in the middle or tail end of the rainy season and oats, sorghum or another suitable annual fodder sown to replace pasture that is no longer productive, help its decomposition and provide dry season fodder, without extra land being set aside.

The most powerful way of intensifying forage production in smallholder farms is through use of fertilizer nitrogen. It is the most effective, and also the most flexible, tool a smallholder farmer can use to match forage supply with demand. Nitrogen is effective in prolonging growth into the early dry season and also in accelerating it at the start of the rains. Under zero grazing, nitrogen fertilizer is always necessary, although many farmers do not use it frequently because of its high price.

Planted fodders also play a significant role in smallholder dairy production. They are mostly identified with zero grazing and stall feeding in smallholder dairying. Of all the planted fodders used by smallholders, Napier grass is the most popular. It forms up to 40 percent of the dry matter in the diet of dairy cattle, the rest coming from other cultivated grasses, fodders, crop by-products, crop residues and purchased concentrates. Table 18.4 gives a list of the most common fodders used by smallholder dairy farmers in eastern Africa.

INTENSIFICATION OF FEED SUPPLY AND CHANGES IN PRODUCTION PRACTICES

A survey of smallholder dairy farms in Kenya showed a changing trend in feeding and production practices in both intensive and semi-intensive systems. There is a clear evidence of increased intensification of dairying by smallholder farmers (Staal *et al.*, 1998). More than 75 percent of smallholders who practise intensive dairy production exclusively stall-feed their dairy cattle, a practice consistent with the small size of farms. In semi-intensive systems, farms are on average

Table 18.4 Suitable forages and farm by-products for smallholder systems of eastern Africa.

Zone elevation Annual precipitation	Suitable feeds			
	Legumes	Ley grasses	Crop by-products	Fodders
1. Cold and wet high-altitude areas 2400–3000 m 1000–2500 mm	Clover (*Trifolium repens*) Kenya white clover (*T. semipilosum)* Vetch *(Vicia villosa)* Lucerne (*Medicago sativa*)	Kikuyu grass *(Pennisetum clandestinum)* Perennial ryegrass *(Lolium perenne)* Cocksfoot *(Dactylis glomerata)* Tall fescue *(Festuca arundinacea)*	Vegetable wastes (kales and carrots) Maize stover Banana pseudostems	Oats, Brassicas (kale), Fodder beets, Turnips Guatemala grass *(Tripsacum laxum)*
2. Cool and wet medium-altitude areas 1850–2400 m 1000–2500 mm	Desmodiums Glycine *(Neonotonia wightii)* *Stylosanthes guianensis* Njahi *(Lablab purpureus*) Lupin (*Lupinus albus*) Lucerne (*Medicago sativa*) Vetch	Rhodes grass, Setaria, Coloured guinea *(Panicum coloratum)*, Star grass, Kikuyu grass, Congo signal (*Bracharia ruziziensis*)	Maize stover Bean stover Banana pseudostems Sweet potato vines	Napier grass Giant panicum *(Panicum maximum)* Guatemala grass Sudan grass Columbus grass *(Sorghum almum)* Maize, Oats, Sweet potato Russian (Quaker) comfrey (*Symphytum peregrinum*)
3. Warm and wet medium-altitude areas 1200–1850 m 1000–2500 mm (bimodal or unimodal) or <1000 m 1000-1800 mm (bimodal or unimodal)	Desmodiums Glycine *(Neonotonia wightii)* *Stylosanthes* Njahi *(Lablab purpureus)* *Leucaena, Calliandra,* Sesbania *(Sesbania sesban, S. grandiflora)*	Rhodes grass Setaria grass Panicum *(Panicum maximum)* Star grass Kikuyu grass Congo signal (*Bracharia ruziziensis*)	Maize stover Banana pseudostems Wheat straw Bean stover Sweet potato vines Sugar cane tops Cowpea	Napier grass, Giant setaria *(Setaria splendida)* Giant panicum Guatemala grass (*Tripsacum laxum* or *andersonii)* Sudan grass *(Sorghum sudanense)* Sweet potato vines Edible canna *(Cannaedulis)* Russian comfrey
4. Hot and humid coastal strip low altitude areas (Kenya and Tanzania coast)	Clitoria (*Clitoria ternatea*) Centrosema *(Centrosema pubescens)* Siratro *(Macroptilium atropurpureum)* *Desmodium* spp. *Stylosanthes* spp. *Leucaena* Gliricidia *(Gliricidia sepium, G. maculata)*	Rhodes grass Setaria Coloured guinea *Eragrostis superba*	Cashew nut Vegetable waste Sweet potato vines Cassava Bananas Cowpea	Giant panicum Napier grass Giant setaria Sudan grass

larger and some smaller-scale farmers have access to communal land where they can graze their dairy herds. An increasing number of dairy farmers now stall-feed their cattle compared with the situation ten years earlier. The proportion of smaller-scale dairy farmers putting cattle to grazing had almost halved in ten years. This is clear evidence of increased intensification of smallholder farming, with intensified dairying as an important part of the change. Although stall feeding is generally predominant, even on the farms where cattle graze, pastures were not usually the main source of feed; rather it was fodders (including by-products and crop residues) gathered from farms or from public land, or purchased. More smallholder dairies grow high yielding fodder, Napier grass being the major forage crop grown by over 70 percent of farmers.

In intensive systems with zero grazing, nearly half of the farmers purchase fodder as the main source of feed, a third (35 percent) grow fodder on their own land, and one in eight (13 percent) gather it from public land. Over the past ten years, dairy feeding strategies in smallholder farms have changed in that there has been an increased number of smallholder dairies using concentrate feeds, minerals, crop residues and agro-industrial by-products. There has been a reduction in the number of smallholder dairy farmers using roadside grazing and natural salt licks. These resources from outside underpin the strategies of smallholder dairy farmers addressing seasonal feed shortages. The main strategies reported were purchase of fodder (60 percent) and purchase of concentrates (15 percent). These changes reflect an increasing dependency on purchased feed, both concentrate and fodder, and the reduced availability and dependency of communal feed resources gathered at no cost.

There has been an increasing demand for forage, which now commands high prices. Smallholders who do not own dairy cattle often serve as a source of fodder, mainly Napier grass, which is commonly grown for sale. Planted forages have grown in importance, particularly as smallholder dairy feeding shifts from grazing to stall feeding. In contrast to the omnipresent Napier grass, other forages, including fodder trees and shrubs and herbaceous legumes have not been widely adopted on most smallholders farms. A few smallholders plant *Sesbania* spp. or *Calliandra* spp., and herbaceous legumes, such as *Desmodium* spp.

MANURE IN SMALLHOLDER DAIRY SYSTEMS

In eastern Africa, many smallholders fertilize their food and forage crops with manure, either as compost, slurry or fresh dung. Some store the bedding and faeces before applying it to Napier grass on the farm. Sometimes, many of the nutrients in the faeces and urine are lost. Much improvement in nutrient management could be made through better handling and storage of faeces and urine. A few farmers give away or sell their manure; some use it as fuel; and others to plaster their houses. Manure sustains soil through increased pH, increased water-holding capacity, increased hydraulic conductivity and infiltration and decreased bulk density. Manure has the long-term effect of raising soil organic matter levels. Some farmers confine their animals at night and free them during the day. The amount of recoverable manure also depends on the efficiency of its collection and management The diet of the animals producing the manure and the conditions under which it is stored and handled are major factors affecting quality. In intensive cut-and-carry systems, without recycling manure and use of fertilizer the soil is depleted of minerals, and an initially high yield may quickly decline.

Under grazing, most of the consumed minerals are returned to the soil. Distribution of minerals through faeces and urine by grazing cattle is poor, however, and high losses of nitrogen may occur due to leaching, denitrification or volatilization. Cows producing 20–25 kg of milk daily only convert about 20 percent of the nitrogen consumed into protein in milk and meat. Napier grass yielding 10–20 tonnes DM ha^{-1} yr^{-1} may remove as much as 300 kg N, 50 kg P and 600 kg K. In contrast, a well managed maize crop in high-potential areas of eastern Africa (double cropped), only the grain being harvested, and assuming a yield of 4000 kg ha^{-1} for each crop, may not remove more than 120 kg N ha^{-1} yr^{-1} from the soil. Recycling animal manure to Napier grass for forage will at least largely replenish the quantities of P and K removed by cut-and-carry systems for zero-grazing and stall feeding; dung and urine could be effectively collected, stored and returned to the Napier grass or fodder at the right time and in an efficient manner. At the same time, in Kenya, fertilizer is much more commonly used on maize than on fodder crops. A successful "self-contained" zero-grazing system, therefore, requires a high level of management.

In cut-and-carry systems, animals are confined, and urine and faeces are deposited on a small area. Without proper collection, large losses of minerals may occur. In the zero-grazing unit, the manure can easily be collected as slurry if the floor of the zero-grazing unit is made of concrete. To avoid severe depletion of soil minerals, this slurry should be returned to Napier grass as efficiently as possible (Snyders, Orodho and Wouters, 1992). However many smallholders do not have cemented floors, nor slurry collection pits. Some may not even have a roof on their zero-grazing unit. In such cases there is great loss of nutrients. Generally, small-scale farmers with zero-grazing units collect the manure daily; most bring it straight back to the Napier grass field and work it into the soil. In this way the losses of nutrients are kept to a minimum. Moreover, no big storage facilities are needed, only a small pit in which the urine and the dung of the day can be collected before transporting it to the field. The use and handling of manure is an important aspect of the zero-grazing package. If no recycling is done, the soil will be overexploited, soil fertility will decline and the yield of Napier grass and other fodders will be reduced.

NUTRITIVE VALUE OF FORAGES

The low quality of forages, particularly in the dry season, is a major constraint to dairy production. The ultimate value of forage plants must be considered mainly in terms of milk. The nutritive value of most forages is influenced by age and varies according to the phenological stages of the crop. As forages develop in age, their acid detergent fibre (ADF), neutral detergent fibre (NDF), lignin, stem, dead material and dry matter (DM) increase, while their crude protein (CP), leaf-to-stem ratio, ash and digestibility decrease. The chemical compositions of hay from common grasses range from 3.7 to 9.2 percent CP and 32 to 45.5 percent crude fibre (CF). In general the chemical composition of forage depends on the plant species or cultivar, the plant part, water status and age. Crowder and Chheda (1982) reported that the genetic differences in pasture species at variety level contribute towards difference in *in vivo* digestibility. CP is higher in leaves than stems, while stems have higher CF. Napier grass, Rhodes grass and setaria are among the most nutritious and productive grasses grown in eastern Africa. Napier grass produces 20–25 tonnes DM ha^{-1} yr^{-1} and the production of chloris and setaria ranges from 14 to 18 tonnes DM ha^{-1} yr^{-1}. Yields and quality of forages are not constant over seasons and years. Grasses are highly seasonal, delivering most of their potential in the first months of the growing season. Forage yields of 15 tonnes DM ha^{-1} in the year of sowing or planting are not unusual without N fertilization. Unless topped up with N, however, yields drop to a third or less by the third year, especially in a cut-and-carry system.

In the wet season, the concentration of minerals and CP is higher than during the dry season. In young grasses the initial high CP levels drop considerably with maturity. The total amino acid content of grass correlates highly with the total N contents and declines during growth. Grasses have protein degradability varying between 12 and 33 percent and for most grasses the proportion of non-lignin N to total protein is 15 percent higher than the digestible coefficient of the total N. In general, K and Fe appear to be adequate in most grasses, but P, Na, Cu, Co, Mg and Ca are normally deficient in grasses, particularly during the wet season. In terms of mean percentage DM, the P levels in legume forage range from 0.13 to 0.38 percent, the Ca level range from 0.88 to 1.45 percent, the Mg levels range from 0.25 to 0.67 percent and the Na levels from 0.02 to 0.23 percent (Skerman *et al.*, 1988). Most forages grown by smallholder farmers have CP of 10–12 percent and digestibility (D-vitro) of 55–60 percent. After the first year of production, this is rarely possible without application of N fertilizer. The CP of Napier grass and many other forage crops is normally less than 14 percent, which is insufficient for efficient dairy production, hence the need for supplementation to increase dairy productivity. The *in vivo* digestibility of Napier grass, Guatemala, and giant setaria range between 47 and 56 percent, but can be as high as 71.1 percent when cut young. Smallholder dairy farmers can increase the quality of forages by using the recommended forage management practices. Smallholder farmers can also improve the quality of their forage crops by growing grass and legumes mixtures. Some farmers grow and harvest the grasses and legume separately, but mix them when stall feeding their dairy cattle. In some cases, growing and managing grasses and legume separately, increases yield. The CP of most legumes ranges between 14 and 18 percent and CF ranges between 27 and 35 percent. *Leucaena leucocephala* has CP of 21.6 and CF 18.3 (Skerman *et al.*, 1988). Although legumes have higher CP than grasses, some tropical legumes have lower digestibility and palatability because they have higher tannin contents.

FORAGE SEED PRODUCTION

Lack of seed and planting materials of recommended forages is a major constraint to increased forage production on smallholder farms in eastern Africa. Frequently, only very small amounts of seed of recommended forages are available to smallholders. Once a forage cultivar has proved to be promising, its seed must be produced in large quantities to be used by smallholders. If seed production is inefficient, then seed costs are high and this reduces the incentive for pasture improvement. There is a need to ensure that adequate seed of recommended forages is available to smallholders in the various ACZs. Seed production potential for recommended forages needs to be assessed and technologies developed to exploit their genetic potential. Even though species and varieties behave similarly in response to agronomic practices, they vary markedly in seed yield potential. To maximize forage seed yields it is necessary to carefully select sites on which the seed crop is grown and to optimize the establishment and management of the seed crop, and also to follow proper harvesting procedures. Nitrogen fertilizer and row width are important tools to manipulate heading and subsequent seed setting, and thus increase seed yields of tropical forages. Forage grasses produce more seed under conditions of closer tiller density and appropriate N dressing. Seed setting, which is the major variable determining pure germinating seed (PGS) content, is greatly affected by weather conditions in the pre-heading period. Rainfall is of critical importance. Studies showed that seasons with good seed yields had more rains, which were better distributed, without prolonged drought. A good rainfall distribution is a major prerequisite to achieving high PGS percentage.

Government has the responsibility for producing Breeder or Basic seed from its forage breeding programmes at appropriate research centres. The Basic seed is supplied to seed companies for multiplication, processing, distribution and sale. Some seed companies have their own breeders involved in breeding and selection of forages. Seed companies normally produce seed of forages that are widely grown and can be sold extensively in the nation and some exported; they are not keen on producing forages that are zone- or area-specific, or handling vegetatively propagated planting materials such as Napier grass, Guatemala grass, sweet potato, edible canna, giant setaria and giant panicum. Smallholders normally get vegetative material and forage seed that are zone specific from government institutions or from other farmers.

An independent national body provides plant health inspectorate services, and is responsible for seed inspection and certification, ensuring that the quality of seed produced meets international quality standards. This body also works closely with breeders to maintain the distinctness, uniformity and stability (DUS) of the seed produced. Imported pasture seed, particularly of legumes, is very expensive for smallholders. Such seed can be produced locally and cheaply by smallholders after appropriate training in production technologies that have been developed locally. Some smallholder farmers have taken seed production as a business and are producing forage seed for sale. Legume seed production by smallholder farmers is more profitable than producing maize as a commercial crop. Farmers who have no dairy cattle are growing crops such as Napier grass and selling the forage and vegetative propagating planting materials to livestock farmers. Some smallholders set aside surplus forage in the year of sowing to harvest farm-saved seed. This provides a low-cost source of seed to establish more pastures. It is partly through this system that improved forages such as Rhodes grass and setaria have spread rapidly. More seeds can move into markets, which are otherwise inaccessible. This informal seed production sector is likely to develop. Smallholder seed producers could come together and form cooperatives and associations to enable them to market forage seeds throughout eastern Africa. Forage production on smallholder farms should increase as a result of both the informal and formal seed production sectors.

CROP RESIDUES, AGRO-INDUSTRIAL BY-PRODUCTS AND THEIR LIMITATIONS

Forages on most smallholder farms in eastern Africa are deficient in protein, energy and certain minerals, such as P and S, particularly in the dry season. Some dairy farmers supplement these from other sources of feed. Most protein supplements are expensive and the majority of smallholder farmers are unable to afford them. Besides, concentrates are not readily available to smallholders and, when available, their quality is questionable. In arable areas of mixed farming, the use of crop residue and by-products to supplement forages as livestock feeds is commonest in dairying. Smallholders supplement their livestock with purchased grain, concentrates and by-products such as bran, wheat pollard and dairy meal. Because of the high cost of dairy meal, many farmers opt for cereal by-products. Variable amounts of concentrates are usually fed to dairy cows at milking, with many smallholders feeding a flat rate of about 1 to 2 kg daily throughout lactation. Most commonly used are crop by-products. Crop residues in smallholder dairy-crop systems include cereal straw and stover (e.g. rice, sorghum, millet and wheat); horticultural crop and vegetable waste (e.g. cowpea and chick pea vines); sugar cane tops and leaves; maize, wheat, and barley wastes; pyrethrum marc; pigeon pea leaves; green maize strippings; banana leaves and pseudostems; potato and banana peelings; residues from soya beans, peas, groundnuts and lab-lab bean (*Lablab purpureus*, syn. *Dolichos lablab*); coconut, simsim and cotton seed cake; and field bean, soya bean and groundnut hulls.

In some parts of Kenya, crop residues, mainly maize and bean stover, provide an average of 35–45 percent of the total livestock feed requirement in smallholder farms. The contribution of crop by-products to feeding dairy cattle depends on farm size. The smaller the farm, the larger the proportion of feeds drawn from crop residues and crop by-products compared with forages. In the highlands of Mount Kilimanjaro and Mount Meru in Tanzania, crop residues, including banana, maize and other cereals, are commonly fed to cattle, especially in the dry season, to supplement forage. In Uganda, crop residues are available as feed for the dry season. Banana peel and potato vine are the most commonly used in intensive systems. Cassava foliage could be a valuable feed in smallholder farms, but is rarely used for fear of cyanide toxicity. The major agro-industrial by-products are cereal bran (maize and rice), oil cake (soya bean, cotton seed and sunflower) and brewers' grains. All cereal straws contain a large pool of structural carbohydrates that can potentially be used as an energy source by the ruminant microbes. The minimal contents of crude protein (2–4 percent of DM), the imbalance in essential minerals contained in the straw, their low digestibility as well as low intake by ruminants are disadvantages in using straw as feed (Wanapat, 1986).

By-products such as maize stover, wheat and rice straw are fibrous and in most cases lignified, making them less digestible and acceptable to livestock. However, these deleterious properties may be reduced through chemical and physical treatment. Treating roughage with sodium hydroxide has proved effective in improving feed conversion and animal performance, and fermentation with urea is a well-known technique. The level of use of a by-product is limited by the cost of collection and transportation from source, storage and processing, in addition to other intrinsic factors, and seasonality in availability. Fresh cassava leaves contain high levels of hydrocyanic acid (HCN), which is toxic. Sun drying the leaves reduces the HCN contents to levels that are less toxic (Wanapat *et al*., 1989). However, not all cultivars of cassava are poisonous and ruminants may be able to consume the fresh leaves without ill-effect. Crop residue are usually deficient in at least one nutrient and cannot support animal production without adequate supplementation. Owing to their bulky nature, most crop residues and by-products must be used at or close to where they are produced, to reduce transport cost. The commonly available farm by-products are sugar cane tops, sugar cane rejects, maize stalks and maize cobs; these were found to be high in crude fibre, calcium and potassium but low in protein and phosphorous (NARS, 1977). They are good roughage but require supplementing with crude protein and energy in the form of readily digestible carbohydrates and minerals when fed to dairy cattle.

DISEASES AND PESTS OF FORAGES

There is an increase incidence of diseases and pests affecting forages in eastern Africa. In the past, forage diseases and pests caused minimal economic losses, so little research was directed towards solving forage disease and pest problems. However, in the past two decades, there have been important outbreaks of diseases and pests, some of which have caused serious economic losses to smallholders. Some of the important forage legumes affected by diseases and pests included *Stylosanthes,* attacked by anthracnose, a fungal disease caused by *Colletotrichum gloeosporioides*; leucaena is attacked by Leucaena psyllids (*Heteropsylla cubana*); siratro is attacked by *Rhizoctonia* fungus; and lupins succumb to a root rot caused by *Sclerotia rolfsii*. Among the forage grasses affected by diseases and pests are *Chloris,* which is attacked by *Fusarium graminearum* fungus;

Cynodon is attacked by *Helminthosporium cynodontis* fungus; *Panicum* is attacked by *Cerebella andropogonis; Paspalum* is attacked by *Claviceps purpurea*, with the danger of animal poisoning; and Kikuyu grass is attacked by the root knot nematode, *Meloidogyne kikuyuyensis*.

Napier grass is also seriously attacked by diseases and pests, causing serious economic loss and concern to farmers. Some of the important diseases and pests are considered below.

Snowmould fungal disease Most Napier grass clones are susceptible to the fungus, *Beniowskia sphaeroidea*, which causes a white mould on the leaves and stems. Research in the early 1970s developed a variety – clone 13 – which was a selection from seedling progenies of French Cameroon seeds. Clone 13 is resistant to the fungal disease (Van Wijk, 1974). This disease has also been reported to attack Kikuyu grass and Nandi setaria.

Napier grass head smut This is a serious Napier grass disease, first reported in cooler parts of central Kenya in 1992. Affected plants develop fungal symptoms that look like flower structures but open up releasing black spores. The stems then become smaller and the total DM of the affected crop is drastically reduced. After 2–3 cuttings the entire stool dries. The causative organism is a fungus, *Ustilago kamerunensis*. This disease, which spreads rapidly, has caused serious concern to smallholders, who report losses of 90–100 percent of their Napier grass and have been forced to drastically reduce their livestock number. Through collaborative research in screening and evaluation, a Napier grass variety, Kakamega 1, developed by the author, has been identified as high yielding and resistant to Napier grass head smut (Farrel, 1998). This variety is being multiplied and bulked in government institutions and has been distributed to over 10 000 smallholders in just one year. The farmers usually come together and contribute money to hire transport to collect Kakamega 1 Napier grass from institutions where the material has been bulked. This cultivar is spreading very fast because smallholders who already have it supply it to their neighbours.

Napier grass stunting disease Another serious Napier grass disease has developed in Western Kenya; it was first reported in Bungoma district, bordering Uganda, in 1997. A similar disease had been reported in Uganda and its cause was suspected to be a virus, probably transmitted by insects (Tiley, 1969). This disease is spreading fast and is present in several districts of Western Kenya, causing serious economic loss to the smallholder dairy industry. Most Napier grass varieties grown in the area are susceptible to the disease, which usually becomes visible in re-growth after cutting or grazing. Affected shoots become pale yellow green in colour and seriously dwarfed, with a densely bunched (fastigiated) appearance on account of repeated proliferation of auxiliary buds. Individual leaves and sheaths of affected plants greatly diminish. Root development is severely impaired and is reduced to a weak fibrous growth. Infection at the early stage results in symptom expression in the whole stool, with complete loss in yield and eventual death. Many smallholders have lost all their Napier crop and are forced to de-stock or sell their herd. The author is collaborating with scientists from KARI and from international research institutes in studies to identify the cause of Napier stunting disease and its mode of transmission. Collaborative tests by KARI and CABI Scientific at CAB International laboratories have confirmed that the stunting disease is caused by a mycoplasma. Further tests are being carried out at CABI International laboratories to confirm whether or not the disease is insect transmitted. Scientists are also screening over 50 Napier grass germplasm lines in order to locate any resistant varieties that could be used in Western Kenya.

THE IMPORTANCE AND FUTURE OF FORAGE-BASED SMALLHOLDER DAIRYING

Milk marketing channels

The dairy industry in eastern Africa has both formal and informal milk market pathways, which are largely dependent on market surplus from smallholder dairy producers. More than 70 percent of milk is marketed through informal channels, which include:

- direct sales by smallholders to consumers, either individually or through institutions;
- sales to farmers' cooperatives, dairy societies or self-help groups that have milk collecting centres, which boil or cool milk before delivering it, either fresh or soured, to consumers or processors; and
- sales to small traders and market agents (milk bars, kiosks or mobile traders), who buy milk from farmers or farmers' cooperatives and sell to consumers or retailers.

Only a few dairy cooperatives pasteurize milk; most milk marketed through informal channels is consumed raw, a system that has public health implications. Until recently, the formal channel was dominated by large parastatal dairy cooperatives, which were instituted to collect milk from surplus areas, process it and sell on to deficit areas. The Cooperative Development of Industry Act in Kenya, under which dairy marketing cooperatives fall, did not, in the past, allow sufficient farmer control of dairy cooperatives, thereby contributing to high incidence of mismanagement and misappropriation of farmers' funds by management. This served as a great disincentive to increased forage and dairy production by smallholder dairy farmers.

Major policy changes are under way in eastern Africa, in line with liberalization. In the dairy sub-sector, the major change has been the liberalization of milk marketing. The immediate impact has been a rapid growth of formal and informal private sectors, resulting in increased participation in milk marketing by formal and informal market agents. This has led to re-distribution of benefits of dairying, including increased employment opportunities, increased milk availability and a rise in real farmgate milk price, thereby increasing producer earnings. Market liberalization has also brought change in relative prices in various regions, which now reflect market access and production costs. There is also a positive shift in concerns of smallholder dairy farmers, from past difficulties with milk marketing to now investigating options for increased production.

Increased farmgate prices by traders and farmers' groups has provided further incentive for conservation of feed for the dry season, when milk price is highest, hence an increase in forage production on smallholder dairy farms. This trend will continue in future and smallholders will intensify forage production and increase dairy productivity by adopting appropriate technologies.

SOCIO-ECONOMIC FACTORS

Smallholder dairying is profitable. Important determinants of profitability, investment and operating costs vary between production systems. As summarized in Table 18.5, engaging in intensive dairying substantially increases both annual per cow investment and the operating cost compared with the semi-intensive and extensive systems utilizing communal grazing. Per-cow annualized monetary returns are very high for zero-grazing systems, compared with unfenced extensive dairying. In Tanzania, production costs were compared for three systems: small-scale intensive rural dairying (Arusha/Kilimanjaro and Southern highlands); small-scale intensive urban dairy

Table 18.5 Summary of annualized per-cow gross returns to dairy production in Uganda

Cost parameters		Dairy production system		
		Extensive	Semi-intensive	Intensive
Total investment costs (a)	US$	4.18	37.18	139.17
	U Sh	4 426	39 335	147 237
Total operating costs (b)	US$	18.66	151.74	276.05
	U Sh	19 751	160 545	292 066
Total gross revenue (c)	US$	185.05	693.53	1 651.71
	U Sh	195 837	733 760	1 747 513
Gross margin, (c) -(b) -(a)	US$	162.20	504.61	1 236.45
	U Sh	171 660	533 880	1 308 210
Cost per litre of sold milk	US$	0.08	0.11	0.16
	U Sh	86	115	164
Cost per litre of total milk	US$	0.06	0.10	0.12
	U Sh	58	101	130

Note: Exchange rate in July 1996 was U Sh 1058 = US$ 1.0
Source: ILRI, 1996.

(represented by Dar-es-Salaam); and smallholder semi-intensive dairy with Zebu cattle (Chalinze). Gross margins per cow differ considerably, depending on the cost of inputs (particularly forage), price of milk and increase in herd value.

Despite the high cost for purchase of forage, urban intensive dairying yields a higher gross margin per cow than similar production systems in rural areas, where the price of milk is lower. This may in part explain the exponential growth of the urban cattle population in recent years. The gross margin per animal from semi-intensive dairy production with zebu is, predictably, low given that this is a low-input, low-output system, as shown in Table 18.6.

Table 18.6 A summary of annual per-cow gross returns to dairy production for the most important smallholder production systems in Tanzania

Cost parameters		Dairy production system		
		Semi-intensive with zebus	Intensive rural with exotic crosses	Intensive urban with exotic crosses
Total variable costs	US$	126	260	747
	T Sh	75 500	156 234	448 348
Total revenue	US$	242	1 020	1 750
	T Sh	145 080	612 217	1 051 330
Gross margin	US$	116	760	1 000
	T Sh	69 580	455 983	602 982
Gross margin per litre	US$	0.2	0.38	0.52
	T Sh	6 159	230	315

Note: Exchange rate (1998, approximate) T Sh 600 = US$ 1.0.
Source: MOAC/SUA/ILRI, 1998.

FODDER-BASED DAIRY PRODUCTION

A study was carried out on the Kenya coast, comparing three smallholder farming systems: (i) an intensive dairy production system where smallholder farmers adopted three related technologies: keeping grade or cross-bred dairy cows; planting forages, mainly Napier Grass; and using the infection and treatment method of immunization against East Coast Fever; (ii) a system where smallholder farmers did not adopt the technology and kept only local cattle; and (iii) where smallholder farmers did not adopt the technology and kept no cattle. The results are summarized in Table 18.7. Results of the survey of 202 households drawn from the three production systems showed that intensive dairying substantially increased household income, generated paid (secondary) employment, and improved the nutritional status of pre-school-age children in the households (Nicholson *et al.*, 1999). The average income per month of farmers practising forage-based intensive dairying was much higher than the income of farmers that did not plant forages (Napier grass).

Differences in cash income from dairying accounted for over 40 percent of the differences in total mean cash income between intensive dairy production and non-dairy farmers. The profits obtained by farmers growing high yielding forages for intensive dairy production is very high. Smallholders who use recommended practices and technologies to grow forage get better returns from milk sales. To further consolidate the system, there is a need both to strengthen forage improvement programmes to develop Napier grass cultivars resistant to the common diseases and to enforce plant quarantine regulations better in order to reduce movement of diseased forage planting materials, especially across borders. There will also be a need to improve and popularize other fodder crops so that smallholder farmers have alternative high yielding forages for intensive dairying.

EMPLOYMENT

Employment generation is another impact of adoption of intensive smallholder dairying, in part because sown forage production and management and the care and feeding of dairy animals require more labour. More than three quarters of households responding to the detailed survey indicated that they hired more labour as a result of taking up intensive smallholder dairy production. Hired labour is usually from the areas surrounding the farms and the financial benefit of dairying is thus shared among the dairy farmers and others in the community. Although smallholders who did not adopt intensive technology hired some labour, their labour received much less total pay because of less work and shorter working hours compared with those employed by smallholder dairy farms.

HOUSEHOLD NUTRITION

Smallholder dairying increases a household's nutritional status; increased milk production usually increases home milk consumption. Milk is a significant source of energy and proteins, including many essential amino acids lacking in many carbohydrates, and it contains essential micronutrients, such as vitamins A and D. Households with dairy cattle can purchase more and a wider variety of foods. The incidence of chronic malnutrition was significantly lower among children of households with dairy cattle than among children of farmers without dairy activity.

Table 18.7 Reported monthly cash income by type of income and adoption status in Kenya

Income type		Households with no cattle and no planted fodder	Households with local cattle but no planted fodder	Households with grade or cross-bred cattle and planted fodder (Napier grass)
Percentage of households with cash income from:				
Dairying		0	14	87
Poultry or eggs		9	8	14
Crops		75	80	73
Wages, salaries or off-farm activities		65	66	65
Remittances		39	35	22
Other income		13	25	18
Mean monthly cash income from dairying				
	US$	0	5.18	109.82
	K Sh	0[ac]	321[ab]	6 809[bc]
	(SD)	–	(1 211)	(7 836)
Mean monthly cash income from poultry or eggs per month	US$	10.50	4.31	31.68
	K Sh	651	267	1 964
	(SD)	(3 261)	(1 682)	(8 687)
Mean monthly cash income from crops				
	US$	12.0	19.94	26.26
	K Sh	744[c]	1 236	1 628[c]
	(SD)	(991)	(2 242)	(2 967)
Mean monthly cash income from wages, salaries or off-farm activities	US$	42.66	48.68	149.32
	K Sh	2 645[c]	3 018[b]	9 258[bc]
	(SD)	(5 585)	(4 169)	(17 079)
Mean monthly cash income from remittances				
	US$	7.29	7.77	4.84
	K Sh	452	482	300
	(SD)	(965)	(902)	(852)
Mean monthly cash income from other sources				
	US$	2.27	2.35	12.82
	K Sh	141	146	795
	(SD)	(823)	(350)	(4 115)
Mean total monthly cash income from all activities				
	US$	75.27	87.78 5	337.29
	K Sh	4 667[c]	439[b]	20.912[bc]
	(SD)	(7 280)	(4 444)	(23 761)

Notes: SD = Standard deviation of the mean. The superscripts a, b and c indicate that the means for the two adoption categories [between adopters, i.e. households with at least one grade/cross bred (G/C) animal and planted fodder, and non-adopters, i.e. household owning no G/C animal and with no planted fodder] with the same letter are statistically different at the 95 percent confidence level. Exchange rate (1999) K Sh 62 = US$ 1.

Source: Nicholson *et al.*, 1999.

MAAIF [Ministry of Agriculture, Animal Industry and Fisheries]. 1993. *Supply and Demand, Master Plan for Dairy Sector*. Vol. IV: Subject Matter Reports. No. 1 – Milk Production and Supply; No. 2 – Support Services to Dairy Farmers; No. 3 – The Market for Milk and Dairy Products. MAAIF, Entebbe.

Mdoe, N. & Wiggins S. 1997. Returns to smallholder dairy in the Kilimanjaro Region, Tanzania. *Agricultural Economics,* **17**: 75–87.

Mlozi, M.R.S. 1995. Information and the problems of urban agriculture in Tanzania. Intentions and realizations. Ph.D. thesis, University of British Columbia, Vancouver, Canada.

MOAC [Ministry of Agriculture and Co-operatives] / **SUA** [Sokoine University of Agriculture] / **ILRI**. 1998. *The Tanzania dairy sub-sector : A Rapid Appraisal.* Vol.3 – Main report. Funded by the Swiss Agency for Development and Cooperation, Switzerland.

Mougeout, L.J.A. 1994. African city farming from a world perspective. *In:* A.G. Egziabher, D. Lee-Smith, D.G. Maxiwell, P.A. Memon, L.J.A. Mougeout and C.J. Sawio (eds). *Cities feeding people: An examination of urban agriculture in East Africa.* International Development Research Centre (IDRC), Ottawa, Canada.

Mullins, G., Wahome, L., Tsangari, P. & Maarse, L. 1996. Impact of intensive dairy production on smallholder farm women in coastal Kenya. *Human Ecology,* **24**: 231–253.

Muriuki, H.G. & Thorpe, W. 2001. Smallholder dairy production and marketing in Eastern and Southern Africa. Regional synthesis. *In:* D. Rangnekar and W. Thorpe (eds). *Smallholder dairy production and marketing opportunities and constraints.* Proceedings of the South-South workshop held at NDDB, Anand, India, 13–16 March 2001.

NARS [National Agriculture Research Station]. 1977. NARS, Kitale, Annual Report, 85p.

Nicholson, C.F., Thornton, P.K., Mohammed, L., Muinga, R.W., Mwamachi, D.M., Elbasha, E.H., Staal, S.J. & Thorpe, W. 1999. *Smallholder Dairy Technology in Coastal Kenya. An adoption and impact study. ILRI Impact Assessment Series,* No. 5. 59p.

Orodho, A.B. 1990. Dissemination of research technology on forages and agriculture by-products in Kenya. *In:* B.H. Dzowela, A.N. Saidi, Asrat Webden-Agenehu and J.H. Kategile (eds). *Utilization of research results on forage and agricultural by-product materials as animal feed resources in Africa.* Proceedings of the First Joint Workshop of the Pasture Network for Eastern and Southern Africa (PANESA) and the African Research Network for Agriculture By-products (ARNAB). Lilongwe, Malawi, 5–9 December 1998.

Pratt, D.J. & Gwynne, M.D. 1977. *Range Management and Ecology in East Africa.* London: Hodder & Stoughton.

Price Waterhouse. 1990. The role of women in the National Dairy Development Programme, Kenya. Price Waterhouse, Nairobi, Kenya.

Skerman, P.J., Cameron, D.G. & Riveros, F. 1988. *Tropical forage legumes. FAO Plant Production and Protection Series,* No. 2. 692p.

Snyders, P.J.M, Orodho, A.B. & Wouters, A.P. 1992. Effects of manure application methods on yield and quality of Napiergrass. National Animal Husbandry Research Centre, Naivasha publication, Kenya Agricultural Research Institute (KARI), Kenya.

Sombroek, W.C., Braun, H.M.H. & van der Pauw, B.J.A. 1982. Explanatory soil map and agro-climatic zone map of Kenya. *Kenya Soil Survey Report,* No. E1. National Agriculture Laboratories, Soil Survey Unit, Nairobi, Kenya. 56p.

Staal, S.J., Chege, L., Kinyanjui, M., Kimani, A., Lukuyu, B., Njubi, D., Owango, M., Tanner, J., Thorpe, W. & Wambugu, M. 1998. Characterization of dairy systems supplying the Nairobi milk market. A pilot survey in Kiambu District for the identification of target groups of producers. Smallholder Dairy (R&D) Project. KARI, ILRI and Livestock Production Department (Ministry of Agriculture).

Tiley, G.E.D. 1969. Elephant grass. *Kawanda Technical Communication,* No. 23.

Van Wijk, A.J.P. 1974. Breeding of Napier Grass. Agricultural Research Station, Kitale, Maize and Pasture Annual Report, 1974.

Wanapat, M. 1986. Better Utilization of Crop-residue for Buffalo Production. *In:* C. Chantalakhana (ed). *Proceedings of the Buffalo Seminar.* International Buffalo Information Centre, Kasetsart University Library, Bangkok, Thailand.

Wanapat, M., Uriyapongson, S., Chanthai, S., Wanapat, S., Wachirapakoin, C. & Thummasang, K. 1989. The utilization of dried cassava leaves and urea-treated rice straw for draught swamp buffaloes during the dry season at village level. *In:* Proceedings of the 27th Annual Conference, Kasetsart, Bangkok, Thailand.

Wandera, J.L. 1996. Pasture and Cover Crops. *KARI Review of Kenyan Agricultural Research,* Vol. 23.

CHAPTER 19

Rural Depopulation and Grassland Management in Patagonia

Pablo Borrelli[1] and Andrés Cibils[2]

ABSTRACT

Patagonia's grasslands and pastoral systems are briefly described. Three historical periods of sheep farming development are recognized and described. The first stage (1880–1930) of land occupation, population growth, capital accumulation and increase of sheep stocks could be called "the golden years". A five-decade period followed, characterized by full land occupation, a progressive loss of profit, stabilization of sheep stocks, and the beginning of a rural depopulation process. A long declining stage started in the last two decades of the twentieth century. Sheep stocks declined by half, decapitalization was evident, and about 500 farms were abandoned. Rural depopulation from pastoral systems became important during this period in all Patagonian provinces, but the process was more evident in the southern provinces.

Significant structural changes are required to attain sustainable, scientifically based utilization of Patagonian grasslands, with competitive and diversified products of high quality and improved quality of rural life. Rural depopulation cannot be stopped or reversed in any other way. After decades without sustainable development policies, the situation has become critical. There is no doubt that, without leadership, technical assistance and long-term public funding, the changes needed are unlikely to occur in time. The Patagonian Sheep Sustainable Development Programme (PSSDP) is described, being a policy proposal for the next decade. PSSDP could change the scenario to the better and accelerate innovation. The proposed long-term, systemic, participatory approach could solve the current crisis and lead to sustainable pastoral systems in Patagonia.

[1]Former Secretaría de Agricultura Ganadería Pesca y Alimentación. Argentina. (Now a private consultant www.ovis 21.com.ar)

[2]Department of Animal and Range Sciences, Box 30003, MSC 3-1, New Mexico State University, Las Cruces, NM 88003, USA.

INTRODUCTION

General description of Patagonia

The vast area of southern Argentina and Chile that spans the South American continent between latitudes 39° and 55° S is usually referred to as Patagonia. Strictly speaking, however, almost all of Patagonia's grazing lands are on the cool semi-arid steppes of most of the extra-Andean territory of southern Argentina (approximately 750 000 km^2), extending into Chile in the area surrounding the Straits of Magellan (Paruelo, Jobbágy and Sala, 1998; Villamil, 1997).

Patagonia is mostly made up of sedimentary landscapes that blend with volcanic (basalt) deposits from the Mesozoic and Cenozoic eras, unfolding as a series of plateaus that lose elevation as one moves east from the Andean foothills (Soriano, 1983). The Patagonian mesa landscape is interrupted only by a series of important rivers that flow from west to east from the Andes into the Atlantic Ocean, such as the Colorado, Negro, Chubut, Chico, Santa Cruz and Coyle (Plate 19.1) rivers. Irrigated floodplains in some of these river valleys have supported the development of important agricultural oases.

Patagonia's climate is influenced mostly by Pacific Ocean air masses that are forced inland across equatorward-moving ocean currents (that are warmer than the land at these latitudes) by the prevailing westerlies of the southern hemisphere (MacArthur, 1972). The Andes mountain range stands between these moist air masses and the Patagonian steppes, creating an extensive rain-shadow and thus dominating climatic patterns in the entire region (Paruelo *et al.*, 1998).

Plate 19.1 Sheep being trailed to a typical *estancia* along the Río Coyle, on the Magellanic Steppe, Santa Cruz.

A very steep decreasing gradient of mean annual precipitation (MAP) is observed east of the Andes, ranging from 4 000 mm at the foothills (at about 42° S) down to 150 mm in the central plateau, approximately 180 km east of the mountains (Soriano, 1983). At the same time, inter-annual variation in precipitation increases exponentially with decreasing annual rainfall, reaching coefficients of variation (standard deviation/mean) exceeding 45 percent at the drier end of the gradient (Jobbágy, Paruelo and León, 1995).

The eastern coastline of most of Patagonia is influenced by moist air coming from the Atlantic Ocean, which creates local conditions of somewhat higher annual precipitation (200 to 220 mm). Rainfall is evenly distributed throughout the year in the coastal region, as opposed to the prevailing winter rainfall pattern of most of Patagonia (Paruelo *et al.*, 1998; Soriano, 1983). The MAP/PET (potential evapotranspiration) ratio of the Patagonian steppes fluctuates between 0.45 and 0.11, with overall marked moisture deficits in spring and summer (Paruelo *et al.*, 1998). Hence, here as in other arid areas of the world, water is the single most important factor regulating primary productivity.

Mean annual temperatures in Patagonia range from 15.9°C in the north (Cippoletti) to approximately 5.4°C in the far South of Tierra del Fuego (Ushuaia) (Soriano, 1983). Mean temperatures of the coldest month of the year (July) are above 0°C, although absolute minimum temperatures can exceed –20°C in several areas of the region (Paruelo *et al.*, 1998). Temperatures may be rising in some areas of Patagonia. Cibils and Coughenour (2001) reported a significant increase in mean annual temperatures over the last 60 years for Río Gallegos, one of the towns with the longest weather records on the steppes surrounding the Straits of Magellan. Although this trend is apparently consistent with predictions of climate change based on Global Circulation Models simulating enhanced atmospheric CO_2 concentration (Hulme and Sheard, 1999), conclusions from weather station data analysis are still preliminary.

Strong and persistent westerly winds are one of the most outstanding characteristics of Patagonia's climate. Because there is relatively little land mass in the Southern Hemisphere, westerlies between the latitudes of 40° and 50° S gain impressive momentum, with mean annual intensity values varying between 15 and 22 km/h. Strong gusts of over 100 km/h are common in spring and summer (MacArthur, 1972; Paruelo *et al.*, 1998; Soriano, 1983). Strong winds affect water balance on the steppes by increasing evaporation losses significantly, and can have a considerable influence on sheep performance through their wind-chill-factor effect (Borrelli, unpublished data; Soriano, 1983).

More than half of Patagonia's soils have been classified as Aridisols (desert soils), with Entisols (soils with little development) and Mollisols (dark-coloured, base-rich steppe soils) being, respectively, the second and third most important soil types of the region (del Valle, 1998). Over 70 percent of Patagonia's topsoil is coarse-textured, ranging from sand to sandy-loams (del Valle 1998). Soil textures can explain a large portion of the variation in dominant plant life form (grasses versus shrubs) across the region (Noy-Meir, 1973; Sala, Lauenroth and Golluscio, 1997). Small-scale spatial heterogeneity of soils tends to increase with aridity (Ares *et al.*, 1990); important changes in leaching and salinity conditions occur over short distances, possibly causing soils within the same taxonomic group to function differently (del Valle, 1998 and references therein). Over 90 percent of Patagonian soils exhibit some degree of degradation, mostly as a consequence of improper land use; severe desertification affects over a quarter (19 to 30 percent) of the region (del Valle, 1998). Some of the most dramatic expressions of erosion processes in Patagonia occur in the form of sand macro-accumulations (dune tongues or *lenguas medanosas*) that, in the early 1970s, covered an area of approximately 85 000 km^2 (Monteith [1972]). Both aerial photography and satellite image analyses indicate that many of these accumulations are about 100 years old, suggesting that the rate of wind-driven erosion processes may have been accelerated by the introduction of domestic livestock in the region (Soriano, 1983).

Guanacos (*Lama guanicoe*) are the only large native ungulate of Patagonia (Soriano, 1983) and although the region has generally been considered to have evolved under light grazing pressure by large herbivores (Milchunas, Sala and Lauenroth, 1988), pre-European colonization numbers of guanacos may have been higher than previously assumed (Lauenroth, 1998). Recent guanaco counts throughout the region show that populations are fairly stable at approximately 500 000 individuals (Amaya, von Thüngen and Delamo, 2001). Overall, the number of native vertebrate faunal species within the region is considered to be relatively poor (Soriano, 1983). The lesser rhea (*Ñandú petiso* or *choique*; *Pterocnemia pennata pennata*) and the upland goose (*cauquén* or *avutarda*; *Cloephaga picta*) are possibly the most conspicuous native birds. The *mara* or Patagonian hare (*Dolichotis patagonum*) and the *piche* or small armadillo (*Zaedyus pichyi*), together with the lesser rheas, are considered to be important Patagonian zoogeographical indicators (Soriano, 1983). There is a significant number of predator species, including red foxes (*Dusicyon culpaeus*), grey foxes (*Dusicyon griseus*), pumas (*Felis concolor*) and skunks (*Conepatus humboldtii*) (Soriano, 1983). Red foxes, together with pumas, are possibly responsible for most of the livestock losses due to predation; in some instances, lamb losses to red fox predation can be as high as 75–80 percent of all lambs born in a given season (Manero, 2001).

Natural vegetation

Extra-Andean Patagonia is mostly covered by treeless shrub and grass steppes that give way to dwarf-shrub semi-deserts in the drier areas of the central plateaus (Roig, 1998). Overall, vegetation is characterized by the dominance of xerophytic species that have evolved remarkable adaptations to cope with severe water deficit conditions (León *et al.*, 1998). Shrub species, for example, commonly exhibit either very small sclerophyllous leaves with abundant glandular hairs or leaves with thick cuticles, and in many instances dwarf cushion-growing habits. Grasses, in comparison, commonly have leaves with thick cuticles, convoluted laminae and bunch growth habits with fairly large accumulations of standing dead biomass (León *et al.*, 1998). Blended in the steppe landscapes are small areas associated with rivers or other permanent water sources, with more mesic plant communities comprising mostly grasses, sedges and rushes, and commonly referred to as riparian meadows (*vegas* or *mallines*) (Golluscio, Deregibus and Paruelo, 1998; Roig, 1998). Although riparian meadows account for a very small proportion of the total area of Patagonia, they frequently play a key role in livestock production and, in many instances, are areas that suffer the effects of bad land management (Golluscio, Deregibus and Paruelo, 1998).

Two phytogeographic provinces cover all of arid and semi-arid Patagonia, namely the Patagonian Province, and the Monte Province (Cabrera, 1971). The latter occupies most of the arid west of Argentina; its southern tip enters Patagonia, occupying approximately the northern third of the region. The remaining two thirds of extra-Andean Patagonia correspond to the Patagonian Province (Cabrera, 1971). Paruelo, Jobbágy and Sala (1998) classified vegetation of the region on the basis of functional characteristics using productivity-related indices derived from NOAA satellite imagery analysis (Figure 19.1). One advantage of this approach is that it provides a functional up-to-date classification of vegetation reflecting the current state of rangelands, rather than the potential expected vegetation types that occur in most structural classification maps. Although here the authors mostly follow the grouping of biozones proposed by Paruelo, Jobbágy and Sala (1998), an

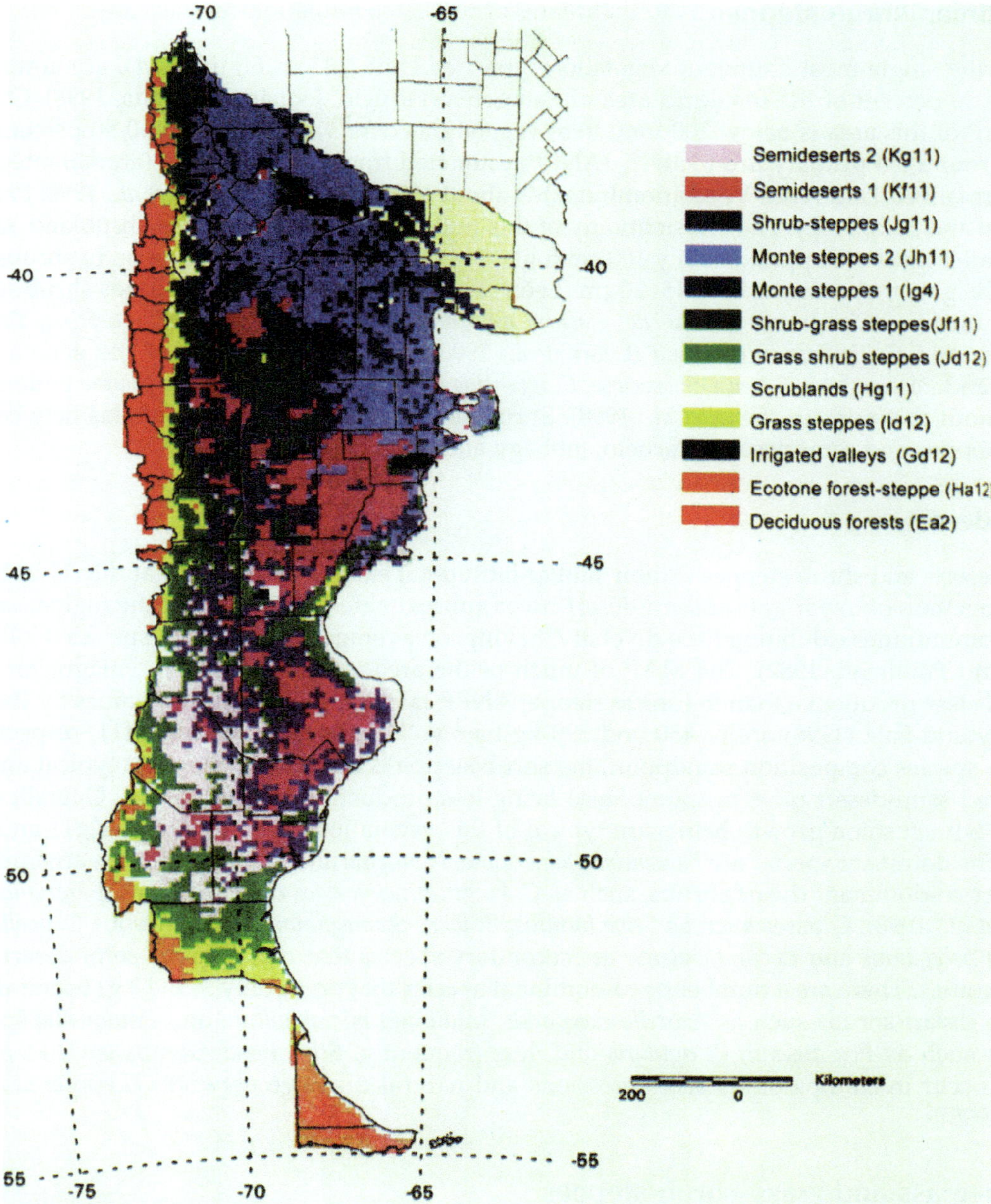

Figure 19.1 Biozones of Patagonia
Source: From Paruelo, Jobággy and Sala (1998), reproduced by permission of authors and the Editors of Ecología Austral.

attempt has been made to produce a synthesis of both structural and functional aspects of the vegetation of each biozone.

Patagonian shrub steppes

This is the single most abundant vegetation type (see Unit Jg11 in Figure 19.1), accounting for close to 20 percent of the semi-arid area of Patagonia (Paruelo, Jobbágy and Sala, 1998). Overall, the MAP of this area is below 200 mm; total vegetation cover varies between 30-50 percent; and aboveground net primary productivity (ANPP) estimated from NDVI-I values (annual integral of the Normalized Difference Vegetation Index) is about 490 kg ha^{-1} yr^{-1} (León *et al.*, 1998; Paruelo, Jobbágy and Sala, 1998). The physiognomy of this unit is that of a two-layered shrubland: shrubs in the taller layer are approximately 100 cm high, while the lower layer is made up of shrubs with crown heights that rarely exceed 15–20 cm (León *et al.*, 1998). The most conspicuous shrub species of this unit are *Chuquiraga avellaneda, Lycium ameghinoi, L. chilense, Verbena ligustrina, Prosopis denudans* and *Colliguaya integerrima* (León *et al.*, 1998; Roig, 1998). Grasses of the genera *Stipa, Festuca* and *Poa*, such as, *S. neaei, S. speciosa, F. argentina* and *P. ligularis*, occur as a sparse understorey throughout this domain (León *et al.*, 1998). Shrub steppes occur in transition areas between the grass steppes and semi-deserts (Paruelo, Jobbágy and Sala, 1998).

Semi-deserts

Semi-deserts and shrub steppes exhibit similar latitudinal extents (Units Kg11 and Kf11 in Figure 19.1); together, both variants of semi-desert cover approximately 22 percent of the region, and are plant communities exhibiting little diversity, having, on average, only 19 plant species (Golluscio, León and Perelman, 1982). The MAP of much of the area is below 150 mm. Patagonian semi-desert is less productive than the shrub steppe; ANPP calculated from NDVI-I values by Paruelo, Jobbágy and Sala (1998) are of 450 and 390 kg ha^{-1} yr^{-1} for units Kg11 and Kf11, respectively. From a species composition standpoint, the shrub steppe is considered the most typical and less degraded semi-desert type, in spite of also being less productive (Soriano, 1983). Overall, dwarf shrubs with cushion-growth habits are typical of this vegetation type in both the Kg11 and Kf11 units. The dominant species are *Nassauvia glomerulosa, N. ulicina* and *Chuquiraga aurea*, accompanied by other co-dominant dwarf shrubs, such as *C. kingii, Brachyclados caespitosus* and *Perezia lanigera* (León *et al.*, 1998). Grasses such as *Stipa humilis, S. ibari, S.ameghinoi* and the shrubs *C. avellaneda, Schinus polygamus* and *Lycium chilense* are secondary species that make up the semi-desert plant communities. There are a number of co-dominant species that occur only in the Kg11 semi-desert, namely dwarf-shrubs such as *Azorella caespitosa, Mullinum microphyllum* and *Franquenia* sp., and grasses such as *Poa dusenii, P. ligularis* and, less frequently, *Stipa neaei*. Shrubs such as *Junellia tridens* occur in clumps on palaeodepressions and natural drainage networks (León *et al.*, 1998; Roig, 1998).

Shrub–grass and grass–shrub steppes

These units (units Jf11 and Jd12 in Figure 19.1) span almost the entire latitudinal extent of Patagonia, and jointly cover approximately 20 percent of the arid area. According to the analysis by Paruelo, Jobbágy and Sala (1998), both vegetation types exhibit similar ANPP levels (650 kg ha^{-1} yr^{-1}) although NDVI values peak somewhat later in the growing season in the grass–shrub steppes. Total vegetation cover of the shrub–grass steppes is approximately 47 percent, with plant communities that, on average, have about 26 plant species (Golluscio, León and Perelman, 1982). *Adesmia campestris, Mullinum spinosum, Senecio filaginoides, Berberis heterophylla, Colleguaya integerrima,*

Trevoa patagónica and *Schinus polygamus* are the most conspicuous shrub species of this biozone (León *et al.*, 1998). The most important grasses are *Stipa speciosa*, *S. humilis*, *Poa ligularis*, *P. lanuginosa*, *Festuca argentina* and *F. pallescens*, which occur together with *Carex* sedges (León *et al.*, 1998). The grass–shrub steppes are a transition between the shrub steppes and the grass steppes on the one hand, and the grass steppes and the *Nothofagus* sub-Antarctic forest ecotone on the other (Figure 19.1) (Roig, 1998). Paruelo, Jobbágy and Sala (1998) collapse the latter (*Chiliotrichum diffusum* grass–shrub steppes) and the steppe-forest ecotones of southern Santa Cruz and northern Tierra del Fuego into a single biozone (unit Ha12). Even though physiognomically these are clearly different domains, they may exhibit similar functional attributes and, consequently, should be significantly more productive than the grass–shrub steppes that occur at the opposite end of the moisture gradient mentioned above. Dominant species of the grass–shrub steppes are the same as those mentioned above, except for the shrubs *Junellia tridens*, *Nardophyllum obtusifolium*, *Berberis buxifolia* and *Chillotrichum diffusum*, and the grasses *Stipa ibari*, *Poa dusenii*, *Festuca pallescens*, *F. gracillima*, *F. pyrogea*, among others (León *et al.*, 1998; Roig, 1998).

Grass steppes

This biozone (unit Id12 in Figure 19.1) also spans the entire latitudinal extent of mainland Patagonia as a narrow belt that runs along the Andean foothills and widens in the south, reaching the Atlantic Ocean. This gives way to the Magellanic steppes that occupy all of the Argentine and Chilean area surrounding the Straits of Magellan, both on the mainland and on the northern tip of the Island of Tierra del Fuego (Figure 19.1) (Paruelo, Jobbágy and Sala, 1998; Cibils and Coughenour, 2001). Overall, this vegetation unit occurs in areas where the MAP is greater than 250 mm. Average ANPP of this vegetation unit is about 900 kg ha^{-1} yr^{-1} according NDVI-I-derived estimates reported by Paruelo, Jobbágy and Sala (1998). Plant species richness is about 34 and 40 species in the sub-Andean grass steppes and the Magellanic steppes, respectively (Golluscio, León and Perelman, 1982; Boelcke, Moore and Roig, 1985). Vegetation cover is about 65 percent on the sub-Andean grass steppes, where *Festuca pallescens* accounts for up to 70 percent of total plant cover, and occurs along with *F. magellanica*, *F. pyrogea*, *Deschampsia elegantula*, *D. flexuosa*, *Phleum communatum*, *Elymus patagonicus* and *Rytidosperma virescens* (León *et al.*, 1998).

Monte shrublands and Monte ecotone

Close to 33 percent of semi-arid Patagonia is covered by the Monte vegetation units (units Jh11, Ig4 and Hg11 in Figure 19.1), with ANPP levels ranging from 650 to 730 kg ha^{-1} yr^{-1} (Paruelo, Jobbágy and Sala, 1998). The Hg11 scrubland unit is the most productive and is considered an ecotone between the Monte and Espinal phytogeographic provinces (Cabrera, 1971; Paruelo, Jobbágy and Sala, 1998). Rainfall across this area rarely exceeds 200 mm and, due to the influence of moist air masses coming inland from the Atlantic Ocean, precipitation tends to be evenly distributed throughout the year (León *et al.*, 1998; Paruelo, Jobbágy and Sala, 1998). The most conspicuous species of these biozones is the shrub *Larrea divaricata*, which occurs in association with other shrub species of this genus, such as *L. cuneifolia*, and *L. nitida* and with other shrubs such as *Prosopis alpataco* and *P. flexuosa* and species of *Lycium*, *Chuquiraga*, *Ephedra* and *Atriplex*. Grasses such as *Stipa tenuis*, *S. speciosa*, *S. neaei*, *Poa ligularis* and *P. lanuginosa* make up the herbaceous stratum of these shrublands (León *et al.*, 1998).

The description of units corresponding to sub-Antarctic forests and forest ecotones are beyond the scope of this chapter. For detailed descriptions of these ecosystems, see Veblen, Hill and Read (1996) and references therein.

BRIEF DESCRIPTION OF PASTORAL AND AGRICULTURAL SYSTEMS

Sheep farming is almost a monoculture in the arid and semi-arid steppes of Patagonia. Although intensive agricultural activities, such as fruit and horticultural crop production, are important in a few irrigated valleys, they are almost absent from sheep farms (Borrelli *et al*., 1997). Sheep farming systems are extensive, with each farm having, on average, three to four 5 000 hectare fenced paddocks. No supplemental feeding is used, as on-farm hay or silage production is negligible and bought-in feeds are too expensive in relation to product prices. Animals graze freely over large areas and are not housed at any time, in spite of the severe winter storms, that in some areas may be frequent. Due to this, there are significant mortalities during big snowfall events. Sheep are gathered three or four times per year. Water provision comes from natural springs and lagoons, rivers, artificial ponds and windmill-pumped groundwater.

Patagonia is free of bovine spongiform encephalopathy (BSE) disease (as is the rest of Argentina), and free of foot-and-mouth disease (Robles and Olaechea, 2001). Environmental conditions constrain the development of internal parasites, so the use of anthelmintics is often unnecessary (Iglesias, Tapia and Alegre, 1992); no antibiotics or hormonal treatments are applied. Under these conditions, Patagonian farms supply wool and meat that can naturally reach the highest standards of quality in terms of food safety and lack of contaminants.

Three kinds of farms can be recognized in Patagonia in relation to farm size (Casas, 1999), namely: *large commercial farms*, with flocks of more than 6 000 head. These are usually derived from the first farm settlements, and are located predominantly on the best-suited rangelands; *small- and medium-sized commercial farms* are located in the drier areas of the region, with flock sizes ranging from 1 000 to 6 000 head. This category has serious financial problems under the current wool price scenarios; and *subsistence farms*, with less than 1 000 sheep, located mainly in northwestern Patagonia. These farms belong mostly to aboriginal family groups that graze their flocks on unfenced public lands.

More than half the sheep farmers in Patagonia are extremely poor and own less than 10 percent of the total sheep flock (Casas, 1999). The other entire half is in effect limited by farm size, so under the current conditions, net farm income does not satisfy their economic expectations. There are about 400 farms that run sheep monoculture operations that could be considered economically viable, and own approximately 40 percent of all sheep in Patagonia. There is a marked gradation from north to south, with subsistence farms most common in the northwest (Neuquén), ranging to Tierra del Fuego, where large, company-owned farms are more frequent than anywhere else in Patagonia. The combination of biozone and farm size (which correlates with the kind of farm operation) results in a wide range of pastoral systems, which differ in terms of objectives, productivity and sustainability.

Farmers make grazing management decisions based on a combination of subjective criteria and previous experience (Golluscio *et al*., 1999). Paddocks are grazed continuously all year round, with the exception of the high-elevation ranges, which are seasonally grazed in summer (Borrelli and Oliva, 1999). Vast grazing areas, divided into only a few large paddocks, restrict the potential for controlling grazing impact, especially when paddocks include grazing sites with contrasting forage availability (e.g. both meadows and arid steppes) (Golluscio *et al*., 1999).

The vegetation of Patagonia is generally considered to have only limited capacity to cope with grazing by domestic herbivores, as the entire region is thought to have evolved under conditions of light grazing pressure from native ungulate herbivores (Milchunas, Sala and Lauenroth, 1988). Although Lauenroth (1998) has recently challenged this notion, there is general consensus that vegetation throughout most of Patagonia has been modified significantly by sheep grazing over the last century, and particularly in the last 40–50 years (Golluscio, Deregibus and Paruelo, 1998; Paruelo *et al.*, 1993).

Deterioration of vegetation under grazing in Patagonia is demonstrated by the widespread replacement of palatable grass species with unpalatable woody life forms (Bertiller, 1993). The degree of plant life form displacement varies among biozones depending on abiotic constraints within each ecosystem (Sala, Lauenroth and Golluscio, 1997; Perelman, León and Bussacca, 1997).

HISTORICAL DEVELOPMENT

Three historical stages can be recognized in Patagonian sheep farming from the beginning of European colonization to the present (Figure 19.2). The first stage involved land occupation, population growth, capital accumulation and increase of sheep stocks. This period lasted about three decades in each province, and five decades at the regional scale, and could be called "the golden years". This stage was followed by a stable period of almost five decades, characterized by

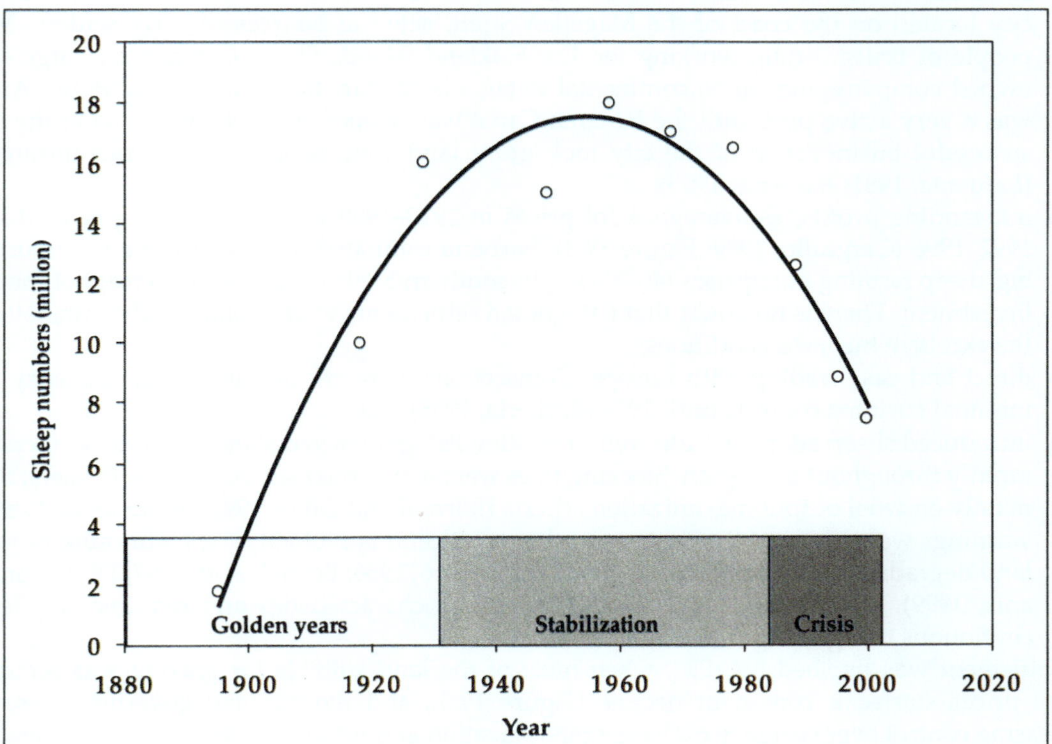

Figure 19.2 Stocks and development stages of sheep farming in Patagonia.
Source: From SAGPyA, 2001a.

full land occupation, a progressive loss of profit, stabilization of sheep stocks, and the beginning of a rural depopulation process. A long decline started during the last two decades of the twentieth century. Sheep flocks were reduced to half (SAGPyA, 2001a), decapitalization was evident, and about 500 farms were abandoned (Borrelli *et al*, 1997). Rural depopulation from pastoral systems became important in all Patagonian provinces (Mendez Casariego, 2000).

THE GOLDEN YEARS: 1880–1930

Barbería (1995) described this stage well. Many factors promoted the settlement of sheep farmers from 1880, including:

- increased demand for wool fibre, sheepskins and frozen meat generated by the industrial revolution in Europe (Lafuente, 1981);
- liberalization and internationalization policies in Argentina started under the leadership of "the 1880's generation". As a result of these policies, the country joined global commerce as a commodity supplier, and development boomed;
- Argentina was in dispute with Chile concerning sovereignty over Patagonian territories. Chile offered public land to settlers in Tierra del Fuego (Lafuente, 1981). The government in Buenos Aires responded in a like manner, and colonization of southern Patagonia started with attractive offers of land to settlers;
- some previously colonized areas, such as the Falkland Islands or Punta Arenas, a Chilean city located on the coast of the Magellan Strait, acted as sources of early settlers. Many people of British origin working for The Falkland Islands Sheep Company, a large state-owned company, moved to continental Patagonia to start their own farms. Punta Arenas was a very active port until the Panama Canal was opened in 1914, and some of the most successful businessmen of the city took grant land concessions on the Argentinean side (Lafuente, 1981; Barbería, 1995);
- outstanding profits, as average wool prices in 1910–1930 were five times greater than in 1992–1996 (Gargiullo, 1999; Figure 19.3). Barbería estimated that in 1910 the net return for big sheep farming enterprises (40 000 ha) in southern Santa Cruz was 50 percent of the total investment. There is no doubt that Patagonian settlements were economically stimulated by the excellent business conditions;
- direct and easy trading with Europe. Transactions were effected in foreign currency with minimal customs controls until 1916 (Barbería, 1995); and
- introduced sheep adapted quite well to pristine Patagonian grasslands, and flocks increased rapidly throughout the region. Stocking rates were established subjectively by farmers, based mainly on wool output maximization criteria (Borrelli and Oliva, 1999). From an early stage, warnings were given of the dangers of this practice: it is probably the main cause of grassland degradation in Patagonia (Willis, 1915; Soriano, 1956; Borrelli *et al*, 1997; DHV Consultants, 1999). Lack of knowledge about rangeland characteristics and requirements led to continuous overgrazing.

Settlement was finished by 1930, when most of the land suitable for grazing was occupied. Wool prices started a consistent decline (Figure 19.3), and the national government exerted increasing control over currency exchange rates, taxation and internal costs. As a result, the sheep farming industry reached an economic plateau, and remained almost unchanged for the next five decades (Lafuente, 1981; Barbería, 1995).

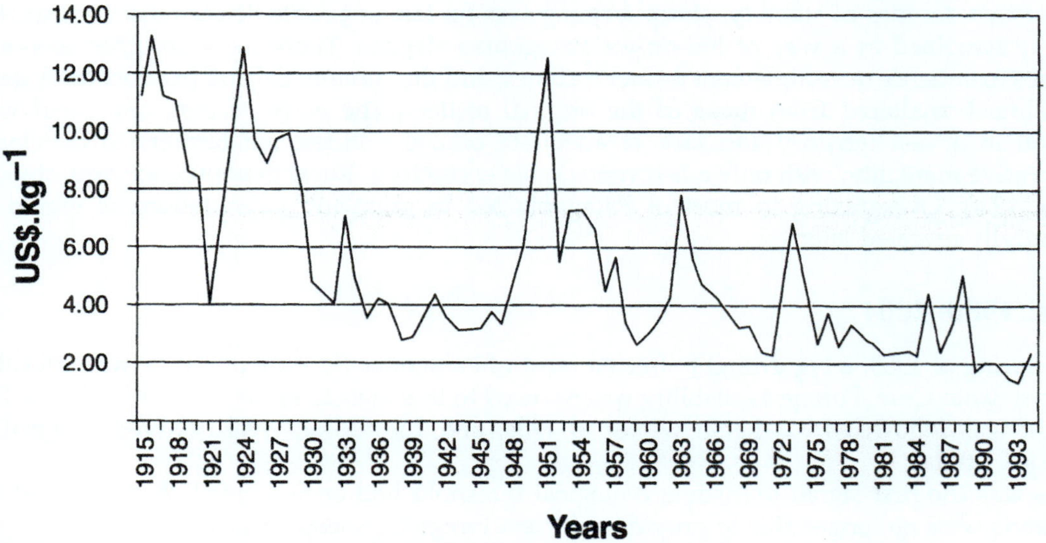

Figure 19.3 Evolution of wool prices (adjusted to equivalent US$ of 1996).
Source: Reproduced by permission from Huerta, 1999.

Stabilization: 1930–1980

Traditional sheep farming was a well-established way of life by 1930, and second-generation farmers perceived it in that way. They learnt from the first settlers, with their profitable sheep farming operations, and tried to continue in the same way. However, important changes occurred within production systems and in external constraints.

Sheep stocks and production remained quite stable, but profitability fell progressively. Small-scale farmers could no longer achieve profit. Progressively greater flocks were needed to obtain a profit. Big families meant that second-generation farmers included many inheritors. High grazing pressure was maintained over all the grasslands to maintain income. Farm size or scale became one of the most important factor in attaining economic sustainability in the region.

Little technological change occurred during this period. The introduction of the Corriedale breed from New Zealand and the use of artificial insemination could be considered the most significant. In 1958, the national government created the National Institute of Agricultural Technology (INTA), a research and extension organization that soon established three research stations in the region. During the 1970s, a joint INTA/FAO project introduced sheep farming technology from Australia and New Zealand. The most important advances were flock improvement techniques based on quantitative genetics, and range improvement by seeding with exotic forage species (Molina Sanchez, 1968). The first natural resource inventories were carried out (Monteith, 1972; Boelcke, Moore and Roig, 1985).

Range management techniques were almost non-existent, and stocking rates were nearly double the estimated carrying capacity of the rangelands. Desertification resulted from uncontrolled grazing at continuous fixed stocking rates (Oliva, Rial and Borelli, 1995).

There was no specific development policy in place, and desertification was an ignored problem that was absent from discussions, political speeches and the mass media. There was little social awareness of problems, even among landowners.

After five decades of stability, sheep farming was no longer profitable for most farmers, but certainly remained as a way of life on the Patagonian steppes. There were no other economic activities to enhance or complement farmer incomes, and the commercial and production systems were almost unaltered from those of the original settlers. The early success, combined with isolation in a vast territory and lack of adequate official policies, led farmers to develop a conservative mentality, with only a few remarkable exceptions. Rural depopulation was obvious (Figure 19.4). Overgrazing in most of Patagonia led to grassland degradation, as would be dramatically revealed later.

Crisis: 1980–2001

In the spring of 1983, a big drought affected most of Patagonia. Scarcity of forage was critical in northern Santa Cruz. Forage availability was reduced to less than 20 kg ha^{-1} in many places. This drought was followed by a harsh winter in 1984, and 1.5 million weak, starved sheep died (SAGPyA, 2001a).

This was the first signal that some ecological threshold had been crossed. A big part of the grasslands were no longer able to provide sufficient forage for sheep survival.

In the 1980s, INTA developed range evaluation methods for Patagonian grasslands. Grazing trials were conducted to study the effect of grazing intensity on soil, vegetation and animal variables (Borrelli, 1999; Oliva *et al.*, 1998; Sifreddi *et al.*, 1995). A big effort was made in inventorying soils, vegetation and the native fauna at a regional scale. Range deterioration was reported for most biozones of Patagonia (Paruelo *et al.*, 1993). By the end of the decade, INTA and the University of Buenos Aires had released range evaluation procedures and recommendations for grazing management (Borrelli, 1990; Ayesa and Becker, 1991; Elissalde, Escobar and Nakamatsu, 1991; Paruelo, Golluscio and Deregibus, 1992; Cibils, 1993). Few farmers were willing to follow such recommendations before 1990. Other aspects of sheep production, such as shearing date, breed comparisons, animal health and reproductive efficiency were studied by INTA. Scientists realized that desertification was a serious problem, but politicians and farmers did not (Table 19.1). Consequently, no sustainable development policy was implemented in this period.

The eruption of Hudson Volcano in Chile covered 7 million hectares of north Santa Cruz with ash in August 1991. One million sheep died or had to be slaughtered. Most of the affected farms (about 350) were abandoned. Many small farms that were not affected by the ash were also abandoned, due to low productivity and lack of profit.

Two big snow events occurred in successive winters (1994 and 1995). Another million sheep died in southern Santa Cruz and Tierra del Fuego, due to snow drifting, extreme cold and starvation.

In 1989, INTA started *Sustainable Development of Patagonia* (PRODESAR), a joint project with the German International Cooperation Office (GTZ), that had a significant effect during the 1990s. There were many achievements that derived from this international cooperation (del Valle *et al*,, 1995), including:

- increased social awareness concerning desertification. It was established as a societal issue at local, regional and national scales. The mass media, schools, universities, political forums and farmers began to recognize desertification as the biggest environmental problem for Patagonia;
- about 250 farmers adopted the recommended range management practices, and range management became a practical discipline (Golluscio *et al.*, 1999).

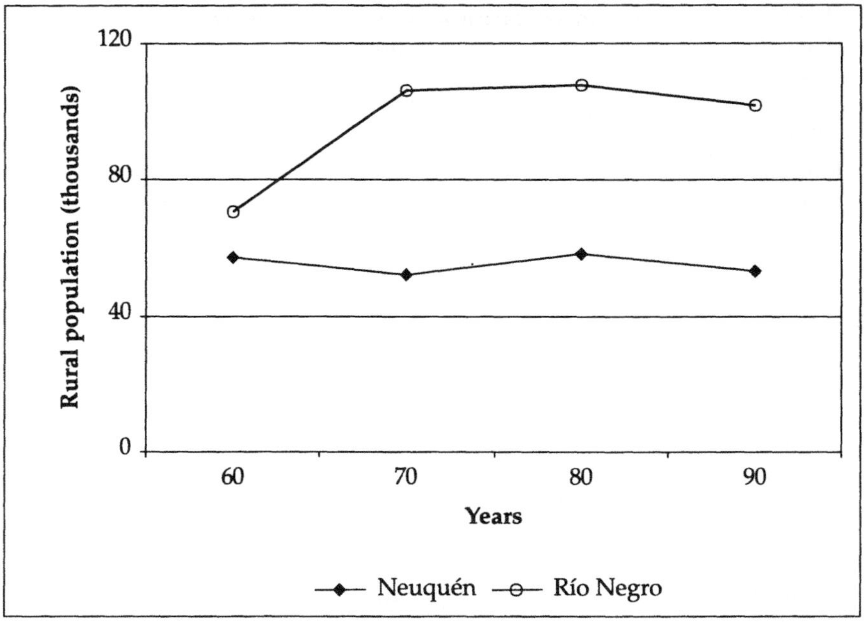

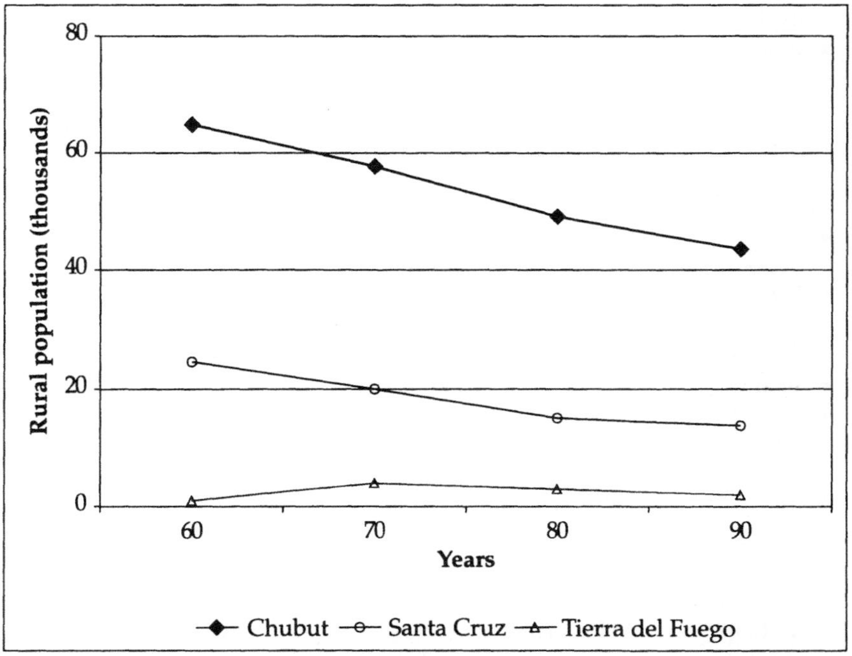

Figure 19.4 Rural population in Patagonia. Above: Río Negro and Neuquén. Below: Chubut, Santa Cruz and Tierra del Fuego

Note: Rural population numbers include villages of less than 2000 inhabitants. The presence of intensive agriculture in irrigated valleys in Río Negro and Neuquén skews the available data.

Source: From Mendez Casariego, 2000.

Table 19.1 Main activities and results associated with the sustainable development strategy for Patagonian grasslands in recent past, and future plans

Period	Research and development	Technology adoption and education	Social awareness	Sustainable development policies
1980–1988	Range evaluation methods developed • Grazing trials • Land resources inventory	7 farms (0.05%) started range evaluations and grazing management	Desertification not an issue	None implemented
1989–2001	Satellite imagery processing and GIS Technology • First Decision-Support Systems • Adaptive management and Extensive Management Technology (EMT) • Diversification and intensification alternatives were explored. • Integration with International Scientific community.	Courses, workshops and field days • Publications • 250 farms evaluated and grazing planned (2%) • 10% of planned farms start adaptive management and provide ongoing information.	Desertification recognized as a societal problem • Mass Media • Schools and universities • Politics and farmers	Sheep Reconversion Credits • Argentina signed the International Convention against Desertification • National Action Program against desertification (NAP) formulated but not implemented.
2002–	Resources monitoring at regional and farm level • Better Decision-Support Systems • Better knowledge of diversification alternatives • Biodiversity • Sustainability of pastoral systems	Systemic and permanent education for capacity building, at personal and organization levels • Better communication system • 50% of the farmers working to quality assurance protocols, including sound grazing management practices.	Increased consumer awareness about environmental quality of products. • Promotion of organic and other quality assured sheep products.	Sheep Law Patagonian Sheep Sustainable Development Programme (PSSDP) • Selective help to farmers that adopt sustainable grazing practices • Quality and environmental care linked • NAP implemented with external funding (GEF).

- individual techniques and recommendations concerning extensive sheep management were integrated into a package called Extensive Management Technology (EMT), based on an adaptive management approach (Borrelli and Oliva, 1999), and 25 farmers, mostly in southern Patagonia, applied EMT during the 1990s, with yearly monitoring of rainfall, vegetation, sheep production and farm economics. Information provided by these farms helped to calibrate grazing recommendations derived from grazing trials;
- many short courses, field days, workshops and publications provided farmers and others with new information, and publicized EMT, desertification and sustainable use;

- facilities for processing satellite imagery and application of Geographical Information System (GIS) technology increased the local capacity for natural resources assessment and management. These facilities were used both for research purposes and as a tool for grazing management at farm level (Oliva *et al.*, 1995; Oliva, Rial and Borelli, 1995; Rial, Vázquez and González, 1999);
- training of scientific staff was continuous, and, during the 1990s, PRODESAR promoted capacity building of the small staff of scientists working in Patagonia. In 1994, an International Workshop held at Río Gallegos (Montes and Oliva, 1995) brought together well known ecologists and range scientists to discuss the sustainability problems of Patagonia. This increased communication led to cooperative activities with centres of excellence in Israel, South Africa and the United States of America; and
- diversification options were explored. The sheep monoculture paradigm was challenged, and new opportunities for Patagonian farmers were proposed and studied. The most promising alternative activities were intensive crops production for off-season niche markets (Quargnolo, 1996), agritourism, and breeding of native fauna (guanacos and rheas).

Scientific progress in the 1990 was significant, despite the ever increasing budget restrictions at national government level. Policies for sustainable development at national and provincial level did not advance at the same rate (Table 19.1). Adjustment policies were non-selective, which implied that many projects that were necessary for sustainable development were subject to the same budget cuts as unproductive, non-specific public expenditure.

Nevertheless, some development policies were launched in the 1990s, as a response to the severe crisis among Patagonian farmers. In 1992, a credit line called "Sheep Reconversion" was offered by the national government to farmers through the National Bank. For the first time, farmers were required to prepare a grazing plan as a requirement for access to loans. This was the first attempt to link farmer support with sustainable management practices. Although US$ 27 million were destined for medium- and large-scale farmers in Patagonia, the *Sheep Reconversion* project failed to produce the expected results, for several reasons:

- the grazing plan requirement became merely a bureaucratic requirement, and consistency and application were not monitored;
- most of the projects merely involved an increase in sheep flock numbers. There was no conversion at all, as some farms bought more sheep at high interest rates but managed them in the traditional way. There was no change in the marketing system, in product quality, in diversification, or in the extent of EMT application; and
- projects were based on deterministic models of sheep production, which did not match real high-risk operations. Variability in sheep production and market prices were ignored.

Sheep Reconversion loans led to seriously increased debt levels among medium-scale farmers, who had (and still have) extreme difficulties in repaying. It is now recognized that those sheep farmers were unable to recapitalize by conventional loan approaches based on high interest rates. It was also clear that no single tool would lead to sustainability – a systemic approach was required. When a process is constrained by many simultaneous problems, a broad-based multifaceted programme is required to cope with them all at the same time.

The National Secretary of Agriculture, Fisheries and Food (SAGPyA) implemented two projects related to Patagonian products. A "Programme for Wool Quality Improvement" (PROLANA) was launched in 1994, with the aim of improving the shearing, handling, grading and packing processes in order to prevent fibre contamination and improve wool appearance. The Programme introduced wool quality assurance procedures, and provided training to shearers, wool classifiers and supervisors. This Programme was considered successful, and 37 percent of all Patagonian wool was marketed with the PROLANA quality label in 2000 (SAGPyA, 2001a).

Carne Ovina Patagónica [Patagonian Lamb Meat] was a programme that introduced a label of geographical origin. It was an attempt to differentiate Patagonian lamb meat, and to gain access to speciality niche markets. The programme was established quite well in the local market, but was interrupted after the big snow storms of 1994 and 1995. Some of its weaknesses became evident: lack of supply strategy, absence of quality control procedures, and little participation by farmers in the marketing process.

On the environmental side, Argentina signed the United Nations Convention Against Desertification, and subsequently, in 1998, formulated a National Action Plan (NAP), although the plan has not been implemented, due to lack of funding.

In the 1990s, Argentina controlled inflation rates, and began a non-selective, unconditional opening of its economy. Most state-owned companies were privatized. The economy concentrated its resources, and most local industry disappeared. Privatized companies had fewer employees. As a result, unemployment rose countrywide, and dramatically so in Patagonian cities, rising to 30 percent of the active population in some places (Salvia, 1999). This was a new problem for Patagonia, a region that had the reputation of providing opportunities for newcomers. Simultaneously, rural depopulation in Patagonia increased (Figure 19.4).

Rural depopulation is almost a worldwide process. In most cases it has been driven either by farm mechanization or by fast industrialization processes in the cities that offered better opportunities for rural people and promoted migration to urban zones. This was not the case for Patagonia, where it was driven mostly by the crisis in the sheep industry and by the increased imbalance availability of services and social infrastructure between rural and urban environments. Most of those displaced from pastoral systems moved to the cities as unemployed, living in marginal conditions, and looking for government aid. They were not pulled by other economic activity.

Rural depopulation of pastoral systems was most evident in the southern provinces of Patagonia, namely Chubut, Santa Cruz and Tierra del Fuego (Figure 19.4), but less evident in Neuquén and Río Negro. There were several contributing causes for this.

- Subsistence farmers predominated in Neuquén and Río Negro. They tended to stay on their farms, due to their lower expectations and less probability of obtaining city jobs.
- Small- and medium-scale farmers of the southern provinces were more educated and had greater profit expectations, so many of them moved to the cities and became part-time farmers or abandoned the farms when they realized that there would be no profit.
- Hired labour had featured strongly on the medium- and large-sized farms of the southern provinces. The drop in wool prices and the relative increase in salary cost led to massive job losses in the 1990s.
- Available statistics do not allow disaggregation of the rural population in pastoral systems from those in irrigated agriculture production. Neuquén and Río Negro both have considerable irrigated agriculture production, and this may be masking rural depopulation from pastoral systems.

The critical situation was finally recognized by the national government, which tried to maintain the farmers and their personnel in the rural environment. Farmers received payments in relation to their labour use, as an emergency policy, with no requirement for sustainable land use, technology adoption or marketing system. Gargiullo (1999) estimated that about US$ 50 million were distributed as farmer payments in the 1990s. This was used as an analgesic, but the disease was not treated. Progressive farmers received a negative signal: "no matter what you do, you receive the same as your neighbour".

If crises are opportunities for change, these policies tended to attenuate the signal and maintain the status quo. As a natural result, sheep stocks declined continuously, more farms were abandoned, and rural depopulation accelerated.

The causes of the crisis were described and analysed by Regional Decision Support Systems promoted by the PRODESAR Project (Borrelli *et al.*, 1997; Mendez Casariego, 2000) and by DHV Consultants (1999). A participatory, interdisciplinary approach was used to integrate existing information and to propose sustainable development strategies. These studies supported the conclusions and recommendations of Noy-Meir (1995), and sustainability of pastoral systems was questioned. The crisis was described as the result of a complex combination of political, economic, ecological and social problems.

SUSTAINABLE DEVELOPMENT OF PATAGONIAN GRASSLANDS

The role of pastoral systems in the future

Sheep farming is often considered as having a negative environmental impact. This way of thinking confuses the impact of continuous overgrazing that occurred in some traditional farms with the intrinsic effect of sheep grazing. There is no evidence that carefully managed sheep flocks produce significant damage to the environment (Oliva *et al.*, 1998). Therefore, the continued operation of pastoral systems may be compatible with rangeland conservation and other alternative uses if proper grazing is imposed in time (Golluscio *et al.*, 1999).

At the same time, there are many roles that pastoral systems of Patagonia could play in the next decades, including those below.

Supply quality products for selected market niches

Fine wool and natural lamb meat are valuable products that consumers appreciate. There are interesting niche markets that are accessible to the industry, both within the country and overseas. (DHV Consultants, 1999).

Provide goods for processing industries

The drop in stock numbers has seriously affected the local wool and meat industry. For many small companies, one strategy to overcome decreased supply of raw material will be to increase quality and value-added of products. To this end, increasing integration can be expected between primary producers and processors. Industry demands will be more specific, and farmers will specialize to supply them. Vertical integration would also allow farmers to process and market their own goods.

Provide a baseline for new economic activities

There are new activities that have potential for diversifying pastoral enterprises, including agritourism, forestation, some niche-oriented intensive agriculture, and breeding of some native fauna with promising scope, such as guanacos and rheas. Sheep farming can provide financial support and operating structure for new businesses.

Reactivating the local economy

Sheep farming has a greater impact on the local economy than might be inferred from gross production statistics. Farmers are distributed over all the region, and most of them operate locally. Therefore, sheep farming plays a significant role in driving commerce and services, especially in small villages and country towns. Most of these places are outside the influence of the oil and mining industry, and have a direct relationship with sheep farming.

Job generation

Noy-Meir (1995) concluded, "Further major economic and demographic development of Patagonia cannot be expected from the primary pastoral industry". Nevertheless, when unemployment rates exceed 30 percent in some cities of Patagonia, the role of retaining and generating genuine jobs in rural environments is not insignificant. The contribution of sustainable, diversified pastoral systems to the solution cannot be ignored, especially when alternative activities that could drive future economic and demographic progress are poorly developed.

Land and culture stewardship

There is increasing awareness of a stewardship role for sheep farmers in the future. They can be seen as land stewards, taking care of all ecosystem resources. A wider approach to land resources would lead to multiple activities in the same geographical space, which in turn is expected to increase landcare practices and protection of endangered flora and fauna.

Patagonia has had only a short history since European colonization. But there is a Patagonian heritage, and certainly sheep farmers represent it. From the roots of their European background, they have had five generations of learning the skills and lifestyle that most suits the Patagonian environment.

THE NEED FOR A LONG-TERM POLICY OF SUSTAINABLE DEVELOPMENT

Short-term palliatives have proven to be ineffective in promoting the structural changes that are required to counteract the long process of decline in the sheep industry (DHV Consultants, 1999).

The complex nature of Patagonian sustainability problems demands sound planning and management strategies, with a participatory approach. A programme is required to integrate diverse strategies – marketing, quality improvement and certification, technology adoption, desertification control, genetic improvement, farmer organization, education and training. A decade is considered the minimum period, due to the slow rate at which sustainability strategies can be implemented and before verifiable results can be expected. As Gargiullo (1999) pointed out, a *new decision culture* is necessary to establish a clear long-term policy, with explicit objectives, serious commitment of public and private organizations, and the introduction of scientific knowledge into the decision-making process.

The "Sheep Law": a lifetime opportunity for Patagonian farmers

Law 25.442, Sheep Farming Recovery, the so-called "Sheep Law", was enacted in May 2001. The National Congress of Argentina made law a regime that aims to improve the competitiveness and sustainability of sheep farming systems nationwide. The main operating instrument is a Fiduciary Fund, supported by the national government, of US$ 6.8 million per year for ten years. This Fund was available from 2003, and has provided financial support to five Sustainable Sheep Development Programmes countrywide.

After a long period with erratic and short-term policies, the Sheep Law provides the opportunity for participatory long-term policies that could solve many of the problems iterated earlier.

From a Patagonian perspective, this is the first time that there is an explicitly stable political horizon, with commitment of government and farmers to eliminate the human-caused factors in the sheep farming crisis.

The Patagonian Sheep Sustainable Development Programme

The Patagonian Sheep Sustainable Development Programme (PSSDP), embracing a participatory approach, was formulated by a team coordinated by the National Secretary of Agriculture, Fisheries and Food, in which the farmers, provincial governments, INTA and other institutions are collaborating in the planning stage of the Programme.

The PSSDP is funded by the Fiduciary Fund created by the Sheep Law. Fifteen workshops were organized across all Patagonia to elicit ideas and suggestions from farmers. The resulting conclusions were summarized in a Base Document (SAGPyA, 2001b). This report reflects many lessons learned from failures and mistakes of past decades, and also the future vision of all parties. It also reflects recommendations from previous reports (Borrelli *et al.*, 1997; DHV Consultants, 1999; Mendez Casariego, 2000) that increased the comprehension of sustainability problems in Patagonia.

The general objectives of PSSDP are to:

- Increase the competitiveness and sustainability of Patagonian sheep farming systems.
- Control and prevent desertification.
- Maintain the rural population in their environment, and to generate new, quality jobs.
- Improve the quality of life in the rural environment.

The approach endorsed for sustainable development in this context has a number of elements, and these are touched on briefly below.

A participatory and decentralized approach

A deep and continuing change process can not be sustained without true commitment of all the parties that must be involved. However, commitment only comes when there is effective participation of farmers, provincial governments and scientific organizations, in both the planning and the execution processes. The vast region in which the Programme is implemented demands decentralized execution, with effective management decisions.

Prioritizing development over palliatives

The Programme focuses on strategies that will change the marketing, processing and production concepts of the pastoral industry. The re-engineering of many processes at farm and industry scale will open an opportunity for innovation and technology adoption. There is commitment over careful orientation of funding to sustainable development strategies.

From the market to the rangeland

The Programme is market oriented. The strategies will be focused on supplying niche markets that have been identified as opportunities for Patagonian farming systems. Profitability and competitiveness should be based on market prices and conditions, without government subsidies. An integrated approach will be used to combine strategies across the value chain: from consumer demands, through the marketing system, the processing sector and the primary producers, to the natural resources involved. Integration will search for synergy and internal consistency among different strategies.

Linking economic aid with proper rangeland management

The Programme provides financial support for development proposals submitted by farmers, mostly in joint Projects, either cooperatives or farmer associations. Funding access is selective and conditioned to the adoption of some basic technologies. There is agreement that aid should be given only to farmers that commit themselves to:

- adopting the recommended grazing practices,
- maintaining their grazings free from scrub infestation, and
- adopting PROLANA practices for wool management.

After more than a century of uncontrolled grazing, often promoted by public policies, this can be considered a milestone for Patagonian rangelands conservation.

Developing alternatives for abandoned farms

About half of the farms in Santa Cruz Province (550 farms, 10 million hectares) have been abandoned in the last decade (Borrelli *et al.*, 1997). Almost half the abandoned farms could re-initiate sheep farming if they combined activities into large sheep operations, or if they included some other economic activities, such as agritourism or intensive agriculture. However, the land of the other half of these farms has been degraded to such an extent that it cannot supply enough forage to cover sheep nutritional requirements for production, or even for maintenance (Borrelli *et al.*, 1997). Although vegetation recovery has not received sufficient scientific study, the authors hypothesize that the combined effect of loss of topsoil and unfavourable climatic conditions implies that range recovery will be very slow, probably a matter of decades.

A land stewardship system is proposed to convert these farms into private protected areas, to allow the farmer to stay on the farm, with a commitment to prevent sheep grazing on degraded rangelands. Protection of the floristic, faunistic, archaeological and historical resources would be their role until other economic activities could be introduced. The concept of sustainable occupation could be applied, increasing awareness and perceptions of environmental resources and cultural heritage.

A human development process

Sustainability, either in economic, ecological or social terms, is the greatest challenge for Patagonian farmers in the future. Far beyond the adoption of new technologies or the funding of the investments that are urgently required, the process is fundamentally one of human development. The key is to build up the capacity of farmers, politicians and scientists to compete successfully in a global market, conserving or improving the land resources involved while satisfying social needs with equity. This requires a maximal effort in education, training and communication, at both individual and organizational levels.

Main strategies and projects

A programme for quality

Forage production in most of Patagonia's grasslands is severely restricted by rainfall. This limitation might be overcome by irrigation, but in only a very a small part of the area. In practical terms, forage availability imposes an overall limit on sheep production.

Lamb production could be increased with the adoption of sound management practices (Borrelli and Oliva, 1999; Mendez Casariego, 2000), but this would be irrelevant in terms of the global market. Therefore the focus must be on product quality and associated marketing strategies. PSSDP can be defined primarily as a Certified Quality Programme. There is increasing demand for high quality lamb meat and wool. Consumers are looking for safe, traceable, pure and lean lamb meat, and they are prepared to pay a premium for products that guarantee those characteristics.

High quality wool: a challenge for Patagonian Merino breeders

Wool prices increase drastically when fibre diameter is below 20 microns (Argentine Wool Federation, 2001), but simultaneously the quality requirements increase, especially in terms of number of coloured fibres, colour and resistance to breaking (Mueller, 2001).

The objectives for the fine-wool programme are:

- to produce finer, high quality branded wool to supply the most exigent textile markets. By doing this. it is expected that wool prices received by the farmers would increase significantly (by 30–40 percent); and
- to increase the local value-added to wool. New links between primary producers and the textile industry will be developed, looking for integrated approaches to production and marketing.

The strategies to attain the objectives include:

- A genetic improvement project that has been developed between the Argentine Merino Breeders Association, INTA and the national government. Great effort will be put into reducing the average fibre diameter of the Merino clip by one micron in ten years, without losing fleece weight or body size. Stud breeders that agree with these objectives and methods will receive financial aid to buy semen from progeny tested top sire rams. Commercial breeders will have financial aid to buy rams that have been produced under this improvement scheme and objectively evaluated.
- Better grazing management and eradication of scrub should lead to better fibre growth.
- Generalization of good manufacture practices proposed by PROLANA will reduce wool contamination and increase wool quality.
- New marketing strategies will be developed for branded wool, which will often be organic. The identification of niche markets and the development of specific supply contracts with textile processors should bring new marketing alternatives to the existing commodity routes.

Patagonian lamb: capturing the value of a pure environment

For more than a century, Patagonia has supplied lamb meat that can be considered as an organic product, free from pesticides, additives or contaminants. Native rangelands are the only source of sheep feed, drinking water is pure, and agrochemicals are not used. These important attributes are lost in the marketing process, and therefore prices do not reflect the competitive advantage of the product.

Patagonian sheep farming can obtain organic or other quality certification system accreditation with lower opportunity costs compared with other intensive sheep production systems.

The objectives of the organic lamb project are to:

- supply traceable, differentiated organic lamb meat,
- increase lamb prices and farm profit, and
- use market incentives to promote sound grazing management practices.

To achieve these objectives, the strategies foreseen are to:

- create a farmer-owned marketing company to identify and enter access niche markets for Patagonian Organic Lamb. Re-engineering of the marketing process is a first priority;
- update abattoirs and processing plants to meet the highest quality standards;
- promote whole-farm planning and good management practices to increase lamb production and quality;
- help farmers face the incremental costs of transition from conventional to organic farming (technical assistance and initial certification costs); and
- create a range-monitoring network to follow soil and vegetation changes over time and space.

Strengthening small-scale farmers: Association and training

The improvement of wool and meat prices would not be sufficient for small farmers to attain profit or at least to satisfy their expectations. Association of small farmers is needed, not only for marketing purposes, but also as a strategy to lower fixed costs and to increase economic efficiency of primary producers. The concept builds on scaling up farming or commercial operations, by means of cooperatives or other association forms. In the past, the approaches used resulted in many failures, so new approaches are imperative.

The objectives are to:

- promote association of small farmers to increase effect farm unit size, reduce fixed costs and increase their group negotiating capacity; and
- increase small-scale farmers' skills in business management, at both farm and integrated levels.

The chosen implementation strategies are that:

- some programme benefits can only be accessed through associations of farmers. Individual small-scale farmers would not be eligible for PSSDP assistance;
- the Programme will help with financial support for technical assistance and management costs of farmer associations in their first three years; and
- intensive training is provided to management staff and farmers to increase their skills for integrated business in a global market.

A private land stewardship system

As a part of PSSDP, a project is now being formulated with the National Secretary for Sustainable Development and Environmental Policy. A private-reserve network is proposed, focusing on the valorization of the natural resources and the maintenance of sustainable occupation of farms.

The Project would also provide a range-monitoring system, at both regional and farm scales, with funding for environmental restoration in the most critical areas, and support for the development of better Decision-Support Systems for farm and PSSDP managers and policy-makers. The project will be submitted for international funding.

The objectives will be to:

- protect and valorize natural resources on farms that are currently out of sheep farming;
- maintain the rural population in their own environment;
- allow for a transition period until new economic activities can be developed;
- install a range-monitoring system, at both farm and regional levels; and
- develop better Decision-Support Systems.

The strategies towards these objectives would be to:

- develop a private land stewardship system by offering a minimum payment to farmers who accept the responsibility of staying on their farms and protecting the natural resources involved according to standard protocols;
- increase existing knowledge about natural resources and their state of conservation by means of farm-scale inventories;
- promote agritourism based on the natural resources and recreational values of Patagonian rangelands; and
- create a range-monitoring system to integrate weather records, multitemporal satellite imagery processing, paddock-installed benchmark stations, and records and conclusions from farmer experiences from adaptive management.

The need for diversification

It is obvious that sheep monoculture is an undesirable fact. Climatic and market risks are too high for farmers, and profitability is low (Borrelli *et al.*, 1997). Even in the event of successful modernization of the sheep industry – an industry that could attain the objectives described above – it would not be enough to satisfy ever-increasing farmer and society expectations. Diversification is as imperative as sheep farming modernization. Consequently, greater efforts are needed to develop non-traditional economic activities that could have a beneficial impact on farm profitability, risk reduction and job generation.

Activities that seem to have good potential are:

- sheep cheese production (in some areas);
- breeding of native species (guanacos and rheas);
- agritourism; and
- irrigated agriculture for niche markets. Best opportunities seem to be in off-season fruits, vegetables and ornamental crops (Quargnolo, 1996; Gargiullo, 1999).

Most of these activities are unaffordable by individual farmers, due to problems associated with marketing scale, lack of capital and cost of technology access. Sustainable development requires specific programmes whereby most of these limitations can be overcome.

The Sheep Law has restrictions on funding targets. Only the first two activities could be supported by PSSDP. Specific projects will provide proper planning and funding for innovative farmers. There is a need for development programmes for agritourism and intensive agriculture.

CONCLUSIONS

Fundamental structural changes are required to attain sustainable, scientifically based utilization of Patagonian grasslands, with competitive, high quality diversified products and improved quality of rural life. Rural depopulation cannot be stopped or reversed in other ways. After decades without sustainable development policies, crises tend to be more extreme. There is no doubt that without leadership, technical assistance and long-term funding from government, the changes needed are unlikely to occur in time.

The Sheep Law and PSSDP may positively change the scenario and speed up innovation. The proposed long-term, systemic and participatory approach has potential for achieving permanent solutions. Patagonian farmers have a great opportunity and face an extraordinary challenge: to jump from a declining inertia to a sustained improvement of their decisional culture.

ACKNOWLEDGEMENTS

Dr Gabriel Oliva, of INTA Santa Cruz, reviewed the manuscript and provided helpful comments and suggestions. The unconditional commitment to sustainable development of Patagonian grasslands by INTA staff is duly recognized. An anonymous reviewer provided suggestions that greatly improved earlier versions of this manuscript.

REFERENCES

Amaya, J.N., von Thüngen, J. & Delamo, D.A. 2001. Relevamiento y distribución de guanacos en la Patagonia: Informe Final [*Survey and distribution of guanacos in Patagonia: Final Report*]. Comunicación Técnica No.111. Area Recursos Naturales-Fauna 111. INTA- Bariloche

Ares, J.O., Beeskow, A.M., Bertiller, M., Rostagno, M., Irisarri, M., Anchorena, J. & Deffossé, G. 1990. Structural and dynamic characteristics of overgrazed lands of northern Patagonia, Argentina. pp. 149–175, *in:* A. Breymeyer (ed). *Managed Grasslands*. Amsterdam, The Netherlands: Elsevier Science Publishers.

Argentine Wool Federation. 2001. Argentine Wool Statistics. Vol. 512. 22p.

Ayesa, J. & Becker, G. 1991. Evaluación forrajera y ajuste de la carga animal [*Forage evaluation and stocking rate adjustments*]. Proyecto de Prevención y Control de la desertificación en la Patagonia. PRECODEPA. Comunicación Técnica Recursos Naturales (Pastizales) No. 7. 17p.

Barbería, E. 1995. Los dueños de la tierra de la Patagonia Austral: 1880–1920 [*The owners of the land in Southern Patagonia*]. Universidad Federal de la Patagonia Austral, Río Gallegos.

Bertiller, M. 1993. Conclusiones. p. 110, *in:* J.M. Paruelo, M.B. Bertiller, T.M. Schlichter and F.R. Coronato (eds). Secuencias de deterioro en distintos ambientes patagónicos. Su caracterización mediante el modelo de estados y transiciones [*Range deterioration sequences in different Patagonian environments. Its characterization by the state and transition model*]. Ludepa, Bariloche. INTA-GTZ

Boelcke, O., Moore, D.M. & Roig, F.A. (eds). 1985. Transecta botánica de la Patagonia Austral [*Botanical transect of Southern Patagonia*]. Buenos Aires: Conicet (Argentina), Royal Society (Great Britain), and Instituto de la Patagonia (Chile).

Borrelli, G., Oliva, G., Williams, M., González, L., Rial, P. & Montes, L. (eds). 1997. Sistema Regional de Soporte de Decisiones. Santa Cruz y Tierra del Fuego. Prodesar (Proyecto de Prevención y Control de la Desertificación en Patagonia), Río Gallegos. INTA, Santa Cruz

Borrelli, P. 1990. Manejo racional de pastizales: una cuestión de supervivencia. pp. 77–92, *in:* VI Jornadas Cooperativas de Lanas. FECOLAN Ed. Buenos Aires

Borrelli, P. 1999. Informe ampliado del efecto de la intensidad de pastoreo sobre variables del sistema suelo-planta-animal y limitantes del pastizal para la producción ovina (1991–1997) [*Extended report of the effect of grazing intensity on variables of the soil-plant-animal system and grassland limitations for sheep production*]. in: El ensayo de pastoreo de Moy Aike Chico (1986-1999). Una experiencia compartida entre Estancia Moy Aike Chico y la Estación Experimental Agropecuaria Santa Cruz. INTA Río Gallegos. INTA, Santa Cruz

Borrelli, P. & Oliva, G. 1999. Managing grazing: experiences from Patagonia. pp. 441–447 (Vol. 1) *in:* D. Eldride and D. Freudenberger (eds). *People and Rangelands: Building the future.* Proceedings of the VI International Rangeland Congress. Townsville, Queensland. Aitkenvale, 1999.

Cabrera, A.L. 1971. Fitogeografía de la República Argentina [*Argentine Republic Phytogeography*]. *Boletín de la Sociedad Argentina de Botánica,* 14: 1–42.

Casas, G. 1999. Recomendaciones y estrategias para compatibilizar el desarrollo productivo en la Patagonia con la prevención y el control de la desertificación [*Recommendations and strategies to compatibilize productive development and desertification control in Patagonia*]. *In:* DHV Consultants, 1999, q.v.

Cibils, A. 1993. Manejo de pastizales [*Range management*]. *In:* Catálogo de Prácticas. Tecnología disponible. Cambio Rural-INTA EEA Santa Cruz (Eds.)INTA, Santa Cruz.

Cibils, A. & Coughenour, M.B. 2001. Impact of grazing management on the productivity of cold temperate grasslands of Southern Patagonia – a critical assessment. pp. 807–812, *in:* Proceedings of the 19th International Grassland Congress. Sao Paulo, Brazil, 11–21 February 2001.

del Valle, H.F., Eiden, G., Mensching, H. & Goergen, J. 1995. Evaluación del estado actual de la desertificación en áreas representativas de la Patagonia. Informe Final Fase I [*Evaluation of current status of desertification in representative areas of Patagonia. Final Report of Phase I*]. Proyecto de Cooperación Técnica entre la República Argentina y la República Federal Alemana. "Lucha contra la Desertificación en la Patagonia a través de un sistema de Monitoreo Ecológico". 182p. INTA–GTZ.

del Valle, H.F. 1998. Patagonian soils: a regional synthesis. *Ecologia Austral,* **8**: 103–122.

DHV Consultants [Consorcio DHV Consultants-Swedforest]. 1999. Desertificación en la Patagonia. Informe Principal [*Desertification in Patagonia. Main Report*].

Elissalde N., Escobar, J. & Nakamatsu, V. 1991. Metodología expeditiva para la evaluación de pastizales de la zona árida del Chubut [*An expeditious method for grassland evaluation in arid Chubut*]. pp. 217–218, *in:* Actas de la X reunión Nacional para el Estudio de las regiones Aridas y Semi-áridas (Bahía Blanca).

Gargiullo, G. 1999. Diagnóstico Económico [*Economic Diagnosis*]. *In:* DHV Consultants, 1999, q.v.

Golluscio, R.A., León, R.J.C. & Perelman, S.B. 1982. Caracterización fitosociológica de la estepa del Oeste del Chubut. Su relación con el gradiente ambiental [*Phytosociological characterization of West Chubut steppe. Its relationship with the environmental gradient*]. *Boletín de la Sociedad Argentina de Botánica,* **21**: 299–324.

Golluscio, R.A., Deregibus, V.A. & Paruelo, J.M. 1998. Sustainability and range management in the Patagonian steppes. *Ecologia Austral,* **8**: 265–284.

Golluscio, R., Giraudo, C., Borrelli, P., Montes, L., Siffredi, G., Cechi, G., Nakamatsu, V. & Escobar, J. 1999. Utilización de los Recursos Naturales en la Patagonia. Informe Técnico [*Natural Resources Utilization in Patagonia*]. *In:* DHV Consultants, 1999, q.v.

Huerta, G. 1999. Cuadros e información complementaria para el diagnóstico económico [*Tables and complementary information for economic diagnosis*]. *In:* DHV Consultants, 1999, q.v.

Hulme, M. & Sheard, N. 1999. *Climatic Change Scenarios for Argentina.* Climatic Research Unit, Norwich, Great Britan. 6p. See: http://www.cru.uea.ac.uk/~mikeh/research/wwf.argent.pdf (accessed 2/5/2004)

Iglesias, R., Tapia, H. & Alegre, M. 1992. Parasitismo gastrointestinal en ovinos del Departamento Guer Aike. Provincia de Santa Cruz, Argentina [*Gastrointestinal parasitism in sheep of Guer Aike Department, Santa Cruz Province, Argentina*]. pp. 305–322, *In:* III Congreso Mundial de Ovinos y Lanas. Buenos Aires, Argentina.

Jobbágy, G.E., Paruelo, J.M. & León, R.J.C. 1995. Estimación de la precipitación y de su variablilidad interanual a partir de información geográfica en el NW de la Patagonia, Argentina [*Rainfall and interannual variability estimation from geographical information in NW Patagonia*]. *Ecologia Austral,* **5**: 47–53.

Lafuente, H. 1981. La región de los Césares. Apuntes para una historia económica de Santa Cruz [*The Caesar's Region. Notes for a history of the Santa Cruz economy*]. Editorial de Belgrano. 175p.

Lauenroth, W. 1998. Guanacos, spiny shrubs and the evolutionary history of grazing in the Patagonian steppe. *Ecologia Austral,* **8**: 211–215.

León, R.J.C., Bran, D., Collantes, M., Paruelo, J. M. & Soriano, A. 1998. Grandes unidades de vegetación de Patagonia extra andina [*Major vegetation units of extra-Andean Patagonia*]. *Ecologia Austral,* **8**: 125–144.

MacArthur, R. 1972. Climates on a rotating Earth. pp. 1–19, *in:* R. MacArthur (ed). *Geographical Ecology.* New York NY: Harper & Row.

Manero, A. 2001. El zorro colorado en la producción ovina [*Red fox in sheep production*]. pp. 243–252, *in:* P. Borrelli and G. Oliva (eds). Ganadería ovina sustentable en la Patagonia Austral [*Sustainable sheep farming in Southern Patagonia*]. INTA, Reg. Pat. Sur.

Méndez Casariego, H. 2000. Sistema de soporte de decisiones para la producción ganadera sustentable en la Provincia de Río Negro (SSD-Río Negro) [*Decision-Support System for sustainable animal husbandry in Río Negro Province*]. INTA-GTZ. Centro Regional Patagonia Norte. EEA Bariloche. EEA Valle Inferior. Proyecto Prodesar. [CD-ROM].

Milchunas, D.G., Sala, O.E. & Lauenroth, W.K. 1988. A generalized model of the effects of grazing by large herbivores on grassland community structure. *The American Naturalist,* **132**: 87–106.

Molina Sánchez, D. 1968. Pasturas perennes artificiales en la Provincia de Santa Cruz [*Sown perennial pastures in Santa Cruz Province*]. pp. 80–82, *in:* Actas de la III reunión Nacional para el Estudio de las Regiones Aridas y Semi-áridas. Trelew.

Monteith, N. 1972. Estudios sobre la erosión en Patagonia. Informe final [*Erosion studies in Patagonia. Final Report*]. Proyecto INTA-FAO para el desarrollo ovino en la Patagonia. 62p. (unpublished).

Montes, L. & Oliva, G.E. 1995. Patagonia. Actas del Taller Internacional sobre Recursos Fitogenéticos, Desertificación y Uso Sustentable de los Recursos Naturales de la Patagonia [*Patagonia. Proceedings of the International Workshop on Plant Genetic Resources, Desertification and Sustainability*]. Río Gallegos, 7–11 November 1994. INTA. 238p.

Mueller, J. 2001. Mejoramiento genético de las majadas patagónicas [*Genetic improvement of Patagonian sheep*]. pp. 211–224, *in:* P. Borrelli and G. Oliva (eds). Ganadería Ovina Sustentable en la Patagonia Austral. Tecnología de Manejo Extensivo. INTA, Reg. Pat. Sur.

Noy-Meir, I. 1973. Desert Ecosystems: Environment and producers. *Annual Review of Ecology and Systematics,* **4**: 25–51.

Noy-Meir, I. 1995. Working Group on Sustainability and Management of Natural Resources in Patagonia. Conclusions and Recommendations. pp. 204–205, *in:* Montes & Oliva, 1995, q.v.

Oliva, G., Rial, P. & Borrelli, P. 1995. Desertificación y posibilidades de uso sustentable en la Provincia de Santa Cruz [*Desertification and sustainable use possibilities in Santa Cruz Province*]. pp. 48–55, *in:* Montes & Oliva, 1995, q.v.

Oliva, G., Rial, P., González, L. Cibils, A. & Borrelli, P. 1995. Evaluation of carrying capacity using Landsat MSS images in South Patagonia. pp. 408–409, *in:* N. West (ed). *Rangelands in a sustainable biosphere.* Proceedings of the V International Rangeland Congress. . Salt Lake City, Utah. 1995. Denver, CO: Society for Range Management.

Oliva, G., Cibils, A., Borrelli, P. & Humano, G. 1998. Stable states in relation to grazing in Patagonia: a 10-year experimental trial. *Journal of Arid Environments,* **40**: 113–131.

Paruelo, J., Golluscio, R. Deregibus, V.A. 1992. Manejo del pastoreo sobre bases ecológicas en la Patagonia extra andina: una experiencia a escala de establecimiento [*Ecologically sound grazing management in extra-Andean Patagonia: a farm-scale experience*]. *Anales de la Sociedad Rural Argentina,* **126**: 68–80.

Paruelo, J.M., Jobbágy, E.G. & Sala, O.E. 1998. Biozones of Patagonia (Argentina). *Ecologia Austral,* **8**: 145–154.

Paruelo, J.M., Bertiller, M.B., Schlichter, T.M. & Coronato, F.R. (eds). 1993. Secuencias de deterioro en distintos ambientes patagónicos. Su caracterización mediante el modelo de estados y transiciones [*Range deterioration sequences in different Patagonian environments. Its characterization by the state and transition model*]. LUDEPA, Bariloche.

Paruelo, J.M., Beltrán, A., Jobbágy, E.G., Sala, O.E. & Golluscio, R.A. 1998. The climate of Patagonia: general patterns and controls on biotic processes. *Ecologia Austral,* **8**: 85–102.

Perelman, S.B., León, R.J.C. & Bussacca, J.P. 1997. Floristic changes related to grazing intensity in a Patagonian shrub steppe. *Ecography,* **20**: 400–406.

Quargnolo, E. 1996. Programmea para el desarrollo de la agricultura en la provincia de Santa Cruz. [*Programme for agricultural development in Santa Cruz Province*]. EEA, Santa Cruz.

Rial, P., Vázquez, M. & González, L. 1999. Evaluación de la capacidad productiva de la Meseta Central de Santa Cruz mediante Sistemas de Información Geográfica [*Production capacity evaluation in the Central Plateau of Santa Cruz using Geographical Information Systems*]. VII Conferencia Iberoamericana sobre SIG. Mérida, Venezuela.

Robles, C. & Olaechea, F. 2001. Salud y enfermedades de las majadas [*Health and disease in sheep flocks*]. pp. 225–243, *in:* P. Borrelli and G. Oliva (eds). *Ganadería Ovina Sustentable en la Patagonia Austral. Tecnología de Manejo Extensivo.* INTA, Reg. Pat. Sur.

Roig, F.A. 1998. La vegetación de la Patagonia [*Patagonian vegetation*]. pp. 48–166, *in:* M.N. Correa (ed). *Flora Patagónica*: Parte I. INTA, Buenos Aires.

SAGPyA. 2001a. Boletín ovino. Web Page, see www sagyp.mecon.gov.ar. Accessed 3 October 2001.

SAGPyA. 2001b. Programmea para el Desarrollo de la Ganadería Ovina Sustentable en la Patagonia. Documento Base [*Programme for Sheep Sustainable Development in Patagonia. Base Document*]. Foro Lanero Comodoro Rivadavia. February 2001. 19p.

Sala, O.E., Lauenroth, W.K. & Golluscio, R. 1997. Plant functional types in temperate semi-arid regions. pp. 217–233, *in:* T.M. Smith, H.H. Shugart and F.I. Woodward (eds). *Plant functional types: Their relevance to ecosystem properties and global change.* Cambridge, UK: Cambridge University Press.

Salvia, A. 1999. La Patagonia de los noventa. Sectores que ganan, sectores que pierden [*Patagonia in the nineties. Sectors that win, sectors that lose*]. Universidad Nacional de la Patagonia Austral. 256p.

Siffredi, G., Ayesa, J., Becker, G.F., Mueller, J. & Bonvisutto, G. 1995. Efecto de la carga animal sobre la vegetación y la producción ovina en Río Mayo (Chubut) a diez años de pastoreo [*Effect of stocking rate on vegetation and sheep production in Río Mayo (Chubut) after ten years of grazing*]. *In:* R. Somlo and G.F. Becker (eds). Seminario Taller sobre Producción, Nutrición y Utilización de pastizales. Trelew.

Soriano, A. 1956. Aspectos ecológicos y pastoriles de la vegetación patagónica, relacionados con su estado y capacidad de recuperación [*Ecological and pastoral aspects of Patagonian vegetation related with its state and recovery capacity*]. *Rev. Inv. Agr.*, **10**: 349–372.

Soriano, A. 1983. Deserts and Semi-Deserts of Patagonia. pp. 423–460, *in:* N.E. West (ed). *Ecosystems of the World – Temperate Deserts and Semi-Deserts.* Amsterdam, The Netherlands: Elsevier Scientific.

Veblen, T.T., Hill, R.S. & Read, J. (eds). 1996. *The ecology and biogeography of Nothofagus forests.* New Haven, USA: Yale University Press.

Villamil, C.B. 1997. Patagonia. Data Sheet-CDP South America-Site A46- Diversity Conservation Project- NMNH Smithsonian Institution [Web page, accessed 19 October 2001]. See: www.nmnh.si.edu/botany/projects/cpd/sa/sa46.htm

Willis, B. 1915. El Norte de la Patagonia. Naturaleza y Riquezas [*Northern Patagonia: Nature and Richness*]. Tomo I. Ministerio de Obras Públicas de la República Argentina. Buenos Aires. 500p.

CHAPTER 20

Australian Temperate Grasslands: Changing Philosophies and Future Prospects

D.R. Kemp[1] and D.L. Michalk[2]

ABSTRACT

Australia has vast areas of grazing lands covering around two-thirds of a continent larger than the continental United States of America. Since European settlement after 1788 the area of grasslands has increased due to land clearing and management of those grasslands has progressed from an *exploitive* through *reactive* periods. The *exploitive* phase saw the rapid expansion of livestock across the continent, often resulting in overexploitation of natural resources as farmers tried to establish the optimal stocking rate. The *reactive* phase was first characterized by moving on to new areas as existing grasslands were overused, then, when much of the available land had been cleared, rehabilitation was attempted by sowing introduced species. As the terms of trade for livestock production continued to become less positive, farmers then focused more on trying to make better use of the naturalized mixtures that developed under livestock grazing, and only resorting to pasture re-sowing where economic benefits from so doing were clear.

Coming to terms with a unique and drier landscape characterized much of the early period of European settlement. Early grazing research tended to be focused on livestock performance and did not always provide solutions to the longer-term management of grasslands. More recently, research has developed more efficient management practices to rehabilitate or maintain the grassland resource base. Farming practices are changing, as

[1]The University of Sydney, Centre for Rural Sustainability, Leeds Parade, Orange, New South Wales, Australia 2800. Email: David.Kemp@orange.usyd.edu.au

[2]NSW Agriculture, Orange Agricultural Institute, Forest Rd, Orange, New South Wales, Australia 2800.

better rules have been developed to define when and why animals are moved between paddocks. Examples are discussed of where the subtle use of strategic rest and graze periods can rehabilitate grasslands and enhance production.

The future management of Australia's grasslands will, however, require a more *proactive* approach based upon the application of a broad definition of sustainable systems and a more rational consensus as to how producers need to manage grazing ecosystems over the longer term. Increasing social pressures require a more environmentally friendly approach rather than an exploitive, degradation pathway. To achieve such a preferred future will require considerable research into sustainable grazing systems and extension programmes that deliver the necessary tools to farmers. It is argued that solutions are possible that would improve the productivity of grasslands while satisfying the needs for more sustainable, biodiverse environments. Future markets could include payments to farmers for environmental services and maintaining the resource base that will affect how grasslands are managed. Many of the impacts and lessons learnt in Australia are paralleled in numerous other countries.

INTRODUCTION

Grasslands are one of the world's greatest natural assets – providing the bulk of nutrients for ruminant production and revitalizing soils for crop production. However, in many regions, degradation has occurred, grassland productivity has lessened, environmental problems have increased and the communities directly dependent on those grasslands have declined. The need to develop more appropriate management systems is acknowledged as a major issue for pastoral livestock production in Australia (Kemp and Michalk, 1993) as well as in major grassland areas such as northern Asia (Tao, 2001) and elsewhere (Reynolds and Frame, this volume). Unless effective rehabilitation practices are developed to maintain grassland productivity in a sustainable manner, the viability of many communities, industries, the environment and the world's capacity to support an increasing population will be in jeopardy. Two hundred years of European settlement in southern Australia achieved considerable success in livestock production, but at an environmental cost that is now becoming more obvious. As a consequence, adaptation of grassland practices to ever-changing conditions within this old, relatively dry and highly climatically variable continent is needed to maintain livestock production and repair environmental damage.

This chapter considers some of the common issues in grassland management that have occurred over recent decades in southern Australia and management solutions that have been developed. The emerging trends for grassland management over the next few decades are then discussed. The lessons being learnt in Australia have application in similar ecosystems elsewhere in the world; for example, they have been recently applied in northern China (Michalk, Chen and Li, 2001). An important element in these developments is to consider the framework within which grasslands are managed. More solutions are being developed from considering grasslands as ecosystems rather than as linear, factory-like, production systems. The principles that underpin the sustainable solutions being developed are considered in the later parts of this chapter.

MAN AND AUSTRALIAN GRASSLANDS

The history of grassland management in Australian can be considered as a continuum from *exploitation* to *reactive* to *proactive* strategies. There is evidence that this pattern has applied to many areas settled by man (Flannery, 1994). Australian aborigines probably went through the same process, coming to terms with the continent over a period of more than 60 000 years.

When Europeans first settled in Australia, around 1800, they found a vast continent with low numbers of native grazing macropods, but did not realize how rapidly man could degrade these finely balanced grassland ecosystems. The available forage and nutrients that had accumulated over millennia enabled a rapid increase in livestock across vast expanses of the landscape. However, the initial fertility was quickly mined and the underlying low-fertility soils were unable to support the large numbers of sheep and cattle without external inputs. With minimal nutrient inputs and intense grazing pressure, grasslands rapidly deteriorated from the initial dominance of perennial species to ephemeral swards of annual species and weeds (Moore, 1970). These changes were widely recorded as early as 1850 (Bride, 1898). after only 50 years of European settlement. Overgrazed systems lost fertility through severe soil erosion, particularly up to the mid-twentieth century, when wind-borne dust was deposited on New Zealand, more than 2 000 km away across the Tasman Sea. Dust storms have been part of Australia since the continent became drier millions of years ago, but possibly the frequency of such storms did increase during that period. Frequent droughts and highly variable rainfall created further problems in determining and maintaining appropriate stocking rates. During the initial exploitive phase, stocking rates clearly exceeded the longer-term carrying capacity. For example, in western New South Wales (the drier 40 percent of that state), the number of livestock at the end of the nineteenth century was twice that in the same region at the end of the twentieth century. In hindsight, the initially low numbers of native macropods represented an ecosystem in balance that accommodated the constraints of limited surface water at critical times of the year, frequent droughts, the predominance of grassy woodlands over grasslands, and hunting by aborigines. In general, the native grasses that dominated the grasslands when Europeans arrived proved to be poorly adapted to continuous, intensive grazing by cloven-hoofed livestock.

As Europeans came to terms with Australia, much experience was gained in managing grasslands in response to the constraints imposed by the unique environment, and some of the management approaches that emerged in this reactive phase are discussed here, where attempts were made to rehabilitate the landscape using introduced species, fertilizer, herbicides and other tactics. The use of fertilizer, often in combination with sown introduced pasture species, has increased carrying capacity of some land several-fold. However, the planning horizon tended to be short and driven by reacting to changing conditions.

A continual decline in the terms of trade (often around 3 percent per annum) has limited the continued application of a simplistic approach where the aim was to re-sow grasslands, often with introduced species. Research has turned to ecology to provide solutions to grassland problems in an attempt to attain genuinely sustainable systems through a more *proactive* approach. This has challenged producers to adopt new thinking about how grassland systems function and how grazing management provides the key to sustainable management. At present, few managers have become genuinely proactive within a medium- to long-term planning horizon, but many are moving in that direction. Planning for sustainable solutions requires a broader vision that considers not only the financial viability of a grazing enterprise, but also takes account of the environmental context within which the system operates and the social and community impacts. Australian research, farmers and advisory services are moving down the path to more proactive, integrative strategies, and this is likely to be a major trend over the next twenty years.

In temperate Australia (Map 20.1), permanent naturalized grasslands comprising mixtures of native, introduced and volunteer plant species, are widespread and support more than half of the

Map 20.1 The higher rainfall (> 550 mm per annum) temperate zone of southern Australia.

nation's cattle and sheep. These grasslands are from the annual cereal-cropping zone to the coast. Despite their higher productivity, there has been considerable concern about the sustainability of these perennial-based grasslands (Wheeler, Pearson and Robards, 1987). These grasslands provide the main focus of this chapter because they have a major impact on Australian livestock industries because of the grasslands' potential to produce large quantities of high quality livestock products. As a consequence, they have been more intensively studied in recent years. These grassland types are also expanding through the cereal-cropping zone.

With escalating costs and declining commodity prices, farmers claim that they cannot afford to sow new grassland or provide the inputs needed to readily maintain or increase production. This has led to increasing interest in alternative, low-cost solutions, where grazing management is used to maintain or rehabilitate these grasslands (Kemp and Dowling, 2000). Such management tactics aim to develop a desirable plant species composition over the longer-term in order to maintain reasonable levels of production and to avoid the costs of frequently re-sowing grasslands. In addition, there are increasing social pressures for producers to include sustainability in the broad sense as a management goal. These trends will continue as farmers and the wider community seek more sustainable agricultural production systems.

THE STATUS OF SOUTHERN AUSTRALIAN TEMPERATE GRASSLANDS

The total area in Australia where temperate species are major components in grasslands (Map 20.1) or could be sown is estimated to be around 85 million hectares, of which approximately 60 percent is found in New South Wales (Table 20.1). These estimates exclude rangelands, but include those grasslands dominated by annual species and where regular cropping is employed as part of a ley farming system. The actual area of permanent grasslands is probably approaching 75 percent of these estimates, based upon the areas of native, naturalized and sown perennial species (Table 20.2). There are areas dominated by C4 (tropical) grasses within the more temperate zones of southern Australia.

The temperate grasslands across southern Australia have gone through a series of successional stages since European settlement. Under the intense grazing pressure applied by high numbers of cloven-hoofed ruminants, the original vegetation changed from a suite of palatable C4 perennial bunchgrasses (Moore, 1970), minor forbs and a few legumes (most Australian legumes tend to be shrubs or trees – Davidson and Davidson, 1993) to less diverse mixtures dominated by exotic

Table 20.1 The estimated area of temperate pastures in each Australian state based upon farm surveys. The Australian Capital Territory is included in the data for New South Wales

State	Area (million ha)
Queensland	10.5
New South Wales	48.7
Victoria	9.1
Tasmania	3.1
South Australia	5.4
Western Australia	8.5
Total	85.3

SOURCE: Hill and Donald, 1998

Table 20.2 The estimated areas over which different temperate pasture species and groups occur within Australia, and estimates of the potential area they could occupy, based upon climatic limitations

Species	Estimated area[(1)] (million hectare)	Potential area[(2)] (million hectare)	Percent of potential
Subterranean clover (*T. subterraneum*)	29.3	71.1	41.2
All annual *Medicago* spp.	23.6	n.a.	n.a.
Burr medic (*M. polymorpha*)	8.3	n.a.	n.a.
Barrel medic (*M. truncatula*)	4.6	39.5	11.7
Balansa clover (*Trifolium balansae*)	1.3	24.7	5.2
Serradella (*Ornithopus compressus*)	1.2	64.6	1.9
Red clover (*T. pratense*)	0.5	19.8	2.6
Persian clover (*T. resupinatum*)	0.2	31.2	0.6
White clover (*T. repens*)	7.8	16.0	49.0
Alfalfa (*M. sativa*)	3.5	81.2	4.4
Perennial ryegrass (*Lolium perenne*)	6.0	14.4	41.4
Phalaris (*Phalaris aquatica*)	4.7	30.8	15.4
Cocksfoot (*Dactylis glomerata*)	4.1	20.3	20.4
Tall fescue (*Festuca arundinacea*)	1.1	16.9	6.5
Unimproved native	37.9	n.a.	n.a.
Fertilized and oversown native	5.9	n.a.	n.a.

Note: n.a. = data not available.
Sources: (1) from Hill and Donald, 1998. (2) Hill, 1996.

annual species. These botanical shifts were compounded by a series of protracted droughts,especially in the 1890s and 1940s, during which severe overgrazing occurred. The first century of European influence can be characterized as an exploitative phase, where the initial soil fertility declined, useful species were grazed out and the settlers were coming to terms with practising pastoral agriculture in a dry and ancient landscape.

Little phosphate fertilizer was applied to the grossly phosphorus-deficient soils until the mid-twentieth century, when legumes such as subterranean clover (*Trifolium subterraneum*) became more widely used. The major gains in pasture productivity in Australia can be attributed to the great success of herbaceous legumes and phosphatic fertilizer (Donald, 1982). However, as soil nitrogen levels increased, volunteer annual grasses such as vulpia (*Vulpia* spp.), barley grass

(*Hordeum leporinum*), a range of bromes (*Bromus* spp.) and broadleaf weed species invaded, especially where high grazing pressure was applied. The rapid demise of native perennial grasses lead to the need to sow a range of introduced perennial grasses to restore stability to these grasslands. Introduced grasses and legumes have contributed much to the pastoral development of Australia, and in many districts are often the only productive species present.

Unfortunately, the economics of sowing introduced pasture species across medium- to lower-fertility sites for sheep and beef cattle is now often marginal due to worsening terms of trade for livestock producers (Vere, 1998; Vere, Campbell and Kemp, 1994: Figure 20.1). A decision as to whether or not to sow introduced species will depend upon ones view as to the longer-term (e.g. 10-year) prices expected for livestock products (Vere, Jones and Campbell, 2001). Managing the better quality naturalized grasslands can be now as profitable as sowing introduced species because while productivity may not be as high, costs are less (Vere, Jones and Campbell, 2001). These conclusions are supported by a survey of the status of grasslands (Pearson *et al.*, 1997) that found the carrying capacity of many sown phalaris (*Phalaris aquatica*) grasslands was only 5–10 dry sheep equivalents (DSE) ha^{-1}, which is below the economic break-even point for sown grasslands (Vere, Campbell and Kemp, 1994).

The economic environment confronting livestock producers in Australia, where farm subsidies are almost non-existent, is illustrated by the worsening terms of trade, i.e. the ratio of prices received to prices paid (Figure 20.1).

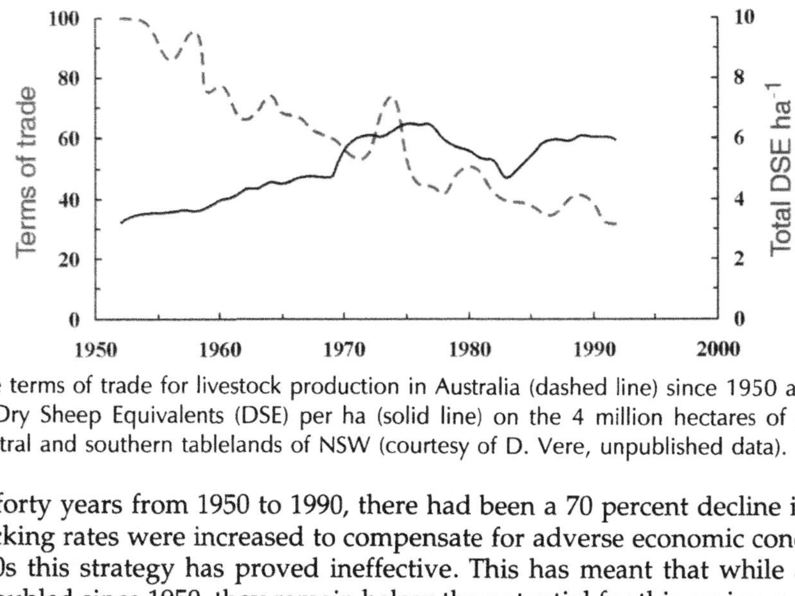

Fig. 20.1 The terms of trade for livestock production in Australia (dashed line) since 1950 and the stocking rate in Dry Sheep Equivalents (DSE) per ha (solid line) on the 4 million hectares of grazing land on the central and southern tablelands of NSW (courtesy of D. Vere, unpublished data).

Over the forty years from 1950 to 1990, there had been a 70 percent decline in terms of trade. Initially, stocking rates were increased to compensate for adverse economic conditions, but since the mid-1970s this strategy has proved ineffective. This has meant that while average stocking rates have doubled since 1950, they remain below the potential for this region, namely 12–14 DSE ha^{-1} that newly-sown pastures are known to sustain on better soils (Kemp and Dowling, 2000). It is likely that the better soils have now been "improved" and the potential productivity of those that remain is low. Trends show that stocking rates for "unimproved" grasslands halved from 1950 to 1990, implying that fertility in these areas had declined. The over-riding problem is that producers need to improve their productivity by 2–3 percent per annum to simply maintain the profitability of their enterprise. However, environmental constraints have come into play limiting the simple solution of simply increasing stocking rates.

Changing strategies

While these species have made a significant contribution to livestock production in temperate areas of southern Australia, there has been a reappraisal of how to best use them in the current economic and environmental circumstances. For example, all introduced perennial grasses are now considered "weeds" in natural areas in New South Wales, and farmers will need to employ practices that limit their spread. A major issue that emerged with highly productive pastures dominated by legumes and annual grasses was the realization that soils were becoming progressively more acid (Williams, 1980; Ridley *et al.*, 1990). A key to the successful use of pasture legumes is management that maintains a good complement of desirable perennial grasses with a capacity to utilize soil nitrate and provide stability to mixtures throughout the year. Without a good balance of perennials, the build-up of soil nitrogen stimulates an explosion of annual grasses and nitrophilous broadleaf weeds, leading to increasing instability and exposure to erosion. This has lead to the view that a range of species in pastures is needed to utilize all the available resources to maximize productivity and minimize the negative impacts of grassland-based livestock production systems, i.e. a more ecological approach (Kemp *et al.*, 1999).

Lime applications can effectively counter acidity when it is confined to the topsoil, but is not always economic under the low to moderate levels of productivity of many regions. Two alternative approaches are available to counter soil acidity. First, acid-tolerant species such as serradella (*Ornithopus compressus* and *O. sativa*) have been selected for adaptation to high levels of exchangeable aluminium and manganese, both toxic metals associated with low soil pH (Michalk and Revell, 1994). Second, pastures can be managed in ways that build up soil organic matter. This has proved to be beneficial in tropical areas, but little research has been done to investigate this approach in temperate Australia.

Alfalfa provides an interesting example of some of the trends in attitudes to grassland management. Alfalfa has a well-deserved reputation as a highly productive species that is prized by many farmers. However, it requires rotational grazing to persist and needs to be re-sown at intervals because no effective means of regenerating stands have been developed, even when seed set is adequate. Over time, more pest and disease problems have emerged, requiring more vigilant management. The commercial pressures on farmers and reduced rural labour supply have meant that the implementation of well-known management practices for alfalfa has been less than ideal. Farmers often comment that they prefer species that are simpler to manage, or are more accommodating of errors in management than alfalfa. This may partly explain why the area sown to alfalfa is considerably less than its potential (Table 20.2).

Chicory (*Cichorium intybus*), a forage herb, has recently emerged as an alternative to alfalfa in some circumstances (Rumball, 1986; Alemseged, 1999). In contrast to alfalfa, chicory is more acid tolerant and it can be economically managed to rejuvenate stands by tactical rest periods to allow seed to be set and for seedlings to establish. However, like alfalfa, chicory requires rotational grazing to ensure plant persistence (Alemseged, 1999). Chicory probably will not replace alfalfa over its whole range. Chicory has an immediate role where more marginal soil conditions and management practices limit the productivity of alfalfa. Chicory has a deep taproot and new uses are emerging for it in managing soil water tables and soil nutrients. A limitation with chicory could prove to be its requirement for a high level of nitrogen to sustain production.

Programmes now exist to domesticate native perennial grasses for wider use. This is driven by the realization that large areas of "native" grasslands (Table 20.2) still support large numbers of livestock and many native grasses have survived often intense grazing over decades. Initially it was thought that native species were not adapted to grazing or higher fertility, but clearly that was not the case. *Microlaena stipoides*, for example, seems to have evolved in similar circumstances to perennial ryegrass, i.e. within the margins of woodlands that are frequently and intensively grazed by kangaroos and where accumulated dung and urine has produced higher fertility. Today, *Microlaena stipoides* is often found around stockyards, sheep camps and other sites where fertility and grazing pressure is high, where it forms dense competitive lawns in which other species cannot survive. The forage quality of many (unselected) native species tends to be lower than selected introduced species (Archer, 1989), but future work may show that this deficiency can be overcome. Certainly the productivity of several native species is of a similar order to European introductions. The main genera in domestication programmes are *Microlaena, Austrodanthonia, Elymus* and *Themeda*. These programmes have raised an interesting dilemma concerning the status of native species ecotypes developed through selection, with claims that they should be regarded as "introductions" when sown outside their area of origin, and are therefore inappropriate for restoring native grasslands. The strength of this argument will need to be balanced against strong evidence that in some cases native species are likely to be more useful than any introduced species. At present, few native species selections are being sown, as it has proved difficult and costly to obtain adequate quantities of harvestable seed.

Australian agronomists have been immensely successful at finding productive pasture species for the vast array of environments across the continent. However, funding for such programmes has declined in recent years, probably because of the success of the research that has produced a large number of suitable species and cultivars that are capable of producing forage to near the limit for the environment. It is also reasonable to argue that since the current forage grown on many farms is not being utilized as efficiently as it could be, the existing cultivars are not the limiting factor for production. Only when more farmers are managing their grasslands to ensure the more desirable species dominate and they are efficiently utilizing more of the forage grown will there be a strong case for seeking additional productive plants. Knowledge that the existing suite of species and cultivars are capable of greater levels of production when well managed led to a recent shift in research and development towards devising better ways of managing these species. Selection work is currently restricted to identifying species and ecotypes suitable for problem sites, such as saline areas.

Trends in grazing management research

With vast areas of grasslands, Australian research has been important in the development of more efficient grazing systems for livestock production. However, grazing management research had traditionally focused on the grazing animals, with the assumption that plant species would cope with whatever management was applied. The aim was often to measure animal response to fertilizer inputs or different plant species mixtures (e.g. Moore, Barrie and Kipps, 1946). Such research was often short term (<3 years) even though farmer experience indicated that important changes in plant species composition occur over a longer term. Where different grazing methods were tested, these tended to be fixed and did not take into account the patterns of plant phenological development or seasonal conditions. As a result, a lot of information was collected on the productivity of grassland communities (i.e. changes in biomass), but little emphasis was placed on measuring plant species dynamics that specify community structure and stability (Kemp and Dowling, 2000) or on how the grassland ecosystem functioned (i.e. water, nutrient and energy flows). The end result was that this research often concluded that continuous grazing was better than rotational grazing. This was at odds with the common farmer practice of moving their livestock around a series of paddocks in a flexible deferred rotation system. Despite considerable investment in grassland research, losses of desirable perennial grasses (both sown and native), declining productivity and a deterioration in the condition of the agro-ecological system have been persistently identified as grassland problems in southern Australia over the last twenty years (Wheeler, Pearson and Robards, 1987; Pearson *et al.*, 1997). It was realized that grazing practices were a common major contributor to the decline in composition and productivity in a range of grassland types (Beattie, 1993). Research then changed radically to focus on tailoring grazing management practices to enhance the proportion of specific desirable perennial grasses within grassland systems (Kemp and Michalk, 1993; Kemp and Dowling, 2000). Better management practices were required for both native and introduced perennial grass species (Kemp, 1993b).

Manipulating composition with grazing

Although the early research on grazing management was rarely directed at managing the composition of grasslands, it helped develop a set of principles that need to be employed to successfully manipulate plant species composition (Table 20.3, based upon Wilson and Hodgkinson, 1990; Kemp, 1991, 1993a). These principles are based upon field observations and research results.

Consideration of the principles set out in Table 20.3, reinforced by the pressures from farmers to produce results of more immediate use on commercial farms, led to a new approach in experimental designs. An *open communal grazing design* was developed (Kemp and Dowling, 2000; Kemp *et al.*, 2001 – based upon a closed communal design developed by Michalk and McFarlane (1979)). In this design, small plots were fenced off within a larger commercial paddock so that each plot could be opened or closed to grazing without interfering with adjoining plots or in the general operation of the paddock. This design was used across 23 on-farm sites in southeastern Australia to test a range of ideas on pasture management. This design was similar to that often used in rangelands with exclusion plots (Michalk, Chen and Li, 2001).

The open communal design enabled the impact of resting and other strategies to be tested against a continuously grazed control in commercial paddocks at minimal cost compared to traditional designs. The control treatment was defined by normal farmer practice and hence was familiar and more meaningful to them. The rests were designed to help identify the sensitive periods in the life cycles of desirable perennial grasses to grazing and to assess any associated impacts on less desirable species. In the past, grazing management research often focused almost exclusively on exerting extra grazing pressure or developing fixed multi-paddock grazing routines. Such tactics can work, but they are more difficult for farmers to use and hence adoption rates were low. Further, such tactics are usually designed to exert pressure on less desirable species as a method of control. In practice, it is often easier for farmers to exclude stock from an area by simply closing the gate, and they are therefore more likely to adopt resting strategies provided they have alternative paddocks where they graze their livestock. Since the emphasis needs to be on enhancing the desirable plant species, *rests* are arguably more important than *grazing* periods for rehabilitating degraded grasslands.

Recent grazing management research

Two national research programmes, the Temperate Pasture Sustainability (TPS) programme (Kemp and Dowling, 2000) and its successor, the Sustainable Grazing Systems (SGS) (Mason and Andrew,

Table 20.3 The principles for successfully managing pasture composition with grazing.

Principle	Explanation
The pasture needs to contain desirable species.	The greater the proportion of desirable species, the easier and faster it is to change composition. In some instances desirable grass species may be too dominant and the aim can be to reduce their incidence to improve the legume content.
Undesirable species need to be considered.	If these species are unpalatable, e.g. woody shrubs, very spiny or toxic, then some mechanical or chemical control may be required to reduce their incidence. Often, these species are not totally undesirable, but less desirable then others (e.g. annual grasses versus perennial grasses) and often palatable and utilized by stock for part of their growth cycle. Annual species may be palatable and productive, but can occupy space that limits the establishment of perennial species.
Fertility of the paddock is important.	The higher the fertility, the faster plants will grow, respond to management and the quicker any changes in composition will occur, whereas on degraded soils it may take some years to achieve desired results.
Stocking rate is the key to manipulating composition.	Considerable influence is exerted on grassland by varying grazing and rest periods to regulate when animals can be selective and when they cannot. This means that the timing of grazing and rest periods needs to be linked to an understanding of the phenology of the dominant and desirable species.
All pasture plants are sensitive to grazing.	This sensitivity varies with stages in their growth cycles, soil fertility and seasonal conditions. In general, grassland plants are more sensitive to defoliation when germinating, regenerating from buds and during flowering. When germinating and regenerating from buds, plants run down their reserves. When flowering, vegetative growth is often suppressed and reserves are used for stem, flower and seed production. Plants are more sensitive to stress and management near the limits of their area of adaptation.
Desirable species benefit from rests.	Desirable species benefit from rests during their more sensitive phases of growth, while less-desirable species need to be more heavily utilized during their sensitive growth phases in order to influence changes in composition in the required direction.
Changing pasture composition can take time.	It may take a few years to significantly increase the perennial grass com ponent of a pasture where it is necessary to allow plants to flower, set seed and become established plants from the seed rain. Wait for an appropriate season to amend composition – it is difficult to make much progress in a drought, but drought periods are often when the most damage is done.
Compromises are common.	In mixed pastures management, difficulties can arise where both desirable and less desirable species have similar growth cycles. In these cases decisions may require some risk as to which species are favoured most by management tactics.

1998), have been instrumental in developing innovative approaches for management of grasslands in southern Australia. TPS focused on screening the impact of tactical rests as a management option for a range of common perennial grass species *in situ* on farms (Kemp and Dowling, 2000) whereas SGS concentrated more on broader issues of sustainable grassland management, such as water use. Both programmes included a wide range of naturalized or sown grasslands, common protocols were used for all measurements (Lodge and Garden, 2000), and new statistical procedures were developed to analyse the performance of individual species components (Orchard *et al*., 2000). For TPS, the results from 23 experiments located strategically across southeastern Australia were scrutinized to determine the general patterns of response to strategic resting in different regions at different times of the year and under a range of grazing practices. These results are summarized in Table 20.4.

Table 20.4 The general response across all 23 Temperate Pasture Sustainability (TPS) experiments of perennial grass content to core treatments. Effects are rated as + (positive effect), – (negative effect) and 0 (no effect)

Species	Core treatments							
	Continuous	Spring rest	Summer rest	Autumn rest	Winter rest	Spring short	Fodder cut	Mob stock
Dactylis glomerata	– –	–0+	++	++	–0+	0	0	0
Phalaris aquatica – Northern NSW[1]	– –	+	– –	+	– –	– –	– –	– –
P. aquatica – Southern[2]	– –	–0+	–0	+	++	-0	-0	++
Lolium perenne[3]	– –	++–	–+	+0	+0	0	0	+0
Themeda australis	– –	++	0	0	0	—	+–	– –
Aristida ramosa	– –	– –	– –	– –	– –	– –	– –	– –
Austrodanthonia spp.	0	0	0	0	0	0	0	0
Microlaena stipoides	0	0	0	0	0	++	0	0

Notes: Seasonal rests extended for 12 weeks in each case; 'spring short' involved extra grazing pressure to prevent excessive growth; fodder cuts were done in late spring; mob stocking involved short periods of grazing at 20 times normal stocking rates at variable intervals, where rest periods varied from 3 weeks in spring to 12 weeks in drier or colder seasons. Northern sites have more summer rain than southern. Additional effects: (1) spring/autumn rest combination was most effective; (2) rotational grazing all year increased phalaris; (3) rotational grazing improved ryegrass persistence.
Source: from Kemp, Michalk and Virgona, 2000.

These results showed that most species used on commercial farms were sensitive to grazing at some stage during their annual growth cycles. The only exceptions were *Austrodanthonia* spp., (wallaby grasses) which tended to be less productive native species adapted to lower fertility soils (and which were not grazed very low in the continuously grazed controls, limiting the experimental contrast) and *Microlaena stipoides*, which responded positively to extra grazing pressure. Extra grazing pressure on other species could be implicated in the latter case, thereby reducing their productivity, and providing more space for *Microlaena* to grow, which under these conditions can form mats that are below grazing height. The sensitivities to grazing for other species were sometimes ambiguous, reflecting the varying conditions across all the experiments, but patterns were evident (Table 20.4). In this research, 12-week rests were used to test major contrasts over the typical length of season. Further work may show that rests do not always need to be this long to be effective.

The general conclusion from the TPS programme was that, in most cases, some form of deferred rotational grazing is needed to enhance or sustain the perennial grasses used across southern Australia for livestock production. The form of rotational grazing is being resolved, but it is clear that the same practice does not need to be applied throughout the year. In a sense, this result supports the on-farm practice of paddock subdivision, though most farmers are uncertain as to rules to use to determine when to move their livestock. The management practices employed now also take a more ecosystem approach than in the past, as often solutions are found from considering how all the species in the grassland interact. Examples developed over recent years include the use of summer resting strategies to increase the proportion of cocksfoot in grasslands (Dowling *et al.*, 1996) and the control of *Aristida ramosa,* a native grass that has seeds that are injurious to livestock (Lodge and Whalley, 1985).

DEVELOPING IDEAS FROM CURRENT RESEARCH

The change in grassland management in Australia over the last couple of decades has been an evolutionary process in response to changed circumstances, as illustrated in the examples outlined above. Many producers in southern Australia have already made the major shift from an animal to a plant+animal management focus (see Reeve *et al.*, 2000, for some early trends that have strengthened in more recent unpublished surveys carried out by Meat and Livestock Australia). Recent surveys indicate that some 80 percent of producers now rest paddocks strategically, based on the needs of the grassland, and move stock more frequently, compared with only 20 percent in the early 1990s. However, this change is probably still more *reactive* than *proactive* management. As the implications of refocusing management criteria are being worked through, additional directions for research are developing and these will probably be an influence on the future trends in grasslands. Some of the directions for grassland management are briefly outlined below.

- **Rest or reduce stocking rates?** The rehabilitation of grasslands will involve more use of tactical rests, although it may be possible in many instances to reduce the grazing pressure without removing all livestock to achieve desirable botanical change (Kemp, 1999). This would probably apply where desirable species are a moderate component of the sward, e.g. 30 percent.
- **What is the minimum herbage mass for management?** There is increasing evidence that herbage mass is a common factor related to other criteria for grassland management, including plant growth rates, livestock intake and legume content. Current work aims to better define the level of herbage mass below which grasslands should not be grazed at critical times. This would provide farmers with an easy-to-use tool for better grassland management.
- **When to rotationally graze?** More farmers now use some form of rotational grazing, but they do not have a consistent set of rules to decide when and why they rotate. It is likely that in many instances a continuous rotation around paddocks is not necessary, but rather rotational grazing may be more appropriate during particular times of the year or for specific management purposes, e.g. grassland manipulation or livestock rationing (Clark, 1993; Doyle, Grimm and Thompson, 1993; Young, 1993).
- **Manage existing grasslands or re-sow?** Faced with a continual decline in the terms of trade and the need to become more efficient, the re-sowing rate of grasslands remains low (Kemp and Dowling, 2000). Grasslands that would have been re-sown in the past now remain in

use and there is increasing interest in how best to rehabilitate them. Clear guidelines for when intervention should occur to rehabilitate grasslands have still to be determined. Intervention should occur before there is any drop in livestock production.

- **Managing recruitment.** Rehabilitation of grasslands requires better management of the recruitment processes of more desirable species, particularly perennial plants that do not regenerate as well from seed as do annuals. Information is now available for cocksfoot (Avery *et al.*, 2000) and chicory (Alemseged, 1999) to foster recruitment, but not for most critical species. This information will be fundamental to achieving better long-term management of grasslands, as it will lead to more cost-effective means of maintaining productivity.
- **Better links between grassland and livestock management.** Major gains are likely to come from more closely matching animal demand to forage supply, including better use of forage systems and supplementary feed to reduce pressure on grasslands at critical times. This will challenge farmer practices, but important gains have already been made by more spring lambing and calving compared with twenty years ago. Further refinements in timing and use of forages and supplements can be made.
- **What is the role for plant improvement?** With a reduced rate of re-sowing of grasslands, more farmers are relying on the plants they already have. Many valuable grassland species have been introduced, selected, bred and widely used in Australia, but those programmes are now somewhat reduced. Genetic engineering proposals have not produced plants with any widespread impact in grasslands. It is likely that, in the future, emphasis will be more on selecting plants for niche conditions where clear benefits from introductions can be demonstrated. Better management of the existing suite of species and cultivars is arguably a greater limitation than the need for new plants.
- **How best to management weeds and pests?** Weeds are among the more important issues farmers face (Reeve *et al.*, 2000), probably reflecting an inadequate level of grassland management (Friend and Kemp, 2000). The frontline for weed management will probably be better grassland management aimed at maintaining a more competitive sward (Michalk *et al.*, 1999; Kemp and King, 2002), rather than relying on herbicides. Biological control practices will be increasingly investigated because they are more selective and environmentally friendly. Pest management will also incorporate the encouragement of beneficial predators.
- **Appropriate management tools and concepts for farmers.** The new directions discussed here mean that some new concepts and approaches need to be developed to enable farmers to understand and then utilize better grassland management practices. Training courses, such as Prograze (Bell and Allan, 2000), have been highly successful and will be further developed. Concepts such as the Pasture Management Envelope (Kemp, 1991; Michalk and Kemp, 1993) have helped build understanding and focus management ideas, such as on the importance of minimal herbage mass and minimal plant species contents as criteria for management. The pasture species composition matrix has effectively quantified state and transition models for grasslands that identifies desirable states for management and pathways to get there (Kemp *et al.*, 1997). These tools take an ecosystem approach to grassland management, in contrast to the production-line concepts of the past.

GRASSLAND MANAGEMENT AND THE RESOURCE BASE

Grasslands are a major part of the landscape, and have many other roles beyond livestock production. Across Australia, there has been growing community concern about the ways grasslands are being used, on a range of environmental issues and on the resource base that sustains grassland productivity. Optimal solutions to the various problems are not yet available, but work in the recent national Sustainable Grazing Systems programme (Mason and Andrew, 1998) provides a better understanding of the issues involved and of practical solutions needed. The current understanding is that:

- Perennial grasses have a crucial role in grassland ecosystem rehabilitation (Michalk *et al.*, 2003). An increased perennial grass content helps to reduce soil acidification rates by more efficient nitrogen capture, as well as helping to control water tables and salinity by transpiring more water, sustaining native organisms over longer periods than ephemeral grasslands and restricting soil erosion.
- Perennial grasses are unlikely to be as effective as trees in higher rainfall areas across southern Australia in transpiring all the incident rainfall, due to the high winter rainfall (White *et al.*, 2000). However, as the whole landscape will not be planted to trees to reduce rising water tables and reduce salinity problems, more lateral thinking is required concerning how best to use perennial grasses in the ecosystem. It may be possible to manage grasslands at different times of the year to partition water flows more usefully between surface runoff (without soil erosion), lateral flow within the soil (especially in duplex soils) and deep drainage.
- Perennial plants can capture more soil nitrate at depth and thereby reduce soil acidification (Ridley *et al.*, 1990; Ridley and Windsor, 1992). However problems often arise during dry seasons, when roots die, reducing the effectiveness of perennial species for this purpose. Understanding is poor about how best to manage grasslands to maintain viable root systems throughout the year to optimize the capture of soil nitrate. Solutions may include the use of special purpose species, e.g. chicory, in appropriate parts of the landscape. Chicory has the potential to extract more nitrogen from the soil than grasses such as phalaris (Kemp *et al.*, 2002). A consequence of using non-traditional species that are more efficient at extracting nutrients is a reduced need for fertilizer, as argued by Elliott (1908) many years ago.

Biodiversity

Biodiversity is an area of increasing social concern and farmers are faced with the challenge to find ways to maintain the diversity of their biological resources. This issue can be considered from two perspectives. First, the diversity of native plants and animals, and, second, the total species diversity – both exotic and native. In the first case, society is often more interested in conservation of rare and endangered species than in protecting common species. Since such species are most often confined to remnant grasslands than distributed throughout the landscape, management of those remnants is the important issue. Native species that are still prominent within existing grasslands have arguably adapted to current practices and can be retained, provided practices do not dramatically change. Overgrazing and weed invasion pose major threats to retaining native plant species. Managing to a minimum herbage mass is a feasible tactic to retain native species diversity within grasslands (Kemp *et al.*, 2003). Most soil insects also appear to be surviving under the "normal" range of management practices likely to be imposed (A. Reid, pers. comm.).

The linkage between total species biodiversity and grassland productivity may provide an incentive for producers to consider biodiversity in their management programmes. Initial evidence indicates that the maximum productivity of grasslands was reached with 6–10 species (Tilman *et al.*, 1997). This finding has important implications for the way introduced forage species are used, as most improved pasture mixtures include only 2–3 species (cultivars are not relevant in this case), yet within a few years volunteer species have invaded and most pastures contain 15–20 species. Although the functional roles of these additional species is poorly understood, there is growing interest in redesigning pasture to include some forbs and other components that may be more useful than the volunteer species. Such mixtures may be more stable over the long term and more easily maintain a dominance of desirable components. Recent efforts to increase the diversity of sown pastures have come with the release of cultivars of chicory and plantains (Rumball, 1986). This approach could be considered a revival, as a wider range of species was included in the original mixtures that were sown in the UK (Elliot, 1908). The optimal number of species groups needed to maximize productivity within a paddock will probably depend upon the size of the paddock and the number of different resource niches available.

Ecosystem services

Within western countries, there is a growing interest in acknowledging the contribution of plants and animals to environmental quality, such as clean air and water, and considering whether farmers should be compensated for the ecosystem services they provide (e.g. Daily *et al.*, 2000). One mechanism through which compensation could be made is accreditation verified by Environmental Management Systems (EMS). Implementation of an EMS could attract payments through price premiums for livestock products in lieu of direct payment for environmental services. Alternatively, payment for management that improves water quality, reduces salinity, increases carbon sinks or optimizes biodiversity could be rewarded through a credit system. No consensus has emerged as to the better mechanism. Depending upon how markets for ecosystem services develop, this could have a significant impact on the ways grasslands are managed.

Water quality is a major issue in Australia. All grasslands have some significance as catchment areas, yet in Australia there has been limited interest in managing grazing lands for quality, i.e. for clean water production. Future markets may compensate farmers for the clean water delivered from their properties and penalize them if contaminated. Irrigators may have to pay farmers in recharge zones in order to obtain the water they require.

Salinity is an inherent feature of the Australian landscape, arising from the geological processes that have resulted in salt accumulating over millennia. Typically, rising water tables result in salt being transported to the soil surface and into river systems. Grassland management will become an increasingly important mechanism for salinity management as grasslands dominate much of the landscape. Farmers might trade "salinity credits" to ensure better quality water in river systems.

Carbon accumulation in the atmosphere is considered to be responsible for recent trends in global warming. Permanent grasslands have a higher ability to store organic carbon than new tree plantations (Davis and Condron, 2002). Australian grasslands could make a substantial contribution to storing carbon and should be eligible for "carbon credits" in the future under the Kyoto Protocol if it is implemented by the Australian Government. This could provide an additional source of income for farmers and could mean an increase in the area of permanent grasslands.

Biodiversity is an important issue and grassland research has only just started to investigate the impacts of management practices on the diversity of herbaceous species and of total organisms in grazed ecosystems. Two hundred years of European settlement has already had significant impact on most Australian grasslands, with native grasslands probably only containing those species that have adapted to livestock grazing. Society is more likely to support the recovery of rare and endangered species than native species in general, hence protection of remnant areas will attract priority funding for biodiversity rather than disclimax grasslands. In turn, this will limit the ability of farmers to attract biodiversity credits for grasslands to remnant areas of native vegetation on their farms.

FUTURE GRASSLANDS

The various themes outlined above have all contributed to the recent changes in direction in grassland research, development and practice in southern Australia, based on lower-cost ecological approaches that restore and maintain grasslands instead of relying on the traditional replacement approach. The success of the ecological approach also focuses on the production of higher value products by tuning animal production to grassland growth cycles, and striving for a broad sustainability approach that encompasses economic, biophysical and social parameters. All these trends are important and in contrast to 40–50 years ago, when all native species were discarded as unproductive and there was an extensive search for exotic species that were more persistent and productive as part of a replacement philosophy.

The shape and use of Australian grasslands will continue to change. Most grasslands now differ significantly from those present when Europeans first arrived around 1800. While many native grasses and the dominant trees and shrubs remain, they now coexist with exotics – predominantly volunteers or deliberate introductions from Europe. Other areas are sown to introduced pasture and forage species that need to be re-sown at regular intervals.

The management of grasslands is a complex science, but with increasing information available on the more important species present, research has developed cost-effective approaches that can fit in with commercial practice. The development of simple, low-cost solutions to grassland degradation problems is likely to continue as a major emphasis of Australian grassland for some time. As indicated in this chapter, the application of these low-cost solutions may require farmers to accept a short-term loss in income. However, the alternative to accepting a small penalty is the complete re-sowing of the grasslands at considerable expense. The overall aim is to improve the sustainability and profitability of grasslands over the medium to long term, with minimal capital outlay in the improvement process.

The new directions in grazing management research that have emerged over the last decade are starting to show benefits for farmers as they become more conscious of the species within their grasslands and the link between grassland composition and productivity. However, much has to be done to encourage producers to adopt new management philosophies and practices. At present, most farmers are making the transition from a reactive to a more active management approach. To become proactive will require considerably more knowledge about grassland ecosystems, how they function and how their utilization with animals can be optimized within a sustainability framework. It is possible that higher levels of utilization could be obtained and hence higher levels of animal production within sustainable systems. The controlled grazing systems developed in New Zealand (Clark, 1993) show that this is possible and can be adapted to Australian conditions (Young, 1993).

The big issue facing southern Australian grasslands is how to achieve systems that are sustainable over the long term in competitive world markets. Sustainability means maintaining productivity, maintaining the natural resource base, providing satisfactory incomes for land users, providing food, fibre and fuel for communities and helping to maintain the wider community. These broad goals are hard to achieve to equal levels of satisfaction, and the best that can probably be achieved is some optimal solution within a range of values, as argued here.

It is anticipated that society will exert a greater influence on the management practices employed in grazing systems. Farmers will need to be proactive in this debate in order to achieve a workable balance between environmental concerns of the wider society (in Australia and among the customers of Australian agricultural products) and the practicality of running a viable livestock enterprise. There are a host of dilemmas in this debate, which will need to be identified and resolved. For example, consumers want clean, green and cheap food, rather than paying the real costs of production or premiums for food produced according to consumer wishes, or acknowledging that some environmental risks need to be taken in producing any commodity. It is possible that these debates may be treated as public health issues, possibly limiting the ability of farmers to propose compromises. The attitudes of urban dwellers regarding what is an acceptable landscape may also constrain grazing practices. Public education campaigns may be needed to point out what a "sustainable" landscape implies for different regions.

Maintaining production is a farmer issue that requires careful consideration. In southern Australia, clear gains in production can be demonstrated over the last fifty years. This has been achieved at the expense of the natural environment and with a requirement for increasing inputs. The earlier increases in production may have exceeded the capacity of the environment (Figure 20.1) and current levels may be the limit that the environment can sustain. Inputs to sustain those current levels of production will increase in cost. High inputs per unit output also imply an increasing level of risk to sustain current activities (Kemp Michalk and Charry, 2001; Kemp, Michalk and Goodacre, 2002) and this will either mean that farmers take a more conservative approach, or improve their managerial skills to cope with higher levels of risk.

Within the frameworks discussed here, two main trends are emerging in the ways grazing systems are managed. For pastoralists producing standard livestock products for the bulk (main) markets, their focus on economic efficiency is likely to result in only satisfying minimum standards for more sustainable grazing systems. Those who take a more exploitive approach will face more regulation by the community. Producers wishing to secure premium markets will need to be more conscious of consumer requirements for a higher standard of environmental management. To achieve this, producers may have to reduce livestock densities to consistently implement an environmental management system to quality assure their products. This group would also include some large-scale producers in rangeland areas who have minimal inputs for animal production and can claim "organic" status. Among those chasing greater economic efficiency and those chasing premium markets, a wider use of more rotational grazing systems is likely, where the rules used to decide when to move animals are more objective than is currently the case. These new rules will be based upon a greater understanding of the plant species found within different grassland systems and when to vary grazing practices to ensure that desirable species remain major components within the sward.

A major continuing challenge in Australia continues to be how to deal with the highly variable climate. Livestock systems suffer from a limited ability to adjust to seasonal conditions compared with crop production. Crop yields go up and down with seasonal rainfall, which means yields and profits can be high in good years – unless there are considerable surpluses. In contrast, livestock do not adjust as quickly to good conditions and hence the difference in profits between good and bad years is not as great as with crops. Tactics that enable producers to capitalize on good conditions have met with only limited success because the time frame for response is usually too short to breed additional animals. Further, prices paid for additional animals at the start of good seasons can be unprofitable due to market shortages and lower prices are received when the animals are then sold at the end of the season, when market surpluses are more common. However, if one accepts a sustainability approach that considers livestock production as part of the grassland ecosystem, "good" years provide an opportunity for grassland rehabilitation by enabling plant species to restore seed banks, increase litter levels and soil organic matter to support production in poorer years. This depends on setting an upper limit to livestock numbers for good years and to have lower limits that help maintain the system in dry years. The tactics discussed here for managing to a minimal herbage mass provide a ready tool to limit overgrazing in dry years.

Taking an ecological and integrative approach to grassland management to solve the multiple problems that beset livestock production offers better prospects for achieving sustainable solutions than by seeking high technology solutions, such as genetic engineering. While genetic engineering and the development of some specific plant traits do have a role, it is likely to be confined to a minor part of the landscape. Grasslands are complex mixtures, where one species is rarely likely to determine the system's productivity and sustainability under all circumstances.

In Australia, grazing management has been one of the underutilized tools for rehabilitating and maintaining grasslands. This is now changing, with more livestock producers across temperate southern Australia resting paddocks and exploring ways to improve the sustainability of their systems. To be successful, farmers and researchers need to develop a clearer understanding of the ecosystem processes regulating plant population dynamics within grasslands. When that is achieved and appropriate management practices developed, grasslands will then be managed on a more sustainable basis.

ACKNOWLEDGEMENTS

The ideas in this paper have been developed through discussions with many colleagues, especially Drs P. Dowling, W. King, G. Lodge and others involved in the TPS and SGS programmes funded by Meat and Livestock Australia, Land and Water Australia, The Woolmark Company, and research agencies. The excellent technical support from the team led by Mr G. Millar is gratefully appreciated and we thank our many colleagues for allowing us to use current and unpublished work.

REFERENCES

Alemseged, Y. 1999. The competitiveness, productivity and management of chicory (*Cichorium intybus* L.) for pastures. PhD Thesis, University of New South Wales, Australia. 232p.

Archer, K.A. 1989. Managing pastures for better utilisation and animal production. *Proceedings of the Annual Conference, Grassland Society of NSW*, **4:** 34–41.

Avery, A.L., Michalk, D.L., Thompson, R.P., Ball, P., Prance, T., Harris, C.A., FitzGerald, D.W., Ayres, J.F., & Orchard, B.A. 2000. Effects of sheep grazing management on cocksfoot herbage mass and persistence in temperate environments. *Australian Journal of Experimental Agriculture*, **40:** 185–206.

Beattie, A.S. 1993. Grazing for pasture management in the high-rainfall perennial pasture zone of Australia. pp. 62–70, *in:* D.R. Kemp and D.L. Michalk (eds). *Pasture management: technology for the 21st century*. Melbourne, Australia: CSIRO.

Bell, A.K. & Allan, C.J. 2000. Prograze – an extension package in grazing and pasture management. *Australian Journal of Experimental Agriculture*, **40:** 325–330.

Bride, T.F. 1898. *Letters from Victorian Pioneers*. Melbourne, Australia.

Clark, D.A. 1993. Grazing for pasture management in New Zealand. pp. 53–61, *in:* D.R. Kemp and D.L. Michalk (eds). *Pasture management: technology for the 21st century*. Melbourne, Australia: CSIRO.

Culvenor, R.A. 1993. Effect of cutting during reproductive development on the regrowth and regenerative capacity of the perennial grass, *Phalaris aquatica* L., in a controlled environment. *Annals of Botany*, **72:** 559–568.

Daily, G.C. *et al*. 2000. The value of nature and nature of value. *Science*, **289:** 395-396.

Davidson, B.R. & Davidson, H.F. 1993. *Legumes: The Australian Experience. The botany, ecology and agriculture of indigenous and immigrant legumes*. UK: Research Studies Press and John Wiley and Sons.

Davis, M.R. & Condron, L.M. 2002. Impact of grassland afforestation on soil carbon in New Zealand: a review of paired-site studies. *Australian Journal of Soil Research*, **40:** 675–690.

Dellow, J.J., Wilson, G.C., King, W.M. & Auld, B.A. 2002. Occurrence of weeds in the perennial pasture zone of New South Wales. *Plant Protection Quarterly*, **17:** 12–16.

Donald, C.M. 1982. Innovation in Australian agriculture. pp. 55–82, *in:* D.B. Williams (ed). *Agriculture in the Australian Economy*. Sydney, Australia: Sydney University Press.

Dowling, P.M., Kemp, D.R., Michalk, D.L., Klein, T.A. & Millar, G.D. 1996. Perennial grass response to seasonal rests in naturalised pastures of central New South Wales. *Rangelands Journal*, **18:** 309–326.

Doyle, P.T., Grimm, M. & Thompson, A.N. 1993. Grazing for pasture and sheep management in the annual pasture zone. pp. 71–90, *in:* D.R. Kemp and D.L. Michalk (eds). *Pasture management: technology for the 21st century*. Melbourne, Australia: CSIRO.

de Fégely, C. 1997. Grasslands Productivity Program 1993-1997. pp. 111–115, *in: Proceedings 38th Annual Conference, Grassland Society of Victoria*. July 1997, The Grassland Society of Victoria.

Elliot, R.H. 1908. *The Clifton Park System of Farming and laying down land to grass*. 4th edition. London: Simpkin, Marshall, Hamilton, Kent.

Flannery, T.F. 1994. *The Future Eaters*. Melbourne, Australia: Reed Books. 423p.

Friend, D.A. & Kemp, D.R. 2000. Grazing management methods. pp. 139–159, *in:* B. Sindel (ed). *Australian Weed Management Systems*. Melbourne, Australia: R.G. and F.J. Richardson.

Helyar, K.R. & Porter, W.M. 1989. Soil acidification, its measurement and the processes involved. pp. 61–101, *in:* A.D. Robson (ed). *Soil Acidity and Plant Growth*. Sydney, Australia: Academic Press.

Hill, M.J. 1996. Potential adaptation zones for temperate pasture species as constrained by climate: A GIS-based modelling approach. *Australian Journal of Agricultural Research*, **47**: 1095–1117.

Hill, M.J. & Donald, G.E. 1998. Determination of benefits from pasture improvement. Final Report to Australian Meat Research Committee. 251p.

Kemp, D.R. 1991. Defining the boundaries and manipulating the system. pp. 24–30, *in: Proceedings of the Sixth Annual conference, Grassland Society of NSW*, Orange NSW Australia, June 1991, The Grassland Society of New South Wales.

Kemp, D.R. 1993a. Managing pastures by grazing. *Proceedings of the Eighth Annual Conference, Grassland Society of NSW*, **8:** 32–38. July 1993, Orange NSW Australia, The Grassland Society of NSW.

Kemp, D.R. 1993b. Pasture management research priorities. pp. 149–154, *in:* D.R. Kemp and D.L. Michalk (eds). *Pasture management: technology for the 21st century*. Melbourne, Australia: CSIRO.

Kemp, D.R. (1999). Managing pasture mass: the minimum threshold concept. Proceedings of Friends of Grasslands, ACT, Workshop, *Pasture Management for Production, Catchment and Biodiversity*. Queanbeyan, August 1999. 3p.

Kemp, D.R., Charry, A.A., Whiteley, W.J. & Gardner, M.W. 2002. Australian agricultural ecosystems: searching for biophysical sustainability performance indicators. *Proceedings 13th International Farm Management Congress*. Arnhem, The Netherlands, July 2002, The International Farm Management Association.

Kemp, D.R. & Culvenor, R.A. 1994. Improving grazing and drought tolerance of temperate perennial grasses. *New Zealand Journal of Agricultural Research*, **37**: 365–378.

Kemp, D.R. & Dowling, P.M. 1991. Species distribution within improved pastures over central NSW in relation to rainfall and altitude. *Australian Journal of Agricultural Research*, **42:** 647–659.

Kemp, D.R. & Dowling, P.M. 2000. Towards sustainable temperate perennial pastures. *Australian Journal of Experimental Agriculture*, **40:** 125–132.

Kemp, D.R. & King, W.McG. 2002. Plant competition in pastures – implications for management. pp. 85–102, *in:* P. Tow and A. Lazenby (eds). *Competition and Succession in Pastures*. Wallingford, UK: CABI Publishing.

Kemp, D.R., King, W.McG., Michalk, D.L. & Alemseged, Y. 1999. Weed-proofing pastures: How can we go about it? pp. 138–143, *in: Proceedings 12th Australian Weeds Conference*, Hobart, Australia, September 1999, Tasmanian Weed Society.

Kemp, D.R., King, W.McG., Lodge, G.M., Murphy, S.R., Quigley, P., & Sanford, P. 2003. SGS Biodiversity Theme: The impact of plant biodiversity on the productivity and stability of grazing systems. *Australian Journal of Experimental Agriculture*, **43:** 962–975.

Kemp, D.R. & Michalk, D.L. (eds). 1993. *Pasture management: technology for the 21st century*. Melbourne, Australia: CSIRO. 177p.

Kemp, D.R., Michalk, D.L. & Charry, A.A. 2001. The development of performance indicators for sustainable systems. pp. 4c, 202, in: *Proceedings 10th Australian Agronomy Conference*. February, 2001. Hobart, Australia. The Australian Society of Agronomy.

Kemp, D.R., Michalk, D.L., Dowling, P.M. & Millar, G.D. 2001. The use of open communal grazing designs to screen options for grazing management. pp. 868–869, *in: Proceedings XIXth International Grassland Congress*. Brazil, February 2001.

Kemp, D.R., Michalk, D.L., Dowling, P.M.& Klein, T.A. 1997. Analysis of pasture management tactics using a pasture composition matrix. p. 1090, *in: Proceedings 18th International Grassland Congress*. July 1997, Canada.

Kemp, D.R., Michalk, D.L. & Goodacre, M. 2002. The productivity of pasture legumes and chicory in central New South Wales. *Australian Journal of Experimental Agriculture*, **42:** 1–11.

Kemp, D.R., Michalk, D.L. & Virgona, J.M. 2000. Towards more sustainable pastures: Lessons learnt. *Australian Journal of Experimental Agriculture*, **40:** 343–356.

Lodge, G.M. & Garden, D. 2000. Grazing management studies in the Temperate Pasture Sustainability Key Program: site descriptions, treatments and data collection. *Australian Journal of Experimental Agriculture*, **40:** 133–141.

Lodge, G.M. & Whalley, R.D.B. 1985. The manipulation of species composition of natural pastures by grazing management on the northern slopes of New South Wales. *Australian Rangelands Journal*, **7:** 6–16.

Mason, W. & Andrew, M. 1998. Sustainable grazing systems (SGS) – developing a national experiment. pp. 314–317, *in: Proceedings Ninth Australian Agronomy Conference*, January, 1998.

McWilliam, J.R. & Kramer, P. J. 1968. The nature of the perennial response in mediterranean grasses. I. Water relations and summer survival in *Phalaris*. *Australian Journal of Agricultural Research*, **19**: 381–395.

Michalk, D.L., Chen Liang & Li Qingfeng. 2001. Development of sustainable grazing systems for degraded grassland in Xingan League, Inner Mongolia. *Desert Technology VI International Conference*. Urumqi, China, 16–24 September 2001.

Michalk, D.L. & Kemp, D.R. 1993. Pasture management, sustainability and ecosystems theory: Where to from here? pp. 155–169, *in:* D.R. Kemp and D.L. Michalk (eds). *Pasture management: technology for the 21st century*. Melbourne, Australia: CSIRO.

Michalk, D.L. & McFarlane, J.D. 1979. A low-cost, communal grazing design for preliminary evaluation of grazing systems. *Journal of the British Grassland Society*, **33:** 301–306.

Michalk D.L. & Revell, C.K. 1994. Serradella (*Ornithopus* spp.) research: An Australian overview. pp. 39–46, *in:* D.L. Michalk, A.D. Craig and W.J. Collins (eds). *Alternative Pasture Legumes 1993*. Proceedings of the Second National Pasture Legumes Workshop. *Primary Industries South Australia Technical Report*, No. 219.

Michalk, D.L., Kemp, D.R., Campbell, M.H. & McLaren, D. 1999. What we don't know about the control of serrated tussock. pp. 20–24, *in: Proceedings 12th Australian Weeds Conference*. September 1999, Hobart, Australia.

Michalk, D.L., Dowling, P.M., Kemp, D.R., King, W.McG., Packer, I.J. Holst, P.J. Priest, S.J. Millar, G.D. Brisbane, S. & Stanley, D.F. 2003. Sustainable grazing systems for the Central Tablelands of New South Wales. *Australian Journal of Experimental Agriculture*, **43:** 861–874.

Moore, R.M. 1970. Australian Grasslands. pp. 87–100, *in:* R. Milton Moore (ed). *Australian Grasslands*. Canberra: Australian National University Press.

Moore, R.M., Barrie, N. & Kipps, E.H. 1946. A study of the production of a sown pasture in the Australian Capital Territory under three systems of grazing management. *CSIRO Bulletin*, No. 201.

Oram, R.N. 1983. Ecotypic differentiation for dormancy levels in oversummering buds of *Phalaris aquatica* L. *Botanical Gazette*, **144:** 544–551.

Orchard, B.A., Cullis, B.R., Coombes, N.E., Virgona, J.M. & Klein, T. 2000. Grazing management studies within the Temperate Pasture Sustainability Key Program: experimental design and statistical analysis. *Australian Journal of Experimental Agriculture*, **40:** 143–154.

Pearson, C.J., Brown, R., Collins, W.J., Archer, K.A., Wood, M.S., Petersen, C. & Bootle, B. 1997. An Australian temperate pastures database. *Australian Journal of Agricultural Research*, **48**: 453–465.

Quigley, P.E., Ward, G.N. & Morgan, T. 1992. Botanical composition of pasture in south-western Victoria. *Proceedings of the Sixth Australian Agronomy Conference*, **6:** 533. February 1992, Armidale Australia.

Reeve, I.J., Kaine, G., Lees, J.W. & Barclay, E. 2000. Producer perceptions of pasture decline and grazing management. *Australian Journal of Experimental Agriculture*, **40:** 331–341.

Reynolds, S. & Frame, J. 2005. Introduction. This volume.

Ridley, A.M., Slattery, W.J., Helyar, K.R. & Cowling, A. 1990. Acidification under grazed annual and perennial grass-based pastures. *Australian Journal of Experimental Agriculture and Animal Husbandry*, **30:** 539–544.

Ridley, A.M. & Windsor, S.M. 1992. Persistence and tolerance to soil acidity of phalaris and cocksfoot in north east Victoria. *Australian Journal of Experimental Agriculture*, **32:** 1069–1075.

Rumball, W. 1986. 'Grasslands Puna' chicory (*Cichorium intybus* L.). *New Zealand Journal of Experimental Agriculture*, **14**: 105–107.

Tao, W. 2001. Rangeland desertification in China. pp. 953–955, *in: Proceedings of the XIXth International Grassland Congress*, February 2001, Sao Paulo, Brazil.

Tilman, D., Knops, J., Wedin, D., Reich, P., Ritchie, M. & Siemann, E. 1997. The influence of functional diversity and composition on ecosystem processes. *Science,* **277**: 1300–1302.

Vere, D.T. 1998. Investigating improved pasture productivity change on the New South Wales tablelands. *Agricultural Economics,* **18:** 63–74.

Vere, D.T., Campbell, M.H. & Kemp, D.R. 1994. Pasture improvement budgets. *NSW Agriculture, Economic Services Unit Bulletin.*

Vere, D.T., Jones, R.E. & Campbell, M.H. 2001. The economics of temperate pasture systems on the central and southern tablelands of New South Wales. *Rangeland Journal,* **23:** 159–72.

Wheeler, J.L., Pearson, C.J. & Robards, G.E. (eds). 1987. *Temperate Pastures: their production, use and management.* Melbourne, Australia: Australian Wool Corporation, and CSIRO. 609p.

White, R.E., Helyar, K.R., Ridley, A.M., Chen, D., Heng, L.K., Evans, J., Fisher, R., Hirth, R., Mele, P.M., Morrison, G.R., Creswell, H.P., Paydar, Z., Dunin, F.X., Dove, H. & Simpson, R.J. 2000. Sustainability and productivity of perennial and annual pastures in the high rainfall zone of southeastern Australia. *Australian Journal of Experimental Agriculture,* **40:** 267–283.

Williams, C.H. 1980. Soil acidification under clover pasture. *Australian Journal of Experimental Agriculture and Animal Husbandry,* **20:** 561–567.

Wilson, G. & Hodgkinson, K. 1990. The response of grasses to grazing and its implications for the management of native grasslands. pp. 47–57, *in:* P.M. Dowling and D.L. Garden (eds). *Proceedings of the Native Grass Workshop.* Melbourne, Australia: Australian Wool Corporation.

Young, R.L. 1993. Pasture management and productivity in practice: the perennial pasture zone. pp. 125–129, *in:* D.R. Kemp and D.L. Michalk (eds). *Pasture management: technology for the 21st century.* Melbourne, Australia: CSIRO.

CHAPTER 21

The Future for Savannah and Tropical Grasslands: A Latin American Perspective

Raúl R. Vera[1]

ABSTRACT

This chapter describes trends in grassland management and utilization in subtropical and tropical Latin America and the Caribbean in the 1990s, and attempts to identify some of the main issues likely to dominate the sector in the early twenty-first century.

It is posited that during the 1980s and 1990s there have been major changes in grassland-based ruminant production systems, largely induced by macroeconomic policies of that period. The evidence reviewed shows the horizontal expansion of the pasture frontier that characterized earlier decades decreased significantly, notwithstanding reconversions within extant production systems. At the same time, the available data show that grassland-based ruminant animal production increased and has become more efficient. This is particularly true in the case of cow milk production, which has witnessed major technological advances; the process of intensification and "modernization" is still ongoing, and although most clearly visible among medium to large producers, the phenomenon applies also to a large fraction of the small-scale farming sector, which has become increasingly market oriented.

Grassland management and utilization are becoming gradually more sophisticated, and traditional extensive grazing systems are yielding to more management- and knowledge-based systems, a trend that will probably become more marked during the next few years. In parallel, land, particularly grassland, degradation has become more noticeable. Depending upon the specific ecosystem considered, economically feasible alternatives to pasture renovation are available and are being adopted in favoured areas, whereas, in more marginal regions, complex combinations of policy and technology may be required for successful grazing land reclamation. In some of the latter areas, as well as elsewhere, new market opportunities for grassland products are being actively sought, such as for organic meat,

[1]Private consultant. 2 Norte 443, Viña del Mar, Chile.

milk and cheese; agroecotourism; and many other activities. The picture that emerges is of an active sector that, in contrast to earlier decades, is actively exploring economic and technological alternatives, with an emphasis on short- to medium-term economically viable solutions. Longer-term issues, such as those involving societal and environmental concerns, are also being raised by some sectors of Latin American society, and are likely to affect grassland management in future years.

INTRODUCTION

Grazing lands constitute the largest single land use system in tropical and subtropical Latin America and the Caribbean (LAC) – a use that is rooted in the history and culture of the subcontinent. In effect, the ratio of grazing lands to cultivated crop lands in the region is 4:1. The utilization of native and sown grasslands has been influenced during the last decade by socio-economic forces that have led to changes in the technologies applied to these lands. This chapter aims to analyse these changes and the resulting trends in grassland utilization that are still continuing.

THE SOCIO-ECONOMIC CONTEXT

Major macroeconomic reforms have taken place throughout tropical and temperate LAC from the mid-1980s up to the present. Numerous "structural adjustment" plans have been put into effect that have changed the ground rules that all economic sectors had been accustomed to in the previous, more sedate, decades. Thus, abandoning of protectionist policies, opening up of economies to international markets, reducing public expenditures and privatizing numerous public institutions, *inter alia*, have characterized this process (CEPAL/FAO/GTZ, 1998). Grasslands and grasslands-based products have become subject to new rules, with important consequences, as discussed below, but probably the major impact has been the new policies introducing an incentive to increase the competitiveness of the sector. This in turn has produced, and is still producing, a notable, although uneven, rate of technical progress in all matters related to grassland and pasture utilization, despite the historically skewed land distribution that characterizes the region (CEPAL/FAO/GTZ, 1998). These changes in production and modernization are an integral part of what has been called "the livestock revolution" by IFPRI, a process that still continues (Delgado, Rosegrant and Meijer, 2001), and best illustrated by the observation that the market value of the worldwide increase in meat and milk that took place between the early 1970s and mid-1990s was more than double the market value of the increased cereals consumption that resulted from the well known "green revolution" of the 1970s.

LAC has a relatively high GNP (US$ 3940 per capita in 1997), which is substantially higher than many other developing countries, although its distribution is highly skewed. It is estimated that 24 percent of its population is poor (i.e. the percentage living on less than US$ 1/day), and 15 percent is undernourished (USDA, 2000).

The population of LAC derives a higher proportion of its calories and protein from animal products compared with the rest of the developing world (18 and 46 percent, respectively, compared with the average of 11 and 26 percent for the developing world (Delgado, Rosegrant and Meijer, 2001; CAST, 1999)). Not surprisingly, annual per capita meat consumption (46 kg/year) is also much higher than the average of the developing countries (21 kg; CAST, 1999). Milk consumption in LAC in 1996–98 averaged 112 kg per capita, compared with 43 kg for the developing world (Delgado, Rosegrant and Meijer, 2001), a fact that has been associated with the high degree of urbanization of the region (see below). For the period 1993–2020, projected annual growth rates of meat and milk for LAC are 2.3 and 1.9 percent, respectively; figures that are lower than the respective averages for the developing world (2.7 and 2.3 percent, respectively) but, as indicated above, starting from a much higher base.

The crop and livestock sectors of LAC have experienced large changes for the reasons mentioned above, and the contribution of the agricultural sector to the economy has decreased in relative terms. Nevertheless, the sector grew at an acceptable rate of 2.9 percent per annum over the period 1990–95 (Muchnik, Morales and Vargas, 1997), but differences between countries are large, and varied between the –4 percent of Haiti to the +13 percent of Guyana. Large variations can also be observed between regions within countries, so that a technologically sophisticated farming sector co-exists with more traditional ones in varying states of "modernization".

The 1990s were characterized by a sustained increase in agricultural output. For example, average yields increased at a rate of 3.3 percent per year compared with only 1.3 per cent in the 1980s (Muchnik, Morales and Vargas, 1997), and simultaneously the portfolio of commodities produced in the region changed and diversified. Probably the most important qualitative change during the 1990s was the emergence of labour-intensive fruits and vegetables as the most valuable agricultural exports of the region, displacing traditional commodities such as coffee and sugar (Diaz-Bonilla, 1999); a substantial fraction of the area cultivated to fruits corresponds to the savannah biome.

A complementary dimension in this analysis is the continued rate of urbanization across tropical LAC; detailed analyses of this and of the differences between Central and South America are available (e.g. Bilsborrow and Carr, 2001). The urban population in the mid-1990s was estimated at 75 percent, and it has been predicted that it will reach 85 percent by 2005 (CEPAL, 1999). Nevertheless, strong linkages between the urban and rural sectors appear to have developed, since there is a rapidly growing proportion of urban residents (particularly in small- to medium-sized urban centres) working in rural activities. It has been estimated that 20 percent or more of the urban residents fall into this category in Argentina, Brazil, Chile and Venezuela (Dirven, 1997). Similarly, rural non-farm employment has increased, as has the phenomenon of rural urbanization (Reardon, Berdegué and Escobar, 2001; da Silva and del Grossi, 2001). These aspects are likely to be important in the near future as urban values and ethics begin to influence decision-making at the farm level, particularly with regard to the role of grasslands in environmental conservation and the management of grazing livestock. Of more immediate consequence is the fact that urbanization is changing demand for grassland products, most notably milk and beef, so that consumers are increasingly demanding products of reliable and standardized quality and hygiene, conveniently packaged and pre-processed, and easily and rapidly cooked (Vera, 2001). Similarly, the demand is growing for more sophisticated products, including "natural", "organic" and "nutraceutical" ones in some urban sectors.

LAND RESOURCES

LAC is characterized by being relatively well endowed with land resources in comparison with the average for all low- and middle-income countries of the world (Table 21.1), in part helped by relatively low population densities. Thus, Central America and the Caribbean have a population density of 57 inhabitants/km^2, whereas tropical South America averages 20 persons/km^2 (data

Table 21.1 Land use indicators for LAC in comparison with the average of all low- and middle-income countries

Indicator	LAC	All low and middle income countries
Arable land – ha per capita in 1996	0.3 ha	0.2 ha
Arable land – proportion of total land area	7%	10%
Arable land – annual change (1965–89)	1.3%	0.5%
Permanent pasture – proportion of total land area	29%	27%
Permanent pasture – annual change (1965–89)	0.5%	0.1%
Labour force in agriculture – proportion of total labour	25%	58%

Source: USDA, 2000.

for 1999 from FAO databases; see also Table 21.2 for data specific to the grasslands regions). In recent decades, agricultural land increased by 0.3 percent annually worldwide, whereas the rate of growth in the LAC region averaged 1.3 percent per annum (USDA, 2000). The relatively ample availability of land per capita, and historical traditions dating back to the settlement by Spaniards and Portuguese, help explain the inordinately large cattle population. In effect, whereas LAC contains only 8 percent of the world's human population, it has 26 percent of the world's cattle stock. Similarly, the ratio of grazing lands to croplands is 4:1 across LAC, compared with the world's average of 2.3:1. Thus, land use systems in LAC are shaped by grasslands and ruminant stock to a much larger degree than in most other continents.

Table 21.2 Grasslands of Central and South America in perspective

Region	Land area (million km^2)	Agricultural mosaic area (million km^2)	Population (million)	Population (per km^2)
Global grasslands	53	7.2	792	14.8
Central America and Caribbean	1.1	0.02	31	27.0
South America	5	1.4	58	11.5
LAC, as proportion of world	11.5%	20.1%	11.1%	—

Source: White, Murray and Rohweder, 2000.

GRASSLANDS-BASED PRODUCTION SYSTEMS

The characteristics of the major LAC farming systems have recently been briefly reviewed by Dixon, Gulliver and Gibbon (2001). Only one (the maize-beans system of Meso-America) of the 15 production systems identified by them does not include livestock and grasslands as important components, which therefore leaves close to 90 percent of the region as having a grassland-based element as an integral part of most farming systems. For a good description of the grasslands available in LAC, see Maraschin (2001).

In the context of the present chapter, "grasslands" are defined as open lands that are used for livestock production, following the functional definition advanced by House and Hall (1999). Table 21.2 puts the Latin American grasslands in a world perspective, and shows the considerable differences that exist between South America, and Central America and the Caribbean in terms of grassland resources and population pressure. Seré, Steinfeld and Groenewold (FAO, 1996) classified the world's livestock systems, and their classification applies well to tropical LAC.

The 1990s have seen major changes in land use and crop, livestock and grassland management practices in tropical LAC. In aggregate, arable and permanent crops have in general grown modestly (see Figure 21.2), but the aggregated data mask large displacements of areas sown to specific crops within certain countries, and a decrease in traditional perennial crops such as coffee and cacao. The most dramatic change has been the rapid expansion of soybean production in both tropical (Figure 21.1) and temperate LAC. In effect, the area of soybean harvested grew 20 percent between 1990 and 2001. The expansion of cereals and oil crops, and soybean in particular, has taken place in the grassland ecosystems of LAC, namely the savannahs of the Cerrados of Brazil and Bolivia, eastern Paraguay, the Pampas and subtropical grasslands of southern Brazil, Uruguay and northern Argentina, and, to a smaller extent, the Llanos of Colombia and Venezuela.

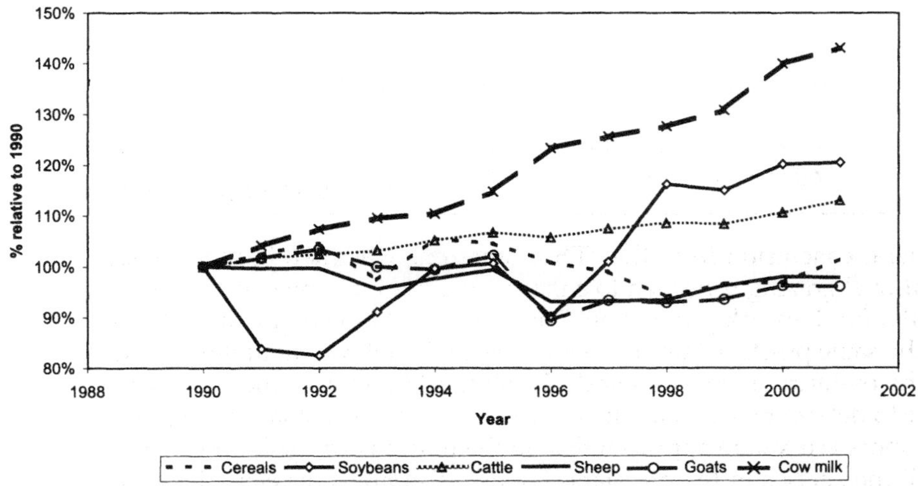

Fig. 21.1 Changes in land use, ruminant stock and cow milk production in tropical LAC, 1990–2001, relative to 1990. Data for cereals and soybean refer to area harvested; the data for stock refers to number of head; and for cow milk is based on fresh milk produced. Tropical LAC was considered as the sum of all countries listed in the FAO database, with the exception of Argentina, Chile and Uruguay.
Source: FAO databases

This expansion of grain crop area has implied a horizontal displacement of part of the stock of cattle, from the increasingly agronomic areas of the central regions of the neotropical savannahs towards more marginal areas (Mattos and Uhl, 1994), but most notably it has also led to intensification of production in those areas where cattle continue to have a significant presence, as discussed below. For the same reasons, large portions of the neotropical savannahs have been subject to large-scale conversion and modification. As an example, it has been estimated that 75 percent of the Brazilian Cerrados have been effectively occupied for human activity (CLAES, 2000) and 50 percent cleared of its original vegetation (Klink, 2000). As for the immediate future of neotropical savannahs, IFPRI projections (Rosegrant *et al.*, 2001) predict that although growth in demand for cereals is slowing down, an additional 8 million hectares will be cultivated with cereals in LAC over the period 1997–2020, and it is clear that, if this expansion occurs, it will be largely at the expense of grasslands.

To a large extent, land owners in the savannahs and derived savannahs of tropical LAC have introduced increasingly high degrees of mechanization, and increased crop yields in mixed farming systems (e.g. see David *et al.*, 1999). They are becoming technologically sophisticated relative to the recent past, and, regardless of farm size, are highly market-oriented. These concepts apply relatively more to the crop sector than to the grassland-based ruminant sector, but both have intensified and have become more sophisticated.

The utilization of neotropical grasslands and pastures intensified in the 1990s (e.g. David *et al.*, 1999). This is contrary to the trend observed in earlier decades, when there was a large expansion of the area of sown pastures, with little evidence of intensification. As shown in Figure 21.2, the area of permanent pastures stayed essentially constant, with only a 1 percent change between 1990 and 1999, whereas cattle production grew substantially. This growth in production in both temperate and tropical LAC was particularly marked for cow milk. Whereas the cattle stock of tropical LAC grew only 0.97 percent per year, milk production grew at a rate of 3.22 percent per year during the period 1990–2001. If a longer time perspective is adopted, cumulative changes are marked, but vary depending upon the period considered. For example, in Central America (excluding El Salvador), the area under pastures increased from 3.5 million hectares in 1950 to 10.5 million hectares in 1983, but it was 9.5 million hectares in 1991. This compares with a stable cropping area of 6.2 million hectares. Similarly, the cattle population increased from 4.2 to 10.1 million head in the period 1950–1978, but it had already reached 9.6 million head by 1992 (Kaimowitz, 1996).

The figures for tropical LAC are dominated by growth of the Brazilian sector (3.74 percent per year over the last decade), since Brazil represents about 55 percent of tropical LAC production, but the growth rate of the rest of tropical LAC was also large, at 2.77 percent per year, with a few countries growing at much higher rates of between 5 and 7 percent (e.g. Bolivia, Honduras, Costa Rica). It is difficult to document at the country level what this process of intensification has entailed since there is scant documentation based on case studies, but one example from

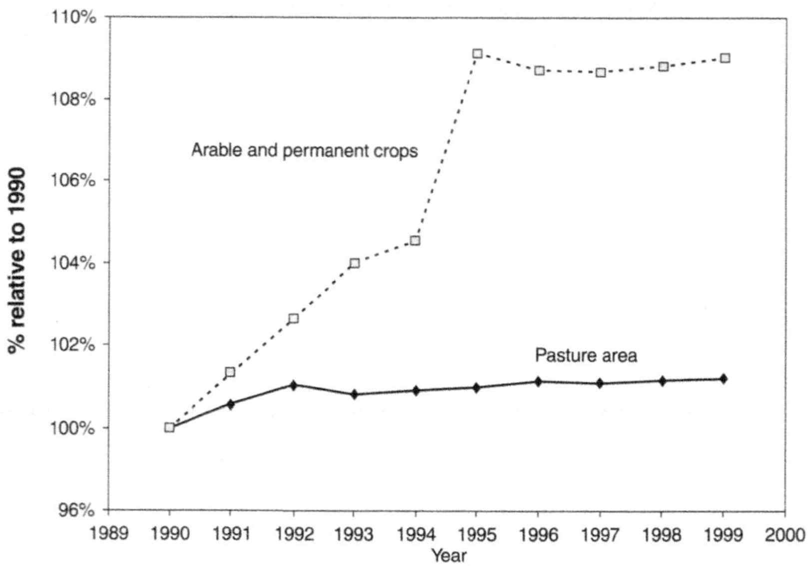

Fig. 21.2 Changes in the area planted to arable and permanent crops, and in the pasture area of tropical LAC over the period 1990–1999.
Source: FAO databases and calculated by the author

(subtropical) Uruguay exemplifies trends that also may be observed elsewhere (Table 21.3) and that show increased efficiency of production per worker, per animal and per hectare, as well as large increases in the use of conserved forages and smaller increases in concentrate feeding. These changes imply not only significant technological adjustments in a sector that in recent past decades was deemed as highly conservative and traditional, but it has raised also the need for much more sophisticated management of all production resources.

Table 21.3 Changes in milk production systems of southern Uruguay, 1987–1997

	Annual rate of change
Farm milk production	
total	6.1%
per worker	17.1%
per hectare	7.5%
per cow	5.2%
Conserved forage fed per litre	10.7%
Concentrates fed per litre	5.1%

Source: Vera, 2001.

Lastly, and notwithstanding all of the above, it should be noted that no country in tropical LAC is a significant exporter of dairy products and thus the growth of the dairy sector has been solely in response to demands of internal markets.

Milk in tropical LAC comes mainly from so-called dual-purpose production systems (Seré and Vaccaro, 1985), and to a smaller extent from specialized dairy systems, depending upon the predominant environmental conditions. In the context of the region, dual-purpose systems are those that produce milk as well as beef in the form of calves and culled mature animals. These systems are heavily based on *Bos indicus* crossbred to Holstein, Brown Swiss or Creole breeds, where milking cows are most frequently milked once a day with the calf at foot to stimulate milk letdown. Depending upon the existing infrastructure and the weather, hand milking of cows can be interrupted by periods of days to weeks when milk cannot be transported to markets, and for this and other reasons, they are considered to be "low-risk systems in high-risk situations" (Vera, Reategui and Loker, 1997).

Depending upon the region, the feeding base is constituted by sown and naturalized pastures and grasslands subject to minimum management, frequently supplemented during the dry season with cut-and-carry forages (sugar cane and Napier grass (*Pennisetum purpureum*)), silage (maize, Napier grass, various other mixtures), and, less frequently, concentrates. In general terms, dual-purpose systems predominate in tropical LAC at altitudes below 2000 m, thus covering all of the savannah areas as well as most of the low- to mid-altitude hillsides region. At the same time, specialized Holstein- and Brown Swiss-based systems are found at higher tropical altitudes (or, alternatively, at higher latitudes), and are most frequently based on pastures sown to temperate (e.g. *Lolium* spp., *Trifolium* spp.) and tropical species (e.g. *Pennisetum clandestinum*), regularly supplemented with concentrates. It has been estimated that dual-purpose systems contribute 40 percent of the saleable milk produced in the region and use 78 percent of the existing stock of milking cows (Rivas, 1992). As an example, the dual-purpose system in Colombia occupies 11 million of the 26 million hectares of grasslands, and represents 33 percent of the stock of cattle (as opposed to 13 percent of specialized dairy cattle; Suárez Gómez, 2001).

As indicated above, all of these systems, but the dual purpose one in particular, are under economic pressure to improve biological and economic performance, and to produce a more regular flow of milk of reliable quality. Without doubt this implies intensification of feeding strategies and the use of improved animal genotypes. In terms of feeding, improved pasture management and utilization, continued diversification in terms of forage resources, supplementation with higher quality conserved or cut-and-carry forages, and concentrate feeding are generally considered essential (e.g. Holmann, Kerridge and Lascano, 1999). Similarly, defined crosses between the base population of *Bos indicus* and Creole cattle with *Bos taurus*, and stabilization of the resulting animal genotypes or other appropriate breeding strategies are equally important (Madalena, 2001). Nevertheless, the major and still incompletely resolved issue is the strong genotype-environment interaction that has been documented for the lowland tropics. In effect, it has clearly been demonstrated, at least in short-term experiments, that realization of the potential afforded by improved tropical forages can only be attained by animals of better genotypic potential than the current average population (Lascano and Avila, 1991, 1993). In contrast, long-term on-farm monitoring of milk production based on improved pastures (e.g. tropical grass-legume mixtures such as *Brachiaria* spp. + *Stylosanthes* spp. and/or *Centrosema* spp.) has shown disappointing results in a variety of environments when there is no simultaneous upgrading of animal genotypes (Vera *et al.*, 1994; Vera, Reategui and Loker, 1997; Ullrich, Vera and Weniger, 1994). Similarly, crossbred animals of higher potential require improved environmental conditions, implying superior feeding and management strategies (Vaccaro *et al.*, 1997; Madalena, 2001).

The evidence thus far available clearly suggests that given the current price structure of inputs and outputs, intermediate levels of milk yield per crossbred cow (somewhere around 2500 to 3500 kg per milked cow lactation, depending upon the country and region considered) obtained from reasonably good, managed tropical pastures (generally based on *Brachiaria* spp.) and forages, and with low levels of purchased supplementary concentrates, have the lowest cost of production per litre of milk and the highest profitability (Madalena, 2001). In essence, the data available support the view that systems based on maximizing the utilization of relatively low cost, abundant and potentially high yielding tropical pastures (if well managed) of medium quality by crossbred cattle are biologically and economically the most feasible. The extent to which these systems need to be complemented with special purpose forages (e.g. tropical maize, Napier grass, forage sugar cane, legume shrubs and trees) depends very much upon the length and severity of the dry season and the market requirements (and economic incentives) for a steady supply of fresh milk.

In the higher altitude (>2000 m) slopes of the Andes ("Sierras"), grassland-based production systems differ considerably, depending upon rainfall and altitude. In the well-watered Andean highlands of Venezuela, Colombia and Ecuador, productive and high quality derived grasslands can be found where species such as *Lolium* spp. and *Trifolium repens* are naturalized. Also, but less frequently and confined mainly to parts of the Colombian and Ecuadorian highlands, *Pennisetum clandestinum* is common. These and still other grasslands and cultivated forages (e.g. lucerne) are part of traditional mixed systems whose prototype is the potatoes-cereals-milk system found across much of the Andes (Estrada, Paladines and Quiros, 1997). These pastures constitute the base of year-round grazing systems that use Holstein and cattle of other European breeds for relatively intensive milk production systems. These mixed systems have been evolving during the last decade with the increasing importance of milk (and therefore pastures) vis-à-vis crops, in response to marked changes in the relative prices of milk, grains and potatoes (Estrada and Paladines, undated) that favour milk, with the result that, for example, milk production rose 90 percent in Ecuador between 1980 and 1996.

At higher altitudes, native grasslands become dominant, although interspersed with small areas of sown pastures, which become particularly important as rainfall diminishes and when irrigation is available. These areas of extensive native grasslands and small areas of irrigated forages (e.g. alfalfa) are the basis for mixed crop-livestock (sheep and native camelids) in the Puna regions of Peru and Bolivia, and probably represent the oldest, and also most degraded, grassland-based production areas. Heavy grazing pressure has been associated with soil erosion, loss of fertility and loss of biodiversity.

Lastly, the large areas of semi-arid grasslands of tropical LAC constitute another important phytogeographical province that supports a substantial ruminant herd and a relatively small and generally poor human population. These areas are extremely diverse and it is difficult to make significant generalizations about them. In tropical South America, there are at least three important areas: the northern Argentinean, Paraguayan and Bolivian Chaco; the Caatinga of northeast Brazil; and the area that extends across the northeast tip of the continent from Venezuela into Colombia. In general, these grasslands include trees and shrubs of variable density, and a herbaceous layer that combines a variety of grasses and forbs. Extensive grazing systems varying from beef cattle in the Chaco (Fretes and Dwyer, 1969) to various combinations of goats and hairless sheep in parts of the Caatinga, characterize the above regions (Cándido, 1998).

NEOTROPICAL GRASSLANDS

Tropical LAC includes a great diversity of native and derived grasslands and pastures (Maraschin, 2001). Detailed descriptions and classifications of the various biomes, with varying objectives, are available (Nabinger, de Moraes and Maraschin, 2000; Sarmiento, 1984, 1994; Sarmiento and Cabido, 1996; Joly *et al.*, 1999; FAO, 2001; Olson *et al.*, 2001; Young and Solbrig, 1993), but, in general, most neotropical grasslands incorporate trees and woodlands to various degrees, so that nomenclature tends to be confusing (for definitions and further references, see House and Hall, 1999). Table 21.4 shows a recent classification of these biomes in South America, emphasizing the complex mosaic of grasslands, tree savannahs, woodlands and forests that characterizes the region.

Central America and southern Mexico have an equally complex mosaic in a much smaller area, due to greater human intervention, a variable topography and the proximity of the Atlantic and Pacific oceans, with the resulting climatic variability. These physiognomic features gave rise to high biodiversity, and also high landscape variability; for example, Holdridge identified 14 "life zones" in Guatemala, 13 in Costa Rica, 12 in Panama and 9 in Belize (Godoy, cited by FAO, 2001).

What follows is therefore only a brief summary of the main formations found throughout tropical LAC, aiming to show the broad diversity that exists and to summarize the current status.

Savannahs

A large proportion of the neotropical grasslands is conditioned by the underlying soils. In effect, 58 percent of the lowlands (below 1000 m) are covered by old, leached soils that in the USDA soil classification fall in the categories of *ultisols* and *oxisols*. Ultisols tend to predominate in regions that have or have had tropical lowland forests, and in foothills, whereas oxisols dominate the savannah areas. These two soil categories, together with others, are associated with a variety of climatic and topographic features that have been classified into *land use systems* by Cochrane *et al.* (1985).

Table 21.4 Important forest areas in tropical South America that include grasslands, as proportion of ecological zone

	Tropical					Subtropical
	Rainforest	Moist	Dry	Shrub	Mountain	Humid
All Tropical South America	82%	27%	89%	15%	26%	19%
Area of zone (million ha)	665	397	133	9	156	46
Types and location of grasslands included	Derived grasslands in the Amazon and seasonal grasslands along river borders	Cerrados; parts of Venezuelan Llanos; Pantanal	Caatinga; Caribbean coast; Chaco; Parts of the Pacific coast of Peru and Ecuador	Pacific coast south of Guayaquil to Tropic of Capricorn; Western Venezuela	Derived and naturalized grasslands of northern Andes of Colombia and Venezuela; Eastern Andes; inter-Andean valleys of Colombia, Ecuador, north Peru	Plateaus and lowlands of the Atlantic coast of southern Brazil, Uruguay and Argentina

Source: FAO, 2001, with modifications.

Other classifications of natural vegetation, with emphasis on grasslands and with different objectives, exist (e.g. Veloso *et al.*, 1991, cited by Joly *et al.*, 1999), but they all attest to the diversity of grassland and other vegetation forms found in the neotropics. The floristic diversity of these grasslands is extremely rich: it has been estimated that the Cerrados of Brazil, parts of Bolivia, Paraguay and patches of Venezuela include 6000 to 7000 vascular plants (Joly *et al.*, 1999). In a relatively small area of approximately 1500 km^2 surveyed in the Llanos of Colombia, Serna-Isaza *et al.* (2001) reported the presence of 197 vascular species from 114 genera and 43 families; a similar number was found in another site of roughly 200 km^2. Most of these estimates are larger than those listed by Groombridge (1992), and little is still known about the ecophysiology of many of these species and plant communities. In fact, few long-term studies on factors affecting ecosystem functions have been conducted in the neotropical savannahs (see reviews by House and Hall, 1999; Solbrig, Medina and Silva, 1996).

Wetlands

The neotropical lowlands include one of the world's largest wetlands, that of the Pantanal, in western Brazil and shared with Bolivia and Paraguay. Ironically, the Pantanal is probably the most conserved grassland phytogeographical province in tropical America, despite being almost all privately owned by cattle ranchers (Joly *et al.*, 1999), where nature, man and cattle have co-existed harmoniously for over 200 years. Its vegetation has been studied (e.g. Pott and Pott, 1997), and appears to be as rich as that of other grasslands. In addition, it is home to an extremely rich fauna.

Anthropogenic grasslands

As one moves from the lowlands to the mid-altitude hillsides, there is a transition from grasslands as climax vegetation, to derived grasslands. In effect, probably most of the mid-altitude neotropics have had some form of forest as climax vegetation, but it is those areas, found throughout Central America, along the Andes of South America and parts of the Atlantic Coast, that have been subject to very intense deforestation, much greater than that of the Amazon (e.g. the Atlantic Rain Forest of Brazil, reduced to 8 percent of the original cover (Joly *et al.*, 1999)). Depending upon the soil and the rainfall regime, these derived grasslands are now covered with introduced, naturalized grasses such as *Digitaria decumbens* and *Cynodon nlemfuensis* on the better soils (e.g. some coffee-growing areas on well watered hillsides), with, more commonly, species such as *Melinis minutiflora* and *Hyparrhenia rufa* on poorer and drier soils (Pacific coast of Central America, many inter-Andean valleys, etc.).

Semi-arid lowland grasslands

The prototypical semi-arid steppes of tropical South America are probably best represented by the Caatinga of northeast Brazil. These grasslands are varied, but in general are characterized by a hot, semi-arid climate, dominated by plants that are well adapted to long periods of water deficiency and that include deciduous species, annual herbs, shrubs and trees with an abundance of thorns, and in general a discontinuous cover. Numerous vegetation subgroups are present, as discussed at length by Joly *et al.* (1999).

Similarly, most of the Central American Pacific coastal area extending from Panama to Central America (roughly 520 000 km^2), originally under dense dry tropical forest, is covered by derived grasslands and remnants of the dry forest. The original biome is considered to be a highly threatened one, since it has been estimated that only 0.1 percent of the habitat remains (Janzen, 1988).

Andean grasslands

Grasslands occupy 58 percent of the area in the Andes, a region of approximately 2 million km^2 over six countries, and are located above 2000 m, namely northern Venezuela, Colombia, Ecuador, Perú, Bolivia, Chile and Argentina (León-Velarde and Izquierdo, 1993; López, 1993). Livestock contribute 14–31 percent of the agricultural production of the Andean countries (León-Velarde and Izquierdo, 1993). Nevertheless, the tropical Andean countries (Chile and Argentina excluded) are net importers of livestock products.

At the farm level, livestock, and in particular ruminant, production is an important contributor to cash incomes of small-scale farm households, as exemplified by case studies carried out in contrasting areas of Ecuador (Table 21.5; Ramírez, Izquierdo and Paladines, 1996).

The changes in macroeconomic policies alluded to earlier had major consequences for land use across the Andes. Many basic agricultural crops (barley, maize, wheat) typical of the region's mixed farming systems (León-Velarde *et al.*, 2000) decreased dramatically. In the case of Colombia, the cropping area decreased by 1 million hectares between 1990 and 1998 (Estrada and Quiroz, 2000), leaving livestock production as almost the only economically viable alternative.

The region is characterized by large variations in altitude and climatic conditions. The Andean highlands of Colombia and Ecuador are similar, and differ significantly from those of Perú and

Table 21.5 Household cash income derived from livestock activities in Ecuador. Data from case studies

Area	Altitude (masl)	Proportion of household income from livestock	Farm size (ha)	
			Average	Range
Humid and subhumid Pacific coast, Manabí Province	< 500	84%	n.a.	n.a.
Eastern Amazonian piedmont, Pastaza and Morona Santiago Provinces	< 1500	95%	122	50–186
Inter-Andean valleys of Gonzanama, Loja Province	1700-2300	92%	53	23–95
Inter-Andean valleys, Chimborazo Province	2500-3500	77%	26	4–93
"Páramo" of Zumbahua, Cotopaxi Province	3500-4500	18%	Communal areas	

Source: Ramírez, Izquierdo and Paladines, 1996.

Bolivia. In general terms, the former are characterized by higher rainfall, lower altitudes and better soils (Lotero, 1993), factors that influence grassland productivity and species adaptation. These grasslands, generally at altitudes above 2000 m, tend to be dominated by naturalized *Pennisetum clandestinum*, frequently associated with the also-naturalized *Trifolium repens*. Many other species are of course present (Lotero, 1993; Ramírez, Izquierdo and Paladines, 1996), including native species such as *Paspalum humboldteanum* and others. Given current management practices, such as overstocking and low to no maintenance fertilization, grasslands based on the naturalized species tend to degrade in 4–6 years (Ramírez, Izquierdo and Paladines, 1996). Pasture degradation in these ecosystems implies reduced soil cover, invasion by less desirable species, such as *Holcus lanatus*, *Paspalum* spp. and others, and consequent reduction in carrying capacity and animal productivity.

The *Páramos* ecosystem is found at somewhat higher altitudes of Colombia and Ecuador, namely 3800–4400 m, where native grasslands are subjected to higher rainfalls and dense cloudiness. Mismanagement of these areas constitutes an important concern for the respective countries, given that the most important rivers have their sources in this ecosystem.

Further south along the Andes, in Perú and Bolivia, rainfall and temperature diminish significantly at equivalent altitudes. In very general terms, the region known as *Sierra* in Perú, is characterized by native grasslands, where *Festuca dolichophylla* and other grasses make a major contribution, while shrubs and trees are sparse (Florez, 1993); crops (potato, barley, quinoa), covering less area than the grasslands, make an important economic contribution. Higher up, and generally above 3800 m, the *Puna* grasslands, characterized by *Festuca* sp. and *Calamagrostis* sp., are used for sheep and camelid production, with only nominal presence of agricultural crops (<10 percent of the area) (Florez, 1993).

General

A different and generally less well-known aspect of the neotropical grasslands is that they cover major South American tropical and subtropical catchments. These catchment areas (Depetris and Paolini, 1991) include the basins of the Orinoco River (third-largest river in the world in terms of water discharged, and 18th in terms of area drained), and the Paraná-Paraguay system (9th and 5th, respectively). The largest South American catchment, the Amazon (the world's largest in terms of water discharge and area drained), also includes substantial areas of grasslands, since 16 percent of the Cerrados lies in its basin (IBGE, 1994). It is well known (e.g. White, Murray and Rohweder, 2000) that grasslands have an important influence on water cycling, erosion control, climate regulation, wildlife habitat and recreational and leisure activities, so their condition, management, utilization and conversion not only matter in economic terms for the individual farmers and ranchers, but also have regional and global consequences. These consequences have little been investigated in tropical LAC, and little is therefore known about the consequences of grassland use and conversion in terms of environmental implications.

SOWN PASTURES AND GRASSLAND IMPROVEMENT

Tropical LAC has witnessed the massive development of sown pastures based exclusively on introduced grasses of African origin, a process that accelerated in the late 1960s and that has continued unabated (Maraschin, 2001). Although no reliable data are available, it can be estimated with considerable confidence that the area sown to these species is about 80 to 100 million hectares (Oliveira *et al.*, 2001). The reasons for the success of the introduced species relative to native species have been summarized by Solbrig, Medina and Silva (1996), *inter alia*, and include larger photosynthetic rates under favourable soil water conditions, higher N use efficiency and higher mineral nutrient contents. Nevertheless, at least the latter two aspects imply that human intervention is required for their maintenance, so that with some exceptions (e.g. *Hyparrhenia rufa, Melinis minutiflora*), the introduced grasses have not been too invasive and have not displaced native species, particularly in the savannahs. Indirectly, these traits have brought about different management practices that have contrasting environmental consequences: in effect, the low mineral content and generally lower nutritional value of native savannah species lead to reduced consumption by animals and the build up of substantial amounts of fuel for periodic fires, whereas introduced species are not regularly burnt.

Pasture development has been most intense in the lowland savannahs and has clearly been dominated by the establishment of grass-only pastures based mostly on *Brachiaria* species, and to a smaller extent *Panicum* spp. and others. As indicated above, soil nutrient limitations are common in the neotropical savannahs, such that inadequate maintenance of soil fertility and mismanagement are blamed for the widespread phenomenon of pasture degradation (Lascano and Spain, 1991). For example, at least 50 percent (Oliveira *et al.*, 2001) of the introduced pastures in the Brazilian Cerrados are considered to be in various stages of degradation. Additionally, large losses of *Brachiaria* spp. pastures, due to the incidence of diseases caused by *Fusarium* spp. and *Rhizoctonia* spp., have been reported (Pizarro, 2001). Introduction of legumes has been considered essential to prevent at least N deficiencies, and so the selection and breeding of tropical forage legumes has received a great deal of attention throughout tropical LAC (Pizarro, 2001), but legume adoption is still very limited.

Complementary strategies that have been investigated are regular pasture fertilization (Oliveira *et al.*, 2001) and the rotation of crops and pastures such that the latter make use of the residual fertilizers applied to the former, while crops benefit from the pasture phase; these benefits have been amply documented (Guimarães *et al.*, 1999; Thomas and Ayarza, 1999), and are well supported by dramatic results observed on-farm (Ayarza *et al.*, 1999). Adoption of these and similar systems is variable and although commonly used in irrigated rice-growing areas of subtropical LAC, the bulk of the tropical lowlands have not yet witnessed a comparable development. Nevertheless, degraded pastures are frequently converted to crops such as soybean (Brannstrom, 2001) and cereal crops as well, using no-till or minimum tillage techniques, at times affecting over 30 percent of the cropping area in the Cerrados of Brazil (Denardin and Kochlann, 1998; Thomas, Ayarza and Lopes, 1999); the extent to which these practices will give rise to regular crop-pasture rotations remains to be seen.

In most of the coffee-growing countries of LAC, pasture-based milk, and to a lesser extent beef, production expanded in the 1990s at the expense of coffee, particularly where the crop is more marginal. Although the expansion of milk production is welcomed, given the still low rates of milk consumption per capita in most tropical countries, that trend has also negative implications, since the coffee sector employs much more intensive labour than either milk or beef production. In Colombia, for example, 37 percent of agricultural jobs are in coffee. Sown pastures constitute the mainstay of milk production in these regions, and species such as *Cynodon nlemfuensis* and *Brachiaria* spp. have rapidly expanded. These same species extend to hillsides areas associated with, but not under, coffee plantations, throughout much of Central America and the well-watered inter-Andean valleys of South America. The introduction of sown legumes is very promising throughout most of these regions, including herbaceous species such as *Arachis pintoi*, and various legume trees, such as *Leucaena leucocephala*, *Gliricidia sepium* and *Cratylia argentea*, among others (Pizarro, 2001; Holmann, 1999; Holmann, Kerridge and Lascano, 1999), constituting incipient silvipastoral systems, which are particularly suitable for small-scale farmers.

Plant improvement

Pasture plant improvement through germplasm collection and screening, classical plant breeding and novel biotechnology is limited in the tropical lowlands of LAC. Breeding activities in the 1990s were, and still are, largely confined to the efforts of EMBRAPA, the Brazilian national research organization, and CIAT (Centro Internacional de Agricultura Tropical). This has been reviewed recently by do Valle (2001), Miles (2001) and Pizarro (2001). Nevertheless, resources for these activities appear to be dwindling, at least at the international level.

During the last few decades there have been successful efforts concentrating on introduction, selection and breeding of African grasses, mainly genera of *Brachiaria* and *Panicum*; collection, selection and breeding of native legumes, such as *Stylosanthes* spp. and *Centrosema* spp. (do Valle, 2001; Miles, 2001); and screening of introduced *Desmodium ovalifolium* and other *Desmodium* spp. (Pizarro, 2001). Several cultivars of these species were released in the late 1980s and early 1990s, but with little real adoption. In more recent years, the exploration of the potential range of adaptation of the native legume *Arachis pintoi* showed a wide range of suitable environmental and production system conditions, leading to release of a number of commercial cultivars (Pizarro, 2001), with more promising prospects than in the above cases. Among the many native grasses, only *Paspalum atratum* has been subject to systematic evaluation (Pizarro, 2001), whereas among the native legume trees and shrubs, other than *Leucaena* spp., *Cratylia argentea* has received considerable attention as a potential component of small-scale mixed farm systems (Holmann, Kerridge and Lascano, 1999).

Without doubt, these improved grass species have been incorporated successfully into a variety of production systems in the lowlands and mid-altitudes of the LAC tropics, whereas legume use has generally not fulfilled the expectations created.

From this very brief summary, it can be inferred that the potential contribution of the ample diversity of grass species present in the neotropical savannahs has largely been ignored.

In the cooler environments of tropical LAC there are currently few systematic, long-term initiatives to screen improved forages, and even less to breed improved cultivars. These activities include the screening of species of *Hemarthria, Lolium, Festuca, Bromus, Phalaris, Trifolium, Dactylis* and *Cajanus* (Euclides Filho, 1999). In these regions, ample use is also made of cultivars of various grass and legume species imported from Europe, New Zealand and elsewhere.

CHALLENGES FOR THE NEOTROPICAL GRASSLANDS

In the late 1990s and early 2000s, the grassland sector of tropical LAC has been confronted with various concerns that, to a large extent, are shared by both market-oriented farmers and – increasingly – the small-scale farming sector that in the past has traditionally been associated with subsistence activities. Many of these concerns are shared by the majority of government institutions as well, and are summarized in what follows.

Economic efficiency

At the turn of the millennium, there is no doubt that the primary concern of market-oriented graziers and government research and development institutions of the region is how to increase economic efficiency and increase production of animal products for the markets, including export markets, as previously discussed. These issues imply complex, debatable and ideology-laden views on conservation versus development (for a brief review of different paradigms, see Brannstrom, 2001) that in the late 1990s and early 2000s have begun to be addressed by some sectors of LAC society. In the meantime, emphasis will probably continue to be placed on technological alternatives and supportive policies that are bound to concentrate on short-term results.

As indicated previously, the new macroeconomic policies that came into being in the 1990s had numerous effects on the grassland sector, and most of these will continue to be felt in the near term. Changes in the ratio of input to output prices, together with increases in the cost of living, lead producers to seek increased production per hectare, with increased efficiency, and to mitigate the historical incentives to speculate with land and cattle ownership. Competition between countries and regions for market share provides additional incentives for the intensification of grassland use, and changes in the stock using them (e.g. moving from sheep to dairy cattle). The need to intensify is, in turn, leading to the adoption of more sophisticated technologies than in the past, such as the continued introduction of forage species and cultivars, and the expansion of forage conservation (e.g. maize silage, round bales, strategically irrigated forages, strategic fertilization). In this context, it is anticipated that large and small ranchers will have to face the challenge of evolving from being traditional livestock raisers to becoming proficient grass growers. In effect, whereas until the very recent past, grass growth and production was taken almost for granted, now cattle and sheep managers will need to extend their abilities with the skills and knowledge required for efficient grassland production and utilization, as shown by the results of a recent informal survey of Uruguayan small-scale ranchers (IPA, 2001).

Grassland-based sheep production systems have suffered the most drastic changes in the last decade. Where these systems have survived, the dominant trend will be towards the production of super-fine wool in temperate areas, complemented with fat lamb production (frequently as contract production; for other forms of contract production in tropical LAC, see CEPAL/FAO/GTZ, 1998) and sheep milk in both temperate and tropical environments. Both alternatives will require significant adjustments in grassland management and in feeding and reproduction strategies, with increased use of information and knowledge.

Healthiness of grassland products

Health and hygiene, as witnessed by the recent events surrounding the outbreak of foot-and-mouth disease, and control and the avoidance of other potentially devastating diseases, will probably increase in importance as intensification proceeds and as markets become more demanding. Improved grassland management can contribute to some extent, but clearly other management issues are more important.

In the long term, an equally important aspect is the extent to which beef, lamb and milk composition can be influenced by grassland-based production systems, as opposed to industrial feeding systems. Aspects such as the content of conjugated linoleic acid (CLA), and other components with human health implications, are beginning to be addressed in the region (Rearte, 2001), and they may have consequences in terms of market prices and desirability of grassland-fed animal products. To some extent, preservation and enhancement of those qualities may run counter to intensification of grazing systems via concentrate supplementation, but these are issues that remain to be examined through further research.

A somewhat less tangible concern is the reclamation of different kinds of grassland-based production rooted in local traditions and folklore, which may provide some added-value as a way to rescue traditional cultures and ways of life, and would provide an added incentive to the production of certified products with denomination of origin.

Lastly, the issue of traceability of animal products, and meat and cheese in particular, is being actively discussed in much of the region, and will probably contribute to further modification of grassland-based production systems.

Reclamation of degraded pasturelands

It is clear that the native grassland area in tropical LAC will continue to decrease as a consequence of conversion to croplands, urbanization and other uses. Similarly, the area sown to introduced pastures, although large to start with, is now nearly stagnant, yet stocking rates seem to have increased, probably indicating an intensification in the use and management of such pastures (David *et al.*, 1999). In this context, and assuming that demand for livestock products will continue to increase, it seems imperative to reclaim degraded grasslands. For a detailed discussion of the issues regarding pasture "degradation", see Lascano and Spain (1991).

Tropical pasture and grassland degradation is an extremely important but unresolved issue, with different implications depending upon the ecosystem under consideration. The semi-arid highlands (e.g. *Puna*) have been overgrazed for centuries, forest resources have been seriously depleted, and loss of biodiversity and soil erosion are pervasive. Technical solutions to pasture and range degradation have been widely investigated but their adoption has been marginal. In this context, the prospects for that ecosystem are dim, and there is controversy as to the extent to which technical solutions can address the issue in the short term (Florez, 1993; Estrada, Paladines and Quiros, 1997; Pomareda, 1997), in contrast to other policies, including the generation of off-farm employment opportunities. In the Andean countries, institutional efforts have been, and

Another aspect of policies applied to the sector, and an aspect that will surely be re-visited in the near future, is the issue of subsidies, direct and indirect payments and equivalent policy tools. The Cairns group, to which several countries of the region belong, has ruled out and strongly opposes the use of agricultural subsidies. Notwithstanding this, societies in the Northern Hemisphere, and to a lesser extent in LAC, increasingly see farmers and ranchers as stewards of the land, managing resources that affect the quality of life of all peoples. The so-called multifunctional aspects of agricultural land use (MFCAL, 1999) are currently receiving increasing attention in Europe and Japan, regions that constitute important current and potential markets for the agricultural products of the neotropical grasslands. Thus, it is likely that importing countries will increase their demands that attention be given to the social and environmental aspects of agriculture by the exporting countries. In that context, restoration of the large areas of degraded grasslands in tropical LAC is bound to eventually receive international attention in an effort to preserve and reclaim ecosystem functions and benefits. Possible models to achieve this aim are the European initiatives that relate to "less favoured areas" (LSIRD, 2000), or the North American initiatives that aim to preserve and recover catchments and riparian areas (CFDA, undated), all of which imply some form of payment to land managers. A recent electronic conference held for the region has raised this issue in regard to the development and adoption of silvipastoral systems and carbon sequestration (LEAD, 2001), signalling that the issue will receive further attention in the near future.

REFERENCES

Angelsen, A. & Kaimowitz, D. (eds). 2001. *Agricultural Technologies and Tropical Deforestation*. Wallingford, UK: CABI.

Ayarza, M.A., Vilela, L., Pizarro, E.A. & da Costa, P.H. 1999. Sistemas agropastoriles basados en leguminosas de usos múltiples. pp. 175–194, *in:* Guimarães et al., 1999, q.v.

Berretta, E.J., Risso, D.F., Montossi, F. & Pigurina, G. 2000. Campos in Uruguay. pp. 377–395, *in:* G. Lemaire, J. Hodgson, A. Moraes, P.C. De F. Cavalho and C. Nabinger (eds). *Grassland Ecophysiology and Grazing Ecology*. Wallingford, UK: CABI.

Bilsborrow, R.E. & Carr, D.L. 2001. Population, agricultural land use and the environment in developing countries. pp. 35–55, *in:* D.R. Lee and C.B. Barrett (eds). *Tradeoffs or Synergies*? Wallingford, UK: CABI.

Brannstrom, C. 2000. *The Brazilian environment: new issues and challenges*. Downloaded November 2001 from http://www.brazilnetwork.org/environ_intro.htm

Brannstrom, C. 2001. Conservation-with-development models in Brazil's agro-pastoral landscapes. *World Development*, **29**: 1345–1359.

Cándido, M.J.D. 1998. *Caatinga- importante recurso forrageiro do Nordeste brasileiro*. See: http://www.tdnet.com.br/domicio/Caatinga.pdf.

Cardoso, I.M., Guijt, I., Franco, F.S., Carvalho, A.F. & Ferreira Neto, P.S. 2001. Continual learning for agroforestry system design: university, NGO and farmer partnership in Minas Gerais, Brazil. *Agricultural Systems*, **69**: 235–257.

CAST [Council for Agricultural Science and Technology]. 1999. *Animal agriculture and global food supply.*. New York NY: CAST. 105p.

CEPAL [Comisión Económica para América Latina y el Caribe, Santiago, Chile (=Economic Commission for Latin America and the Caribbean (ECLAC))]. 1999. El nuevo perfil demográfico de América latina y el Caribe. *Notas de la CEPAL*, No. 6.

CEPAL/FAO/GTZ. 1998. *Agroindustria y pequeña agricultura: vínculos, potencialidades y oportunidades comerciales*. Santiago, Chile.

Cerri, C.C., Bernoux, M. & Cerri, C.E.P. 2001. Soil carbon and nitrogen dynamics after pasture installation in the Amazon region. pp. 157–159, *in: Proceedings of the 19th International Grassland Congress*. São Pedro, SP, Brazil, February 2001.

CFDA [Catalog of Federal Domestic Assistance]. (undated) 10.062 *Water Bank Program*. See: http://www.cfda.gov/public/

CLAES [Centro Latino Americano de Ecologia Social]. 2000. *Observatorio Ambiental Agropecuario MERCOSUR*, No. 16 (12 Sept. 2000). CLAES, Montevideo.

Cochrane, T.T., Sánchez, L.D., de Azevedo, L.D., Porras, J.A. & Garver, C.L. 1985. *Land in Tropical America*. Cali, Colombia: CIAT, and CPAC-EMBRAPA. 3 vols.

da Silva, J.G. & del Grossi, M.E. 2001. Rural nonfarm employment and incomes in Brazil: patterns and evolution. *World Development*, **29**: 443–453.

David, M.B. de A, Waniez, P., Brustlein, V., de Biaggi, E.M., de A. Rollo, P. & dos Santos Rodrigues, M. 1999. *Transformaciones Recientes en el Sector Agropecuario Brasileño. Lo que muestran los censos*. Santiago: CEPAL. 128p.

Delgado, C.L., Rosegrant, M.W. & Meijer, S. 2001. *Livestock to 2020: the revolution continues*. Paper presented at the Annual Meeting of the International Agricultural Trade Research Consortium (IATRC), Auckland, New Zealand. 38p.

de Miranda, E.E., Mangabeira, J.A., Miranda, J.R. & Dorado, A.J. 1998. Pobreza y medio ambiente: seguimiento de 400 campesinos a lo largo de 10 años en Amazonía brasileña. VII Encuentro Internacional De RIMISP. Impacto ambiental de la pobreza rural, impacto social del deterioro ambiental. Rol de los instrumentos de desarrollo agrícola. See: http://www.rimisp.cl

Denardin, J.E. & Kochlann, R.A. 1998. Fast no-till adoption in Brazil without subsidies: a successful partnership. 1998 Northwest Direct Seed Intensive Cropping Conference. (Downloaded 21 June 1999 from http://pnwsteep.wsu.edu/DirectSeed/world.htm)

Depetris, P.J. & Paolini, J.E. 1991. Biogeochemical aspects of South American.rivers: the Paraná and the Orinoco. *In:* E.T. Degens, S. Kempe and J.E. Richey (eds). *Biogeochemistry of major world rivers*. (SCOPE John Wiley Series No. 42). International Council of Scientific Unions, Scientific Committee on Problems of the Environment. Hoboken, NJ: Wiley.

Diaz-Bonilla, E. 1999. Macroeconomic, trade and sectoral issues in the agriculture of Latin America and the Caribbean. *USDA Agricultural Outlook Forum*, 1999.

Dirven, M. 1997. El empleo agrícola en América Latina y el Caribe pasado reciente y perspectivas. *CEPAL, Serie Desarrollo Productivo*, No. 43.

Dixon, J., Gulliver, A. & Gibbon, D. 2001. Global Farming Systems Study: Challenges and Priorities to 2030. Synthesis and Global Overview. Rome: FAO.

do Valle, C.B. 2001. Genetic resources for tropical areas: achievements and perspectives. pp. 477–482, *in: Proceedings of the 19th International Grassland Congress*. São Pedro, SP, Brazil, February 2001.

Ellyard, P. 1998. Rangelands in a planetist future. *Rangeland Journal*, **20**: 41–60.

Euclides Filho, K. 1999. Retrospectiva e desafios da produçâo de ruminantes no Brasil. Sociedad Brasileira de Zootecnia (SBZ). See: http://www.sbz.org.br/eventos/PortoAlegre/homepagesbz/kepler.htm.

Estrada, R.D., Paladines, O. & Quiros, R. 1997. Pobreza y degradación de suelos en los Andes altos. La experiencia de CONDESAN. VIII Encuentro Internacional de RIMISP: Impacto ambiental de la probreza rural, impacto social del deterioro ambiental. Rol de los instrumentos de desarrollo agrícola. See: http://www.rimisp.cl

Estrada, R.D. & Paladines, O. (undated). El impacto de las políticas en el desarrollo lechero de los Andes. El caso de la Sierra ecuatoriana. Informe CONDESAN.

Estrada, R.D. & Quiroz, R. 2000. Technological and institutional changes affecting mixed crop-livestock production systems in the Andes. pp. 83–94, *in:* P.M. Tulachan, P.A.M. Saleem, J. Mak-Hokkonen and T. Partap (eds). *Contribution of Livestock to Mountain Livelihoods*. Kathmandu: ICIMOD.

Evia, G. & Gudynas, E. 2000. *Agropecuaria y Medio Ambiente en Uruguay*. Montevideo: CLAES. 199p.

FAO. 1996. World Livestock Production Systems. Current status, issues and trends. Edited by C. Seré, H. Steinfeld and J. Groenewold. *FAO Animal Production and Health Paper*, No. 127.

FAO. 2001. Global Forest Resources Assessment 2000. Main Report. *FAO Forestry Paper*, No. 140 (available at http://www.fao/org/forestry/fo/fra/main/menu.sjp)

Ferguson, J.E. & Sauma, G. 1993. Towards more forage seeds for small farmers in Latin America. *pp.* 1751–1756, *in: Proceedings of the 17th International Grassland Congress*. Palmerston North, New Zealand, and Rockhampton, Australia, 14-21 February 1993.

Fisher, M.J., Lascano, C.E., Vera, R.R. & Rippstein, G. 1992. Integrating the native savannah resource with improved pastures. pp. 75–100, *in: Pastures for the Tropical Lowlands. CIAT's contribution*. Cali, Colombia: CIAT.

Fisher, M.J., Rao, I.M., Thomas, R.J., Ayarza, M.A., Lascano, C.E., Sanz, J.I. &Vera, R.R. 1994. Carbon storage deep in the soil by introduced pastures in the South American savannahs. *Nature*, **371** (15 September 1994): 236–238.

Florez, A. 1993. *Producción y Utilización de los Pastizales AltoAndinos del Perú*. Quito: REPAAN. 202p.

Fretes, R.A. & Dwyer, D.D. 1969. Range and livestock characteristics of Paraguay. *Journal of Range Management*, **22**: 311–314.

Gastó, J. 2001. Ordenación territorial para la gestión animal. pp. 22–52, *in: Simposio Internacional en Producción Animal y Medio Ambiente. Proceedings*. Departamento de Zootecnia, Facultad de Agronomía e Ingeniería Forestal, Pontificia Universidad Católica de Chile, Santiago.

Groombridge, B. (ed). 1992. *Global biodiversity. Status of the earth's living resources*. London: Chapman & Hall.

Guimarães, E.P., Sanz, J.I., Rao, I.M., Amézquita, M.C. & Amézquita, E. (eds). 1999. *Sistemas Agropastoriles en Sabanas Tropicales de América Latina*. Cali, Colombia: CIAT. 313p.

Holmann, F. 1999. Análisis ex-ante de nuevas alternativas forrajeras en fincas con ganado en sistemas de doble propósito en Perú, Costa Rica y Nicaragua. *Pasturas Tropicales*, **21**(2): 2–17.

Holmann, F., Kerridge, P.C. & Lascano, C.E. 1999. Sistemas mejorados de alimentación basados en leguminosas forrajeras para el ganado de doble propósito en fincasa de pequeños productores de América Latina y Tropical. *Informe de progreso 1999*. Centro Internacional de Agricultura Tropical (CIAT).

House, J.I. & Hall, D.O. 1999. Net primary production of savannahs and tropical grasslands. *In:* H. Mooney, J. Roy and B. Saugier (eds). *Terrestrial Global Productivity: Past, Present and Future*. New York NY: Academic Press. Also available at http://www.savannahs.org.uk/

IBGE [Instituto Brasileiro de Geografia e Estatístca]. 1994. *Anuario Estatístico do Brasil*. Brasilia: IBGE.

IPA [Instituto Plan Agropecuario]. 2001. *Revista del Plan Agropecuario*, No. 100 (Edición Especial). Montevideo: Instituto Plan Agropecuario.

Janzen, D.H. 1988. Management of habitat fragments in a tropical dry forest growth. *Annals of the Missouri Botanical Garden*, **75**: 105–116.

Joly, C.A., Aidar, M.P.M., Klink, C.A., Mcgrath, D.G., Moreira, A.G., Moutinho, P., Nepstad, D.C., Oliveira, A.A., Pott, A., Rodal, M.J. & Sampaio, N. 1999. Evolution of the Brazilian phytogeography classification systems: implications for biodiversity conservation. *Ciencia e Cultura*, **51**: 331–348.

Kaimowitz, D. 1996. *Livestock and deforestation in Central America in the 1980s and 1990s: a policy perspective*. Special publication. Bogor: Center for International Forestry Research. 95p.

Klink, C.A. 2000. Bit by bit, the Cerrado loses ground: human occupation and land-use of Brazil's tropical savannahs. Downloaded November 2001 from http://www.brazilnetwork.org/cerrado.htm

Kluthcouski, J., de Oliveira, I.P., Yokoyama, L.P., Dutra, L.G., de A. Portes, T., da Silva, A.E., da S. Pinheiro, B., Ferreira, E., da M. de Castro, E., Guimarães, C.M., de C. Gomide, J. & Blabino, L.C. 1999. Sistema Barreirâo: recuperación/ renovación de pasturas degradadas utilizando cultivos anuales. pp. 195–231, *in:* Guimarâes et al., 1999, q.v.

Lascano, C.E. & Avila, P. 1991. Potencial de producción de leche en pasturas solas y asociadas con leguminosas adaptadas a suelos ácidos. *Pasturas Tropicales*, **13**(3): 2–10.

Lascano, C.E. & Avila, P. 1993. Milk yield of cows with different genetic potential on grass and grass-legume tropical pastures. pp. 2006–2007, *in: Proceedings of the 17th International Grassland Congress*. Palmerston North, New Zealand, and Rockhampton, Australia, 14-21 February 1993.

Lascano, C.E. & Spain, J.M. (eds). 1991. *Establecimiento y Renovación de Pasturas*. Cali, Colombia: CIAT. 425p.

LEAD [Livestock Environment and Development Initiative]. 2001. *Conferencia electrónica: Sistemas silvopastoriles para la generación de servicios ambientales en Latino América*. Electronic conference hosted by CATIE and FAO, October–November 2001. See: http://lead.virtualcentre.org/es/ele/conferencia3/conferencia.htm

Lee, D.R., Ferraro, P.J. & Barrett, C.B. 2001. Introduction: Changing Perspective on Agricultural Intensifcation, Economic Development and the Environment. pp. 1–16, *in:* D.R. Lee and C.B. Barrett (eds). *Tradeoffs or Synergies*? Wallingford, UK: CABI.

León-Velarde, C. & Izquierdo, F. 1993. *Producción y utilización de los pastizales de la zona altoandina. Compendio*. Quito: REPAAN. 228p.

León-Velarde, C., Quiroz, R., Zorogastúa, P. & Tapia, M. 2000. Sustainability concerns of livestock-based livelihoods in the Andes. pp. 183–202, *in:* P.M. Tulachan, P.A.M. Saleem, J. Mak-Hokkonen and T. Partap (eds). *Contribution of Livestock to Mountain Livelihoods*. Kathmandu: ICIMOD.

López, A. 1993. *Producción y utilización de los pastizales altoandinos del Perú*. Quito: REPAAN. 203p.

Lotero, J. 1993. *Producción y utilización de los pastizales de las zonas altoandinas de Colombia*. Quito: REPAAN. 155p.

LSIRD [Livestock Systems in Integrated Rural Development]. 2000. Collected papers of the European Network for Livestock Systems in Integrated Rural Development. A Concerted Action of the Commission of the European Union, DG VI FAIR1 CT95-0114. See: http://www.mluri.sari.ac.uk/~mi361/lsird.htm

Madalena, F. 2001. Produçâo de leite por animais puros e mestiços. pp. 39–52, *in: Simposio Manejo e Nutriçâo de Gado de Leite*. Goiania, Goias, Brazil: CBNA,

Maraschin, G.E. 2001. Production potential of South American grasslands. pp. 5–15, *in: Proceedings of the 19th International Grassland Congress*. São Pedro, SP, Brazil, February 2001.

Mattos, M.M. & Uhl, C. 1994. Economic and ecological perspectives on ranching in the Eastern Amazon. *World Development*, **22**: 145–158.

MFCAL [Multifunctional Character of Agriculture and Land]. 1999. *FAO/Netherlands Conference on the Multifunctional Character of Agriculture and Land*. Maastricht, The Netherlands.

Miles, J.W. 2001. Achievements and perspectives in the breeding of tropical grasses and legumes.pp. 509–516, *in: Proceedings of the 19th International Grassland Congress*. São Pedro, SP, Brazil, February 2001.

Muchnik, E., Morales, C. & Vargas, G. 1997. *Desk study of CGIAR involvement in Latin America*. Washington, D.C.: CGIAR.

Nabinger, C., de Moraes, A. & Maraschin, G.E. 2000. *Campos* in Southern Brazil. pp. 355–377, *in:* G. Lemaire, J. Hodgson, A. de Moraes, C. Nabinger and P.C. de F. Carvalho (eds). *Grassland Ecophysiology and Grazing Ecology*. Wallinsford, UK: CABI.

Oliveira, O.C. de, de Oliveira, I.P., Ferreira, E., Alves, B.J.T., Miranda, C.H.B., Vilela, L., Urquiaga, S. & Booley, R.M. 2001. Response of degraded pastures in the Brazilian Cerrado to chemical fertilization. *Pasturas Tropicales*, **23**: 14–17.

Olson, D.M., Dinerstein, E., Wikramanayake, E.D., Burgess, N.D., Powell, G.V.N., Underwood, E.C., D'Amico, J.A., Itoua, I., Strand, H.E., Morrison, J.C., Loucks, C.J., Allnutt, T.F., Ricketts, T.H., Kura, Y., Lamoreux, J.F., Wettengel, W.W., Hedao, P. & Kassem, K.R. 2001. Terrestrial ecoregions of the world: a new map of life on earth. *Bioscience*, **51**(11): 933–938. (see also http://www.wordlwildlife.org/ecoregions/)

Pizarro, E.A. 2001. Novel grasses and legumes germplasm: advances and perspectives for tropical zones. pp. 93–100, *in: Proceedings of the 19th International Grassland Congress*. São Pedro, SP, Brazil, February 2001.

Pomareda, C. 1997. En zonas ambientalmente degradadas la poca viabilidad de resolver la probreza via la agricultura y la ganadería. *VII Encuentro Internacional de RIMISP: Impacto ambiental de la probreza rural, impacto social del deterioro ambiental. Rol de los instrumentos de desarrollo agrícola*. Santiago, Chile; RIMISP. See: http://www.rimisp.cl

Pott, A. & Pott, V.J. 1977. *Plants of Pantanal*. Brasilia: EMBRAPA. 320p.

Ramírez, P., Izquierdo, F. & Paladines, O. 1996. *Producción y utilización de pastizales en cinco zonas agroecológicas del Ecuador*. Quito: MAG/GTZ/REPAAN. 235p.

Reardon, T., Berdegué, J. & Escobar, G. 2001. Rural nonfarm employment and incomes in Latin America: overview and policy implications. *World Development*, **29**: 395–409.

Rearte, D. 2001. Beef quality on grazing systems in temperate regions. pp. 259–273, *in: Proceedings, International Symposium on Animal Production and the Environment*. World Association of Animal Production and Sociedad Chilena de Producción Animal, SOCHIPA. Santiago, Chile, 25–27 July 2001.

Rivas, L. 1992. El sistema ganadero de doble propósito en América Latina Tropical. Paper presented to the Simposio Internacional sobre Alternativas y Estrategias en Producción Animal. Universidad Autónoma de Chapingo, México.

Rosegrant, M.W., Paisner, M.S., Meijer, S. & Witcover, J.U. 2001. 2020 *Global Food Outlook. Trends, Alternatives and Choices*. Washington, D.C.: International Food Policy Research Intsitute-IFPRI. 24p.

Sarmiento, G. 1984. The savannahs of tropical America. pp. 245–288, *In:* F. Bourliere (ed). *Tropical Savannahs*. Amsterdam, The Netherlands: Elsevier.

Sarmiento, G. 1994. Sabanas naturales, génesis y ecología. pp. 17–56, *in: Sabanas Naturales de Colombia*. Bogotá: Banco de Occidente.

Sarmiento, G. & Cabido, M. (eds). 1996. *Biodiversidad y Funcionamiento de Pastizales y Sabanas en América Latina*. Caracas: Ediciones CYTED & CIELAT. 318p.

Schipper, R.A., Janse, H.G.P., Bourman, B.A.M., Hengsdijk, H., Nieuwenhuyse, A. & Sáenz, F. 2001. Integrated bioeconomic land-use models: an analysis of policy issues in the Atlantic Zone of Costa Rica. pp. 267–284, *in:* D.R. Lee and C.B. Barrett (eds). *Tradeoffs or Synergies*? Wallingford, UK: CABI.

Seré, C., Steinfeld, H. & Groenewold, J. 1996. World Livestock Production systems. Current status, issues and trends. *FAO Animal Production and Health Paper* 127, 98 p.

Seré, C. & Vaccaro, L. 1985. Milk production from dual purpose systems in tropical Latin America. pp. 459–475, *in:* A.J. Smith (ed). *Milk Production in Developing Countries*. Centre for Tropical and Veterinary Medicine, University of Edinburgh, Scotland, UK. 459p.

Serna-Isaza, R.A., Rippstein, G., Grollier, C. & Mesa, E. 2001. Biodiversidad de la vegetación de sabana en la altillanura plana y la serranía de los Llanos Orientales. pp. 46–63, *in:* G. Rippstein, G. Escobar and F. Motta (eds). *Agroecología y Biodiversidad de las Sabanas en los Llanos Orientales de Colombia*, Cali, Colombia: CIAT.

Solbrig, O.T., Medina, E,. & Silva, J. 1996. Biodiversity and tropical savannah properties: A global view. pp. 185–211, *in:* H.A. Mooney, J.H. Cushman, E. Medina, O.E. Sala and E.D. Schulze (eds). *Functional Roles of Biodiversity: A Global Perspective*. SCOPE 55. New York NY: Wiley.

Suárez Gómez, R. 2001. El complejo productivo lácteo en Colombia. pp. 209–262, *in:* M. Dirven (comp). *Apertura Económica y (Des)Encadenamientos Productivos*. Santiago: CEPAL.

Sunderlin, W.D. & Rodríguez, J.A. 1996. Cattle, broadleaf forests and the agricultural modernization law of Honduras. The case of Olancho. *CIFOR Occasional Paper*, No. 7(E). Bogor, Indonesia: CIFOR.

Thomas, R. & Ayarza, M.A. (eds). 1999. *Sustainable Land Management for the Oxisols of the Latin American Savannahs*. Cali, Colombia: CIAT. 231p.

For Product Safety Concerns and Information please contact our EU representative GPSR@taylorandfrancis.com Taylor & Francis Verlag GmbH, Kaufingerstraße 24, 80331 München, Germany

Batch number: 10732778

Printed by Printforce, the Netherlands